OTHER TITLES IN THIS SERIES

National Audubon Society Birds of North America

NATIONAL AUDUBON SOCIETY FIELD GUIDES

African Wildlife
Butterflies
Fishes
Fossils
Insects and Spiders
Mushrooms
North American Birds—Eastern Region
North American Birds—Western Region
North American Trees—Eastern Region
North American Trees—Western Region
North American Seashore Creatures
North American Wildflowers—Eastern Region
North American Wildflowers—Western Region
Night Sky
Reptiles and Amphibians
Rocks and Minerals
Shells
Tropical Marine Fishes
Weather
Marine Mammals of the World
North American Mammals

National Audubon Society

TREES

OF NORTH AMERICA

National Audubon Society

TREES

OF NORTH AMERICA

The complete identification
reference to trees—with
full-color photographs displaying
leaf shape, bark, flowers, and
fruit; updated range maps;
and conservation status

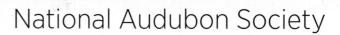Audubon

CONTENTS

POND CYPRESS (*TAXODIUM ASCENDENS*)

TABLE OF CONTENTS

INTRODUCTION

Humans and animals have always had an important relationship with trees. Besides being aesthetically pleasing, trees provide other living organisms with some of their most basic needs—including food and oxygen. Humans have long made tree bark into medicines in addition to spices and dyes, and timber is harvested to build houses and a multitude of wooden products. The fruits and nuts that trees produce are widely enjoyed and provide an important part of the diets of both people and wildlife. Trees provide food and shelter to birds, bears, insects, and many other animals. There are an estimated 3.04 trillion trees around the world. About 15 billion of them are cut down each year. As human population increases and the effects of climate change are realized, how trees will be impacted in the future remains unclear.

About National Audubon Society

The National Audubon Society protects birds and the places they need, today and tomorrow. Audubon works throughout the Americas using science, advocacy, education, and on-the-ground conservation. State programs, nature centers, chapters, and partners give Audubon an unparalleled wingspan that reaches millions of people each year to inform, inspire, and unite diverse communities in conservation action. A nonprofit conservation organization since 1905, Audubon believes in a world in which people and wildlife thrive. The organization introduces children, family, and nature lovers of all ages to the world around them, while Audubon experts, including scientists and researchers, guide lawmakers and agencies in developing conservation plans and policies. Audubon also works with several domestic and international partners, including BirdLife International based in Great Britain, to identify and protect bird habitat. Promoting the growth of native North American plants to support birds and other wildlife is a key part of Audubon's conservation strategy, and for more than a century Audubon has protected birds and their habitat for the benefit of humanity as well as the earth's biodiversity. Audubon promotes the use of native plants and maintains a native plants database as part of its Plants for Birds program. This database allows people to search for plants that are native to their region by entering their ZIP code. You can learn more at **audubon.org/nativeplants**.

WESTERN HEMLOCK (*TSUGA HETEROPHYLLA*)

Names and Classification of Trees

All trees belong to the kingdom Plantae. Each tree species has both a common or English name, such as Longleaf Pine, and a scientific or Latin name, such as *Pinus palustris*. The common name is what we are generally most familiar with, and is what we use in everyday conversation. Many trees have multiple common names, which often vary regionally. The scientific name is more precise for classification purposes and consists of the genus name and the species name. In *Pinus palustris*, the first word indicates that the tree belongs to the genus *Pinus* (pine trees) and the second part describes the species—in this case, a reference to the swampy or marshy habitat in which this tree grows.

Division

Trees are split into two broad divisions—angiosperms and gymnosperms. Angiosperms are flowering plants with seeds enclosed in an ovary. These trees, sometimes referred to as deciduous trees or hardwoods, lose their leaves for part of the year. In cold climates, they lose leaves in the fall, while in the dry climates, they lose leaves in the dry season. These flowering plants are usually further divided into two classes—monocots (short for monocotyledons, which include grasses, aloes, palms, yuccas, and flowers with three or six parts) and dicots (short for dicotyledons, which includes birches, willows, maples, and oaks). Dicots have branched leaf veins with four, five, or more than six flower parts.

The seeds of gymnosperms are not protected by an ovary or fruit like angiosperms. Gymnosperms include cone-bearing trees such as pines, conifers and junipers. These trees do not lose their leaves.

Promoting the growth of native North American plants to support birds and other wildlife is a key part of Audubon's conservation strategy.

One acre of forest absorbs about six tons of carbon dioxide and puts out four tons of oxygen—enough to meet the needs of 18 people annually.

Functions of a Tree

Each part of a tree serves a different function. The flowers on trees produce fruits, nuts, and cones, which contain seeds to grow the next generation of trees.

Most deciduous trees have flowers or nuts with seeds inside, while most conifer trees have cones to protect their seeds. When the seeds are ready to grow, the cone's scales open and seeds fly out.

The tree trunk holds the branches. The branches hold the leaves and the flowers, fruit, or cones. The roots keep the tree from tipping over. The roots and leaves also grab nutrients from the air or ground and supply it to the tree.

Leaves further use energy from the sun to make food for the tree through a chemical process called photosynthesis. In photosynthesis, tree and plant roots absorb water, minerals and nutrients from the soil. These raw materials travel to the leaves, where chlorophyll is found. At the same time, leaves or needles absorb carbon dioxide from the air. Chlorophyll uses energy from the sun to transform the carbon dioxide and water into oxygen and carbon-based compounds such as glucose, a sugar that helps plants grow.

Sometimes photosynthesis takes place in the bark of a tree—especially in smooth-barked trees, where chlorophyll can be found. Trees release excess oxygen produced as a byproduct of photosynthesis into the atmosphere.

Trees are estimated to produce about half of the Earth's oxygen. The U.S. Department of Agriculture estimates one acre of forest absorbs about six tons of carbon dioxide and puts out four tons of oxygen—enough to meet the needs of 18 people annually.

Ecology

Trees are a vital part of the ecosystem. Birds, mammals, insects, and many other forms of wildlife use trees for nesting sites, shelter, food, and cover to evade predators.

While trees serve wildlife in various ways, trees also rely on wildlife for pollination and reproduction. Many animals eat the fruit and nuts that trees produce, which contain seeds needed for the plant's reproduction. The seeds pass intact through an animal's digestive tract and germinate where they are deposited.

Meanwhile, tree roots improve water quality and prevent erosion and runoff by breaking up soil to allow space for air and water.

In addition, trees help to regulate our climate and lessen the intensity of the greenhouse effect by moderating the sun, rain, and wind. Leaves absorb the sun's rays, providing shade and cooling the air temperature in the summer. Trees also preserve warmth in the winter by providing a screen against harsh wind. The U.S. Department of Agriculture estimates the net-cooling effect of a single, healthy tree is equal to 10 air conditioners running 20 hours a day.

BALD CYPRESS (*TAXODIUM DISTICHUM*)

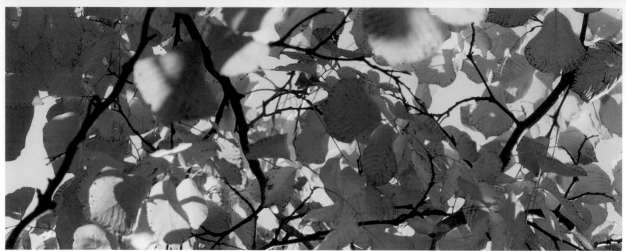

YELLOWWOOD (*CLADRASTIS KENTUKEA*)

Human Use and Personal Value

Aside from their ecological value, trees add to our quality of life in many ways. Trees line city parks, playgrounds, streets, and backyards, creating a peaceful atmosphere for all to enjoy. They provide shade against heat in the summer. Much of the food we enjoy today and treasured recipes we pass down to generations of family members involve delicious fruits and nuts, many of which wouldn't be possible without trees.

Children have long enjoyed climbing trees and building tree forts and tree houses. Trees also hold a sentimental value, marking property lines, holding memories, and even marking burial sites of loved ones. Having healthy, mature trees can also increase property value by as much as 10 percent, according to the U.S. Forest Service.

The availability of trees for lumber, fuel, paper, construction, and manufacturing has provided countless jobs and is an important part of our modern economy. Modern forestry practices are more sustainable than in the past, but the harvesting of trees still must be balanced against the ecological, aesthetic, and community value of the living trees.

Fall Leaf Color

North America's trees are renowned for their bright colors in the fall, when many leaves turn brilliant orange, yellow, and red.

One notices the colors start to emerge just before deciduous trees drop their leaves, when the days become shorter in the fall. When this happens, green chlorophyll in trees breaks down and the green color of the leaves fades as trees stop producing food and start storing nutrients in their roots. Eventually, a tree stops producing chlorophyll altogether, and the yellow or orange carotenoid in the leaves begins to show through.

The red color seen in some tree leaves is caused by a bright red pigment called anthocyanin. Researchers believe anthocyanin is produced by trees as a form of protection to allow trees to recover nutrients before leaves fall off. Trees that display yellow and orange colors in the fall are fairly consistent, but red colors can be more or less vibrant each year depending on the weather. Reds are most vibrant when there are a number of sunny fall days and cool nights above freezing. In the daytime, the leaves can use the warm sun to produce sugar. A cool night prevents sugar sap from flowing through the leaves and down to the branches and trunk. This is where anthocyanins emerge to help the trees and the red color displays.

A severe drought can delay the arrival of fall colors by a few weeks. A warm period with lots of rain will make the colors less vibrant. Severe frost can kill the leaves, turning them brown and causing them to drop early.

Trees that display yellow and orange colors in the fall are fairly consistent, but red colors can be more or less vibrant each year depending on the weather.

Generally, related trees share similar characteristics. Examining these characteristics—especially the leaves and bark—in detail can help to identify a tree to its species.

Identification

Identifying any wild, living organism can be challenging in the field because of natural variation among individuals and the similarity between related species, but knowing what field marks to consider gives an observer a good place to start. Thankfully, trees don't fly, hide, or run away, allowing an observer to take as long as necessary to closely inspect and evaluate a tree's features, including its leaves, bark, flowers, and fruits.

Generally, related trees share similar characteristics. Examining these characteristics—especially the leaves and bark—in detail can help to identify a tree to its species. Here are a few features to use to get started identifying a tree in the wild.

Leaves and Leaf Shapes

Leaf shape—A good first step is to look at the tree's leaf shape. There are three basic leaf types—needles, scales, and broadleaf. Most evergreens have needles or scales, while most broadleaf trees are deciduous, meaning they drop their leaves in the fall.

SCALES OF ALASKA CEDAR
(*CHAMAECYPARIS NOOTKATENSIS*)

BROAD LEAVES OF AMERICAN ELM
(*ULMUS AMERICANA*)

Alternate vs. opposite—Next, determine if the leaves are alternate or opposite. If a plant has alternate leaf arrangements on a branch or twig, then there is only one leaf at each node, and the leaves come off the stem in alternate directions. If leaves are opposite, then there is a pair of leaves at each node going in opposite directions of each other.

NEEDLES OF PACIFIC SILVER FIR (*ABIES AMABILIS*)

Incense Cedar (*Calocedrus decurrens*), for example, has opposite leaves, while American Hazelnut (*Corylus americana*) has alternate leaves. Plants with whorled leaves, a variation of opposite leaves, have three or more equally spaced leaves at a node—an example of this is Mountain Laurel (*Kalmia latifolia*).

SIMPLE LEAVES

ALTERNATE LEAVES

OPPOSITE LEAVES

PALMATELY COMPOUND LEAVES

Simple vs. compound—After determining whether the tree's leaves are alternate or opposite, next determine if they are simple or compound. Simple leaves contain a single leaf blade and middle incision that runs from one side of the leaf to the other. Examples of this include the familiar leaves of maples, sycamores, or oaks.

Compound leaves consist of several or many distinct leaflets joined to a single stem. Compound leaves come in several varieties. Palmately compound leaves radiate from a single point on the leaf stalk, such as in the Ohio Buckeye (*Aesculus glabra*). In pinnately compound leaves, the leaflets are attached to a common axis, called a rachis, such as in California Walnut (*Juglans californica*). Pinnately compound leaves can compound again, branching off into secondary rachises, forming bipinnately and tripinnately compound leaves, as are sometimes seen on Common Honey Locust (*Gleditsia triacanthos*) trees.

PINNATELY COMPOUND LEAVES

BIPINNATELY COMPOUND LEAVES

Margins—After determining if leaves are simple or compound, look at the edge of the leaf, called the margin. Leaf margins may be smooth (called untoothed or entire) or toothed, containing a saw-like edge. Untoothed margins may be lobed or wavy. Toothed margins may be rounded, blunt, or pointed.

LEAVES WITH SMOOTH MARGINS

LEAVES WITH TOOTHED MARGINS

Bark—Next, look at the bark features to determine if the bark is smooth, fissured, furrowed, or scaly. Sometimes bark can peel off in flakes, sheets, or strips in a way that's characteristic of the tree's species.

Fruit and Flowers—Check the tree for flowers or fruit, which sometimes provide important identification clues, especially on leafless trees. If a tree has flowers, examine their shape and form, location on the branch, and whether they are single or crowded into clusters, or inflorescences. Clusters may be simple, with all the flowers on a single axis, or compound, with flowers on secondary axes.

Wood and Bark

Pith is the tissue that forms the central core of a plant's stems and roots, and is the oldest part of the tree. The pith is composed of soft, spongy cells, which store and transport nutrients throughout the plant. Radiating outward from the pith, there are up to five major layers that comprise a tree's wood.

The innermost layer is the heartwood, which keeps the tree balanced and stable. This wood is typically the strongest and most resistant to decay and may be visually different than the surrounding layers when viewed in cross-section such as on a stump or felled log. Heartwood is among the oldest wood on a tree, capable of maintaining its structural integrity sometimes for hundreds or thousands of years.

The next layer is the sapwood, which is younger wood whose main function is to transport water throughout the tree. Sapwood allows water and nutrients to flow between the tree's roots and its leaves. Newer wood on a tree begins as sapwood, which tends to be thicker on fast-growing plants and those in ideal growing conditions.

The next layer is the vascular cambium layer, also called main cambium or wood cambium, which is responsible for tree growth. In many of North America's trees—particularly those that grow in seasonal cycles—each year of growth forms a visible ring in its wood. These rings are the result of new growth in the vascular cambium layer. Counting the rings on a stump reveals the age of the tree. Dicots and gymnosperms have a cambium layer, but monocots such as palms do not.

PACIFIC SILVER FIR (*ABIES AMABILIS*)

LIVE OAK (*QUERCUS VIRGINIANA*)

Bark consists of the two outermost layers of tissue that cover a tree's trunk and branches, much as our skin covers our bodies. Bark forms a barrier that protects trees against extreme temperatures, fires, storms, and attacks from diseases, animals, and insects. Bark also acts like a plumbing system, moving food and water throughout the tree.

The phloem, or inner layer of bark, is where sugars made by photosynthesis in the leaves are transported back down to feed the branches, trunk, and roots. The phloem layer doesn't live long and turns to cork when it dies, becoming part of the outer layer of bark.

The dead outer bark is the protective layer that, like our outermost layer of skin, is constantly being renewed from living tissue underneath. As trees grow and expand, often the outer bark becomes too tight. As this happens, outer bark may split and crack, often in patterns that are characteristic of a particular genus or species.

Roots

Roots are organs of plants that typically, but not always, form under the surface of the soil. It is the part of the plant that does not bear any leaves or nodes. Roots are essential to plants and trees. Although there are many kinds of roots that are specialized to a tree's environment, in general the two major types of tree roots are anchoring roots and feeder roots. Anchoring roots keep the tree upright, while feeder roots absorb water and nutrients from the soil to supply to the rest of the tree.

Anchoring roots can grow deep below the surface and can match the size of the tree's crown. As anchoring roots grow outward, they can send sinker roots down into the soil to provide strength and stability. While anchoring roots may live for years, feeder roots die and are replaced often, similar to skin cells. Most feeder roots occur in the top 6 to 18 inches (15 to 45 cm) of soil, where water and nutrients tend to be most abundant.

Roots take in water and nutrients to feed the tree by root hairs and by mycorrhizae, which are symbiotic fungi that colonize a host tree's root system. Tiny root hairs look much like fringe; the interior of root cells are slightly salty, which attracts water and helps roots take in more liquid. If a plant has associated mycorrhizae, the plant supplies organic molecules such as sugars made by photosynthesis to the fungus. The fungus then supplies water and nutrients to the plant.

Anchoring roots keep the tree upright, while feeder roots absorb water and nutrients from the soil to supply to the rest of the tree.

PACIFIC SILVER FIR (*ABIES AMABILIS*)

BLUE PALOVERDE (*PARKINSONIA FLORIDA*)

Flowers and Fruits

Flowering trees rely on flowers and fruit for reproduction. In many flowering plant species, a single flower produces both eggs and pollen, while in others, the male and female flowers are separate. After flowering seeds are pollinated, they grow into seeds, while the ovary wall around them grows into a fruit.

Coniferous trees, by contrast, produce seeds and pollen inside of cones. Male cones produce pollen and tend to be small and inconspicuous. The familiar brown, woody cones that many people associate with conifers are the female cones, which contain the seeds. The seeds are spread when the female cones are dropped, eaten, or carried away by wildlife.

Tree reproduction has adapted many variations through aeons of evolution. Some seeds have wings, like those of maple trees; these may be commonly called "helicopters," "whirlers," "twisters" or "whirligigs." These seeds are carried by the wind to germinate away from the parent tree. Other trees rely on fires for reproduction; their cones or fruits are sealed with resin, and open to release seeds only after they're exposed to intense heat from a fire. Some plants also rely on smoke chemicals to break seed dormancy and can spend decades waiting in the soil for the next forest fire.

For a more detailed discussion of tree reproduction, see "Forest Regeneration and Reproduction" on page 16.

PEACH (*PRUNUS PERSICA*)

Native trees support an untold number of North American wildlife species ranging from songbirds and mammals to pollinators such as hummingbirds, butterflies, and bees.

Conservation

Trees have evolved as the bedrock upon which entire natural ecosystems are built, but many trees face conservation challenges and threats to their survival. Many of these threats are related to exploitation and other human activity, such as changes in land use and climate change. (For a more detailed discussion, see "Threats to Trees and Forests" on page 25.)

Throughout this book, you will see references to most tree species' conservation status as designated by the International Union for Conservation of Nature's Red List of Threatened Species. These designations, in order of severity, are denoted as follows:

LC Least Concern

NT Near Threatened

VU Vulnerable

EN Endangered

CR Critically Endangered

EW Extinct in the Wild

EX Extinct

Some of the trees in this book are cultivated or domesticated species that have not been assessed in the wild. Generally tree populations planted by humans for horticulture or landscaping are not included in the assessment of a tree species' conservation status in the wild.

Native trees are those that occur naturally in an area. These trees have evolved to the environmental and ecological conditions of the region, and support native wildlife communities that have co-evolved to thrive alongside them. Native plants provide food and shelter for native birds and other widlife.

In North America, a native plant is defined as one that was naturally found in a particular area before European colonization. Native trees support an untold number of North American wildlife species ranging from songbirds and mammals to pollinators such as hummingbirds, butterflies, and bees.

In the modern era, agriculture and urbanization have led to a great deal of intact, ecologically productive land in North America becoming fragmented and altered into monoculture crop fields or lawns and filled with exotic plants. These non-native landscapes do not support functional ecosystems that support our native wildlife. Worse yet, some planted species can become invasive, spreading into remaining natural spaces and degrading or destroying native wildlife habitat.

The importance of native plants to North America's birds and other native wildlife cannot be overstated. Research by entomologist Doug Tallamy has shown that native oaks, for example, support more than 550 species of butterflies and moths alone, compared to the non-native Ginkgo tree that supports just five. These caterpillars are a vital food source for countless migrant and resident birds. Landscapes full of non-native plants mean a shortage of food for birds. Because 96 percent of North America's terrestrial birds feed insects to their young, a shortage of native plants supporting native insects spells disaster for our birds.

For more about native vs. non-native plants, see the discussion in "What Tree Where?" on page 21.

TABLE MOUNTAIN PINE (*PINUS PUNGENS*)

Tree Classification—What Is a Tree?

What is a tree? Although individual definitions may use slightly different wording, most sources usually are quite similar to the definition below, or at least include a combination of these traits to define a tree. Trees, not surprisingly, don't spend much time browsing the internet or reading dictionaries, so in reality many of these defined traits can become muddy. Trees are highly diverse, adaptable, dynamic individuals who develop and respond to their environments in ways that make general definitions difficult.

For this book, we have defined a tree as: A woody perennial plant, typically having a single usually elongate main stem, growing to a considerable height, generally with few or no branches on its lower part.

The first part of this definition starts with the fact that trees are perennial woody plants. Botanists use a series of terms to define the longevity of different plant species. An **annual** is a plant that can go through an entire life cycle (germinate, grow, flower, seed, death) in a single growing season. Therefore, annuals growing in natural habitats don't generally live more than a year, but they bridge growing seasons in the form of seeds. **Biennials** complete their life cycle in two growing seasons and, much like annuals, all parts including the roots die after they produce seed. **Perennial** plants are those that have the capacity to live for more than two seasons and will maintain living organs over dormant seasons. Within perennial plants, there are two main groups: **herbaceous** and **woody plants**, with trees falling in the latter group. Although some annuals, biennials, and herbaceous perennials can grow quite tall, none of these would be considered a tree—size alone isn't enough.

Whether a perennial is considered "woody" is a product of its anatomy and physiology. To be a woody plant, secondary growth must be present, and the "wood" cells are generally strengthened with structural compounds including cellulose and lignin. This secondary growth is what results in the annual rings present in many species. More information can be found in the discussion on "Wood and Bark" on page 11. Based on this definition, technically palm trees aren't trees. Palms are in the family Arecaceae, which are monocots, similar to grasses

and bamboos. While palms have woody-like trunks and are perennials, they lack secondary growth and therefore cannot be considered a true tree from an anatomical perspective. From a functional perspective, palm trees have many important ecosystem and human services, so for the purposes of this book we are including them.

After establishing whether a plant qualifies as a woody perennial, one encounters the chronic confusion about classifying "tree" versus "shrub." Our definition of a tree says this can be determined by whether the plant has a single stem or trunk—but is it really that easy? There are a number of reasons a tree could have more than one stem or trunk. Many trees are regularly found as multi-stemmed individuals (e.g., *Betula populifolia, Acer circinatum, Tamarix ramosissima*). Sometimes this is due to the species' inherent growth and architecture or, more often, it is a result of some sort of stress or disturbance. Fires, flooding, wind damage, and forest harvesting can all cause damage to woody plants, causing them to sprout new stems, and often two or more stems will end up sharing dominance. This is why multi-stemmed species are common in riparian and floodplain areas. This sprouting is also responsible for "fairy rings" in Redwood (*Sequoia sempervirens*) forests—rings of clonal younger Redwoods that have sprouted up from logged stumps. Although this can be confusing, generally shrubs are species that consistently have multiple stems regardless of disturbance or management history, and these stems tend to stay fairly small in diameter over time. Trees tend to have fewer, larger diameter trunks, with the number of stems diminishing over time without repeated disturbance.

Most definitions of a tree also include language about achieving a considerable height. This part of the definition cannot be applied at the individual level, but should be applied to a species as a whole. There are important developmental and climatic considerations. What about seedlings and saplings? Surely young trees are still trees even when they are still quite short. As for climate, consider trees that grow at the timber line. Trees in subarctic and alpine environments regularly manifest as krummholz, which roughly translates from German as "crooked wood." This is the phenomenon in which exposure to extreme weather (i.e. wind, cold) causes trees to become stunted and deformed.

For this book, we have defined a tree as: *A woody perennial plant, typically having a single usually elongate main stem, growing to a considerable height, generally with few or no branches on its lower part.*

Take these individuals to lower latitudes and altitudes and they may be tall, majestic trees, but in these more adverse conditions they tend to hug the ground and look like shrubby ground covers. In both of these instances, the trees have the potential to grow to a considerable height even if they haven't done so yet.

The last part of the definition states that trees generally have few or no branches on their lower part. A lot of the same controls on height and number of trunks also control branching. A variety of species-specific traits including shade tolerance and the ability to self-prune combine with habitat conditions to determine whether a tree has branches close to the ground. A good example is the typical "Christmas Tree," which is shaped like a cone with branches all the way to the ground. A couple of different shade-tolerant conifers (*Abies* spp., *Picea* spp., etc.) are grown for this, generally in open fields. These species are adapted to take advantage of as much light as possible, so in open-grown conditions with ample side light, those trees will hold on to their needles and branches as long as possible to maximize photosynthesis. In forests, in the presence of competition and side shade, their growth habit would be much different.

Here is one final example that demonstrates the challenge in defining trees. There is a grove of Bristlecone Pine (*Pinus aristata*) in the Great Basin of California; some of the most iconic individuals in the grove are more than 5000 years old. These pines have a krummholz-like growth form: they are fairly short, with multiple stems, and branch pretty low to the ground. For all intents and purposes, the "Methuselah" Bristlecone Pine shouldn't be considered a tree by the definitions offered in the dictionary. Yet, you would be hard-pressed to find someone who didn't think these majestic specimens were trees.

BRISTLECONE PINE (*PINUS ARISTATA*) IN A SHRUBBY GROWTH FORM

BRISTLECONE PINE (*PINUS ARISTATA*) GROWING AS A TALL TREE

By this same logic, we've included a few species throughout this book that, although they may not meet the *technical* definition of a tree, can be large and tree-like enough—at least in some conditions—to be commonly thought of as "trees." Acknowledging that the lines can be rather blurry, and that some authorities might quibble about their inclusion in this book devoted to trees, the accounts of these species include notes about whether they are usually large shrubs or otherwise tree-like but not true trees.

■ *Marlyse Duguid*

Forest Regeneration and Reproduction

Trees are remarkable organisms that can live an incredibly long time, but individual trees are not immortal. To sustain their species into the future it is important for trees to reproduce themselves; in forests this process is called regeneration. Individual tree species have a suite of different strategies to ensure their progeny continue into the future; these strategies are highly tied to environment, disturbance, and species traits, but the diversity of these strategies proves there isn't just one way to successfully reproduce.

Although many people may think of plant reproduction as simply a seed germinating and growing into an adult, sexual reproduction methods are only half of the story. Less commonly considered but no less prevalent in trees is asexual reproduction. Asexual reproduction is when a tree produces new individuals (referred to as ramets) that are genetically identical to the parent plant.

Individual tree species have a suite of different strategies to ensure their progeny continue into the future; these strategies are highly tied to environment, disturbance, and species traits, but the diversity of these strategies proves there isn't just one way to successfully reproduce.

Although there are a couple of different asexual reproduction strategies, they are generally similar functionally—dormant buds or meristems (plant stem cells) are triggered to make new and novel organs through hormonal intra-plant communication. Examples include stumps sprouting following logging, or a sapling sprouting after being grazed by an herbivore. The loss of the top of the tree alerts the base to send new sprouts from the root collar. This allows a plant with much older roots to send a new young sapling. This type of growth, referred to as "coppice," has long been utilized by people to manage trees for multiple small stems, for charcoal production, fuelwood, or forage.

Clonal growth can also happen from the roots in some species. This phenomenon can be seen on a grand scale following large disturbances such as fires. There is a large grove of Quaking Aspen (*Populus tremuloides*) in Utah, nicknamed "Pando," which is a clonal colony of more than 40,000 stems, but with a single genetic origin. Although each individual "tree" in the colony is a little more than 100 years old, the roots are estimated to be 80,000 years old.

Certain species also have the ability to produce roots from branches, either while still connected to the parent tree or when broken off. This is an excellent adaptation for floodplain residents and alpine areas that experience landslides. Willow trees (*Salix* spp.) are well-known for this. If a tree falls over and touches the ground it can put down roots, and another upright portion can form. If stems get broken in floods, the branches float downstream, come to rest on newly deposited soil, and quickly establish new vegetation by rooting.

Although asexual reproduction can maintain a tree's space on the landscape for a long time, it limits the ability of the species to evolve and adapt to changing conditions. Considering the earth's climate is constantly changing and trees are relatively slow to mature and develop, sexual reproduction—genetic recombination to produce genetically unique individuals (referred to as genets)—is necessary for the long-term demographic stability of a species. The more genetic diversity there is in a species, the more likely it is that resistance traits will exist in the population that could help it to overcome adver-

QUAKING ASPEN (*POPULUS TREMULOIDES*)

sity. Take the Pando example mentioned previously: That aspen grove in Utah is in decline, and as only one genetic individual, if there is a forest pathogen that attacks it, the population isn't likely to have any surviving individuals. Genetically diverse populations that are a result of sexual reproduction tend to be much more resilient.

Sexual reproduction requires eggs and pollen, but reproductive parts can be arranged in a number of ways. Although here we are referring to these organs as flowers, technically only the angiosperms have flowers. Gymnosperms (conifers and ginkgo) don't have true flowers but instead have cone-like structures called *strobili*. While some species bear flowers with both pollen-producing and egg-bearing parts (so-called "perfect flowers"), it is highly common for trees to have separate male and female flowers. Some trees have both types of flowers, but separately, on the same individual (*monecious*), while other species have separate individuals for pollen and egg bearing flowers (*dioecious*). This is an oversimplification, as many species exhibit intermediate conditions. These strategies are all in place to aid outcrossing, or cross-pollination between different individuals. It is possible for plants to self-pollinate, but this is less advantageous from a genetic diversity standpoint. Separating the eggs

and pollen across both space and time helps to encourage genetic recombination in a population.

Pollination is the process of getting pollen to eggs between individuals, and is primarily achieved in trees by either the wind or by animals. The strategy used is highly tied to climate. In wet tropical rainforests, where climate is fairly predictable and steady, more species tend to rely on animals (insects, birds, and small mammals). By contrast, in boreal forests where there is more wind and seasonality, wind is a more reliable strategy.

Once pollinated, trees will produce some sort of seed. The shape, size, and color of those seeds determine the mode of dispersal. Small, light, winged seeds can get picked up on the wind and travel long distances. Trees with this type of seeds tend to produce a lot of seeds and are often common in early successional habitats. The strategy is to get as many seeds as possible out on the landscape and hope that they land in an area with ideal conditions. Gravity dispersed seeds tend to be round and heavy with hard outer coatings, and are far less numerous because of the parental investment in each seed. Seeds that rely on animal dispersal include berries and fruits that need to be eaten and passed as well as nuts that may be cached by hungry rodents or birds for later consumption, only to go uneaten. Water dispersed seeds float on ocean currents or streams to find new habitats.

Often a single fruit can employ multiple dispersal strategies, and these can be quite specialized. Coconuts (*Cocos nucifera*) on tropical volcanic islands, for example, use gravity to fall to the ground and roll down the mountains to the ocean, and then can float on ocean currents until they find new islands to inhabit. Acorns from oak trees also use gravity for primary dispersal, but then rely on birds and scatter-hoarding rodents to take those nuts away from the parent plant. Some trees even need fire to disperse their seeds: serotinous species require the high temperatures associated with fire to release their seeds (some *Pinus* spp.). This ensures that these fire-adapted species plant their seeds in the ideal habitat for regeneration.

Through all these stages there are many challenges (herbivory, disturbance, habitat mismatch) that often result in reproductive failure. Even if a tree is successful in flowering, pollination, and dispersal, reproduction still isn't guaranteed. That is why many trees employ both sexual and asexual reproduction strategies, and with long lives and many years to try, hopefully they are ultimately successful in reproducing themselves at least once.

■ *Marlyse Duguid*

CRAPEMYRTLE (*LAGERSTROEMIA INDICA*)

Plants with a tree growth form first appear in the fossil record in the middle of the Devonian geological period, about 390 to 380 million years ago.

The Evolution of Trees

During the past few decades, scientists have made great strides towards understanding the evolutionary relationships among plants. All land plants (embryophytes) fall into a single evolutionary lineage, and a major lineage within this group is the vascular plants (tracheophytes). Vascular plants have specialized tissue for conducting water and nutrients, and they are the familiar plants that dominate terrestrial biomes. Research by evolutionary biologists has shown that the tree growth form evolved independently in several major vascular plant lineages. That is, trees (past and present) do not all fall into a single evolutionary group that excludes plants with different growth forms; but instead, the tree growth form is scattered across the vascular plant branch of the tree of life into evolutionary groups that contain species with other growth forms, such as herbaceous plants and shrubs.

All existing North American trees growing in temperate regions are from one major land plant lineage, the seed plants. The tree growth form within the seed plant lineage evolved, was lost, and

GREEN HAWTHORN (*CRATAEGUS VIRIDIS*)

in some cases evolved again. This pattern is reflected in several of the plant lineages mentioned in this book. The rose family (Rosaceae), which has trees such as cherries and hawthorns, also has herbaceous representatives such as strawberry and cinquefoil, plus shrubs, such as roses. Conversely, some lineages that have mostly herbaceous representatives in North America have tree representatives in other areas. The bean family (Fabaceae) has many herbaceous and shrub representatives in North America, but in the tropics the family has many tree species, too, some of which are among the tallest trees of evergreen rain forests. Some other examples include the sunflower family (Asteraceae), which has tree representatives on some islands and in tropical areas; the violet family (Violaceae), which has tropical tree representatives in the genus *Rinorea* that are up to 50 feet tall; and the morning glory genus (*Ipomoea*), which has tree representatives in some dry, tropical forests.

Fossil History

Plants with a tree growth form first appear in the fossil record in the middle of the Devonian geological period, about 390 to 380 million years ago (mya). In addition to seed plants, the tree growth form evolved within the horsetail, fern, and lycophyte lineages (club mosses and relatives), and some other now-extinct groups. Although seed plants rose to predominance during the Permian period (about 290–250 mya), other land plant lineages dominated earlier. During the Carboniferous period (about 360–300 mya), horsetails and lycophytes, along with some early seed plants, formed extensive forests, the fossilized remains of which are a major source of coal. Amazingly, plants in the lycophyte lineage, which are now low-growing, herbaceous plants, reached heights of over 100 feet and had trunks more than 3 feet in diameter. It has been proposed that the evolution and proliferation of lycophytes and other kinds of trees in the Devonian helped to significantly transform the Earth's environment by altering soils, carbon dioxide levels, and climate.

Advantages of Being a Tree

Why did various plant lineages converge on the tree growth form? The tree growth form evolved as a way to deal with various conditions related to reproduction, light acquisition, herbivory, and environmental stress. Being tall allows trees to outcompete smaller-statured species for light, helps reduce herbivory by ground-dwelling animals, provides protection against fire, and facilitates fertilization and dissemination of fruits and seeds. Being a long-lived perennial helps mitigate against negative, short-term environmental conditions (for example, drought). Of course, there are evolutionary trade-offs with the features associated with being a tree, hence the high prevalence of other growth forms among vascular plants.

■ *Patrick Sweeney*

Defining Forested Ecosystems

Ecologists divide and classify vegetative communities into biomes, or communities that share characteristics for the environment in which they live. Biomes can be classified using two simple climatic factors: temperature and precipitation. These two factors can tell us quite a bit about how a vegetative community can develop. In all but the coldest and driest environments around the globe, forests are the common vegetation structure.

At its most basic definition, a forest is an area covered primarily with trees. Forests on this planet are diverse in composition, structure, and origin. A large plantation may be chiefly covered in trees, but these trees may be highly managed like an agricultural field, sprayed with herbicides to keep down competition and thus bare on the ground, with each tree genetically identical and not connected with each other or any other organism (see the discussion of "Plant Communication" on page 23), and harvested and replanted in short rotations. Compare this with a lush old-growth forest, which has compositional and structural diversity and numerous biological connections with other organisms. It is important to realize that these two types of forests can exist in a single climate and that human intervention and land-use greatly influences the nature of these forests. Here we will focus on describing forests in each biome that are not currently intensively managed for timber and are in what many would consider a "natural" state. It should be noted that in many cases this natural condition may be the result of long-standing, low-intensity indigenous human management of ecosystems.

In the warmest, wettest environments we find tropical rainforests. Wet tropical forests are characterized by high tree species diversity, dense canopies, and diverse structure (multiple strata). Warm, wet conditions are optimal for decomposition and nutrient cycling, so most of the nutrients in tropical rainforests are tied up in the vegetation and the soils can sometimes be quite poor. This makes wet tropical rainforests highly susceptible to degradation if they are cleared and burned for land conversion. Also found in the tropics are dry tropical forests; in these forests, distinct wet and dry seasons necessitate species to have unique adaptations. Although dry tropical forests have lower overall species diversity than their wetter counterparts, each of these dry forests is distinct (e.g., Madagascar versus Mexico) and these ecosystems are home to a number of rare and remarkable endemic species. Dry forests worldwide are at risk of land conversion and development.

Boreal forests are the coldest forests transitioning from tundra. These forests stretch over huge expanses across the highest latitudes of the earth that have a relatively low density of people compared to temperate regions. Long, cold winters mean slow decomposition and nutrient cycling, so high amounts of soil carbon and peatland punctuated with areas of permafrost can be found in the boreal zone. Overall, there is fairly low species diversity, with only a few genera dominating at these high latitudes, including conifers—firs (*Abies* spp.), larches (*Larix* spp.), and spruces (*Picea* spp.)—and a few broadleaf species such as aspens (*Populus* spp.) and birches (*Betula* spp.).

The majority of forests in North America are classified as temperate forests. Although they share a similar broad set of climatic requirements, these forests display the influence of local conditions such as physical geography, soils, disturbance regime, and land-use history. Perhaps the most iconic of the temperate forests are the temperate rainforests of the Pacific Northwest of the United States. These forests have high levels of structural diversity, and are dominated by huge conifers such as Western Red Cedar (*Thuja plicata*) and Douglas-fir (*Pseudotsuga menziesii*). These forests also tend to feature a lot of downed wood, undergrowth, ferns, and mosses.

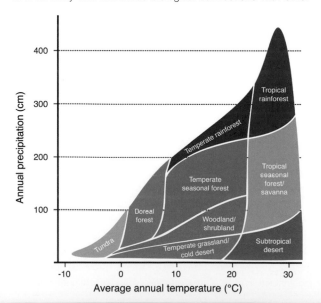

EASTERN HEMLOCK (*TSUGA CANADENSIS*)

In eastern deciduous mixed mesophytic forests, the trees tend to be much smaller, and most of the trees are hardwoods—including oaks (*Quercus* spp.) and maples (*Acer* spp.)—punctuated by relatively few conifers such as White Pine (*Pinus strobus*) and Eastern Hemlock (*Tsuga canadensis*). There is often a much greater diversity of trees in the canopy in eastern forests than in the Pacific Northwest. Most of the eastern deciduous forests are second-growth, recovering from European colonialism, the industrial revolution, and the loss of the American Chestnut (*Castanea dentata*) described in "Threats to Trees and Forests" (page 25).

Not far from these mixed-mesophytic forests are the forests of the coastal plain of the southeastern United States. Although the climate is almost identical in places, the underlying soils of the coastal plain are deep sands. These forests tend to be dominated by conifers, specifically pines, and most are fire-adapted ecosystems. One example is the Longleaf Pine savanna. These forests are open, park-like forests with high levels of herbaceous diversity in the understory. Here the Longleaf Pines (*Pinus palustris*) grow in almost monocultures at wide spacing, a completely different structure than the other two temperate forests. These forests are home to a number of rare species and at one time were much more expansive. Conversion of land for industrial plantation forestry and agriculture are some of the biggest risks to these important ecosystems.

Although the majority of trees in this book can be found in forests, not all trees are. North America is home to many non-forested habitats—chaparral, grasslands, high-plains desert, and urban areas all are home to multiple tree species.

■ *Marlyse Duguid*

What Tree Where?

Throughout this book you will find that most tree species have a distribution map associated with them. These distribution maps represent, to the best of our knowledge, the "native" range of that species. But what does native mean? Nativity for plant species is a complicated concept. We generally think of a "native" species as having evolved in a given area or having arrived without any human assistance, while the presence of "non-native" species in a region is directly attributable to human assistance to overcome a barrier (mountain range, ocean, etc.) that it would be unlikely to disperse across on its own.

Although it is tempting to assume that the native range of a species represents the ideal climate and habitat for that species,

Even when a species is well suited for the climate, it is still continually challenged through a series of environmental filters, both abiotic (light, moisture, nutrients) and biotic (interactions with other species).

TREE BIOLOGY TOPICS

AUTUMN OLIVE (*ELAEAGNUS UMBELLATA*)

this is not always the case. Any wild plant community has been through a sequence of processes. First, each of the species in an area needs to arrive on the site through natural dispersal (see "Forest Regeneration and Reproduction" on page 16). Therefore, the final plant community is constrained from the start, because only a small portion of all potential plant species are likely to find their way to any specific site.

Once those propagules are present, that pool of available species faces a series of filters through which the species need to survive and thrive. The broadest of those filters is the physiological limits in which a plant survive; we call the range of temperatures and moisture conditions a species can survive its climate envelope. Some species have broad tolerances and thus larger envelopes, while others are more sensitive to cold or drought conditions and therefore have smaller envelopes.

Even when a species is well suited for the climate, it is still continually challenged through a series of environmental filters, both abiotic (light, moisture, nutrients) and biotic (interactions with other species).

The species that make it through this succession of filters and can coexist with each other are the ones that ultimately form the native plant community.

Humans, however, have been cultivating and moving plants around for thousands of years, so in most habitats—especially those inhabited by people—plant communities are not exclusively native.

Horticulture and arboriculture use a mixture of knowledge of species life-history traits and trial-and-error to establish trees in new areas. The United States Department of Agriculture (USDA)

puts out plant hardiness zone maps as a guide. These zones are based on the lowest annual temperature in an area, and individual tree species are assigned zone ratings based on the assumed climate envelope from its native range or former horticultural experience. These are broad swathes and don't deal directly with the local habitat conditions. But, in planted conditions we remove a lot of the filters found in native plant communities, specifically competition and other biotic interactions.

Non-native trees can fall into a few different categories. **Casual** non-native species probably represent the largest proportion. These are species that generally stay where you put them and do not maintain reproductively successful populations on their own. Most horticultural and agricultural introductions fall into this category. The persistence of these species on the landscape requires repeated planting and cultivation.

Some non-native tree species become **naturalized**, meaning they are able to self-sustain their populations on the landscape without human cultivation. A good example would be a tree you plant in your garden that tends to seed around a bit, but generally stays on the property. Many trees naturalize only in local areas, building a small, self-sustaining population in pockets where the conditions are favorable for its growth.

The third class of non-native tree species are what people have deemed **invasive**. An invasive tree species is one that not only has readily established a robust population, but also has expanded that population across long distances. Tree species from almost every continent have found themselves deemed invasive in another place. The scale and range of each of these is specific.

If a plant feels a little nibble or set of caterpillar feet on its leaves, it can send chemical signals from cell to cell, and within minutes cells on the opposite side of the plant can produce secondary compounds to protect it from herbivory.

This book includes native, naturalized, and invasive tree species found in North America. Into which of these categories a tree species belongs is often location-dependent.

For example, Monterey Pine (*Pinus radiata*) is an uncommon species of tree found in only a few small native populations in coastal California. Based on the rarity of its native habitat, one might assume that this species has a small climate envelope and is difficult to grow. In fact, Monterey Pine has been introduced as a timber tree in Mediterranean climates worldwide. It can be found in Australia, South America, Europe, and Africa, and is considered an aggressive, non-native invasive species in some localities around the world. Indeed, it is the most commonly found tree in parts of New Zealand after being introduced there by humans. The takeaway is that where we find plants in the wild does not equal where they will best grow and thrive, and that humans have greatly modified plant communities worldwide.

■ *Marlyse Duguid*

Plant Communication

The concept of trees as cognizant beings is not new but continues to be controversial. Sentient plants are a common theme in books and film, and humans love to think of trees as a living and active part of the landscape. Since the controversial best-selling book *The Secret Life of Plants* by Peter Tompkins and Christopher Bird came out in the 1970s and was universally panned by the scientific community, many scientists have been wary of the subject. Relatively recently there has been a bit of resurgence in the field of plant cognizance, sometimes called plant neurobiology. Whatever you choose to call it, it is increasingly clear that plants, including trees, are not just passive parts of the landscape, but dynamic organisms interacting with their environment.

If a plant feels a little nibble or set of caterpillar feet on its leaves, it can send chemical signals from cell to cell, and within minutes cells on the opposite side of the plant can produce secondary compounds to protect it from herbivory. Plants also

MONTEREY PINE (*PINUS RADIATA*)

It is clear that the interactions between trees and other organisms are vibrant and dynamic.

respond to touch, with well-known examples including carnivorous plants like Venus flytraps and sundews. This movement is called thigmonasty, and is primarily a mechanical response, but there is sometimes more to the story. One study showed that Venus flytraps counted how many times their trigger hairs were touched and produced increasing amounts of digestive chemicals in response to larger insects (more touches). Further, if you expose flytraps to lidocaine, a local anesthetic, their leaves will be unresponsive for a couple hours until the chemical wears off—suggesting that both animals and plants react to common anesthetics in similar ways.

Plants can also "hear." After all, sound waves are simply vibrations, and plants clearly have the ability to sense the environment around them through their leaves—but what about their other organs? Recent work has shown that the roots of plants will grow toward the sound of water even when water isn't present. There is also evidence that certain flowers use their petals to sense pollinators flying by and when they are exposed to sound frequencies of flying bees they will produce sweeter nectar within three minutes. Sweeter nectar could keep a pollinator around a little longer and increases the likelihood that the bee will visit more of that species in the future. Although neither of these studies was specifically done in trees, there is every reason to believe that trees could have similar sets of sensitivities to sound.

Plant-to-plant communication is the best-known of these phenomena in trees and happens through a couple of different pathways. Trees appear to have multiple networks for sharing resources and information with one another. Natural root grafts (when the vascular systems merge through direct contact) are common in a number of tree species. We have known for a long time that trees move water, carbon, nutrients, and chemical messages through these grafts, but it was long believed that these connections were limited to other individuals in the same or closely related species. More recent research suggests that trees are communicating between different species through their mutualist partners, mycorrhizal fungi, through what has been deemed the "wood-wide-web." In the 1990s Dr. Suzanne Simard showed that Paper Birch (*Betula papyrifera*) and Douglas-fir (*Pseudotsuga menziesii*), two species not closely related, were sharing resources underground. Since then Dr. Simard and her lab have shown how extremely complex these underground networks are.

Trees also communicate through airborne chemicals. Plants produce a suite of volatile organic compounds (VOCs) for a number of reasons. When a tree is attacked by an herbivore, it will put out a specific set of VOCs, and these compounds travel on the wind. As other trees "smell" those compounds in the air, they respond by producing protective chemicals in anticipation of an attack. Interestingly, there is evidence that trees communi-

PAPER BIRCH (*BETULA PAPYRIFERA*)

cate best with their close kin in this way, even within a species, suggesting that trees can discriminate and pay better heed to the chemical cues from their relatives than from strangers.

This isn't the only case of kin recognition in plants. Studies have shown that plants can recognize if their neighbors are family and choose how they grow their roots or orient their leaves to either share or compete for resources based on relatedness.

Clearly, trees communicate among themselves, but do they also communicate with other organisms? In a couple of cases it has been shown that both predators (specifically birds) and parasitoids can sense some of those VOCs emitted by trees in response to insect herbivory. This could be interpreted as the tree signaling for help when being eaten by an insect, and the insect's predators responding to the distress signal by eating or infecting the offending insect. Although the use of anthropocentric language is probably not accurate for all of these processes, it is clear that the interactions between trees and other organisms are vibrant and dynamic.

So, what about humans? It is clear that spending time in forests is good for us, with recent research showing multiple benefits to mental and physical health. Tree-produced VOCs include a diverse set of chemicals that are difficult to isolate and test. Likely, some of these chemicals can be harmful to humans, and some may have benefits. Modern science doesn't support the idea that trees are communicating with us, though—for now, we will have to leave that to the movies.

■ *Marlyse Duguid*

Threats to Trees and Forests

The earth is home to around three trillion trees distributed across diverse landscapes including forests, savannas, agricultural fields, wetlands, mangroves and cities. These trees bestow untold benefits to the environment and humankind. They clean our air and water, hold our soils and moderate the climate, provide habitat for wildlife, and provide important economic, health, and cultural benefits for people. Yet, across the planet there are significant and imminent risks to trees and forests.

Among the greatest of those threats are humans, responsible for both direct and indirect pressures on trees and forests. The most apparent are human land-use and development. Humans and forests have coexisted since prehistory, but with the growth of urbanization, humans and forests are increasingly competing for space. This manifests across a variety of scales, from the conversion of large tracts of tropical rainforests for industrial grazing land, to fragmentation of forests in previously rural areas to make way for expanding suburban housing, to the removal of forests for mining of energy resources and minerals. Overall, the current estimates of global forest cover lost due to deforestation and land-use conversion are around 192,000 km^2 per year, or about 15.3 billion trees annually.

Indirect negative effects of humans on forest health and tree loss include a suite of interrelated risks including species invasions, pests and pathogens, anthropogenic climate change, and changing disturbance regimes (such as fires and severe storms).

As long as modern humans have traveled, they have moved species around with them, both purposely and inadvertently. The vast majority of these species are relatively harmless, but some are rapidly spreading and have severe ecological and/or economic impacts. A well-known historical example is the loss of the American Chestnut (*Castanea dentata*) from forests across the eastern United States from the invasive fungal pathogen called chestnut blight (*Cryphonectria parasitica*). There are numerous invasive pests and pathogens currently threatening North American forests including Emerald Ash Borer (the insect *Agrilus planipennis*) and sudden oak death (the pathogen *Phytophthora ramorum*). In addition, there are many invasive plant species that displace trees or disrupt forest ecosystems. Examples include Tree-of-heaven (*Ailanthus altissima*) displacing native forest trees across the United States, herbs such as garlic mustard (*Alliaria petiolata*) whose allelopathic tendencies inhibit regeneration, and aggressive vines such as kudzu (*Pueraria montana*) that can strangle and pull down trees.

Although globally we think of climate change as significant warming and increased severe weather events, climate change manifests quite diversely across ecoregions, causing variations not only in temperature, but also precipitation and weather

Overall, the current estimates of global forest cover lost due to deforestation and land-use conversion are around 192,000 km^2 per year, or about 15.3 billion trees annually.

AMERICAN CHESTNUT (*CASTANEA DENTATA*) IS CRITICALLY ENDANGERED SINCE THE INTRODUCED DISEASE CHESTNUT BLIGHT CAUSED ITS POPULATION TO COLLAPSE.

patterns. A changing climate alone is not necessarily bad for trees and forests. Trees in many regions will actually grow better and faster under higher levels of carbon dioxide, warmer temperatures, and potentially increased precipitation. In many other regions, drought and excessive heat can severely stress trees and forests. Individual tree species each have a "climate envelope," the suite of climatic conditions in which that species occurs. Under past climate change scenarios in Earth's past, tree species were able to migrate and change their ranges to accommodate changing climate patterns. Scientists can map species' migrations in the past by examining the historical pollen record in lakebeds and other sediments. They use these historical and current distributions to build models to project future ideal habitat ranges under the anticipated future conditions expected in climate change scenarios.

One of the most challenging aspects of modern anthropogenic climate change for trees is the rate at which the climate is changing now. Trees are relatively slow-growing organisms, taking many years to decades before reaching a reproductive age,

and many having an ability to disperse seeds only over relatively short distances. Along with slow migration, human land-use and forest fragmentation have put additional dispersal barriers—such as urban areas—in place that impede natural migration. Therefore, trees may persist in areas as climate relicts, but may not be able to reproduce, or may be unable to establish progeny in more suitable climatic conditions.

Climate change has the potential to disrupt historical disturbance ecology with often catastrophic consequences for forest health. In many regions, climate change has already changed habitats, initiated record-setting droughts, caused more frequent severe wind events, and shifted fire return intervals and intensity. All of these risk factors can interact with major consequences for forest health. One example is Mountain Pine Beetle (*Dendroctonus ponderosae*), an insect native across much of western North America. Management decisions (e.g., fire suppression) as well as changing climatic conditions have amplified outbreaks of this native forest pest. Denser forests provide more habitat, and longer, hotter summers and milder winters

Climate change has the potential to disrupt historical disturbance ecology with often catastrophic consequences for forest health.

Trees are amazing, resilient organisms, and forested ecosystems have a lot of redundancies in place as back-up plans that have helped them overcome novel challenges for hundreds of millions of years.

increase the lifespan and reproductive rate of these insects and allow their populations to expand into higher altitudes and latitudes. The U.S. Forest Service estimates that between 2010 and 2018 the state of California alone lost 147 million trees due to the combination of drought and bark beetle infestation. Standing dead trees from bark beetle outbreaks combined with record droughts from climate change increases fire susceptibility in these forests. Some of the catastrophic fires in the first couple decades of the 2000s were a result of this combination of forest stressors.

Sea level rise and increases in severe weather events from climate change, such as hurricanes and tornadoes, also have ramifications for trees and forests. Mangrove forests, found along coastlines in Florida, Louisiana, and Texas serve as important barriers for storm surges. Whether these forests will be able to keep pace with sea level rise and increasing hurricane frequency remains to be seen. Wind events can snap and topple trees as well as support establishment of invasive species following disturbance. It is clear that as climate change continues to change ecological relationships between organisms and their environment, we will continue to see novel disturbances and their subsequent effects on trees and forests.

Perhaps the most concerning aspects surrounding the conservation of forests are the threats we cannot predict. New species invasions, pollutants, introductions of genetically modified organisms, and new technologies and the associated pressures on natural ecosystems that accompany them all have the potential to fundamentally change forested landscapes as we know them.

Yet, hope should remain: Trees are amazing, resilient organisms, and forested ecosystems have a lot of redundancies in place as back-up plans that have helped them overcome novel challenges for hundreds of millions of years. Forests grow back following fires and storms, some species develop immunity to pathogens, and humans have undertaken amazing tree planting and restoration activities across the planet.

■ *Marlyse Duguid*

ENGELMANN OAK (*QUERCUS ENGELMANNII*)

WHITE MANGROVE (*LAGUNCULARIA RACEMOSA*)

■ DENSE POPULATION
■ LIGHT POPULATION
■ NATURALIZED

This large, beautiful tree of the Pacific Northwest gets its common name from the silvery lower surface of its foliage. David Douglas (1799–1834), the Scottish botanical explorer and discoverer of this species, named it *amabilis*, meaning "lovely." Although beautiful when young, this tree does not attain as pleasing a shape in maturity.

DESCRIPTION: A large fir with a beautiful conical crown of short, down-curving branches and flat, fernlike foliage. Its needles are ¾–1½" (2–4 cm) long, evergreen, crowded and spreading forward in two rows, curved upward on upper twigs, flat, shiny dark green and grooved above and silvery-white beneath. The bark is light gray and smooth, becoming scaly and reddish-gray or reddish-brown.

FRUIT: Cones: 3–6" (7.5–15 cm) long; cylindrical, upright on the topmost twigs, and purplish; the cone scales have fine hairs, with bracts short and hidden. It has paired, long-winged seeds.

HABITAT: Cool, wet regions, including the coastal fog belt and interior mountain valleys; in coniferous forests.

RANGE: Pacific Coast from extreme southeastern Alaska south to western Oregon; local in northwest California; to 1000' (305 m) in north; to 6000' (1829 m) in south.

USES: The Pacific Silver Fir is used for a variety of commercial purposes including construction, pulpwood, urban landscaping, and Christmas trees.

SIMILAR SPECIES: Western Hemlock, Mountain Hemlock, European Silver Fir.

CONSERVATION STATUS: ⓁⒸ
The population of Pacific Silver Fur is considered stable. Historically this species was logged beyond sustainable levels; clear-cutting and changes in land management in some areas have not been conducive to the regeneration of these forests. This species is highly susceptible to forest fires. Local populations in Washington State and British Columbia have been decimated by the Balsam Woolly Adelgid (*Adelges piceae*), an introduced insect, but some trees have shown a resistance to it.

Diameter: 2–4' (0.6–1.2 m)
Height: 80–150' (24–46 m)

■ DENSE POPULATION
■ LIGHT POPULATION
■ NATURALIZED

This strongly fragrant, small to medium tree is the most common and widespread fir in the East. Balsam Fir is a major pulpwood species, and its strong piney scent makes it a popular choice for Christmas trees and wreaths. Its growth occurs in whorls of branches surrounding an upright leader or terminal, making a symmetrical tree with a broad base and narrow top. It is a relatively short-lived tree, but its numbers naturally regenerate quickly and it is easily grown in nurseries.

DESCRIPTION: This is the only fir native to the Northeast, with a narrow, pointed crown of spreading branches and aromatic foliage. Its needles are ½–1" (1.2–2.5 cm) long, evergreen, spreading almost at right angles in two rows on hairy twigs, curved upward on upper twigs; they are flat with rounded tips (sometimes notched or sharp-pointed), shiny dark green above, with two narrow whitish bands beneath. The bark is brown, thin, smooth, with many resin blisters, becoming scaly.

FRUIT: Cones: 2–3¼" (5–8 cm) long; cylindrical; dark purple; upright on the topmost twigs; the cone scales are finely hairy with bracts mostly short and hidden. It has paired, long-winged seeds.

HABITAT: Coniferous forests; often in pure stands.

RANGE: Alberta east to Labrador and south to Pennsylvania, west to Minnesota and northeastern Iowa; local in West Virginia and Virginia; to the timberline in the north and above 4000' (1219 m) in the south.

USES: This tree is used for Christmas trees, wreaths, and balsam pillows; pulpwood; and sometimes for lumber. The wood is lightweight and limber. This tree also produces a fragrant resin, called Canada balsam, that is used for fragrance, glue, and optical cement.

SIMILAR SPECIES: Fraser Fir, which is restricted to southern Appalachian Mountains where Balsam Fir does not occur.

CONSERVATION STATUS:

Balsam Fir is a common tree with no specific conservation threats known. Natural regeneration methods for these trees are highly effective even in open and disturbed sites, such as heavily cut areas, as long as an adequate seed source exists. It is listed as state-endangered in Connecticut.

Diameter: 1–1½' (0.3–0.5 m)
Height: 40–60' (12–18 m)

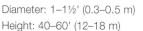

■ DENSE POPULATION
■ LIGHT POPULATION
■ NATURALIZED

The rarest fir in North America, this species is named for its uniquely bristly cones. The entire natural range of this species is limited to a coastal strip about 60 miles (97 km) long within the Los Padres National Forest in California. Aromatic resin from the trunk was used as incense in the early Spanish mission nearby.

DESCRIPTION: This is a small to medium tree with a narrow, conical crown of short, slightly drooping branches. Its needles are 1½–2¼" (4–6 cm) long and ⅛" (3 mm) wide, evergreen, spreading almost at right angles in two rows; they are flat, short-pointed, stiff; and shiny dark green above, with two broad whitish bands beneath. The bark is light reddish-brown, smooth, becoming scaly and slightly fissured at the base. Its twigs are stout, light reddish-brown, and hairless.

FRUIT: Cones: 2½–4" (6–10 cm) long; egg-shaped, purple-brown, upright on the topmost twigs. The cone scales are thin, rounded, hairless, and finely toothed, and each has a bract ending in a very long, spreading, yellow-brown bristle. It has paired, long-winged seeds.

HABITAT: Steep, rocky slopes and canyons; in mixed evergreen forests.

RANGE: Santa Lucia Mountains of southern California; at 2000–5000' (610–1524 m); locally at 600' (183 m).

USES: The Bristlecone Fir is not used for much today, but the fragrant resin was once used as incense in the Spanish missions located near this rare tree's natural range.

SIMILAR SPECIES: This tree's long needles and unique bristles help to separate it from other firs.

CONSERVATION STATUS: NT

This species is not widely used commercially and exists in small stands in areas that are not normally prone to fires. Climate change and the dieback of associated oak species due to Sudden Oak Death, however, have changed the frequency and intensity of fires and increased the amount of fuel available for fires in this species' limited range. Lack of genetic diversity and infrequent regeneration in this species limit its ability to recover from fires or other changes in its environment.

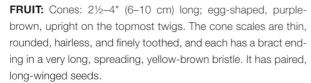

Diameter: 1–3' (0.3–0.9 m)
Height: 40–100' (12–30 m)

■ DENSE POPULATION
■ LIGHT POPULATION
■ NATURALIZED

At least two geographic varieties of White Pine exist in western North America. Rocky Mountain White Fir (var. *concolor*), of the Rocky Mountain region, grows in the warmest and driest climate of all native firs. California White Fir (var. *lowiana* [Gord.] Lemm.), of the Pacific Coast, is grown for Christmas trees and other ornamental purposes. The scientific name, meaning "of uniform color," refers to both needle surfaces. The winged seeds are eaten by a variety of songbirds and mammals, and the needles are eaten by grouse and deer. Porcupines gnaw the bark. Grouse often roost in these trees in winter.

DESCRIPTION: This is a very large fir, widespread in western mountains, with a narrow, pointed crown of short, symmetrical, and horizontal branches. The needles are 1½–2½" (4–6 cm) long, evergreen, spreading almost at right angles in two rows, and curved upward on upper twigs; they are flat, flexible, and almost stalkless, with the tips short-pointed, rounded, or notched. The needles are light blue-green with whitish lines on both surfaces. The bark is light gray and smooth, becoming very thick near the base and deeply furrowed into scaly, corky ridges. The twigs are light brown or gray, stout, and hairless.

FRUIT: Cones: 3–5" (7.5–13 cm) long; cylindrical; greenish, purple or yellow; upright on topmost twigs; cone scales finely hairy, with short, hidden bracts; paired, long-winged seeds.

HABITAT: Moist, rocky mountain soils; in pure stands and with other firs.

RANGE: Extreme southeastern Idaho southeast to New Mexico, west to California, and north to southwestern Oregon; local in northwestern Mexico. At 5500–11,000' (1676–3353 m) in the south; to 2000' (610 m) in the north.

USES: The lightweight wood of the White Fir is used for lumber, crates, doors, and other purposes. It is also a common pulpwood species and is often used for Christmas trees due to its pleasant aroma and sturdy branches.

SIMILAR SPECIES: Grand Fir.

CONSERVATION STATUS: LC
This species is known from many protected areas throughout its range. No range-wide threats have been identified for this species and its population is stable.

Diameter: 1½–4' (0.5–1.2 m)
Height: 70–160' (21–49 m)

■ DENSE POPULATION
■ LIGHT POPULATION
■ NATURALIZED

Sometimes considered a subspecies of Balsam Fir, this rare and declining fir is found in the southern Appalachian Mountains, particularly at Great Smoky Mountain National Park and at Mount Mitchell in North Carolina. With its silvery and green foliage, this species is commonly grown for Christmas trees and ornamental purposes. It is known locally as "She-balsam" because of the resin produced in the bark; in contrast, Red Spruce (*Picea rubens*) in the same forest but without resin blisters is often called "He-balsam." The Scottish explorer John Fraser (1750–1811) discovered this fir and introduced it and many other plants in Europe.

DESCRIPTION: The only native southeastern fir, this is a handsome tree with a pointed crown of silvery white and aromatic foliage. Its needles are ½–1" (1.2–2.5 cm) long, evergreen, and spreading almost at right angles in two rows on slender hairy twigs. The needles are crowded; curve upward on the upper twigs; are flat, with the tip usually rounded; and are shiny dark green above, with two broad silvery-white bands beneath. The bark is gray or brown, thin, and smooth, with many resin blisters, becoming scaly.

FRUIT: Cones: 1½–2½" (4–6 cm) long; cylindrical; dark purple; upright on topmost twigs; cone scales finely hairy, partly covered by yellow-green pointed and toothed bracts; paired long-winged seeds.

HABITAT: Coniferous forests with Red Spruce in high mountains.

RANGE: Appalachian Mountains in southwestern Virginia, western North Carolina, and eastern Tennessee; at 4000–6000' (1219–2012 m).

USES: This small to medium, beautiful, and fragrant tree is widely grown for Christmas trees.

SIMILAR SPECIES: Balsam Fir, but the ranges do not overlap.

CONSERVATION STATUS:

Small, disjunct subpopulations are known from six peaks and their north-facing slopes in the southern Appalachians, including in Smoky Mountains National Park. These trees suffered a massive dieback after the Balsam Woolly Adelgid (*Adelges piceae*) was discovered in 1957 in *Abies fraseri* on Mt. Mitchell and quickly spread to the other subpopulations, sparing only one major population at Mt. Rogers, Virginia. Millions of Fraser Firs died, leaving room for competitors such as *Picea rubra* and *Betula* sp. to establish dominance in several locations in North Carolina. Conservationists have undertaken massive seedling recruitment in some stands that have died, and some of these suffer less damage from subsequent adelgid infestations, raising the hope that some individuals may develop resistance.

Diameter: 1–2' (0.3–0.6 m)
Height: 30–50' (9–15 m)

■ DENSE POPULATION
■ LIGHT POPULATION
■ NATURALIZED

Both the common and scientific names of this species refer to the tree's large size. It often reaches 150–200' (45.7–61 m) tall, with the largest measuring 267' (81.4 m). Like those of related species, the smooth bark of small trunks have swellings or blisters; when pinched or opened, fragrant, transparent resin squirts out.

DESCRIPTION: This is one of the tallest true firs, with a narrow, pointed crown of stout, curved, and slightly drooping branches. The needles are 1¼–2" (3–5 cm) long, evergreen, spreading almost at right angles in two rows, crowded and curved upward on upper twigs. Flat, flexible; shiny dark green above, silvery white beneath. Its bark is brown and smooth, with resin blisters, becoming deeply furrowed into narrow scaly ridges. The twigs are brown and slender, with fine hairs when young.

FRUIT: Cones: 2–4" (5–10 cm) long; cylindrical, upright on the topmost twigs, green or brown; the cone scales are hairy, and the bracts are short and hidden. Paired, long-winged seeds.

HABITAT: Valleys and mountain slopes in cool, humid climate; in coniferous forests.

RANGE: Southern British Columbia south along the coast to California; also south in the Rocky Mountain region to central Idaho; to 1500' (457 m) along the coast; to 6000' (1829 m) inland.

USES: This beautifully symmetric, fragrant species is a popular choice for Christmas trees, especially in the Northwest. The soft wood is commonly used for pulpwood and is sometimes used for plywood and rough construction.

SIMILAR SPECIES: White Fir.

CONSERVATION STATUS: **LC**
The current population is stable and often forms extensive forests. Although logged in many parts of its range, the species regenerates well under uneven-aged management.

Diameter: 1½–3½' (0.5–1 m)
Height: 100–200' (30–61 m)

■ DENSE POPULATION
■ LIGHT POPULATION
■ NATURALIZED

The spires of Subalpine Fir add beauty to the Rocky Mountain peaks. When weighted down to the ground with snow, the lowest branches sometimes take root, forming new shoots. Deer, elk, bighorn sheep, and moose browse the bark of this and related firs; grouse eat the leaves, and many songbirds and mammals consume the seeds. The scientific name, which means "hairy-fruited," refers to the finely hairy cones. Corkbark Fir (var. *arizonica* [Merriam] Lemm.), a southwestern variety, has thin, whitish, corky bark. Some authorities consider the variety *lasiocarpa* ([Hook.] Nutt.) as a separate species, *A. bifolia*, or Rocky Mountain Alpine Fir.

DESCRIPTION: Subalpine Fir is the most widespread western true fir, with a dense, long-pointed, spirelike crown and rows of horizontal branches reaching nearly to base; shrubby at timberline. The needles are 1–1¾" (2.5–4.5 cm) long, evergreen, spreading almost at right angles in two rows, crowded, curved upward on the upper twigs, flat, and dark green, with whitish lines on both surfaces. The bark is gray and smooth, with resin blisters, becoming fissured and scaly. Its twigs are gray and stout, with rust-colored hairs.

FRUIT: Cones: 2¼–4" (6–10 cm) long; cylindrical, upright on the topmost twigs, dark purple; cone scales are finely hairy with short, hidden bracts; paired, long-winged seeds.

HABITAT: Subalpine zone of high mountains to the timberline, forming spruce-fir forests with Engelmann Spruce and with other conifers.

RANGE: Central Yukon and southeastern Alaska southeast to southern New Mexico; at 8000–12,000' (2438–3658 m) in the south; to sea level in the north.

USES: The soft, brittle wood from this tree is used for rough construction, doors, and poles. Smaller trees are used for Christmas trees.

SIMILAR SPECIES: Balsam Fir, but its range does not overlap.

CONSERVATION STATUS: LC
Subalpine Fir is locally common with a stable population. No specific threats have been identified for this species, however, increased fire frequencies and overgrazing by livestock are potential threats.

Diameter: 1–2½' (0.3–0.8 m)
Height: 50–100' (15–30 m)

■ DENSE POPULATION
■ LIGHT POPULATION
■ NATURALIZED

Named for its characteristic bark, this magnificent conifer forms almost pure forests at high altitudes along the western slopes of Sierra Nevada. It is common in Yosemite National Park. Early mountaineers prepared their beds by cutting and overlapping two rows of the plushy, fragrant boughs. In parts of northern California, subpopulations of Shasta Red Fir (var. *shastanensis*) are more scattered than elsewhere.

DESCRIPTION: This large, handsome fir has an open, conical crown rounded at the tip and short, nearly horizontal branches. The needles are ¾–1⅜" (2–3.5 cm) long, evergreen, spreading in two rows, crowded, curved upward on the upper twigs, four-sided, and blue-green with whitish lines. The bark is thick, reddish-brown, and deeply furrowed into narrow ridges. Its twigs are stout and brown, with fine hairs when young.

FRUIT: Cones: 6–8" (15–20 cm) long; cylindrical, purplish-brown, upright on the topmost twigs; cone scales with fine hairs, with yellowish bracts mostly short and hidden (exposed in Shasta Red Fir), pointed and finely toothed; paired, long-winged seeds.

HABITAT: High mountains with dry summers and deep snow in winter; often found in pure stands and also in mixed conifer forests.

RANGE: Cascade Mountains of Oregon south to Coast Ranges of California and through Sierra Nevada to central California and extreme western Nevada. At 6000–9000' (1829–2743 m) in the south; to 4500' (1372 m) in the north.

USES: This tree is used in construction, for paper, and for Christmas trees.

SIMILAR SPECIES: Noble Fir.

CONSERVATION STATUS: **LC**

This species is present in a number of protected areas, including famous national parks, scattered throughout its natural range. The population is large and stable. Although historically logging and forest management practices that favored other conifer species negatively impacted the species' area of occupancy, today California Red Fir forests are either protected from logging or are better managed, allowing regeneration of this species in many areas.

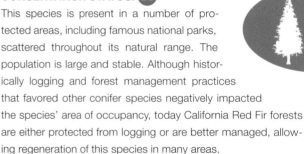

Diameter: 1–4' (0.3–1.2 m)
Height: 60–120' (18–37 m)

■ DENSE POPULATION
■ LIGHT POPULATION
■ NATURALIZED

A handsome tree with large, showy cones mostly covered by papery bracts, Noble Fir was named by the Scottish botanical explorer David Douglas (1798–1834). It is the tallest true fir and one of North America's tallest trees. Aside from its impressive size, this tree is recognized for its important ecological roles as a source of food and shelter for wildlife, and in water purification and carbon capture.

DESCRIPTION: The largest native true fir, this tree has a conical crown rounded at the tip and short, nearly horizontal branches. Its needles are 1–1⅜" (2.5–3.5 cm) long, evergreen, spreading in two rows, flat, grooved above, often notched, blue-green with whitish lines, crowded and curved upward, four-sided, and pointed on the upper twigs. The bark is gray-brown and smooth, becoming brown to red-brown; slightly thickened; and furrowed into irregular, long, scaly plates. The twigs are stout and brown, with rust-colored hairs when young.

Diameter: 2½–4' (0.8–1.2 m)
Height: 100–150' (30–46 m), often much taller

FRUIT: Cones: 4½–7" (11–18 cm) long; cylindrical, upright on the topmost twigs, green becoming purplish brown; cone scales with fine hairs and mostly covered by large papery bracts, finely toothed, long-pointed and bent downward; paired, long-winged seeds.

HABITAT: Moist soils in high mountains with a short, cool growing season and deep snow in winter. It is usually associated with other conifers; not in pure stands.

RANGE: Cascade Mountains and Coast Ranges from Washington south to northwest California; at 3000–7000' (914–2134 m); occasionally at 200–8800' (61–2682 m).

USES: This species is a popular Christmas tree, and is also used in construction and for paper.

SIMILAR SPECIES: California Red Fir.

CONSERVATION STATUS: 🄻🄲
This species is locally abundant. Logging of this valuable fir had a negative effect on the area of occupancy where logged stands were subsequently replaced with other trees or other forms of land use during the past 150 years. These declines have now ceased and the population has become stable.

The genus Cedrus consists of four species of true cedars, all native to the Old World. Atlas Cedar is distinguished by the short, pale blue-green needles—shiny or whitish in some varieties—and by the relatively small cones. The species name refers to the location of the Atlas Mountains of northwestern Africa, near the Atlantic Ocean, where this tree is native.

DESCRIPTION: This large ornamental tree sports a straight trunk, a pointed, pyramidal crown becoming broad and flattened, and pale blue-green or silvery foliage. The needles are ¾–1" (2–2.5 cm) long, evergreen; in clusters of 10 to 20 and crowded on spurs or alternate on leading twigs, three-angled, and pale blue-green or silvery. Its bark is gray and smooth, becoming irregularly furrowed into flat ridges. The twigs are spreading, long, stout, gray, and finely hairy, with many spurs on short side twigs.

FRUIT: Cones: 2–3" (5–7.5 cm) long, 1½–2" (4–5 cm) in diameter; barrel-shaped with a flat top, upright, almost stalkless, light brown; composed of many hard cone scales; maturing in the second year, axis remaining attached; paired, broad-winged seeds.

HABITAT: Moist soils in parks and gardens and around homes; in temperate regions.

RANGE: Native to northwestern Africa. Planted as an ornamental tree in the eastern states and Pacific region.

USES: This species is cultivated as an ornamental in North America. An Atlas Cedar is planted on the South Lawn of the White House in Washington, DC. In 1977 President Carter designed a tree house that was built within the tree for his daughter, Amy. In its native range in Africa, the heavy, aromatic wood is used for construction and cabinetmaking.

SIMILAR SPECIES: Deodar Cedar, Cedar of Lebanon.

CONSERVATION STATUS: **EN**

Many stands are located within National Parks and receive some protection from overgrazing and logging. Programs to monitor the extent and severity of recent die-backs are in place. Native Atlas Cedar forests have been exploited for their timber for centuries and have been subject to overgrazing and repeated burning. Exploitation has increased since the 1940s, with range-wide declines of up to 75% estimated to have occurred between 1940 and 1982. Since the 1980s a series of droughts have led to further decline, especially near the Saharan desert; climate change models suggest a continued decrease in precipitation in this range.

Diameter: 3' (0.9 m)
Height: 80' (24 m)

This large, handsome tree is distinguishable from other true cedars by the slightly drooping branches and long needles. The large developing cones are usually present on lower branches, even through winter. Grown as an ornamental tree in North America, several cultivated varieties differ in having yellow foliage or compact, weeping, and creeping habits. Deodar Cedar is often decorated as a living Christmas tree. In India, the durable, fragrant wood is commercially important for construction and has been used for incense. The name is derived from the Hindi word *devadaru*, meaning "timber of the gods."

DESCRIPTION: This is a large tree with a straight trunk and a regular, pyramidal crown with a curved or drooping tip, and graceful, drooping branches down to the base. Its needles are 1–2" (2.5–5 cm) long, evergreen, in clusters of 10 to 20, crowded on spurs or alternate on leading twigs, three-angled and dark blue-green. The bark is gray, becoming brown and deeply furrowed. Its twigs are gray and densely hairy, with many spurs or short side twigs.

FRUIT: Cones: 3–4" (7.5–10 cm) long, 2–3" (5–7.5 cm) in diameter; elliptical, rounded at the tip, upright, almost stalkless, reddish-brown; composed of many hard cone scales; maturing in the second year, the axis remaining attached. It has paired, broad-winged seeds.

HABITAT: Moist soils in parks and gardens and around homes; in warm, temperate regions.

RANGE: A native of the Himalayas. This species is planted mainly in the southeastern and West Coast states, and along the Mexico border; it is especially popular in California.

USES: This cedar is cultivated as an ornamental tree in North America.

SIMILAR SPECIES: Atlas Cedar, Cedar of Lebanon.

CONSERVATION STATUS: **LC**
The population is stable but deforestation in Pakistan and India along with logging in Afghanistan and other areas within the species' range may pose a threat.

Diameter: 3' (0.9 m)
Height: 80' (24 m)

The national symbol of Lebanon, this large, handsome cedar is native to southwestern Asia and cultivated as an ornamental in North America, particularly in the Southeast and on the Pacific Coast. It is often associated with the Holy Land and is mentioned in the Quran and the Bible. The fragrant, durable wood is used for construction timbers, lumber, furniture, and paneling in its native range.

DESCRIPTION: This large evergreen tree has a straight, stout trunk and a narrow, pointed crown, becoming irregular and broad or flattened with spreading, horizontal branches. Its needles are 1–1¼" (2.5–3 cm) long, evergreen, in clusters of 10 to 15, and crowded on spurs or alternate on leading twigs, three-angled, and mostly dark green. The bark is dark gray, becoming thick and furrowed into scaly plates. Its twigs are abundant, spreading, long and stout, mostly hairless or slightly hairy, with many spurs or short side twigs.

FRUIT: Cones: 3–4½" (7.5–11 cm) long, 1¾–2½" (4.5–6 cm) in diameter; barrel-shaped with a flat top, reddish-brown, upright, almost stalkless, resinous; composed of many hard cone scales; maturing in the second year, the axis remaining attached; paired broad-winged seeds.

HABITAT: Moist soils in parks and gardens and around homes; in temperate regions.

RANGE: Native of Asia Minor from southern Turkey to Lebanon and Syria at high altitudes. Planted as an ornamental mainly in the southeastern United States and on the Pacific Coast.

USES: In its native range of southwestern Asia, this cedar is used for construction, furniture, and paneling. It is cultivated as an ornamental in North America.

SIMILAR SPECIES: Atlas Cedar, Deodar Cedar.

CONSERVATION STATUS:

The species is declining due to loss of habitat. Lebanon and Syria, where there has been a long history of forest destruction, are some of the most threatened areas for the species. An insect called *Cephalcia tannourinensis* also poses a threat in Lebanon. All forests in Lebanon are protected, but there is little enforcement of laws forbidding damage to the cedar trees.

Diameter: 3' (0.9 m)
Height: 80' (24 m)

Grown as a handsome ornamental and in forestry plantations, this species is now naturalized in North America, the most common ornamental larch in the East. In its native range, European Larch is an important timber tree and a source of Venetian turpentine, an aromatic resin once popularly used in medicine.

DESCRIPTION: This introduced, cone-bearing, deciduous tree has a straight trunk and an open, broadly pyramid-shaped crown, becoming irregular with age. Its needles are ¾–1¼" (2–3 cm) long, deciduous, soft, flattened, with a ridge beneath; 30–40 needles of unequal length are crowded on spurs or alternate on leader twigs. The needles are light green, turning yellow in autumn before shedding. The bark is gray turning brown, thick, furrowed, and scaly. Its twigs are gray or yellowish, stout, and hairless, with many dark brown spurs or short side twigs.

FRUIT: Cones: 1–1¼" (2.5–3 cm) long; egg-shaped; upright, stalkless; reddish and like a rosebud when young; maturing in autumn, remaining attached and turning dark gray; cone scales are numerous, rounded, thin, light brown, finely hairy, and longer than 3-pointed bracts.

HABITAT: Moist soil in humid cool temperate regions.

RANGE: Native of northern and central Europe, at high altitudes. Introduced and naturalized in the northern United States and southeastern Canada.

USES: In Europe, this species is used for a variety of purposes, including utility poles, posts, railway cross-ties, shipbuilding, construction, and charcoal. The bark has been used in tanning and medicine. In North America, the species has been planted as an ornamental.

SIMILAR SPECIES: Tamarack.

CONSERVATION STATUS: (LC)

The species is protected in numerous areas, including the Ojców National Park in Poland. There are few threats.

Diameter: 2' (0.6 m)
Height: 70' (21 m)

- DENSE POPULATION
- LIGHT POPULATION
- NATURALIZED

One of our northernmost trees, and one of the most widespread conifers in the world, the Tamarack is a small to medium tree found across northern North America. It can handle extremely cold temperatures, thriving even at the Arctic tree line. It can be told from the nonnative European Larch by its tiny cones (less than an inch long).

DESCRIPTION: This is a deciduous tree with a straight, tapering trunk and a thin, open, conical crown of horizontal branches; it is a shrub at the timberline. The needles are ¾–1" (2–2.5 cm) long, ¹⁄₃₂" (1 mm) wide, deciduous, soft, very slender, three-angled, and crowded in a cluster on spur twigs; they are also scattered and alternate on leader twigs. The needles are light blue-green, turning yellow in autumn before shedding. The bark is reddish-brown, scaly, and thin. Its twigs are orange-brown, stout, and hairless, with many spurs or short side twigs.

FRUIT: Cones: ½–¾" (12–19 mm) long; elliptical; rose red turning brown; upright, stalkless; falling in the second year; several overlapping rounded cone scales; paired, brown, long-winged seeds.

HABITAT: Wet, peaty soils of bogs and swamps; also found in drier upland loamy soils; often in pure stands.

RANGE: Across northern North America near the northern limit of trees from Alaska east to Labrador, south to northern New Jersey, and west to Minnesota; local in northern West Virginia and western Maryland; from near sea level to 1700–4000' (518–1219 m) southward.

USES: The durable lumber of the Tamarack is used as framing for houses, railroad cross-ties, poles, and pulpwood. Native Americans used the slender roots to sew together strips of birch bark for canoes.

SIMILAR SPECIES: European Larch.

CONSERVATION STATUS: (LC)

This species is widespread throughout North America and exists in a constantly renewing ecosystem. The population is stable with few threats.

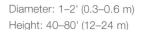

Diameter: 1–2' (0.3–0.6 m)
Height: 40–80' (12–24 m)

■ DENSE POPULATION
■ LIGHT POPULATION
■ NATURALIZED

Our largest larch, the Western Larch is a tall, deciduous tree of the mountain slopes and valleys of the Northwest. It is pale green in spring and summer and yellow in autumn. Pileated Woodpeckers forage and nest in these trees, and grouse eat the buds and leaves. Western Larch often survives fires due to its thick bark.

DESCRIPTION: Western Larch is a large deciduous tree with a narrow, conical crown of horizontal branches. The needles are 1–1½" (2.5–4 cm) long, ¹⁄₃₂" (1 mm) wide, deciduous, crowded in a cluster on spur twigs; also alternate and scattered on leader twigs; 3-angled, stiff, short-pointed, and light green, turning yellow in autumn before falling. Its bark is reddish-brown and scaly, becoming deeply furrowed into flat ridges with many overlapping plates. There are two kinds of twigs: long leaders (orange-brown and hairy when young), and many short spurs.

FRUIT: Cones: 1–1½" (2.5–4 cm) long; elliptical, brown, upright on short stalks; many rounded, hairy cone scales shorter than long-pointed bracts; paired, pale brown, long-winged seeds.

HABITAT: Mountain slopes and valleys on porous, gravelly, sandy, and loamy soils; with other conifers.

RANGE: Southeast British Columbia south to northwest Montana and northern Oregon; at 2000–5500' (610–1676 m) in north; to 7000' (2134 m) in south.

USES: The wood of this tree is used for construction, paneling, flooring, utility poles, plywood, and pulpwood. The natural sugar in the gum and wood resembles a slightly bitter honey and can be made into medicine and baking powder.

SIMILAR SPECIES: Tamarack.

CONSERVATION STATUS: (LC)
This species is widely distributed and is known to regenerate quickly. The population is stable with no identified threats.

Diameter: 1½–3' (0.5–0.9 m), sometimes larger
Height: 80–150' (24–46 m)

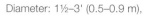

■ DENSE POPULATION
■ LIGHT POPULATION
■ NATURALIZED

Norway Spruce is the common spruce of northern Europe and has been widely cultivated for ornament, shade, shelterbelts, Christmas trees, and forest plantations. Its showy cones are the largest among spruces. Numerous horticultural varieties include trees with a narrow columnar shape, drooping or weeping branches, dwarf habit, and yellowish or variegated needles.

DESCRIPTION: This large, introduced, cone-bearing tree has a straight trunk and a pyramid-shaped crown of spreading branches. Its needles are ½–1" (1.2–2.5 cm) long, evergreen, stiff, four-angled, sharp-pointed, and spreading on all sides of the twig from very short leafstalks. They are shiny, dark green with whitish lines. The bark is reddish-brown and scaly. The twigs are reddish-brown, slender, drooping, mostly hairless, and rough, with peglike bases.

FRUIT: Cones: 4–6" (10–15 cm) long; cylindrical; light brown, hanging down; cone scales numerous, thin, slightly pointed, irregularly toothed, opening and shedding year after maturing; paired long-winged seeds.

HABITAT: Moist soils in humid, cool, temperate regions.

RANGE: Native of northern and central Europe at high altitudes. It is widely planted in southeastern Canada, the northeastern United States, the Rocky Mountains, and on the Pacific Coast. It is naturalized in parts of the Northeast.

USES: The Norway Spruce is widely planted around the world as an ornamental tree. It is also commonly used as a Christmas tree.

SIMILAR SPECIES: White Spruce.

CONSERVATION STATUS:
The population is thought to be stable in its native range; its two varieties have not been formally assessed, but neither is considered threatened.

Diameter: 2' (0.6 m)
Height: 80' (24 m)

- DENSE POPULATION
- LIGHT POPULATION
- NATURALIZED

This species is a large, blue-green evergreen of western North America, closely related to White Spruce and Blue Spruce. It was named after George Engelmann (1809–1884), a German-born physician and botanist of St. Louis and authority on conifers.

DESCRIPTION: Engelmann Spruce is a large tree with dark or blue-green foliage and a dense, narrow, conical crown of short branches spreading in close rows. Its needles are ⅝–1" (1.5–2.5 cm) long and evergreen, spreading on all sides of the twig from very short leafstalks; they are four-angled, short-pointed, slender, and flexible. Its needles, which are dark or blue-green with whitish lines, produce an unpleasant skunklike odor when crushed. The bark is grayish- or purplish-brown and thin, with loosely attached scales. The twigs are brown, slender, hairy, and rough, with peglike leaf bases.

FRUIT: Cones: 1½–2½" (4–6 cm) long; cylindrical, shiny light brown; hanging at the end of a leafy twig. The cone-scales are long, thin, and flexible; narrowed; and may be irregularly toothed. It has paired, blackish, long-winged seeds.

HABITAT: Dominant with Subalpine Fir in the subalpine zone up to the timberline; it is also found with other conifers.

RANGE: Central British Columbia and southwestern Alberta southeast to New Mexico; chiefly in Rocky Mountains; at 8000–12,000' (2438–3659 m) in south; down to 2000' (619 m) in north.

USES: Its resonant qualities make the wood of Engelmann Spruce valuable for piano sounding boards and violins. This species is sometimes cultivated for ornamental purposes.

SIMILAR SPECIES: Blue Spruce, White Spruce.

CONSERVATION STATUS: LC

The population is stable and occupies a vast range in Canada and the Pacific Northwest. The species can be found in protected areas within the U.S. and Canada and is particularly abundant in the northern Rocky Mountains. Modern forestry practices allow this species to regenerate after logging.

Diameter: 1½–2½' (0.5–0.8 m)
Height: 80–100' (24–30 m)

■ DENSE POPULATION
■ LIGHT POPULATION
■ NATURALIZED

This is a common evergreen conifer found across northern North America, doing well in cold climates even north of the Arctic Circle. It is a primary source of pulpwood and generally the most important commercial tree species of Canada. White Spruce and Black Spruce are the most widely distributed conifers in North America after Common Juniper, which rarely reaches tree size. Various kinds of wildlife, including deer, rabbits, and grouse, feed on spruce foliage in winter.

DESCRIPTION: This is a medium to large tree with rows of horizontal branches forming a conical crown; smaller and shrubby at tree line. Its needles are ½–¾" (12–19 mm) long, evergreen, stiff, four-angled, sharp-pointed, spreading mainly on the upper side of twigs from short leafstalks, blue-green with whitish lines. The needles exude an unpleasant skunklike odor when crushed. The bark is gray or brown, thin, smooth or scaly; the cut surface of inner bark is whitish. Its twigs are orange-brown, slender, hairless, and rough, with peglike bases.

FRUIT: Cones: 1½–2½" (4–6 cm) long; cylindrical, shiny light brown, hanging at the ends of twigs and falling at maturity. Cone scales are thin and flexible, with nearly straight margins and without teeth. It has paired brown, long-winged seeds.

HABITAT: Many soil types in coniferous forests; sometimes in pure stands.

RANGE: Across northern North America near northern limit of trees from Alaska and British Columbia east to Labrador, south to Maine, and west to Minnesota; local in northwest Montana, South Dakota, and Wyoming; from near sea level to timberline at 2000–5000' (610–1524 m).

USES: White Spruce is one of the most important commercial tree species of Canada, used for pulpwood, construction lumber, and musical instruments, including pianos and violins. This tree is planted as an ornamental across the United States and Canada.

SIMILAR SPECIES: Engelmann Spruce, Black Spruce.

CONSERVATION STATUS: **LC**

White Spruce and its varieties, var. *glauca* and var. *albertiana,* are the most widespread spruce trees in North America. The population is stable with no identified threats.

Diameter: 1–2' (0.3–0.6 m)
Height: 40–100' (12–30 m)

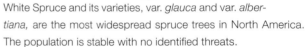

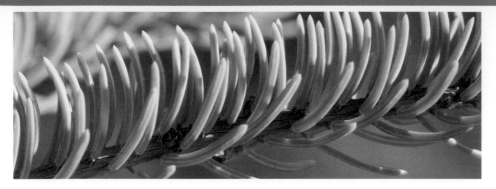

■ DENSE POPULATION
■ LIGHT POPULATION
■ NATURALIZED

Black Spruce is one of the most widely distributed conifers in North America. Its uses are similar to those of White Spruce, but the species' smaller size limits its use in lumber production. The lowest branches of a Black Spruce take root by layering when heavy snow bends them to the ground, forming a ring of small trees around a large one. Spruce gum and spruce beer were made from this species and Red Spruce.

DESCRIPTION: This tree has an open, irregular, conical crown of short, horizontal or slightly drooping branches; a prostrate shrub at timberline. Its needles are ¼–⅝" (6–15 cm) long, evergreen, stiff, 4-angled, sharp-pointed, spreading on all sides of twig from very short leafstalks, and ashy blue-green with whitish lines. The bark is gray or blackish, thin, scaly, and brown beneath; the cut surface of inner bark is yellowish. Its twigs are brown, slender, hairy, and rough, with peglike bases.

FRUIT: Cones: ⅝–1¼" (1.5–3 cm), long; egg-shaped or rounded; dull gray; curved downward on a short stalk and remaining attached, often clustered near the top of the crown. The cone scales are stiff and brittle, rounded and finely toothed. It has paired, brown, long-winged seeds.

HABITAT: Wet soils and bogs including peats, clays, and loams. It is found in coniferous forests, often in pure stands.

RANGE: Across northern North America near the northern limit of trees from Alaska and British Columbia east to Labrador, south to northern New Jersey, and west to Minnesota; at 2000–5000' (610–1524 m).

USES: Black Spruce is primarily used for pulpwood, but also for lumber and Christmas trees.

SIMILAR SPECIES: White Spruce.

CONSERVATION STATUS: (LC)
Widespread and abundant, the Black Spruce population is stable with no identified threats.

Diameter: 4–12" (0.1–0.3 m)
Height: 20–60' (6–18 m)

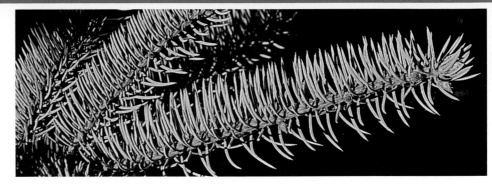

■ DENSE POPULATION
■ LIGHT POPULATION
■ NATURALIZED

The Blue Spruce is native to the Rocky Mountains but has been widely cultivated as an ornamental tree across the continent. The populations tend to be small and isolated in arid Arizona, New Mexico, and parts of Utah, but they are more substantial elsewhere, especially in the more moist areas of the central Rockies. It is a popular Christmas tree and is also used in shelterbelts.

DESCRIPTION: This is a large tree with blue-green foliage and a conical crown of stout, horizontal branches in rows. The needles are ¾–1⅛" (2–2.8 cm) long, evergreen, spreading on all sides of twig from very short leafstalks, 4-angled, sharp-pointed, and stiff, with a resinous odor when crushed; dull blue-green or bluish, with whitish lines. The bark is gray or brown, furrowed, and scaly. Its twigs are yellow-brown, stout, hairless, and rough, with peglike leaf bases.

FRUIT: Cones: 2¼–4" (6–10 cm) long; cylindrical, mostly stalkless, shiny light brown; cone scales long, thin, and flexible, narrowed and irregularly toothed; paired, long-winged seeds.

HABITAT: Narrow bottomlands along mountain streams; often in pure stands.

RANGE: Rocky Mountain region from southern and western Wyoming and eastern Idaho south to northern and eastern Arizona and southern New Mexico; at 6000–11,000' (1829–3353 m).

USES: This species is seldom used for lumber but has been widely used in landscaping as an ornamental planting. It is also used as a Christmas tree.

SIMILAR SPECIES: Engelmann Spruce.

CONSERVATION STATUS: **LC**
Some subpopulations of this species are small and isolated, but the overall population is stable with no identified threats.

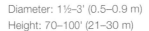

Diameter: 1½–3' (0.5–0.9 m)
Height: 70–100' (21–30 m)

RED SPRUCE *Picea rubens*

■ DENSE POPULATION
■ LIGHT POPULATION
■ NATURALIZED

Red Spruce is an important forest tree in the northeast. It is generally found in pure stands or with Balsam Fir, White Pine, or Black Spruce, having a dark pointed top and reaching to the summits of the eastern mountains, where they can shrink to shoulder height or lower—even to ankle height at the tree line. Except at the highest altitudes, the trunks are straight. The wood is strong in proportion to its weight. The crushed needles have a lovely citrus rind aroma.

DESCRIPTION: A medium-sized evergreen with a narrow cone-shaped crown. Its needles are ½–⅝" (12–15 mm) long, stiff, four-angled, sharp-pointed; spreading on all sides of the twig from short leafstalks; they are shiny green, with whitish lines. The bark is reddish brown, thick, and scaly. The twigs are brown, slender, finely hairy, and rough with peglike bases.

FRUIT: Cones: 1¼–1½" (3–4 cm) long; cylindrical; reddish-brown; hanging down on short, straight stalk; falling at maturity. The cone scales are stiff, rounded, often finely toothed. It has paired brown, long-winged seeds.

HABITAT: Moist sandy loam and rocky mountain soils to the highest altitudes in the southern Appalachians.

RANGE: Ontario east to Nova Scotia; from northern New England south to the Massachusetts shore, through the Adirondacks, the Hudson Valley, in the higher peaks of the Alleghenies, south in the mountains of Virginia, western North Carolina and eastern Tennessee; to 4500–6500' (1372–1981 m) in the southern mountains. Extensive virgin spruce-fir forests are preserved in the Great Smoky Mountains National Park.

USES: A versatile wood with many uses from construction to paper pulp. Its smooth and uniform quality is valued in the manufacture of musical instruments from guitars and organ pipes to the bellies of violins. Spruce gum, a forerunner of modern chew-ing gum made from chicle (gum from a tropical American tree), was obtained commercially from resin of both Red and Black spruce trunks. The young, leafy twigs were boiled with flavoring and sugar to prepare spruce beer.

SIMILAR SPECIES: Black Spruce. Where the ranges overlap, Black Spruce is distinguishable from Red by its smaller, dull gray cones curved downward on short stalks and remaining attached.

CONSERVATION STATUS: LC

Red Spruce is showing damage from air pollution throughout its range, particularly at the higher elevations. Atmospheric pollutants and acid deposition have negatively impacted the population in the past 100 years, with weakened trees becoming more susceptible to insects and disease. The overall number of this locally dominant tree continues to increase, however.

Diameter: 1–2' (0.3–0.6 m), occasionally to 3' (0.91 m)
Height: 50–80' (15–24 m), occasionally to 150' (46 m)

■ DENSE POPULATION
■ LIGHT POPULATION
■ NATURALIZED

This spruce is the world's largest and tallest, recorded at a maximum height surpassing 300' (91 m). Sitka Spruce is common in the coastal rainforests of the Pacific Northwest, where it is one of the most important nesting trees for the endangered Marbled Murrelet, a small seabird that nests inland.

DESCRIPTION: This is the world's largest spruce, with a tall, straight trunk from a wide base, and a broad, open, conical crown of horizontal branches. Its needles are ⅝–1" (1.5–2.5 cm) long, evergreen, spreading on all sides of twig, flattened and slightly keeled, sharp-pointed, and dark green. The bark is gray, smooth, and thin, becoming dark purplish-brown with scaly plates. Its twigs are brown, stout, hairless, and rough, with peglike bases.

FRUIT: Cones: 2–3½" (5–9 cm) long; cylindrical, short-stalked, light orange-brown; hanging at the ends of twigs; opening and falling at maturity. The cone scales are long, stiff, thin, rounded, and irregularly toothed. It has paired, brown, long-winged seeds.

HABITAT: Coastal forests in the fog belt, a narrow strip of high rainfall and cool climate. It may be found in pure stands and with Western Hemlock.

RANGE: Pacific Coast from southern Alaska and British Columbia to northwestern California; to timberline at 3000' (914 m) in Alaska; below 1200' (366 m) in California.

USES: This species is the primary timber tree in Alaska. Its high-grade lumber has been used in many ways, including aircraft construction. It is also used for pulpwood, boats, and piano sounding boards.

SIMILAR SPECIES: White Spruce.

CONSERVATION STATUS: 🅛🅒

The species regenerates well, which has limited its decline despite exploitation for logging. Outdated logging practices may have depleted stands of mature trees and allowed other species to replace them, but today the population is stable.

Diameter: 3–5' (0.9–1.5 m),
sometimes much larger
Height: 160' (49 m)

■ DENSE POPULATION
■ LIGHT POPULATION
■ NATURALIZED

Many Native Americans gathered the cones of the Whitebark Pine and used the seeds as a food source. Clark's Nutcrackers—western birds related to crows and jays—tear open the cones to eat the seeds. This is the primary way in which the trees' seeds are distributed. Whitebark Pine is considered the most primitive native pine because its cones do not open naturally until they decay.

DESCRIPTION: This tree has a short, twisted or crooked trunk and an irregular, spreading crown; it is a shrub at the timberline. The foliage has a sweetish taste and odor. Its needles are 1½–2¾" (4–7 cm) long, evergreen, in bundles of five with the sheath shedding the first year, crowded at the ends of twigs, stout, stiff, short-pointed, and dull green, with faint white lines on all surfaces. Its bark is whitish-gray, smooth, and thin, becoming scaly. Its twigs are stout, tough and flexible, and brown, with fine hairs when young.

FRUIT: Cones: 1½–3¼" (4–8 cm) long; egg-shaped or rounded, almost stalkless, purple to brown; shedding at maturity but not opening; cone scales very thick, with sharp edge ending in raised, stout point. Large elliptical, dark brown, thick-walled, wingless, edible.

HABITAT: Dry, rocky soils on exposed slopes and ridges in subalpine zone to timberline; sometimes forms pure stands and thickets.

RANGE: Central British Columbia, east to southwestern Alberta, south to western Wyoming and west to central California. At 4500–7000' (1372–2134 m) in the north; at 8000–12,000' (2438–3658 m) in the south.

USES: The seed from this species was an important food source for numerous Native American tribes. It is rarely cultivated.

SIMILAR SPECIES: Limber Pine.

CONSERVATION STATUS: 🄴🄽

Whitebark Pine declined due to white pine blister rust, a fungal disease first seen around 1925. The pathogen can kill 35% of the trees in a stand or more, depending on the location. More recently, forest fires and a native pest, the Mountain Pine Beetle (*Dendroctonus ponderosae*), have been responsible for additional mortality. Because Whitebark Pine occurs at higher elevations, it's also vulnerable to suitable habitat shifting as a result of climate change.

Diameter: 1–2' (0.3–0.6 m)
Height: 20–50' (6–15 m)

■ DENSE POPULATION
■ LIGHT POPULATION
■ NATURALIZED

These ancient trees are extremely long-lived, and the oldest ones are an impressive sight, with their twisted trunks and their gnarled roots clinging to windswept mountain ridges. These resilient pines continue to grow slowly even under poor conditions in their exposed habitats, leaving some trees disfigured, twisted, or sheared.

DESCRIPTION: This tree has short needles crowded into a mass suggesting a foxtail, and a broad, irregular crown of spreading branches; it is a low shrub at the timberline. Needles are ¾–1½" (2–4 cm) long, evergreen; five in bundle, with the sheath shedding after the first year; crowded in a long, dense mass curved against the twig; stout, stiff, blunt-pointed; persisting 10–20 years; dark green with white lines on inner surfaces; often with whitish resin dots on the outer surface. The bark is whitish-gray and smooth, becoming reddish-brown and furrowed into irregular, scaly ridges.

FRUIT: Cones: 2½–3½" (6–9 cm) long; cylindrical, dark purplish-brown, almost stalkless; opening at maturity. The cone scales are four-sided with slender, curved bristles. The seeds are brown mottled with black and with a detachable wing.

HABITAT: Exposed, dry, rocky slopes and ridges of high mountains in subalpine zone to timberline; often in pure stands.

RANGE: Arizona, New Mexico, and Colorado; at 7500–11,500' (2286–3505 m).

USES: The wood of Bristlecone Pine is sometimes used locally for fence posts and mine shaft timbers, but many of these trees are located on protected land where harvesting is prohibited.

SIMILAR SPECIES: Intermountain Bristlecone Pine, Foxtail Pine.

CONSERVATION STATUS: **LC**
Although its population is stable, the exotic fungal pathogen *Cronartium ribicola*, which causes white pine blister rust, has been recorded in one subpopulation, and its spread is being closely monitored. Conservation actions include gene conservation to determine rust-resistant seed sources for restoration.

Diameter: 1–2½' (0.3–0.8 m)
Height: 20–40' (6–12 m)

■ DENSE POPULATION
■ LIGHT POPULATION
■ NATURALIZED

The whorls of many knobby, closed cones help to identify this western evergreen. Since the cones may become embedded within the wood of the expanding trunk, this species has been called "the tree that swallows its cones." This pine typically occupies a transitional position between chaparral and woodland and higher elevation forests. When fires kill the trees, cones as much as 30 years old are opened by the heat and shed their seeds. The abundant seedlings then begin a new forest.

DESCRIPTION: This pine has a narrow, pointed crown of slender, nearly horizontal branches turned up at the ends, becoming irregular with age, and with abundant cones remaining closed for many years. Its needles are 3–7" (7.5–18 cm) long, evergreen; three in bundle; slender, stiff, yellow-green. The bark is gray and smooth, becoming dark gray and fissured into large, scaly ridges.

FRUIT: Cones: 3¼–6" (8–15 cm) long; egg-shaped, clustered in many rings or whorls, stalkless and turned back, shiny yellow-brown. The cone scales are raised and keeled, ending in a short, stout spine. Its blackish seeds are about ¼" (6 mm) long and have a narrow wing 1–1¼" (2.5–3 cm) long.

HABITAT: Forms almost pure stands on poor, coarse, rocky, mountain soils.

RANGE: Southwest Oregon south to southern California; local in northern Baja California, Mexico; at 1000–2000' (305–610 m) in the north; 1500–4000' (457–1219 m), sometimes higher, in the south.

USES: This species is sometimes planted for slope protection.

SIMILAR SPECIES: Bishop Pine, Monterey Pine.

CONSERVATION STATUS: LC

The global population is stable, but numbers of mature trees in any given location can fluctuate because fire tends to kill all trees in a locality. Regeneration from seeds normally follows, resulting in numerous seedlings and saplings. Fire prevention and urbanization may lead to a decline in this species.

Diameter: 1–2½" (0.3–0.8 m)
Height: 30–80' (9–24 m)

■ DENSE POPULATION
■ LIGHT POPULATION
■ NATURALIZED

Jack Pine is a pioneer after fires and logging. Its cones usually remain closed for many years until opened by the heat of fires or exposure after cutting. The northernmost New World pine, its range nearly reaches the Arctic Ocean in western Canada. The endangered Kirtland's Warbler (*Dendroica kirtlandii*) depends on Jack Pine for nesting. This rare songbird has a limited breeding range in parts of Michigan, Wisconsin, and Ontario, where it nests only in dense stands of young Jack Pines following forest fires or controlled burns.

DESCRIPTION: This is an open crowned tree with spreading branches and short needles; it is sometimes a shrub. Its needles are ¾–1½" (2–4 cm) long, evergreen; two in bundle; stout, slightly flattened and twisted, widely forking; shiny green. The bark is gray-brown or dark brown and thin, with narrow scaly ridges.

FRUIT: Cones: 1¼–2" (3–5 cm) long; narrow, long-pointed, and curved upward; shiny light yellow; usually remaining closed on the tree many years. The cone scales are slightly raised and rounded, keeled, mostly without prickle.

HABITAT: Sandy soils, dunes, and on rock outcrops; often in extensive pure stands.

RANGE: Northwest Territories and Alberta, east to central Quebec and Nova Scotia, southwest to New Hampshire, and west to northern Indiana and Minnesota; to about 2000' (610 m).

USES: Jack Pine is an important timber species, used for pulpwood, construction lumber, utility poles, fence posts, and in other ways. It is commonly cultivated and used for Christmas trees. The tree also serves as habitat and breeding area for the Kirtland's Warbler, a federally endangered bird.

SIMILAR SPECIES: Lodgepole Pine.

CONSERVATION STATUS:
Jack Pines can be locally abundant. The population is stable with no identified threats.

Diameter: 1' (0.3 m)
Height: 30–70' (9–21 m)

■ DENSE POPULATION
■ LIGHT POPULATION
■ NATURALIZED

This is the common pinyon of Mexico, its range just barely extending into the southwestern United States. The hard seeds are the main commercial pinyon nuts (*piñones*) of Mexico. In the United States, this species has limited distribution and usually bears light cone crops; other species with thin-walled seeds are more common. Rodents often eat the seeds of Mexican Pinyon.

DESCRIPTION: This is a small, resinous tree with a short trunk and a spreading crown of low, horizontal branches and thick-walled, edible seeds; often shrubby. Its needles are 1–2½" (2.5–6 cm) long, evergreen; three in bundle (rarely two), with sheath shedding after first year; slender, flexible; green, with white lines on all surfaces or only two inner surfaces. The bark is light gray and smooth; becoming dark gray or reddish-brown and furrowed into scaly plates.

FRUIT: Cones: ¾–2" (2–5 cm) long; egg-shaped or rounded, almost stalkless, dull orange or reddish-brown, resinous; opening and shedding. The cone scales are thick, slightly four-angled, sometimes with a tiny prickle. The seeds are large, oblong or elliptical, dark brown, and wingless.

HABITAT: Dry, rocky slopes of mesas, plateaus, and mountains; with junipers and evergreen oaks.

RANGE: Central and Trans-Pecos Texas west to southeast Arizona; also Mexico; at 5000–7500' (1524–2286 m); locally to 2500' (762 m).

USES: In Mexico, the hard seeds of this species are the primary source of commercial pinyon nuts.

SIMILAR SPECIES: Colorado Pinyon Pine.

CONSERVATION STATUS: (LC)

This species is widespread, mostly in Mexico, Two subspecies of Mexican Pinyon are threatened, but the overall population is stable with no identified threats.

Diameter: 1' (0.3 m)
Height: 16–20' (5–6 m)

■ DENSE POPULATION
■ LIGHT POPULATION
■ NATURALIZED

This evergreen is endemic to the southeastern United States, found in Florida and parts of Mississippi, Alabama, and Georgia. There are two separate populations. The cones of trees found on the Gulf Coast open at maturity; the cones of the other population, found in central Florida, remain closed until heated by fire.

DESCRIPTION: Florida's small to medium-sized pine, this tree has a rounded or flattened crown. The needles are 2–3½" (5–9 cm) long, evergreen; two in bundle; slender, slightly twisted; dark green. The bark is dark gray or reddish-brown, furrowed into narrow scaly ridges; on small trunks, gray and smooth.

FRUIT: Cones: 2–3½" (5–9 cm) long; narrowly egg-shaped; yellow-brown; short-stalked; bent downwards, often clustered in 2–3 rows or whorls; mostly remaining closed on the tree many years and becoming gray. The cone scales are slightly keeled and raised.

HABITAT: Well-drained sandy soils; often found in pure stands.

RANGE: Florida and the southernmost parts of Mississippi, Alabama, and Georgia, to 200' (61 m).

USES: Sand Pine is widely used for pulpwood. It also provides important habitat for many wildlife species, including the Florida Sand Skink (*Plestiodon reynoldsi*), a lizard endemic to Florida.

SIMILAR SPECIES: Spruce Pine, but these species are found in different habitats.

CONSERVATION STATUS: **LC**
The population is stable with no major threats.

Diameter: 1–1½' (0.3–0.5 m)
Height: 30–70' (9–21 m)

■ DENSE POPULATION
■ LIGHT POPULATION
■ NATURALIZED

Lodgepole Pine is one of the most widely distributed New World pines and the only conifer native in both Alaska and Mexico. Its common name refers to the fact that Native Americans used the slender trunks as poles for conical tents or teepees.

There are three geographic varieties. The Pacific Coast variety, Shore Pine (var. *contorta*), is a small tree with spreading crown; thick, furrowed bark; short leaves; and oblique cones pointing backward, opening at maturity but remaining attached.

Sierra Lodgepole Pine (var. *murrayana*) is a variety found in the Cascade Mountains of southwestern Washington and western Oregon, the Sierra Nevada of central California, and south to northern Baja California. It is a tall, narrow tree with thin, scaly bark, relatively broad leaves, and symmetrical, lightweight cones opening at maturity and shedding within a few years.

Rocky Mountain Lodgepole Pine (var. *latifolia*) of the Rocky Mountain region is a tall, narrow tree with thin, scaly bark, long needles, and cones often oblique and pointing outward. This variety is adapted to forest fires, often with cones that remain tightly closed on the trees many years until a fire destroys the forest. When the heat causes the cones to open, the seeds fall to the bare ground to begin a new forest. This variety is also able to reproduce without fire, and in some areas most of the trees release their seeds without the heat of fire.

DESCRIPTION: This is a widely distributed pine that may grow tall with a narrow, dense, conical crown, or remain small with a broad, rounded crown depending on the variety. The needles are 1¼–2¾" (3–7 cm) long, evergreen; two in bundle; stout, slightly flattened and often twisted; yellow-green to dark green. The bark is light brown, thin, and scaly; or in Shore Pine (the coastal variety), dark brown, thick, and furrowed into scaly plates.

FRUIT: Cones: ¾–2" (2–5 cm) long; egg-shaped, stalkless, oblique or one-sided at the base, shiny yellow-brown; remaining closed on the tree many years, but variable. The cone scales are raised, rounded, keeled, with a tiny, slender prickle.

HABITAT: High mountains on mostly well-drained soils, often in pure stands; Shore Pine in peat bogs, muskegs, and dry, sandy sites.

RANGE: Southeast Alaska and central Yukon south on the Pacific Coast to northern California, south through the Sierra Nevada to southern California, and south in the Rocky Mountains to southern Colorado; also local in the Black Hills of South Dakota and in northern Baja California.

USES: Lodgepole Pine is commonly used for lumber, plywood, paneling, doors, furniture, and fence posts.

SIMILAR SPECIES: Jack Pine.

CONSERVATION STATUS: LC

The overall population is stable; bark beetle outbreaks and forest fire frequency may disrupt some local communities.

Diameter: 1–3' (0.3–0.9 m)
Height: 20–80' (6–24 m)

■ DENSE POPULATION
■ LIGHT POPULATION
■ NATURALIZED

Coulter Pine bears the heaviest cones of any pine in the world, often weighing 4–5 pounds (1.8–2.3 kg). Native Americans once gathered and ate the large seeds, but now it is mostly squirrels and other wildlife that consume the annual crop. White-headed Woodpeckers (*Leuconotopicus albolarvatus*) commonly forage in Coulter Pines, gleaning insects from the bark. This species was described in 1831 by Thomas Coulter (1793–1843), an Irish botanist and physician who collected plants in Mexico and California.

DESCRIPTION: A straight-trunked tree with rows of nearly horizontal branches formed annually. It forms an open, thin, irregular crown and has extremely large, heavy cones. The needles are 8–12" (20–30 cm) long, evergreen, two in bundle, and crowded at the ends of stout, brown twigs; they are stout, stiff, sharp-pointed, and light gray-green with many white lines. The bark is dark gray, thick, and deeply furrowed into scaly ridges; becoming slightly shaggy, blackish-gray, rough, and divided into rectangular plates on branches.

FRUIT: Cones: 8–12" (20–30 cm) long; egg-shaped, bent down on a stout stalk, extremely heavy, slightly shiny yellow-brown, resinous; opening gradually and remaining on the tree. The cone scales are extremely long, thick, and sharply keeled with a long, stout spine flattened at the end.

HABITAT: Dry, rocky slopes and ridges in foothills and mountains; with other conifers.

RANGE: Central and southern California; also northern Baja California; at 3000–6000' (914–1829 m); rarely at 1000–7000' (305–2134 m).

USES: Coulter Pines are cultivated as ornamental trees. The lightweight, soft wood is seldom used for anything other than firewood.

SIMILAR SPECIES: Foothill Pine, Jeffrey Pine, Ponderosa Pine.

CONSERVATION STATUS: **NT**
The species is declining and its subpopulations are severely fragmented. It is highly susceptible to changes in the frequency of forest fires. Fire control, especially around urban areas, can lead to forest succession and the dominance of other species. Where fires occur too frequently, however, seedlings and saplings are destroyed before they mature to reproductive age.

Diameter: 1–2½" (0.3–0.8 m)
Height: 40–70' (12–21 m)

■ DENSE POPULATION
■ LIGHT POPULATION
■ NATURALIZED

Shortleaf Pine is native in 21 southeastern states, the most widely distributed of the "southern yellow pines," which include this species along with Longleaf, Slash, and Loblolly pines. It has been heavily harvested for its lumber. The Red-cockaded Woodpecker (*Leuconotopicus borealis*) relies on old-growth forests of Shortleaf and Loblolly pines for nesting. As these old-growth forests have significantly declined, so have the wood-peckers and other wildlife species.

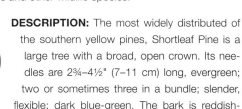

DESCRIPTION: The most widely distributed of the southern yellow pines, Shortleaf Pine is a large tree with a broad, open crown. Its needles are 2¾–4½" (7–11 cm) long, evergreen; two or sometimes three in a bundle; slender, flexible; dark blue-green. The bark is reddish-brown with large, irregular, flat, scaly plates.

FRUIT: Cones: 1½–2½" (4–6 cm) long, conical or narrowly egg-shaped, dull brown, short-stalked; opening at maturity but remaining attached. The cone scales are thin and keeled, with a small prickle.

HABITAT: From dry rocky mountain ridges to sandy loams and silt loams of flood plains, and in old fields. It is often found in pure stands or with other pines and oaks.

RANGE: Extreme southeastern New York and New Jersey south to northern Florida, west to eastern Texas, and north to southern Missouri; to 3300' (1006 m).

USES: Shortleaf Pine is an important timber species, used to produce lumber for construction, millwork, and many other uses, as well as plywood and veneer for containers.

SIMILAR SPECIES: Spruce Pine, Loblolly Pine, Pitch Pine.

CONSERVATION STATUS: (LC)
The population is increasing with no identified range-wide threats.

Diameter: 1½–3' (0.5–0.9 m)
Height: 70–100' (21–30 m)

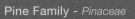

■ DENSE POPULATION
■ LIGHT POPULATION
■ NATURALIZED

This species is known by a multitude of common names, including Two-Needle Pinyon, Two-Leaf Pinyon, New Mexico Pinyon, and simply Pinyon. Its edible seeds—known as pinyon nuts, Indian nuts, pine nuts, and *piñones*—are a wild, commercial nut crop. Eaten raw, roasted, or in candies, these seeds were once a staple food source for southwestern Native Americans. Local residents now harvest them every autumn for local markets, although many of the seeds end up devoured by Pinyon Jays, Wild Turkeys, bears, deer, and other wildlife. This species is the most common tree on the south rim of Grand Canyon National Park.

DESCRIPTION: This is a small, bushy, resinous tree with a short trunk and a compact, rounded, spreading crown. Its needles are ¾–1½" (2–4 cm) long and evergreen; two in bundle (sometimes three or one); stout, and light green. The bark is gray to reddish-brown, rough, and furrowed into scaly ridges.

FRUIT: Cones: 1½–2" (4–5 cm) long; egg-shaped, yellow-brown, resinous or sticky; opening and shedding; with thick, blunt cone scales. The seeds are large, wingless, slightly thick-walled, oily, and edible.

HABITAT: Open, orchardlike woodlands, alone or with junipers; and on dry, rocky foothills, mesas, plateaus, and lower mountain slopes.

RANGE: The southern Rocky Mountain region from Utah and Colorado south to New Mexico and Arizona; local in southwestern Wyoming, extreme northwest Oklahoma, Trans-Pecos Texas, southeastern California, and Mexico; mostly at 5000–7000' (1524–2134 m).

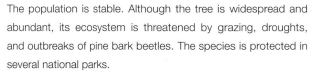

USES: This species ranks first among the native nut trees of the United States that are not cultivated. Local residents harvest the oily seeds for local and gourmet markets. Small pinyons are popular Christmas trees.

SIMILAR SPECIES: Colorado Pinyon Pine is the only pinyon throughout most of its range.

CONSERVATION STATUS: LC
The population is stable. Although the tree is widespread and abundant, its ecosystem is threatened by grazing, droughts, and outbreaks of pine bark beetles. The species is protected in several national parks.

Diameter: 1–2' (0.3–0.6 m) or more
Height: 15–35' (4.6–10.7 m)

■ DENSE POPULATION
■ LIGHT POPULATION
■ NATURALIZED

This is the common native pine of southern Florida. An important lumber species and one of the fastest-growing southern pines, Slash Pine is grown extensively in forest plantations both in its natural range and farther north. Its beauty makes it a popular shade and ornamental tree. South Florida Slash Pine (var. *densa*) is a medium-sized variety of south and central Florida with needles mostly two in a bundle and with a grass-like seedling stage.

DESCRIPTION: Slash Pine is a large tree with a narrow, regular, pointed crown of horizontal branches and long needles. The needles are 7–10" (18–25 cm) long, evergreen; two and three in bundle; stout, stiff; slightly shiny green. The bark is purplish-brown, with large, flattened, scaly plates; on small trunks, blackish-gray, rough and furrowed.

FRUIT: Cones: 2½–6" (6–15 cm) long; narrowly egg-shaped; shiny dark brown; short-stalked; opening and shedding at maturity, leaving a few cone scales on twig; cone-scales flat, slightly keeled, with short stout prickle.

HABITAT: Low areas such as pond margins, flatwoods, swamps or "slashes," including poorly drained sandy soils; also uplands and old fields. In pure stands as a subclimax after fires and in mixed forests.

RANGE: Coastal Plain from southern South Carolina to southern Florida, and west to southeast Louisiana; mostly near sea level, locally to 500' (152 m).

USES: Slash Pine is planted throughout the southeastern United States for reforestation and timber plantations. The wood is used for poles, railroad ties, and a variety of other purposes. This species is also a popular ornamental tree.

SIMILAR SPECIES: Longleaf Pine.

CONSERVATION STATUS: (LC)

The population is increasing with no identified threats. Logging activity has allowed this species to spread to drier sites where it sometimes replaces logged stands of endangered Longleaf Pine.

Diameter: 2–2½' (0.6–0.8 m)
Height: 60–100' (18–30 m)

■ DENSE POPULATION
■ LIGHT POPULATION
■ NATURALIZED

This Mexican pine extends northward into southeastern Arizona and southwestern New Mexico. As the name implies, it grows in Apache Native American homelands. Seedlings of this tree go through a grasslike stage in which they have a short stem and very long needles to 15" (38 cm); the related Longleaf Pine in the southeastern United States goes through a similar stage.

DESCRIPTION: Apache Pine is a medium-sized tree with a straight trunk and an open, rounded crown of few, large, spreading branches—one row added a year—and long needles. The needles are 8–12" (20–30 cm) long, evergreen, three in a bundle (sometimes four to five), crowded at the end of stout twigs, spreading widely or drooping, and dull green. The bark is dark brown or blackish-gray, rough, thick, deeply furrowed into scaly ridges.

FRUIT: Cones: 4–5½" (10–14 cm) long, egg-shaped or conical, almost stalkless, shiny light brown, opening and shedding at maturity, leaving a few cone scales on the twig. The cone scales are four-sided, thick, raised, with a prominent keel and stout, short spine that is straight or curved backward.

HABITAT: Rocky ridges and slopes of mountains; with Arizona Pine and Chihuahua Pine.

RANGE: Southeastern Arizona and extreme southwestern New Mexico; also northern Mexico; at 5000–8200' (1524–2499 m).

USES: This species is rarely harvested due to its limited range, but it is sometimes used for lumber, and locally for firewood.

SIMILAR SPECIES: Ponderosa Pine, Longleaf Pine.

CONSERVATION STATUS: LC
The population is stable with few threats apart from small-scale subsistence harvesting in some areas.

Diameter: 2' (0.6 m)
Height: 50–70' (15–21 m)

■ DENSE POPULATION
■ LIGHT POPULATION
■ NATURALIZED

Both the common and scientific names of this pine refer to its extremely tough and flexible twigs, which can sometimes be twisted into a knot. This flexibility allows the tree to handle the weight of heavy snow. Limber Pines on exposed ridges and at the timberline are shaped by the wind into stunted shrubs with crooked or twisted branches that are bent over and longer on one side. Birds and mammals, especially squirrels, consume the Limber Pine's large seeds.

DESCRIPTION: Limber Pine is medium-sized tree with a short trunk and a broad, rounded crown of annual rows of stout branches nearly down to ground; it can also be a windswept, deformed shrub at the timberline. Its needles are evergreen; five in bundle, with sheath shedding first year; 2–3½" (5–9 cm) long; slender and long-pointed, not toothed; light or dark green, with white lines on all surfaces. The bark is light gray and smooth; becoming dark brown and furrowed into scaly ridges or rectangular plates. The twigs are slender, tough, and flexible.

FRUIT: Cones: 3–6" (7.5–15 cm) long; egg-shaped, yellow-brown, short-stalked; opening at maturity; cone scales are thick, rounded, and end in blunt point. The seeds are large and edible, with short wings.

HABITAT: Dry, rocky slopes and ridges of high mountains up to timberline; often in pure stands.

RANGE: Rocky Mountain region chiefly, from southeast British Columbia and southwest Alberta south to northern New Mexico and west to southern California; also local in northeast Oregon, southwest North Dakota, Black Hills of South Dakota, and western Nebraska.

USES: This species is commonly cultivated as an ornamental and is also used for Christmas trees.

SIMILAR SPECIES: Whitebark Pine.

CONSERVATION STATUS: (LC)

The population is disjunct and decreasing. Limber Pine is susceptible to several diseases and pests, notably to white pine blister rust (caused by the fungus *Cronartium ribicola*) and Mountain Pine Beetles (*Dendroctonus ponderosae*). Limber Pine is often the only tree species able to tolerate some harsh high-mountain and tree line habitats; where there are no replacement species to occupy these areas, the loss of these trees will likely result in the transition of these sites from forests to non-forested landscapes.

Diameter: 2–3' (0.6–0.9 m)
Height: 40–50' (12–15 m)

■ DENSE POPULATION
■ LIGHT POPULATION
■ NATURALIZED

This uncommon, gray-barked pine is found in damp woodlands of the Deep South, often scattered in the shade of mixed forests rather than in pure stands. The wood is not durable and so is rarely used for lumber. Spruce Pine is a species easily killed by forest fires.

DESCRIPTION: This is a large tree with a long, narrow crown and horizontal branches. Its needles are evergreen; 1½–4" (4–10 cm) long; two in bundle; slender, often slightly flattened and twisted; dark green. The bark is gray, smooth, thick; becoming dark gray and furrowed into flat scaly ridges.

FRUIT: Cones: 1¼–2½" (3–6 cm) long; conical or narrowly egg-shaped; reddish-brown; nearly stalkless; pointing backward or downward; opening at maturity, shedding or sometimes persistent. The cone scales are thin or slightly thickened, with a tiny prickle that is usually shed.

HABITAT: Moist lowland soils, especially along rivers; a minor component of mixed swamp forests.

RANGE: Coastal Plain from eastern South Carolina to northern Florida and west to southeastern Louisiana; to about 500' (152 m).

USES: The wood of Spruce Pine is used for pulpwood and infrequently for lumber. The smooth gray bark makes it sought after as an ornamental and sometimes as a Christmas tree.

SIMILAR SPECIES: Shortleaf Pine, Sand Pine.

CONSERVATION STATUS:

The population is stable. Spruce Pines occupy a large geographic range but are usually scattered thinly throughout mixed forests. The species is highly sensitive to fire and does not recover well through regeneration.

Diameter: 2–2½' (0.6–0.8 m)
Height: 80–90' (24–27 m)

■ DENSE POPULATION
■ LIGHT POPULATION
■ NATURALIZED

It is difficult to describe the exact smell that comes from the crushed twigs of a Jeffrey Pine. The scent has been likened to that of lemons and vanilla, but also to violets, pineapples, apples, and butterscotch. The sweet scent is strong in the needles, bark crevices, and resin. This species was named for its discoverer, John Jeffrey, the 19th century Scottish botanical explorer who collected seeds and plants in Oregon and California for introduction into Scotland.

DESCRIPTION: This large tree sports a straight trunk and an open, conical crown of spreading branches with large cones. The bark and the twigs give off a distinctive, sweet odor when crushed. Its needles are evergreen; three in bundle; 5–10" (13–25 cm) long; stout, stiff; light gray-green or blue-green, with broad white lines on all surfaces. The bark is purplish-brown, thick, and furrowed into narrow scaly plates. The twigs are stout, hairless, smooth, and gray-green.

FRUIT: Cones: 5–10" (13–25 cm) long; conical or egg-shaped; light reddish-brown, almost stalkless; opening and shedding at maturity, leaving a few cone scales on the twig. The cone scales are numerous, raised, and keeled, ending in a long, bent-back prickle.

HABITAT: Dry slopes of mountains, especially from lava flows and granite; best developed on deep, well-drained soils; often forming pure stands and with other conifers.

RANGE: Southwest Oregon south through the Sierra Nevada (especially eastern slopes) to western Nevada and southern California; also in northern Baja California; mostly at 6000–9000' (1829–3048 m); less frequently down to 3500' (1067 m) and up to 10,000' (3048 m).

USES: The Jeffrey Pine is primarily used for high-grade lumber for millwork, cabinets, doors, and windows. Pure n-heptane, a component in gasoline, is distilled from Jeffrey Pine resin.

SIMILAR SPECIES: Washoe Pine, a rare tree with a limited range in western Nevada and northeastern California.

CONSERVATION STATUS: 🄻🄲
The population is stable. Natural regeneration is able to counter normal losses to diseases, pests, and forest fires. Air pollution near urban areas, especially around Los Angeles and in some national parks in the Sierra Nevada, weakens these trees enough to constitute a threat to regional populations.

Diameter: 2–4' (0.6–1.2 m), sometimes much larger
Height: 80–130' (24–39 m)

DENSE POPULATION
LIGHT POPULATION
NATURALIZED

One of the most beautiful and largest pines in North America is the Sugar Pine. This tree routinely reaches heights well over 100' (30 m) and has been recorded at more than 10' (3 m) in diameter. The tallest on record is 255' (77.8 m) tall. Sugar Pines also sport the longest cones of any conifer, sometimes reaching 20" (50.8 cm) or more in length.

This species provided early settlers of California with wood for houses, especially shingles, and for fences. Forty-niners made ample use of the wood for flumes, sluice boxes, bridges, and mine timbers. Native Americans gathered and ate the large, sweet seeds. The common name of this pine refers to the sweet resin—also a food source for Native Americans—that exudes from cut or burned wood.

DESCRIPTION: Sugar Pine is a large, tall tree with a straight trunk unbranched for a long span and an open, conical crown of long, nearly horizontal branches, bearing giant cones near the ends; becoming flat-topped. Its needles are 2¾–4" (7–10 cm) long and evergreen; five in bundle, with the sheath shedding in the first year; they are twisted, slender, stiff, sharp-pointed; and blue-green with white lines on all surfaces. The bark is brown or gray; furrowed into irregular scaly ridges; gray and smooth on branches.

FRUIT: Cones: 11–18" (28–46 cm) long; cylindrical, shiny light brown; hanging down on a long stalk near the ends of upper branches. The cone scales are thick and rounded, ending in a blunt point, spreading widely. The seeds are large, long-winged, and edible.

HABITAT: Many kinds of mountain soils; not forming pure stands but occurring in mixed coniferous forests.

RANGE: Mountains from western Oregon south through the Sierra Nevada to southern Califor-nia; also in northern Baja California; at 1100–5400' (335–1646 m) in the north, 2000–7800' (610–2377 m) in the Sierra Nevada, and 4000–10,500' (1219–3200 m) in the south.

USES: Sugar Pine is a major lumber species, used for everything from crates and boxes to interior millwork to piano keys.

SIMILAR SPECIES: Western White Pine.

CONSERVATION STATUS:
The population is stable, and this species seems more tolerant than its relatives of air pollution that affects parts of its range. White pine blister rust, a disease caused by an introduced pathogen (*Cronartium ribicola*), poses a possible threat by limit-ing natural regeneration in some areas.

Diameter: 3–6' (0.9–1.8 m),
sometimes much larger
Height: 100–160' (30–49 m)

■ DENSE POPULATION
■ LIGHT POPULATION
■ NATURALIZED

Also called Smooth-leaf Pine, this tree is primarily found in Mexico. Unlike most pines, this species often produces new shoots or sprouts from cut stumps. The cones mature in three growing seasons instead of the usual two. Cones are usually present in three stages of development, as are many old, open cones. Trees native to the United States and northwestern Mexico are considered a separate variety (var. *chihuahuana*), characterized by three stout needles in a bundle. The typical variety (var. *leiophylla*), of wider distribution in Mexico, has five slender needles in a bundle.

DESCRIPTION: The trunk of this pine bears short, leafy twigs; very thick bark; and a thin, open, spreading crown of upturned branches. Its needles are evergreen; three in bundle, with the sheath shedding the first year; 2½–4" (6–10 cm) long; stout, stiff; blue-green with white lines on all surfaces. The bark is dark brown or blackish, as much as 2–3" (5–7.5 cm) thick, and deeply furrowed into broad, scaly ridges.

FRUIT: Cones: 1½–2½" (4–6 cm) long; narrowly egg-shaped, shiny light brown, long-stalked; usually opening but remaining attached; cone scales flattened, mostly with tiny prickle.

HABITAT: Rocky ridges and slopes of mountains; often found with Arizona Pine and Apache Pine.

RANGE: Southwestern New Mexico, Arizona and Mexico; at 5000–7800' (1524–2377 m).

USES: The dense, strong wood from the Chihuahua Pine is used for construction, railroad ties, and firewood.

SIMILAR SPECIES: Mexican Pinyon.

CONSERVATION STATUS: LC
The population is stable with few threats. Logging has affected parts of its range, but the species is widespread and common.

Diameter: 1–2' (0.3–0.6 m)
Height: 30–80' (9–24 m)

- ■ DENSE POPULATION
- ■ LIGHT POPULATION
- ■ NATURALIZED

These long-living trees are classed among the oldest known living things. The oldest dated living trees are Intermountain Bristlecone Pines more than 4600 years old, protected at Inyo National Forest near Bishop, in eastern California. Other ancient Intermountain Bristlecone Pines are found at Wheeler Peak Scenic Area in the Humboldt National Forest of eastern Nevada. Some shrubs and trees that spread in colonies or clumps from the same root system may be even older. These trees usually grow in large, open stands and are vigorous primary succession trees capable of withstanding harsh or exposed habitat. Intermountain Bristlecone Pine was formerly considered a variety of Bristlecone Pine (*P. aristata*) and is also called Great Basin Bristlecone Pine.

DESCRIPTION: This tree has short needles crowded into a mass suggesting a foxtail, and has a broad, irregular crown of spreading branches; it is a low shrub at the timberline. Its needles are evergreen; five in a bundle, with the sheath shedding after the first year; ¾–1½" (2–4 cm) long; crowded in a long, dense mass curved against twig; stout, stiff, blunt-pointed; persisting 10–20 years; dark green with white lines on the inner surfaces; often with whitish resin dots on the outer surface. The bark is whitish-gray, smooth; becoming reddish-brown and furrowed into irregular, scaly ridges.

FRUIT: Cones: 2½–3½" (6–9 cm) long; cylindrical, dark purplish-brown, almost stalkless; opening at maturity. The cone scales are four-sided, with rounded bases and fine, slender, curved bristle. The seeds are brown mottled with black and with a detachable wing.

HABITAT: Exposed, dry, rocky slopes and ridges of high mountains in subalpine zone to the timberline; often found in pure stands.

RANGE: Utah to Nevada and eastern California; at 7500–11,500' (2286–3505 m).

USES: These trees grow mostly in protected areas and are not used much commercially.

SIMILAR SPECIES: Bristlecone Pine.

CONSERVATION STATUS: LC

The population is stable. The species is known for its slow growth rate and extreme longevity. Stands on mountaintops, at lower elevations, or on other marginal sites may be threatened by climate change, but it is difficult to predict the complex interactions of changes in temperature, precipitation patterns, and insect and disease activity.

Diameter: 1–2½' (0.3–0.8 m)
Height: 20–40' (6–12 m)

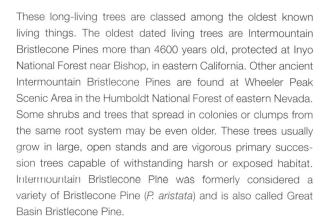

■ DENSE POPULATION
■ LIGHT POPULATION
■ NATURALIZED

This is the only pine in the world with single needles, rather than clusters of two or more as in other pines. The large, mealy seeds of Singleleaf Pinyon are sold locally as pinyon or pine nuts and were once a staple food for Native Americans in the Great Basin region. Many birds and mammals also consume the seeds.

DESCRIPTION: Singleleaf Pinyon is a low-growing, small pine with a spreading, rounded, gray-green crown and low, horizontal branches; it is often shrubby. Its needles are evergreen; one in bundle (rarely two), with the sheath shedding after the first year; 1–2¼" (2.5–6 cm) long; stout, stiff, sharp-pointed; straight or slightly curved; dull gray-green with many whitish lines; resinous. The bark is dark brown or gray, smoothish, becoming furrowed into scaly plates and ridges.

FRUIT: Cones: 2–3" (5–7.5 cm) long; egg-shaped or rounded, dull yellow-brown, almost stalkless, resinous; opening and shedding; with thick, four-angled cone scales, often with a tiny prickle. The seeds are large, wingless, thin-walled, mealy, and edible.

HABITAT: Dry, gravelly slopes of mesas, foothills, and mountains; it is found in open, orchardlike, pure stands and with junipers.

RANGE: Southeastern Idaho and northern Utah south to northwest Arizona and west to southern California; also northern Baja California; at 3500–7000' (1067–2134 m).

USES: The seeds of Singleleaf Pinyon were collected and eaten by Native Americans, and are still often eaten by people today. This species is also cultivated as an ornamental tree, for landscaping, and for Christmas trees.

SIMILAR SPECIES: Parry Pinyon.

CONSERVATION STATUS: ⓁⒸ

Singleleaf Pinyon is widespread throughout the semi-arid southwest; the population is stable with few threats, although drought and engraver beetle outbreaks (*Ips* spp.) have caused some declines.

Diameter: 1–1½' (0.3–0.5 m)
Height: 16–30' (5–9 m)

■ DENSE POPULATION
■ LIGHT POPULATION
■ NATURALIZED

Western White Pine is one of the world's largest pines—the tallest known has reached a height of 239' (72.8 m). An important timber tree, this species is the leading source for matchstick wood, because of its uniformly high grade without knots, twisted grain, or discoloration. White Pine Blister Rust, caused by an introduced fungus (*Cronartium ribicola*), is a serious disease of this and other five-needle white pines.

DESCRIPTION: This is a large tree with a straight trunk and a narrow, open, conical crown of horizontal branches. Its needles are evergreen; five in bundle, with the sheath shedding in the first year; 2–4" (5–10 cm) long; slightly stout; blue-green with whitish lines on inner surfaces. The bark is gray and thin, smooth, becoming furrowed into rectangular, scaly plates.

FRUIT: Cones: 5–9" (13–23 cm) long; narrowly cylindrical, yellow-brown, mostly long-stalked; opening and shedding at maturity. The cone scales are thin and rounded, ending in a small point, and spreading widely. It has long-winged seeds.

HABITAT: Moist mountain soils; in mixed forests and occasionally in almost pure stands.

RANGE: Northern Rocky Mountains from British Columbia southeast to northwestern Montana; also along the Pacific Coast south through the Sierra Nevada to central California; to 3500' (1067 m) in the north; at 6000–9800' (1829–2987 m) in the south.

USES: Western White Pine is an important commercial timber species, a primary source for wooden matchsticks. It is also grown as an ornamental tree.

SIMILAR SPECIES: Eastern White Pine, but the ranges of the two species do not overlap.

CONSERVATION STATUS: **NT**
The population is decreasing due to the effects of overharvesting and pine blister rust. Succession by other species after clear-cutting or die-backs has led to substantial reductions in the overall population.

Diameter: 3' (0.9 m), sometimes much larger
Height: 100' (30 m)

■ DENSE POPULATION
■ LIGHT POPULATION
■ NATURALIZED

This evergreen pine has a limited range, mostly along coastal California. Its cones remain closed for many years until they are opened by the heat of fire. Fossilized cones from the Pleistocene epoch indicate that this tree was associated with extinct vertebrates, including the Woolly Mammoth. Its common name refers to the discovery of this local pine in 1835 near the mission of San Luis Obispo (Saint Louis, Bishop of Toulouse) in California.

DESCRIPTION: Bishop Pine is a tree with a conical, rounded, or irregular crown of stout, spreading branches, and with numerous spiny cones remaining closed for many years. Its needles are evergreen; two in a bundle; 4–6" (10–15 cm) long; stout, slightly flattened, stiff, blunt-pointed, and dull green. The bark is dark gray, thick, furrowed into scaly plates; smoothish on the branches.

FRUIT: Cones: 2–3½" (5–9 cm) long; conical or egg-shaped, shiny yellow-brown, stalkless; oblique or one-sided at the base; many clustered in rings or whorls. The cone scales are raised and keeled, those on the outer part are much enlarged, ending in a stout, flattened, straight or curved spine.

HABITAT: Low hills and plains along coast in the fog belt; found in scattered groves and with other pines.

RANGE: Near sea level along the coast of central and northern California and Santa Cruz and Santa Rosa islands; also local in Baja California; a variety occurs on Cedros Island, Mexico.

USES: This species is sometimes cultivated as an ornamental plant.

SIMILAR SPECIES: Bishop Pine is not likely to be confused with any other species within its range.

CONSERVATION STATUS: VU

This species occurs in eight locations with small, severely fragmented populations. This species depends on periodic fires for successful regeneration; fire suppression and fire fighting in its urbanized coastal habitats benefit competing species.

Diameter: 2–3' (0.6–0.9 m)
Height: 40–80' (12–24 m)

■ DENSE POPULATION
■ LIGHT POPULATION
■ NATURALIZED

Austrian Pine, also known as European Black Pine, is native to that continent where it is a leading timber species. In North America, the species is a popular ornamental tree. It is naturalized in some areas in the northeastern United States and southeastern Canada.

DESCRIPTION: This is a large, upright oval to pyramidal evergreen. The needles are thicker and darker green than those of Red Pine and fold when bent (Red Pine's needles break). Its needles are 4–7" (10–17.8 cm) long, two to a bundle; they are sharp, glossy, flexible, and dark green. The winter bud is silvery (brown in Red Pine).

FRUIT: Cones: 2–4" (5–10 cm) long; yellow-brown to reddish brown, ovoid. Scales are tipped with a tiny prickle, which is lacking in Red Pine.

HABITAT: Introduced in North America and planted for shelterbelts and landscaping in almost every state and province.

RANGE: Naturalized in the Northeast and Great Lakes region but less extensively than Scots Pine. It is native to Europe and northern Africa.

USES: This is a popular ornamental and landscaping tree in North America.

SIMILAR SPECIES: Scots Pine, Red Pine.

CONSERVATION STATUS: LC
The population is stable with no range-wide threats identified.

Diameter: 1–2' (30–60 cm)
Height: 50–60' (15–18 m)

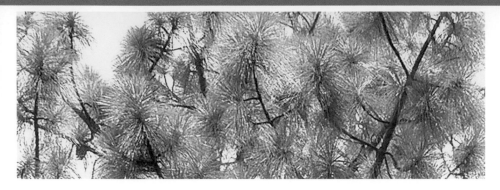

■ DENSE POPULATION
■ LIGHT POPULATION
■ NATURALIZED

This large evergreen is native to the southeastern United States, found from eastern Texas to southern Maryland and south into central Florida. It is the state tree of Alabama. Longleaf Pine is so named because it sports the longest needles of any pine in the world. The seedlings pass through a "grass stage" for a few years, in which the stem grows in thickness rather than height and the taproot develops rapidly. Later, the elongating, unbranched stem produces extremely long needles. Longleaf Pine has been extensively harvested for its wood, and was once the leading timber tree of the South. Remaining old-growth forests of this species serve as important habitat for the declining Red-cockaded Woodpecker (*Leuconotopicus borealis*).

DESCRIPTION: This is a large tree with the longest needles and largest cones of any eastern pine. It has an open, irregular crown of a few spreading branches, with one row added each year. Its needles are mostly 10–15" (25–38 cm) long, and may grow to 18" (46 cm) long on small plants. The needles are evergreen and densely crowded, three in a bundle, slightly stout, and flexible; spreading to drooping; and dark green. The bark is orange-brown, furrowed into scaly plates; on small trunks, gray and rough. The twigs are dark brown and stout, ending in a large white bud.

FRUIT: Cones: 6–10" (15–25 cm) long; narrowly conical or cylindrical; dull brown; almost stalkless; opening and shedding at maturity; cone scales raised, keeled, with small prickle.

HABITAT: Well-drained sandy soils of flatlands and sandhills; often in pure stands.

RANGE: Coastal plains from southeastern Virginia to eastern Florida, and west to eastern Texas. It usually occurs below 600' (183 m); to 2000' (610 m) in foothills of the Piedmont plateau region.

USES: Longleaf Pine is still an important timber species. The trees are tapped for turpentine and resin and then logged for construction lumber, poles and pilings, and pulpwood.

SIMILAR SPECIES: Loblolly Pine.

CONSERVATION STATUS:

Once the most common pine in the coastal plains, large-scale exploitation and conversion of land to agriculture and pastures have reduced its numbers dramatically. The population is still large but decreasing, although the rate of decline has slowed. Its remaining population is severely fragmented. Frequent fire is necessary for its regeneration, and controlled burning programs are in place in an effort to perpetuate the species.

Diameter: 2–2½' (0.6–0.8 m)
Height: 80–100' (24–30 m)

■ DENSE POPULATION
■ LIGHT POPULATION
■ NATURALIZED

This is the most widely distributed and common pine in North America. David Douglas, the Scottish botanical explorer, found this pine in 1826 and named it for its ponderous, or heavy, wood. There are three distinct geographic varieties. The typical variety, Ponderosa Pine or Pacific Ponderosa Pine (var. *ponderosa*), occurs in the Pacific Coast region and has three long needles in a bundle and large cones. Rocky Mountain Ponderosa Pine or Interior Ponderosa Pine (var. *scopulorum*) is found in the Rocky Mountain region, and has two short needles in a bundle and small cones. Arizona Pine or Arizona Ponderosa Pine (var. *arizonica*), found mainly in southeastern Arizona, has five slender needles in a bundle.

DESCRIPTION: This is a large tree with a broad, open, conical crown of spreading branches. Its needles are evergreen; usually two or three in bundle (2–5 in varieties); they are generally 4–8" (10–20 cm) long; stout, stiff, and dark green. The bark is blackish, rough, and furrowed into ridges; on trunks of small trees (blackjacks), becoming yellow-brown and irregularly furrowed into large, flat, scaly plates.

FRUIT: Cones: 2–6" (5–15 cm) long; conical or egg-shaped, almost stalkless, light reddish-brown; opening and shedding at maturity, leaving a few cone scales on the twig. The cone scales are raised and keeled, ending in a short, sharp prickle. It has small, long-winged seeds.

HABITAT: Mostly in mountains in pure stands, forming extensive forests; also found in mixed coniferous forests.

RANGE: Widely distributed; southern British Columbia east to southwestern North Dakota, south to Trans-Pecos Texas, and west to southern California; also found in northern Mexico. It occurs from sea level in the north to 9000' (2743 m) in the south; the best developed stands occur at 4000–8000' (1219–2438 m).

USES: This valuable timber tree is the most commercially important western pine. Its lumber is especially suited for window frames and panel doors. Quail, nutcrackers, squirrels, and many other kinds of wildlife consume the seeds; chipmunks store the seeds in their caches, thus aiding dispersal.

SIMILAR SPECIES: Coulter Pine, Apache Pine, Loblolly Pine, Table Mountain Pine.

CONSERVATION STATUS: LC
Ponderosa Pine is abundant in its range and its population is stable. In many managed forests, selective logging and fire prevention work to the disadvantage of this species.

Diameter: 2½–4' (0.8–1.2 m), sometimes larger
Height: 60–130' (18–39 m)

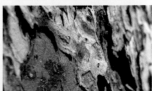

■ DENSE POPULATION
■ LIGHT POPULATION
■ NATURALIZED

This small pine is native to the Appalachian Mountains, easily seen at Shenandoah and Great Smoky Mountains National Parks. This species typically grows to be about 30' (9.1 m) tall, although it has been recorded to reach more than 90' (27.4 m) in height. It is sometimes called Hickory Pine because of its tough hickory-like branches.

DESCRIPTION: This tree has a rounded or irregular crown of stout, horizontal branches and abundant spiny cones in clusters along branches. Its needles are evergreen; 1¼–2½" (3–6 cm) long; two in bundle (sometimes three); stout, stiff, usually twisted; and dark green. The bark is dark brown, thick, and furrowed into scaly plates.

FRUIT: Cones: 2–3½" (5–9 cm) long; egg-shaped; shiny light brown; usually in clusters of three or four; stalkless; pointing backward or downward; opening partly at maturity but remaining attached many years. The cone scales are thickened and keeled, with a stout, curved spine.

HABITAT: Dry gravelly and rocky slopes and ridges of mountains with other pines; sometimes found in pure stands.

RANGE: Appalachian region from Pennsylvania south to northeast Georgia and eastern Tennessee; local in New Jersey, Delaware, and the District of Columbia. It occurs to 4000' (1219 m) and rarely down almost to sea level.

USES: Table Mountain Pine is used locally for firewood, and is sometimes used commercially for pulpwood and timber.

SIMILAR SPECIES: Loblolly Pine, Ponderosa Pine.

CONSERVATION STATUS: **LC**
The population is stable but its subpopulations are scattered. This species regenerates best after fires. Infestations of southern pine beetles (*Dendroctonus frontalis*) can decimate an entire stand.

Diameter: 1–2' (0.3–0.6 m)
Height: 20–40' (6–12 m)

■ DENSE POPULATION
■ LIGHT POPULATION
■ NATURALIZED

This small evergreen has a limited native range in southern California in the United States and northern Baja California in Mexico. Its edible seeds are not gathered commercially because of the tree's limited distribution. Rodents, other mammals, and birds often consume the seeds.

DESCRIPTION: This is a small resinous tree with a spreading rounded crown and low, horizontal branches; it is often shrubby. Its needles are evergreen; four in a bundle (sometimes three or five), with the sheath shedding after the first year; 1–2¼" (2.5–6 cm) long; stout, stiff, short-pointed; bright green with whitish inner surfaces. The bark is light gray and smooth, becoming reddish-brown and furrowed into scaly ridges.

FRUIT: Cones: 1½–2½" (4–6 cm) long; egg-shaped or nearly round, almost stalkless, dull yellow-brown, resinous; opening and shedding. The cone-scales are thick, four-angled, often with a tiny prickle. It has large, wingless, edible seeds.

HABITAT: Dry, gravelly slopes of foothills and mountains; in woodlands or with junipers.

RANGE: Southern California and northern Baja California; at 4000–6000' (1219–1829 m).

USES: Although not gathered commercially, the seeds of the Parry Pinyon are collected and eaten by some people. This tree was historically an important resource to Native Americans in the region. It is sometimes planted as an ornamental pine, and it is also used for Christmas trees.

SIMILAR SPECIES: Singleleaf Pinyon.

CONSERVATION STATUS: **LC**
The population is stable with no identified threats.

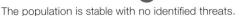

Diameter: 1–1½' (0.3–0.5 m)
Height: 16–30' (5–9 m)

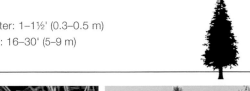

■ DENSE POPULATION
■ LIGHT POPULATION
■ NATURALIZED

Monterey Pine, although rare in its native California range, is one of the world's most valuable pines and is the species most commonly planted in the Southern Hemisphere, where pines do not naturally occur. This species is especially popular in New Zealand, Australia, Chile, and South Africa. Like those of Knobcone and Bishop pines, the cones of Monterey Pine remain closed, sometimes for many years, until opened by the heat of a forest fire; the seeds are then discharged and begin a new forest. The cones sometimes burst open in hot weather with a snapping sound.

DESCRIPTION: Monterey Pine is a tree with a straight trunk; a narrow, irregular, open crown; and many closed cones grouped in rings. Its needles are evergreen; three in bundle; 4–6" (10–15 cm) long; slender, shiny green. The bark is dark reddish-brown, thick, deeply furrowed into scaly ridges or plates.

FRUIT: Cones: 3–6" (7.5–15 cm) long; conical or egg-shaped, oblique or one-sided at the base and pointed at the tip, shiny brown; many clustered in rings or whorls on short stalks and turned back; remaining closed on the tree for many years. The cone scales are thick, slightly raised, and tipped.

HABITAT: Coarse soils, usually sandy loams, on slopes. It is found in pure stands or with Monterey and Gowen cypresses and Coast Live Oak.

RANGE: This tree is rare in the wild, with native stands found at three localities on the coast of central California in the fog belt to about 6 miles (9.7 km) inland; also a variety occurs on Guadalupe Island, Mexico. It is found up to nearly 1000' (305 m).

USES: Monterey Pine is widely used for its wood, including for house construction, in fencing, and sometimes for boats. It is also used for pulpwood, and is commonly used for Christmas trees in Australia and New Zealand.

SIMILAR SPECIES: Bishop Pine, Knobcone Pine.

CONSERVATION STATUS: (EN)

The natural population in its native range is decreasing. Logging, grazing by feral goats, competition from other trees in the absence of periodic fires, and pine pitch canker (a fungal disease caused by the introduced pathogen *Fusarium subglutans pini*) have reduced the tree's area of occupancy, likely permanently. Ironically, where it is cultivated in the Southern Hemisphere, this tree can become weedy or invasive in some habitats if not properly managed. It's listed as a weed in Hawaii.

Diameter: 1–3' (0.3–0.9 m)
Height: 50–100' (15–30 m)

■ DENSE POPULATION
■ LIGHT POPULATION
■ NATURALIZED

This large evergreen is also known as the Norway Pine, a confusing alternate name that probably originated from early Scandinavian immigrants who likened the tree to the Norway Spruce from their home region. Red Pine is a notably long-lived species, with some trees known to be 500 years old.

DESCRIPTION: Red Pine is a common, large tree with small cones and a broad, irregular or rounded crown of spreading branches, with one row added each year. Its needles are evergreen; 4¼–6½" (11–16.5 cm) long; two in a bundle; slender; and dark green. The bark is reddish-brown or gray; with broad, flat, scaly plates; becoming thick.

FRUIT: Cones: 1½–2¼" (4–6 cm) long; egg-shaped; shiny light brown; almost stalkless; opening and shedding soon after maturity. The cone scales are slightly thickened, keeled, and without prickle.

HABITAT: Well-drained soils; particularly sand plains. It is usually found in mixed forests.

RANGE: Southeast Manitoba east to Nova Scotia, south to Pennsylvania and west to Minnesota. It is local in Newfoundland, northern Illinois and eastern West Virginia at 700–1400' (213–427 m) northward; to 2700' (823 m) in the Adirondacks; and at 3800–4300' (1158–1311 m) in West Virginia.

USES: Red Pine is planted as an ornamental and shade tree, and the moderately hard and straight-grained wood is used for general construction, millwork, and pulpwood.

SIMILAR SPECIES: Austrian Pine.

CONSERVATION STATUS: ⓛⒸ
The population is increasing with few threats. This species was extensively logged in the 19th and 20th centuries and is now recovering through regeneration and by planting in the Great Lakes region. It's listed as endangered in Illinois, Connecticut, and New Jersey.

Diameter: 1–3' (0.3–0.9 m), often larger
Height: 70–80' (21–24 m)

Pitch Pine is suited to dry, rocky soil that other trees cannot tolerate, becoming open and irregular in shape in exposed situations. This hardy species is resistant to fire and injury, forming sprouts from roots and stumps. It is the primary pine at Cape Cod, and the New Jersey Pine Barrens are composed of dwarf sprouts of Pitch Pine following repeated fires. The common name refers to the high resin content of Pitch Pine's knotty wood. Early colonists used Pitch Pines as a source of resin, producing turpentine and tar used for axle grease from this species before naval stores were developed from the southern pines. Pine knots fastened to a pole also served as excellent torches in the times before electricity.

DESCRIPTION: Pitch Pine is a medium-sized tree often bearing tufts of needles on the trunk, with a broad, rounded or irregular crown of horizontal branches. Its needles are evergreen; 3–5" (7.5–13 cm) long; three in a bundle; stout, stiff, often twisted; yellow-green. The bark is dark gray; thick, rough, deeply furrowed into broad scaly ridges, exposing brown inner layers.

FRUIT: Cones: 1¼–2¾" (3–7 cm) long; egg-shaped; yellow-brown; opening at maturity but remaining attached. The cone scales are raised and keeled, with a slender, sharp prickle.

HABITAT: Shallow sands and gravels on steep slopes and ridges, also in river valleys and swamps. It forms temporary pure stands, and is gradually replaced by hardwoods; it is also found in mixed forests.

RANGE: Southern Maine west to New York and southwest mostly in the mountains to northern Georgia; local in extreme southern Quebec and extreme southeastern Ontario. From sea level in the Coastal Plain to about 2000' (610 m) in the north; 1400–4500' (427–1372 m) in the upper Piedmont plateau region.

USES: Pitch Pine is now primarily used for rough lumber and pulpwood.

SIMILAR SPECIES: Loblolly Pine, Shortleaf Pine, Pond Pine.

CONSERVATION STATUS:
The population is increasing with no identified range-wide threats.

Diameter: 1–2' (0.3–0.6 m)
Height: 50–60' (15–18 m)

■ DENSE POPULATION
■ LIGHT POPULATION
■ NATURALIZED

This common and widespread pine is found in the foothill woodlands, northern oak woodlands, chaparral, mixed conifer forests, and hardwood forests in California and southwestern Oregon. The soft, lightweight wood of this tree is not durable; the crooked, forking trunks also make the wood impractical to use except as firewood. It was commonly called Digger Pine, in reference to a derogatory term used by settlers to refer to certain Native groups; it is also commonly called Bull Pine and California Foothill Pine.

DESCRIPTION: Gray Pine is a tree with a crooked, forking trunk and branches; an open, thin, irregular, broad, or rounded crown; and large, heavy cones. Its needles are evergreen; three in a bundle; 8–12" (20–30 cm) long; slender and drooping; dull gray-green with many white lines. The bark is dark gray, thick, deeply and irregularly furrowed into scaly ridges; becoming slightly shaggy, light gray, and smooth on branches.

FRUIT: Cones: mostly 6–10" (15–25 cm) long; egg-shaped and slightly one-sided, brown, bent down on long stalks; opening late and remaining on the tree for many years. The cone scales are long, thick, sharply keeled, and four-sided.

HABITAT: Dry slopes and ridges in foothills and low mountains; it is found with oaks and other conifers.

RANGE: Northern to Southern California through the Coast Ranges and the Sierra Nevada; mostly at 1000–3000' (305–914 m); rarely down to 100' (30 m) and up to 6000' (1829 m).

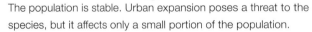

USES: The seeds from this tree were a major food source for some Native Americans.

SIMILAR SPECIES: Coulter Pine.

CONSERVATION STATUS: LC
The population is stable. Urban expansion poses a threat to the species, but it affects only a small portion of the population.

Diameter: 2–4' (0.6–1.2 m)
Height: 40–70' (12–21 m),
sometimes much larger

■ DENSE POPULATION
■ LIGHT POPULATION
■ NATURALIZED

The Latin name *serotina*, meaning "late," refers to this species' cones, which remain closed for years before opening, often following a fire. After fires or other damage, seedlings and trees will produce sprouts from roots. This species is also called Pocosin Pine, based on the Native American word *pocosin* for "pond" or "bog," alluding to this species' habitat.

DESCRIPTION: This is a medium-sized tree with an open, rounded or irregular crown of stout, often crooked branches. Its needles are evergreen; 5–8" (13–20 cm) long; three in a bundle; slender, stiff; yellow-green. The bark is blackish-gray or reddish-brown; furrowed into scaly plates.

FRUIT: Cones: 2–2½" (5–6 cm) long; nearly round or egg-shaped; shiny yellow; almost stalkless; remaining closed on the tree for many years. The cone scales are slightly raised and keeled, with its weak prickle usually shed.

HABITAT: Swamps, shallow bays, and ponds; it is often found in nearly pure stands.

RANGE: Southern New Jersey and Delaware south to central and northwestern Florida, near sea level.

USES: This species is not a popular ornamental, nor is it used much for lumber. Pond Pine is sometimes used for pulpwood.

SIMILAR SPECIES: Pitch Pine.

CONSERVATION STATUS: LC

The population is stable. The largest threat, especially for older pines, is the fungus *Phellinus pini* (or *Porodaedalea pini*) that causes a heartwood rotting disease called red heart. Red heart affects all pine trees in Florida, but only poses a serious threat for mature pine trees.

Diameter: 1–2' (0.3–0.6 m)
Height: 40–70' (12–21 m)

■ DENSE POPULATION
■ LIGHT POPULATION
■ NATURALIZED

The large seeds of Southwestern White Pines are consumed by various wildlife, particularly the Mexican Jay (*Aphelocoma wollweberi*), and were eaten by southwestern Native Americans. This species of the Mexican border region—formerly considered a southern variety of Limber Pine, which has a broader and more northern distribution—is smaller, and has smooth-edged needles with white lines and shorter cones with thick, rounded, blunt-pointed cone scales.

DESCRIPTION: This tree has a straight trunk and a narrow, conical crown of horizontal branches. Its needles are evergreen; five in a bundle, with the sheath shedding in the first year; 2½–3½" (6–9 cm) long; slender, finely toothed at least near tip; bright green, with white lines on inner surfaces only. The bark is gray and smooth, becoming dark gray or brown and deeply furrowed into narrow, irregular ridges.

FRUIT: Cones: 6–9" (15–23 cm) long; cylindrical, yellow-brown, short-stalked; opening at maturity. The cone scales are slightly thickened and long, with the thin, narrow tips spreading and curved back. The seeds are large, edible, and short-winged.

HABITAT: Dry, rocky slopes and canyons in high mountains; a minor component of coniferous forests.

RANGE: Trans-Pecos Texas west to east-central Arizona; also northern Mexico; at 6500–10,000' (1981–3048 m).

USES: This species is grown as an ornamental tree and also used as a Christmas tree. It is sometimes used in making cabinets.

SIMILAR SPECIES: Limber Pine.

CONSERVATION STATUS: (LC)
The population is stable. Unsustainable logging can pose localized threats.

Diameter: 1½–3' (0.5–0.9 m)
Height: 50–80' (15–24 m)

EASTERN WHITE PINE *Pinus strobus*

■ DENSE POPULATION
■ LIGHT POPULATION
■ NATURALIZED

The Eastern White Pine of New England and southeastern Canada was the greatest natural treasure of the American colonies. Once the world's most important timber tree, vast forests of White Pine—straight, light, and strong in proportion to its weight—were the principal material for the building of ships, houses, bridges, and furniture. The virgin White Pine forests, now gone except for isolated old-growth pockets, held millions of tall, straight trees. Efforts by the British crown to reserve the best trees for ships' masts were among the annoyances that sparked the American Revolution. The tree was pictured on the revolutionaries' first flag and is the state tree of Maine.

DESCRIPTION: The tallest tree in the east, Eastern White Pine is a magnificent evergreen tree with a straight trunk and crown of horizontal branches, with one row added per year, becoming broad and irregular. Its needles are 2½–5" (6–13 cm) long, evergreen, five in a bundle, slender, and blue-green. The bark is gray; smooth becoming rough; thick and deeply furrowed into narrow scaly ridges.

FRUIT: Cones: 4–8" (10–20 cm) long, narrowly cylindrical, yellow-brown, and long-stalked. The cone scales are thin, rounded, and flat.

HABITAT: Most well-drained sandy soils; sometimes found in pure stands.

RANGE: Southeastern Manitoba east to Newfoundland, south to northern Georgia, and west to northeastern Iowa; a variety is found in Mexico. The species occurs from near sea level to 2000' (610 m); in the southern Appalachians to 5000' (1524 m).

USES: The largest conifer and formerly the most valuable tree of the Northeast, Eastern White Pine is used for construction, millwork, trim, matches, and pulpwood. Younger trees and plantations have replaced the once seemingly inexhaustible lumber supply of virgin forests.

SIMILAR SPECIES: Shortleaf Pine, Austrian Pine, Jack Pine, Loblolly Pine, Scrub Pine.

CONSERVATION STATUS: LC
Eastern White Pine is widespread and abundant, and its population is increasing. Heavy logging through the 19th century had depleted the species and eliminated most of the old-growth forests, but regrowth has since allowed the species to recover.

Diameter: 3–4' (0.9–1.2 m) or more
Height: 100' (33 m),
formerly 150' (46 m) or more

■ DENSE POPULATION
■ LIGHT POPULATION
■ NATURALIZED

The beautiful pine is native to the Scottish Highlands; in North America it is naturalized locally in southern Canada and in the U.S. throughout the Northeast and upper Midwest. This is the most widely distributed pine in the world and one of the most important European timber trees. In North America, native pines are better adapted for forestry plantations, but Scots Pine is commonly grown for shelterbelts, ornamentals, and Christmas trees.

DESCRIPTION: This is a beautiful, large, introduced tree with a crown of spreading branches that become rounded and irregular, and rich blue-green foliage. The needles are evergreen; 1½–2¾" (4–7 cm) long, two in a bundle; stiff, slightly flattened; twisted and spreading; blue-green. The bark is reddish-brown, thin; becoming gray and shedding in papery or scaly plates.

FRUIT: Cones: 1¼–2½" (3–6 cm) long, egg-shaped, pale yellow-brown, and short-stalked, opening at maturity. The cone scales are thin and flattened, often with a minute prickle.

HABITAT: Various soils from loams to sand. This species tolerates city smoke.

RANGE: Native across Europe and northern Asia, south to Turkey. Naturalized locally in southeastern Canada and the northeastern United States from New England west to Iowa.

USES: Introduced in North America, the Scots Pine is commonly grown for shelterbelts, ornamentals, and Christmas trees.

SIMILAR SPECIES: Shortleaf Pine.

CONSERVATION STATUS: LC
The population is stable with few threats.

Diameter: 2' (0.6 m) and much larger with age
Height: 70' (21 m)

■ DENSE POPULATION
■ LIGHT POPULATION
■ NATURALIZED

A large pine of lowlands and swampy areas, Loblolly Pine is native in 15 southeastern U.S. states. It is one of the most numerous trees in the United States and among the fastest-growing southern pines, and it is extensively cultivated in forest plantations for pulpwood and lumber. The word loblolly means "mudhole" or "mire," which describes the habitats in which this tree often grows. The species is also sometimes colloquially called Bull Pine because of its giant size, and Rosemary Pine because of its fragrant, resinous foliage.

DESCRIPTION: The principal commercial southern pine, Loblolly Pine is a large, resinous, and fragrant tree with a rounded crown of spreading branches. Its needles are evergreen; 5–9" (13–23 cm) long; three in a bundle; stout, stiff, often twisted, and green. The bark is blackish-gray, thick, and deeply furrowed into scaly ridges exposing brown inner layers.

FRUIT: Cones: 3–5" (7.5–13 cm) long, conical, dull brown, almost stalkless; opening at maturity but remaining attached. The cone scales are raised and keeled with a short, stout spine.

HABITAT: From deep, poorly drained flood plains to well-drained slopes of rolling, hilly uplands. It forms pure stands, often on abandoned farmland.

RANGE: Southern New Jersey south to central Florida, west to eastern Texas, north to extreme southeast Oklahoma; to 1500–2000' (457–610 m) .

USES: Perhaps the most important timber species in the southeastern United States, Loblolly Pine is widely cultivated for pulpwood and lumber.

SIMILAR SPECIES: Ponderosa Pine, Slash Pine.

CONSERVATION STATUS: LC
The population is increasing as it reclaims land once cleared for agriculture. The species is considered a weed in Hawaii.

Diameter: 2–3' (0.6–0.9 m)
Height: 80–100' (24–30 m)

■ DENSE POPULATION
■ LIGHT POPULATION
■ NATURALIZED

This is the rarest native pine in North America, its wild range restricted to a few areas in southern California and its total population numbering only about 3000. Most of the remaining trees grow in the Torrey Pines State Natural Reserve in San Diego. When cultivated elsewhere for shade and commercial forestry in moist, warm climates, these trees grow rapidly to a large size, often surpassing 100' (30.5 m) in height.

DESCRIPTION: Torrey Pine is a medium-sized tree with an open, spreading crown of branches; it may be shrubby on exposed sites. Its needles are evergreen; five in a bundle, crowded in large clusters at the ends of stout twigs; 8–13" (20–33 cm) long; stout, stiff, dark gray-green with white lines. The bark is blackish, deeply furrowed into broad, flat, scaly ridges; it is gray and smoothish on branches.

FRUIT: Cones: 4–6" (10–15 cm) long; broadly egg-shaped, nut-brown, and bent down on a stout stalk; maturing in three seasons, opening late and remaining on the tree a few years. The cone scales are thick, keeled, and four-sided, ending in a stout, straight spine. The seeds are very large.

HABITAT: Limited to two areas on dry, sandy bluffs and slopes; found in pure stands.

RANGE: Southern California (San Diego County) and Santa Rosa Island; to 500' (152 m).

USES: An endangered species in the wild, the Torrey Pine is cultivated as an ornamental tree. Native Americans gathered and ate the seeds.

SIMILAR SPECIES: Western White Pine, Coulter Pine.

CONSERVATION STATUS: 🆁
The population is decreasing. A small population is protected in the Torrey Pines State Reserve in California. The species is threatened by fires, pest epidemics, diseases, and urbanization. A small population on Santa Rosa Island in California is not declining due to urbanization, but the risks of fires, pests, and diseases remain.

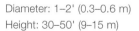

Diameter: 1–2' (0.3–0.6 m)
Height: 30–50' (9–15 m)

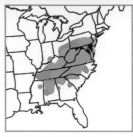

■ DENSE POPULATION
■ LIGHT POPULATION
■ NATURALIZED

Used primarily for pulpwood and lumber, Virginia Pine is more hardy than most pines and great for planting in poor, dry areas. It is commonly used for foresting former coal strip mines.

DESCRIPTION: This is a short-needled tree with an open, broad, irregular crown of long spreading branches; often a shrub. Its needles are evergreen, 1½–3" (4–7.5 cm) long, two in a bundle; they are stout, slightly flattened and twisted, and dull green. The bark is brownish-gray, thin, with narrow scaly ridges, becoming shaggy; on small trunks, smoothish, peeling off in flakes.

FRUIT: Cones: 1½–2¾" (4–7 cm) long; narrowly egg-shaped, shiny reddish-brown; almost stalkless; opening at maturity but remaining attached. The cone scales are slightly raised and keeled, with a long, slender prickle.

HABITAT: Clay, loam, and sandy loam on well-drained sites. It forms pure stands, especially on old fields or abandoned farmland, even in poor or severely eroded soil. Also found in mixed forest types.

RANGE: Southeastern New York (Long Island) south to northeastern Mississippi, and north to southern Indiana; at 100–2500' (30–762 m).

USES: This species is used for lumber, pulpwood, and Christmas trees. It is not a common ornamental tree.

SIMILAR SPECIES: Table Mountain Pine, Shortleaf Pine, Jack Pine.

CONSERVATION STATUS: LC

The population is increasing significantly as Virginia Pine reclaims depleted sites such as strip mined areas and abandoned farmland in the eastern United States.

Diameter: 1–1½' (0.3–0.5 m)
Height: 30–60' (9–18 m)

- DENSE POPULATION
- LIGHT POPULATION
- NATURALIZED

Bigcone Douglas-fir is native only in the mountains of southern California, mostly within national forests, and beyond the range of Douglas-fir. This species is distinguishable from its relative by its much larger cones and by its sharp-pointed, bluish needles. The trees recover from fire damage and injuries by sprouting vigorously from trunks and branches. It is generally shade-tolerant and prefers to grow on slopes. The species is sometimes called Bigcone Spruce.

DESCRIPTION: This tree has an open, broad, conical crown of long, spreading branches. Its needles are evergreen, spreading mostly in two rows, and ¾–1¼" (2–3 cm) long; they are flattened, sharp-pointed, almost stalkless, and blue-green or blue-gray. The bark is dark reddish-brown, thick, and deeply furrowed into broad scaly ridges. The twigs are dark reddish-brown and slender; often drooping; slightly hairy when young; ending in a dark brown, pointed, scaly, hairless bud.

FRUIT: Cones: 4–6" (10–15 cm) long; narrowly egg-shaped, brown, short-stalked; with many thick, stiff, rounded cone scales, each above a short, three-pointed bract. It has paired, long-winged seeds.

HABITAT: Dry slopes and canyons in various soils; with chaparral, Canyon Live Oak, and in mixed coniferous forests.

RANGE: Mountains of southern California; at 900–8000' (274–2438 m).

USES: Due to its limited distribution, this species is not used for any commercial purposes, although it has been used locally for firewood.

SIMILAR SPECIES: It resembles Douglas-fir, but does not overlap in range.

CONSERVATION STATUS: **NT**
The relatively small population of Bigcone Douglas-fir is stable. Long, hot summers make the species vulnerable to forest fires in its limited area of occupancy.

Diameter: 2–3' (0.6–0.9 m)
Height: 40–80' (12–24 m)

■ DENSE POPULATION
■ LIGHT POPULATION
■ NATURALIZED

This ecologically and economically important tree, mainly distributed in the West, comes in two distinct geographic varieties: Coast and Rocky Mountain. Coast Douglas-fir (var. *menziesii*), the typical Douglas-fir of the Pacific Coast, is a very large tree with long, dark yellow-green needles and large cones with spreading bracts. Rocky Mountain Douglas-fir (var. *glauca*), of the Rocky Mountain region, is a medium-sized to large tree with shorter, blue-green needles and smaller cones with bracts bent upward. David Douglas (1798–1834), the Scottish botanical collector who sent seeds back to Europe in 1827, is commemorated in the common name. Many forms of wildlife consume the foliage, including grouse, beavers, rabbits, deer, and elk; many birds and mammals eat the seeds.

DESCRIPTION: This is a large tree with a narrow, pointed crown of slightly drooping branches. Its needles are evergreen; spreading mostly in two rows, ¾–1¼" (2–3 cm) long; they are flattened, mostly rounded at the tip, flexible, and dark yellow-green or blue-green; and have short, twisted leafstalks. The bark is reddish-brown, thick, deeply furrowed into broad ridges; it is often corky. The twigs are orange (turning brown) and slender; hairy; and end in a dark red, conical, pointed, scaly, hairless bud.

FRUIT: Cones: 2–3½" (5–9 cm) long; narrowly egg-shaped, light brown, short-stalked; with many thin, rounded cone scales, each above a long, protruding, three-pointed bract. It has paired, long-winged seeds.

Diameter: 2–5' (0.6–1.5 m),
sometimes much larger
Height: 80–200' (24–61 m)

HABITAT: Coast Douglas-fir forms vast forests on moist, well-drained soils, often in pure stands. Rocky Mountain Douglas-fir is chiefly on rocky soils of mountain slopes; it is found in pure stands and mixed coniferous forests.

RANGE: Central British Columbia south along the Pacific Coast to central California; to 2700' (823 m) in the north and to 6000' (1829 m) in the south; also in the Rocky Mountains to southeastern Arizona and Trans-Pecos Texas; down to 2000' (610 m) in the north and at 8000–9500' (2438–2896 m) in the south. It is also local in the mountains of northern and central Mexico.

USES: One of the world's most important timber species, Douglas-fir ranks first in the United States in total volume of timber, in lumber production, and in production of veneer for plywood. It is also a popular Christmas tree.

SIMILAR SPECIES: Bigcone Douglas-fir, but does not overlap in range.

CONSERVATION STATUS: 🔵LC
The population is stable, but excessive logging has eliminated many old-growth forests. Although Douglas-fir trees tend to regenerate well after logging, the damage to the ecosystem caused by the loss of old-growth forest is often significant.

■ DENSE POPULATION
■ LIGHT POPULATION
■ NATURALIZED

Eastern Hemlock is a slow-growing, long-lived tree that may take 300 years to reach maturity. It is an extremely shade-tolerant tree able to remain in the understory in natural stands for hundreds of years. Hemlock Woolly Adelgid (*Adelges tsugae*), an aphid-like insect, was introduced to eastern North America from Japan in the early 1950s. It feeds on the hemlock's sap, stealing nutrients from the needles, causing them to turn color and eventually fall off. The adult insect is less than 1 mm long and barely visible to the naked eye, but its white, waxy wool-like coat is clearly visible on the stems of trees at the base of the needles. Infested trees slowly weaken until they die after four to six years.

DESCRIPTION: This evergreen tree has a conical crown of long, slender, horizontal branches often drooping down to the ground, and a slender, curved, and drooping leader. Its needles are evergreen; ⅜–⅝" (10–15 mm) long, flat, flexible, and rounded at the tip; spreading in two rows from short leafstalks; they are shiny dark green above, with two narrow whitish bands beneath and green edges often minutely toothed. The bark is cinnamon brown, thick, and deeply furrowed into broad scaly ridges. The twigs are yellow-brown, slender, finely hairy, and rough with peglike bases.

FRUIT: Cones: ⅝–¾" (15–19 mm) long, elliptical, brown, short-stalked; hanging down at the ends of twigs; composed of numerous rounded cone scales. It has paired, light brown, long-winged seeds.

HABITAT: Acid soils, often found in pure stands. Characteristic of moist, cool valleys and ravines; it also occurs on rock outcrops, especially north-facing bluffs.

RANGE: Southern Ontario east to Cape Breton Island, south in the mountains to northern Alabama, and west to eastern Minnesota. To 3000' (914 m) in the north; at 2000–5000' (610–1524 m) in the south.

USES: Eastern Hemlock is a popular ornamental tree. Early settlers used the branches of this tree as brooms.

SIMILAR SPECIES: Carolina Hemlock.

CONSERVATION STATUS: NT

The population is decreasing primarily due to the spread of the introduced Hemlock Woolly Adelgid (*Adelges tsugae*). The pest has spread to 18 eastern states from Georgia to Maine, devastating populations of native Eastern and Carolina hemlocks. It appears to be spreading about 10 miles per year but has not yet reached Eastern Hemlock's main area of occupancy around the Great Lakes. Severe winters appear to limit the spread and abundance of the adelgid, but warming temperatures associated with climate change may allow the pest to expand.

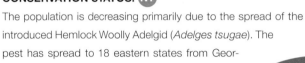

Diameter: 2–3' (0.6–0.9 m)
Height: 60–70' (18–21 m)

■ DENSE POPULATION
■ LIGHT POPULATION
■ NATURALIZED

This uncommon tree has a handsome, symmetrical form. It is sometimes used as an ornamental tree. A local species that went unnoticed by early botanical collectors, Carolina Hemlock was first distinguished in 1850 and named in 1851. It resembles its more widespread relative, Eastern Hemlock, and is susceptible to the introduced Hemlock Woolly Adelgid (*Adelges tsugae*) but to a lesser extent.

DESCRIPTION: This is an evergreen tree with a compact, conical crown of short, stout, often drooping branches and very slender drooping leader. Its needles are evergreen; ⅜–¾" (10–19 mm) long, flat, flexible, slightly notched at the tip, not toothed; and spreading in all directions from short leafstalks; they are shiny dark brown above, with two broad, whitish bands beneath. The bark is red-brown, thick, and deeply furrowed into broad, flat, connected scaly ridges. The twigs are orange-brown, slender, slightly hairy, and rough with peglike bases.

FRUIT: Cones: ¾–1⅜" (2–3.5 cm) long, narrowly elliptical, short-stalked, hanging down at the ends of twigs. They are composed of numerous oblong cone scales spreading widely, almost at right angles at maturity. It has paired, long-winged seeds.

HABITAT: Dry slopes, rocky ridges, ledges, and cliffs; in mixed forests.

RANGE: Southwestern Virginia, northeastern Tennessee, western North Carolina, extreme northwestern South Carolina and extreme northeastern Georgia; at 2500–4000' (762–1219 m).

USES: This species is rarely used commercially.

SIMILAR SPECIES: Eastern Hemlock.

CONSERVATION STATUS: NT

This uncommon tree occurs in scattered, small subpopulations. Climate change may shift its habitats to favor succession by other species, and the impact of the spread of Hemlock Woolly Adelgid (*Adelges tsugae*) is uncertain. The species is listed as threatened in Tennessee.

Diameter: 2' (0.6 m)
Height: 40–60' (12–18 m)

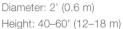

■ DENSE POPULATION
■ LIGHT POPULATION
■ NATURALIZED

Western Hemlock is one of the most common trees in the Pacific Northwest, forming vast, dense groves. It is a shade-tolerant species that tends to thrive in undisturbed climax forests. Young Western Hemlock trees grow slowly in the shade of other conifer species, which they can outlast and eventually replace. Its bark is smooth and reddish-brown when the tree is young, and becomes darker and deeply furrowed with flat-topped scaly ridges as it matures. Western Hemlock needles are short-stalked, flat, finley toothed, irregularly spare, and of unequal length even on the same individual tree.

DESCRIPTION: The largest hemlock, with a long, slender, often fluted trunk; a narrow, conical crown of short, slender, horizontal or slightly drooping branches; and a slender, curved and drooping leader. Its needles are evergreen, spreading in 2 rows, ¼–¾" (6–19 mm) long; they are flat, flexible, rounded at the tip, short-stalked, and shiny dark green above with two broad, whitish bands and indistinct green edges, often with tiny teeth beneath. The bark is reddish-brown to gray-brown, thick, and deeply furrowed into broad, scaly ridges; the cut surface of inner bark is red. The twigs are slender, yellow-brown, finely hairy, and rough with peglike bases.

FRUIT: Cones: ¾–1" (2–2.5 cm) long; elliptical, brown, and stalkless, with many rounded, elliptical cone scales. They are often hanging down at the ends of twigs. It has paired, long-winged seeds.

HABITAT: Moist, acid soils, especially flats and lower slopes; it occurs in dense pure stands and with Sitka Spruce and other conifers.

RANGE: Southern Alaska southeast along the Pacific Coast to northwest California; also southeastern British Columbia south in the Rocky Mountains to northern Idaho and northwest Montana; to 2000' (610 m) along the coast; to 6000' (1,829 m) inland.

USES: This important timber species is one of the best pulpwoods and a source of alpha cellulose for making cellophane, rayon yarns, and plastics. Native Americans of southeastern Alaska used to make coarse bread from the inner bark. Pitch obtained from crevices in the bark has been chewed as a gum, and applied medicinally for a variety of purposes including as a poultice and as a salve to combat sunburn. Western Hemlock leaves and shoot tips can be used to make an herbal tea.

SIMILAR SPECIES: Apache Pine, Atlas Cedar, Austrian Pine.

CONSERVATION STATUS: LC
The population is increasing with no identified threats.

Diameter: 3–4' (0.9–1.2 m)
Height: 100–150' (30–46 m)

- ■ DENSE POPULATION
- ■ LIGHT POPULATION
- ■ NATURALIZED

Mountain Hemlock is a characteristic species of high mountains, varying greatly in size from a large tree at low altitudes to a dwarf, creeping shrub at the timberline. Hemlock groves provide cover, nesting sites, and seeds for birds, as well as foliage for mountain goats and other hoofed browsers. This species is named for its discoverer, German naturalist Karl H. Mertens (1796–1830).

DESCRIPTION: A tree with a tapering trunk, conical crown of slender horizontal or drooping branches, and slender, curved, drooping leader; it is a prostrate shrub at the timberline. The needles are evergreen, usually crowded at the ends of short side twigs, spreading on all sides and curved upward; they are ¼–1" (0.6–2.5 cm) long, short-stalked, flattened above and half-round, stout, blunt and blue-green with whitish lines on both surfaces. The bark is gray to dark brown, thick, and deeply furrowed into scaly ridges. The twigs are light reddish-brown, mostly short, slender, finely hairy, and rough with peglike bases.

FRUIT: Cones: 1–3" (2.5–7.5 cm) long; cylindrical, purplish turning brown, stalkless, hanging down, with many rounded cone scales. It has paired, long-winged seeds.

HABITAT: Moist, coarse or rocky soils, from sheltered valleys to exposed ridges; it is found with firs and in mixed coniferous forests.

RANGE: Southern Alaska southeast along the Pacific Coast to British Columbia and in the mountains to central California; also British Columbia south in the Rocky Mountains to northern Idaho and northeastern Oregon; to 3500' (1067 m) in Alaska and at 5500–11,000' (1676–3353 m).

USES: Mountain Hemlock is grown as an ornamental tree in North America and Europe.

SIMILAR SPECIES: Western Hemlock.

CONSERVATION STATUS: LC
The population is increasing with no identified threats. Mountain Hemlock is protected in Nevada. This species is invading subalpine meadows and late-lying snow basins in some high-elevation areas, likely driven by a warming climate.

Diameter: 1–3' (0.3–0.9 m)
Height: 30–100' (9–30 m)

Also known as Monkey Tail Tree and Chilean Pine, this species is native to Chile—where it is the national tree—and to Argentina. It is cultivated in North America as an ornamental landscaping tree. The common name is said to come from the idea that a monkey would be unable to climb this tree.

DESCRIPTION: This is a medium-sized evergreen tree. The well-spaced, upcurved, ropelike branches are distinctive. The base of a large trunk can resemble an elephant's foot. Its needles are 1–2" (2.5–5 cm) long, leathery, sharp-pointed; in ropelike branches.

FRUIT: Cones: 6" (15 cm) long, brown, rounded to barrel-shaped.

HABITAT: This is an introduced, cultivated species in North America.

RANGE: Cultivated in North America, especially on the Pacific Coast and in the Southeast.

USES: The seeds of this species have been harvested as food in its native range in South America. Monkey Puzzle Tree is a popular ornamental tree.

SIMILAR SPECIES: Monkey Puzzle Tree is unlikely to be confused with any other species in North America.

CONSERVATION STATUS: **EN**
The population is decreasing, partly due to fragmentation of forests by fires and logging. Anthropogenic fires in Chile, where several national parks have been destroyed, are also a threat. The establishment of exotic tree species in Argentina may limit regeneration.

Diameter: 3–5' (1–1.5 m)
Height: 50–80' (15–24 m)

NORFOLK ISLAND-PINE *Araucaria heterophylla*

Despite its common name, this tree is not a true pine. In North America it is a popular houseplant, especially in winter as a potted "Christmas tree." In the wild it is endemic to Norfolk Island, a small island in the Pacific Ocean.

DESCRIPTION: This is a small to medium-sized evergreen tree. The symmetrical, open, conical crown of evenly spaced branches with erect branchlets in rows is distinctive. The trunks of young trees are covered in dead needles. Its needles are ½" (1.3 cm) long, curved, spruce-like, and spirally arranged. In older trees, leaves are scalelike, ¼" (6.4 mm) long, curved inward, densely overlapping. They are bright green to dark green.

FRUIT: Cones: 3–5" (7.6–12.7 cm) long; green, turning brown; roundish; growing at the branch tip. Each scale has a long, curved prickle.

HABITAT: Cultivated in mild climates in North America. It is native to Norfolk Island, east of Australia.

RANGE: Cultivated in North America.

USES: This species is cultivated as an ornamental houseplant and is sometimes used as a Christmas tree.

SIMILAR SPECIES: Cook Pine.

CONSERVATION STATUS:
The population is increasing in its native range due to improvements in land management practices and better control of invasive species.

Diameter: 1–4' (30–122 cm)
Height: 40–70' (12–21 m)

■ DENSE POPULATION
■ LIGHT POPULATION
■ NATURALIZED

An important timber species, Incense Cedar is also the leading wood for the manufacture of pencils, because it is soft but not splintery, and can be sharpened in any direction with ease. The fragrant wood is used for cedar chests and closets. Although stands of young trees are killed by fire, the exceptionally thick bark protects mature trees. Practically no pests attack this tree, but mature tree trunks are vulnerable to dry rot of the heartwood.

DESCRIPTION: This is a large, resinous, aromatic tree with a tapering, irregularly angled trunk and a narrow, columnar crown, becoming open and irregular. Its leaves are ⅛–½" (3–12 mm) long, evergreen, scalelike, and long-pointed, growing opposite in four rows; the side pairs are keeled, each overlapping the next pair, extending down the twig. The shiny green leaves are aromatic when crushed. The bark is light or reddish-brown; thick, deeply and irregularly furrowed into shreddy ridges. The twigs are much-branched and flattish; with wedge-shaped joints longer than broad; composed of scalelike leaves.

FRUIT: Cones: ¾–1" (2–2.5 cm) long; oblong; hanging down at end of slender, leafy stalk; reddish-brown; composed of 6 paired, hard, flattened, pointed cone scales. Seeds: 4 or fewer in cone, paired, with 2 unequal wings.

HABITAT: Mountain soils; it is found in mixed coniferous forests, seldom in pure stands.

RANGE: Western Oregon south to southern California and extreme western Nevada, at 1200–7000' (366–2134 m). It also occurs in Baja California.

USES: Incense Cedar is an important timber species and is one of only a few species whose wood is suitable for making wooden pencils. It is also cultivated as an ornamental tree.

SIMILAR SPECIES: Western Red Cedar.

CONSERVATION STATUS: LC
This species is numerous and the population is stable. Fires could be a threat to isolated subpopulations in southern California and Baja California. The species is protected in Nevada.

Diameter 3–5' (0.9–1.5 m)
Height: 60–150' (18–46 m)

■ DENSE POPULATION
■ LIGHT POPULATION
■ NATURALIZED

Port Orford Cedar is adapted to the humid climate of the Pacific Coast with its wet winters and frequent summer fog. Many horticultural varieties are grown as ornamentals and shade trees, especially in European countries with moist climates. Varieties include columnar, drooping, and dwarf forms and others with foliage of varying shades, ranging from silvery or steel-blue to bright green and yellowish. The names honor Port Orford, Oregon—located in the center of the species' range—and Peter Lawson and his sons, Scottish nurserymen who introduced this species into cultivation in 1854. This species is also known as Lawson Cypress.

DESCRIPTION: This is a large evergreen tree with an enlarged base; a narrow, pointed, spirelike crown; and horizontal or drooping branches. Its leaves are opposite in four rows, $\frac{1}{16}$" (1.5 mm) long, scalelike, dull green above and whitish beneath with a gland-dot. The bark is reddish-brown, thick, and deeply furrowed into long, fibrous ridges. The twigs are slender, flattened, and regularly branched and spreading horizontally in a fernlike spray.

FRUIT: Cones: $\frac{3}{8}$" (10 mm) in diameter; many in clusters, reddish-brown, often with a bloom; with eight or ten blunt cone scales, maturing in one season. It has two to four seeds under a cone scale.

HABITAT: Sandy and clay loams, also rocky ridges; it is found with other conifers and sometimes in pure stands.

RANGE: Southwestern Oregon and northwestern California in a narrow coastal belt; also local in the Mount Shasta area; to 5000' (1524 m).

USES: This species is widely cultivated for its wood and as an ornamental or shade tree. The wood from Port Orford Cedar is known for its rot resistance and is used in Asia, particularly in Japan, for making coffins. It is also used for toys, arrow shafts, aircraft construction, and musical instruments, including guitars.

SIMILAR SPECIES: Atlantic White Cedar.

CONSERVATION STATUS: **NT**

Extensive logging reduced old-growth stands to a fraction of their historical area of occupancy. The population in the wild is increasing, but the spread of the soil-borne pathogen *Phytophthora lateralis* affects successful regeneration in many areas, especially areas accessible by road.

Diameter: 2½–4' (0.8–1.2 m)
Height: 70–200' (21–61 m)

■ DENSE POPULATION
■ LIGHT POPULATION
■ NATURALIZED

This cypress is native to the coastal regions of northwestern North America. The durable wood has a pleasant, resinous odor. It is used for flooring, furniture, interior finish, and boats. Indigenous peoples of the Pacific Northwest Coast made canoe paddles from the wood and carved ceremonial masks from the trunks. The species is also called Yellow Cedar, Nootka Cypress, and several similar variations.

DESCRIPTION: This tree has a narrow crown and horizontal or slightly drooping branches. Its leaves are evergreen, opposite in four rows, ⅛" (3 mm) long, scalelike, pointed and spreading, bright yellow-green, generally without a gland-dot. The bark is gray-brown, thin, with long narrow fissures, fibrous and shreddy. The twigs are slightly stout, flattened or four-angled, regularly branched and spreading horizontally, becoming reddish-brown.

FRUIT: Cones: ½" (12 mm) in diameter, rounded, reddish-brown, with four or six rounded cone scales ending in a long point, maturing in two seasons, with two to four seeds under a cone scale.

HABITAT: Wet mountain soils; mainly in mixed conifer forests, sometimes in pure stands.

RANGE: Pacific Coast region from southern and southeastern Alaska southeast to the mountains of western Oregon and extreme northwest California; it is local farther inland; found at 2000–7000' (610–2134 m), to sea level farther north.

USES: This species is used for flooring, furniture, interior finish, and boats. It is also cultivated as an ornamental.

SIMILAR SPECIES: Port Orford Cedar.

CONSERVATION STATUS: LC
The population is stable, but there is some dieback that could be related to climate change. Less snow cover in some areas gives these trees less root insulation and protection through the winter.

Diameter: 1–4' (0.3–1.2 m)
Height: 50–100' (15–30 m)

This false cypress is native to central and southern Japan and cultivated in North America as an ornamental. A number of cultivars have been selected for garden planting. It is a slow-growing tree, usually reaching about 60' (18 m) in height with a diameter of about 2' (0.6 m).

DESCRIPTION: This is a medium-sized evergreen tree. The trunk is sometimes forked. The leaves are scalelike, tiny, ⅛" (3.2 mm) long, glossy, yellow-green; sharp-pointed; lower surface often has white marks. The leaves on young plants are needlelike.

FRUIT: Cone: ¼" (6 mm) wide; green, becoming brown; rounded with a wrinkled texture; it has 10 scales, each with a tiny point in the center.

HABITAT: In the wild it is found along mountain streams; it is cultivated in North America.

RANGE: Native to Japan; cultivated in North America.

USES: This species is grown for its timber in Japan. It is also a popular ornamental tree, including in North America.

SIMILAR SPECIES: Oriental Arborvitae, Port Orford Cedar, Alaska Cedar, Atlantic White Cedar.

CONSERVATION STATUS: LC
The population is stable with few threats.

Diameter: 1–2' (0.3–0.6 m)
Height: 50–70' (15–21 m)

Chamaecyparis thyoides **ATLANTIC WHITE CEDAR**

- ■ DENSE POPULATION
- ■ LIGHT POPULATION
- ■ NATURALIZED

This medium-sized to large evergreen of the East Coast has long been harvested for its hardy and decay-resistant wood. Ancient logs buried in swamps have been mined and found to be preserved well enough to use as lumber. Pioneers prized the durable wood for log cabins, including floors and shingles. During the Revolutionary War, the wood produced charcoal for gunpowder. One fine forest is preserved at Green Bank State Forest in southern New Jersey. As an ornamental, this species is the hardiest of its genus northward.

DESCRIPTION: This evergreen, aromatic tree has a narrow, pointed, spirelike crown and slender, horizontal branches. Its leaves are evergreen, opposite, ¹⁄₁₆–¹⁄₈" (1.5–3 mm) long, scale-like, and dull blue-green with a gland-dot. The bark is reddish-brown; thin, and fibrous, with narrow connecting or forking ridges; becoming scaly and loose. The twigs are slender, slightly flattened or partly four-angled, and irregularly branched.

FRUIT: Cones: tiny, ¼" (6 mm) in diameter; bluish-purple with a bloom, becoming dark red-brown; with six cone scales ending in a short point; maturing in one season; with one or two gray-brown seeds under each cone scale.

HABITAT: Wet, peaty, acid soils; forming pure stands in swamp forests.

RANGE: Central Maine south to northern Florida and west to Mississippi in a narrow coastal belt; to 100' (30 m).

USES: Atlantic White Cedar has been widely cultivated as an ornamental tree and sometimes as a Christmas tree. It is commonly used for lumber for house construction, siding and paneling, and boat construction. The wood is hardy, durable, and resistant to decay and warping.

SIMILAR SPECIES: Arborvitae, Port Orford Cedar.

CONSERVATION STATUS: ⓛⓒ

The population is increasing. This tree is commonly found in swamp forests in Alabama and Florida and on the Atlantic Coastal Plain. The population is extirpated in Pennsylvania and listed as a special concern in Maine, and it is listed as rare in Georgia and New York.

Diameter: 1½–2' (0.5–0.6 m)
Height: 50–90' (15–27 m)

■ DENSE POPULATION
■ LIGHT POPULATION
■ NATURALIZED

Cryptomeria, also known as Japanese Cedar, has several cultivated varieties and is seen mainly as an ornamental. In Hawaii the trees are also grown in forest plantations and for windbreaks. It is propagated by seeds and is fast-growing. The national tree of Japan, where it is called Sugi, this species is one of the most important native timbers and is widely planted in forests, as a street tree, and around culturally important sites.

DESCRIPTION: This introduced, cone-bearing, evergreen tree has a straight trunk tapering from an enlarged base, a conical or pyramid-shaped narrow crown of horizontal branches, and aromatic resinous foliage. Its leaves are evergreen, spreading on all sides of the twig, ¼–⅝" (6–15 mm) long; they are awl-shaped or needlelike, curved forward and inward, the base extending down the twig; slightly stiff and blunt, somewhat flattened and four-angled, and dull blue-green. The bark is reddish-brown, fibrous, and fissured, peeling off in long strips. The twigs are dull green, slender, and hairless, mostly shedding with leaves.

FRUIT: Cones: ½–¾" (12–19 mm) in diameter; rounded, dull, short-stalked; opening at maturity in fall but remaining attached; they have many cone scales with a bristly point in the center and three to five points at the end. It has two to five slightly three-angled, narrowly winged seeds under each cone scale.

HABITAT: Moist soils in humid, warm temperate regions.

RANGE: A native of Japan and China. It is planted in the eastern United States and in the Pacific states, and is naturalized in Louisiana and North Carolina.

USES: The fragrant, weather- and insect-resistant wood of this species is commonly used in Japan for furniture, paneling, and other construction. It is also a popular ornamental, including in North America.

SIMILAR SPECIES: Giant Sequoia.

CONSERVATION STATUS: (NT)
The population is stable despite extensive logging; plantation forestry and regeneration have replenished the species' numbers, but most of the natural old-growth forests are gone. There has been a recent population decline in lowland forests attributed to climate change. The species is highly susceptible to precipitation changes.

Diameter: 2' (0.6 m)
Height: 50–80' (15–24 m)

■ DENSE POPULATION
■ LIGHT POPULATION
■ NATURALIZED

Arizona Cypress is the only cypress native to the Southwest. It occurs in the wild on dry, sterile, rocky mountain slopes and canyon walls, but it also does well in cultivation when planted in better soils or when irrigated. It generally requires little maintenance but needs deep watering at least every other week for desert planting during the growing season. Arizona Cypress is often grown for Christmas trees. The durable wood is used locally for fence posts. As many as five varieties have been distinguished based on minor differences of foliage and bark.

DESCRIPTION: This is an evergreen tree with a conical crown and stout, horizontal branches. Its leaves are opposite in four rows, 1/16" (1.5 mm) long, scalelike, keeled, and dull gray-green, mostly with a gland-dot that exudes whitish resin. The bark varies from gray or dark brown, rough, and furrowed to reddish-brown, smooth, thin, and peeling. The twigs are four-angled and slightly stout, branching almost at right angles.

FRUIT: Cones: 3/4–1 1/4" (2–3 cm) in diameter, dark reddish-brown with a bloom, becoming gray, with six or eight rounded, sharp-pointed, hard scales, sometimes slightly warty, usually remaining closed and attached. It has many brown seeds 1/8–3/16" (3–5 mm) long.

HABITAT: Coniferous woodlands on coarse, rocky soils; in pure stands or with pinyons and junipers.

RANGE: Local and rare from Trans-Pecos Texas west to southwest New Mexico, Arizona, and southern California; at 3000–5000' (914–1524 m). It is also found in northern Mexico.

USES: This species is widely cultivated as an ornamental landscaping tree. It is also used as a Christmas tree.

SIMILAR SPECIES: Redwood, Baker Cypress.

CONSERVATION STATUS: LC
Arizona Cypress is widespread but found mostly in small, fragmented subpopulations. This species is especially susceptible to fires and needs proper protection; forest fires are a major threat to isolated subpopulations and varieties with limited distribution. The species is protected in Nevada.

Diameter: 1–2' (0.3–0.6 m)
Height: 40–70' (12–21 m)

■ DENSE POPULATION
■ LIGHT POPULATION
■ NATURALIZED

The northernmost and hardiest of the New World cypresses, this rare tree is the only one of its genus native to Oregon. It occurs in small, localized populations in northern California and southwestern Oregon. Its name honors Milo Samuel Baker (1868–1961), the California botanist who described this species in 1898. This species is also known as Modoc Cypress or Siskiyou Cypress.

DESCRIPTION: This is a medium-sized to large evergreen tree with a narrow, conical, open crown. Its leaves are opposite in four rows, more than $\frac{1}{16}$" (1.5 mm) long, scalelike, and dull gray-green with a gland-dot that exudes whitish resin. The bark is reddish-brown, smoothish, often peeling in thin, curling plates. The twigs are four-angled, slender, crowded, and irregularly arranged.

FRUIT: Cones: $\frac{3}{8}$–$\frac{3}{4}$" (10–19 mm) in diameter; nearly round, gray or dull brown, with six or eight rounded, sharp-pointed, hard, warty scales. It has light tan seeds, $\frac{1}{8}$–$\frac{3}{16}$" (3–5 mm) long.

HABITAT: Dry Ponderosa Pine forests, mostly on volcanic soil or serpentine.

RANGE: Southwestern Oregon and northern California, especially the Siskiyou Mountains; at 3800–6000' (1158–1829 m).

USES: Because of its rarity and limited distribution, this species is seldomly used for any commercial purposes.

SIMILAR SPECIES: Arizona Cypress, Gowen Cypress, Monterey Cypress. The only *Cupressus* in Oregon.

CONSERVATION STATUS:
The population is decreasing. Fires that occur too frequently limit reproduction, but the increase of fire suppression policies in the past decades have limited growth. Some cypresses in California are shadowed by competition. Limited light has caused some trees to lose their foliage.

Diameter: 2' (0.6 m)
Height: 30–100' (9–30 m)

■ DENSE POPULATION
■ LIGHT POPULATION
■ NATURALIZED

There are three rare geographic varieties of this endangered New World cypress endemic to California. The typical variety, *Cupressus goveniana goveniana*, is confined to two groves within Point Lobos Reserve and Del Monte Forest in Monterey County. Mendocino Cypress (var. *pigmaea*) varies from a low shrub of 3' (0.9 m) on sandy soil (in Mendocino White Plains) to a large tree 100' (30 m) high (in Mendocino and Sonoma counties). Santa Cruz Cypress (var. *abramsiana*), with large cones to 1¼" (3 cm) long, is local in Ponderosa Pine forests to 2500' (762 m) and in the Santa Cruz Mountains of the Coast Ranges (Santa Cruz and San Mateo counties). This species is named after James Robert Gowen, a 19th century British horticulturist.

DESCRIPTION: This is typically a small evergreen tree with a conical or spreading crown. Its leaves are opposite in four rows, less than ¹⁄₁₆" (1.5 mm) long, scalelike, and mostly bright green and without a gland-dot. The bark is brown to gray, smooth or becoming rough and shreddy, and fibrous. The twigs are slender and four-angled.

FRUIT: Cones: small, usually less than ¾" (2 cm) in diameter; nearly round, gray-brown to dull gray, with six to ten rounded, hard cone scales, remaining closed and attached. It has many dull brown to shiny black seeds, ⅛" (3 mm) long.

HABITAT: Coastal Redwood forests.

RANGE: Northwest and central California; usually near sea level.

USES: Because of its small size and limited distribution, this species is not typically used commercially.

SIMILAR SPECIES: Monterey Cypress, Sargent Cypress, Baker Cypress.

CONSERVATION STATUS: (EN)
The total population of this species is believed to consist of fewer than 2300 mature individuals, and its population trend is decreasing. Gowen Cypress is on the U.S. Endangered Species List and is classified as state-endangered in California. Urbanization and changes in fire management practices threaten the species.

Diameter: ½–1½' (0.15–0.5 m)
Height: 15–30' (4.6–9 m)

MONTEREY CYPRESS *Cupressus macrocarpa*

■ DENSE POPULATION
■ LIGHT POPULATION
■ NATURALIZED

The gnarled, picturesque Monterey Cypresses growing on sea cliffs are a favorite photographic subject. This species is limited to two small relict populations in Monterey County on the central coast of California. The two native groves are protected within Point Lobos Reserve and Del Monte Forest at Point Cypress. It is often found with Gowen Cypress.

DESCRIPTION: This is a medium-sized evergreen tree with large cones and a symmetrical crown when young or where protected; often becoming irregular and flat-topped when exposed to high winds. Its leaves are opposite in four rows, more than ¹⁄₁₆" (1.5 mm) long, blunt, scalelike, and bright green, usually without a gland-dot. The bark is gray, rough, and fibrous. The twigs are four-angled, stout, and irregularly arranged.

FRUIT: Cones: 1–1⅜" (2.5–3.5 cm) long; longer than broad, brown, with eight to twelve rounded, stout-pointed, hard scales; remaining closed and attached. It has many irregular seeds, ¼–⁵⁄₁₆" (6–8 mm) long and shiny, dark brown.

HABITAT: Exposed granitic headlands and sheltered areas on seacoast, subject to salt spray and winds.

RANGE: Confined to two groves near Monterey and Carmel in west-central California.

USES: This species is widely planted as an ornamental, hedge, and windbreak along the California coastline and grown in forest plantations for timber in South Africa, New Zealand, and Australia.

SIMILAR SPECIES: Gowen Cypress, Sargent Cypress, Baker Cypress.

CONSERVATION STATUS: (VU)
This species exists mostly in protected areas, but heavy tourism and recreation in these areas could increase the potential of anthropogenic (human-caused) fires. Trees in the Point Lobos State Natural Reserve in California and other remote locations are at greater risk of fire, but trees in urbanized areas, where fire control is more accessible, are at less risk.

Diameter: 2' (0.6 m); rarely 4' (1.2 m)
Height: 60–80' (18–24 m)

■ DENSE POPULATION
■ LIGHT POPULATION
■ NATURALIZED

Sargent Cypress is endemic to California and limited to the Coast Range mountains. Several stands grow within Mount Tamalpais State Park north of San Francisco. Sargent Cypress was named for Charles Sprague Sargent (1841–1927), the founder and director of Harvard University's Arnold Arboretum and author of the 14-volume *The Silva of North America*.

DESCRIPTION: This evergreen tree has a narrow crown, becoming quite broad where exposed. Its leaves are opposite in four rows, more than ¹⁄₁₆" (1.5 mm) long, scalelike, and dull green, often with a gland-dot. The bark is gray, dark brown, or blackish; rough and furrowed, and thick and fibrous. The twigs are four-angled, stout, stiff, and branching in all directions.

FRUIT: Cones: mostly ¾–1" (2–2.5 cm) long; variable in shape, round or longer than broad, dull brown or gray; with usually six or eight rounded, often pointed, rough, hard cone scales; remaining closed and attached. It has many angular, dark brown seeds.

HABITAT: Exposed and protected slopes of foothills and mountains; it is often found with Gray Pine, oaks, and chaparral.

RANGE: Coast Ranges of California; to 3000' (914 m).

USES: This species is a popular ornamental tree.

SIMILAR SPECIES: Monterey Cypress, Gowen Cypress.

CONSERVATION STATUS:

Although locally abundant, the population is decreasing. Grazing by livestock and frequent fires can stifle regrowth. Because wildfires open the cones and allow the seeds to disperse and germinate, fire suppression can also negatively impact the species. Sargent Cypress is susceptible to cypress canker, a disease caused by fungi in the genus *Seirideum*, which can kill mature trees.

Diameter: 3' (0.9 m)
Height: 30–50' (9–15 m)

■ DENSE POPULATION
■ LIGHT POPULATION
■ NATURALIZED

California Juniper is native to the southwestern United States and Baja California. Able to withstand heat and drought, this species extends farther down into the semidesert zone than other junipers and is important in erosion control on dry slopes. Native Americans used to gather the berrylike cones to eat fresh and to grind into meal for baking.

DESCRIPTION: This is an evergreen shrub or small tree with a broad, irregular crown. Its leaves are usually in threes, ¹⁄₁₆–⅛" (1.5–3 mm) long, scalelike, blunt; forming stout, stiff, rounded twigs; they are yellow-green with a gland-dot. The bark is gray, fibrous, furrowed, and shreddy.

FRUIT: Cones: ½–¾" (12–19 mm) long; berrylike, longer than they are broad, and bluish with a bloom; becoming brown, hard, and dry. They are mealy and sweetish, with 1–2 seeds.

HABITAT: Dry slopes and flats of foothills and lower mountain zones; it is found with pinyons in woodlands and with Joshuatrees in the semidesert zone.

RANGE: Mountains of California, extreme southern Nevada, and western Arizona; also found in Baja California. It occurs at 1000–5000' (305–1524 m).

USES: Native Americans used this species both as a food source and as a medicinal plant. California Juniper is now a popular ornamental plant, and is used for bonsai.

SIMILAR SPECIES: Utah Juniper, Western Juniper.

CONSERVATION STATUS: LC
The population is stable. Forest fires are a threat, but this species is often found in rocky areas where fires are less likely to spread. California Juniper is protected in Nevada.

Diameter: 1–2' (0.3–0.6 m)
Height: 40' (12 m)

- ■ DENSE POPULATION
- ■ LIGHT POPULATION
- ■ NATURALIZED

Common Juniper is only rarely a small tree in New England and elsewhere in the Northeast. In the West, it is a low shrub, often at the timberline. Including geographic varieties, this species is the most widely distributed native conifer in both North America and the world, occurring across the Northern Hemisphere. Juniper "berries" are food for wildlife, especially grouse, pheasants, and bobwhites. They are an ingredient in gin, producing the beverage's distinctive aroma and tang.

DESCRIPTION: Common Juniper is usually a spreading low shrub in North America, sometimes forming broad or prostrate clumps; it is rarely a small tree with an open, irregular crown. The leaves are evergreen; ⅜–½" (10–12 mm) long, awl-shaped, stiff, sharp-pointed, and jointed at the base; they usually occur in threes, spreading at right angles, and are whitish and grooved above and shiny yellow-green beneath. The bark is reddish-brown to gray, thin, rough, scaly, and shreddy. The twigs are light yellow, slender, three-angled, and hairless.

FRUIT: Cones: ¼–⅜" (6–10 mm) in diameter; berrylike, whitish blue with a bloom; they are hard, mealy, aromatic, sweetish, and resinous; maturing in two to three years and remaining attached. It has one to three brown, pointed seeds. Pollen cones are mostly on the same plant.

CAUTION: Consuming any part of this plant is not advised. Some juniper berries are astringent or bitter, and those of a few similar species are toxic. If accidentally ingested, contact your doctor or poison control center.

HABITAT: Rocky slopes in coniferous forests of mountains and plains.

RANGE: Widespread from Alaska east to Labrador and Greenland, south to New York, and west to Minnesota and Wyoming; also south in the mountains to South Carolina and central Arizona; it is also found in Iceland and across northern Eurasia; to 8000–11,500' (2438–3505 m) in the south.

USES: Oil from Common Juniper cones is used to make gin. This species is commonly cultivated as an ornamental tree or shrub.

SIMILAR SPECIES: The lack of scalelike leaves helps to distinguish this species from other junipers in North America.

CONSERVATION STATUS: LC
The population of this widespread species is increasing, but it is struggling in some areas due to changes in land use. It is listed as threatened in Illinois, and a subspecies, Ground Juniper, is listed as endangered in Ohio.

Diameter: to 8" (20 cm)
Height: 1–4' (0.3–1.2 m),
rarely 15–25' (4.6–7.6 m)

■ DENSE POPULATION
■ LIGHT POPULATION
■ NATURALIZED

Creeping Juniper is a native plant of the North that has been widely used in gardens and public plantings as a ground cover. It is slow-growing and trails along the ground but does not run rampant. The berrylike cones are a valuable wildlife food, especially for birds.

DESCRIPTION: This is a prostrate, creeping, mat-forming evergreen shrub that can form ground-covering mats more than 20' (6 m) across. Its leaves are opposite, needle-like when young, and scale-like when mature. More than 100 cultivars are available with foliage varying from grayish-green to yellow.

FRUIT: Tiny, berrylike green cones, ripening to bluish-purple to bluish-black with a whitish bloom; they are juicy when ripe.

HABITAT: Pine and hardwood forests, sagebrush and other shrub habitats, prairies, and grasslands.

RANGE: From Alaska and across Canada south to northern New England, New York, the Great Lakes, the Great Plains, and the Rocky Mountains to Wyoming.

USES: This species is commonly used in landscaping as ground cover.

SIMILAR SPECIES: Common Juniper.

CONSERVATION STATUS:
Creeping Juniper is widespread and can be locally abundant, but it is listed as threatened in Vermont and Iowa and endangered in New Hampshire. A portion of the population is protected in Canada.

Height: 6–18" (15–45 cm)

■ DENSE POPULATION
■ LIGHT POPULATION
■ NATURALIZED

Western Juniper is common at high altitudes in the Sierra Nevada. Giants reach a trunk diameter of 16' (5 m) and an estimated age of more than 2000 years. This species may develop thick, long roots that entwine rock outcrops, mimicking the shape of the branches. Clark's Nutcrackers (*Nucifraga columbiana*), Cedar Waxwings (*Bombycilla cedrorum*), and several other bird species commonly feed on the cones.

DESCRIPTION: Western Juniper is an evergreen tree with a short trunk and a broad crown of stout, spreading branches, becoming ragged and gnarled with age; it can also be a shrub. Its leaves are mostly in threes, ¹⁄₁₆" (1.5 mm) long, and scalelike, forming stout, rounded twigs; they are gray-green with a gland-dot. Its bark is reddish-brown, furrowed, and shreddy.

FRUIT: Cones: ¼–⅜" (6–10 mm) in diameter; short, elliptical, berrylike, blue-black with a bloom, soft, juicy, and resinous; maturing in the second year; it has two or three seeds.

HABITAT: Mountain slopes and plateaus, mostly on shallow, rocky soils.

RANGE: Central and southeastern Washington south to southern California; to 10,000' (3048 m).

USES: Because of its small size, this species is seldom used for anything other than firewood. The cones are an important food source for birds and other wildlife.

SIMILAR SPECIES: California Juniper, Rocky Mountain Juniper.

CONSERVATION STATUS: LC
Western Juniper is locally common and increasing. Western Juniper is an aggressive species, but reproduction and growth takes time; some subpopulations consist only of aging trees that would be vulnerable to events such as a fire or disease.

Diameter: 1' (0.3 m),
sometimes much larger
Height: 15–30' (4.6–9 m)

UTAH JUNIPER *Juniperus osteosperma*

■ DENSE POPULATION
■ LIGHT POPULATION
■ NATURALIZED

The most common juniper in Arizona, it is conspicuous at the south rim of the Grand Canyon and on higher canyon walls. Utah Juniper grows slowly, becoming craggier and more contorted with age. Junipers are also called cedars; Cedar Breaks National Monument and nearby Cedar City in southwestern Utah are named for this tree. Scattered tufts of yellowish twigs with whitish berries commonly found on the trees are a parasitic mistletoe, which is characteristic of this tree.

ALTERNATE NAME: *Sabina osteosperma*

DESCRIPTION: This tree has a short, upright trunk; low, spreading branches; and a rounded or conical, open crown. Its leaves are generally opposite in four rows, forming stout, stiff twigs; 1/16" (1.5 mm) long; they are scalelike, short-pointed, and yellow-green, usually without a gland-dot. Its bark is gray, fibrous, furrowed, and shreddy.

FRUIT: 1/4–5/8" (6–15 mm) in diameter; berrylike, bluish with a bloom; becoming brown, hard, and dry; mealy and sweetish with 1–2 seeds.

HABITAT: Dry plains, plateaus, hills, and mountains, mostly on rocky soils; often found in pure stands or with pinyons.

RANGE: Nevada east to Wyoming, south to western New Mexico, and west to southern California; it is also local in southern Montana. It occurs at 3000–8000' (914–2438 m).

USES: Native Americans used the fibrous bark of this species for cordage, sandals, woven bags, thatching, and matting. They also ate the berrylike cones fresh or in cakes.

SIMILAR SPECIES: California Juniper, Western Juniper.

CONSERVATION STATUS: LC

This aggressive juniper species is increasing; it is the most widespread species of juniper in the southwestern United States. It has little commercial value and is not widely exploited. Because it occurs often on rocky escarpments and plateaus, it is at low risk of competition from other plants.

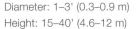
Diameter: 1–3' (0.3–0.9 m)
Height: 15–40' (4.6–12 m)

DENSE POPULATION
LIGHT POPULATION
NATURALIZED

Rocky Mountain Juniper makes a graceful ornamental tree, often with a narrow crown of drooping foliage. Several varieties differ in form and in leaf color. The aromatic wood is especially suited for cedar chests and is also used for lumber, fence posts, and fuel. This species is closely related to Eastern Red Cedar, which has dark green foliage and "berries" that mature in one year.

ALTERNATE NAME: *Sabina scopulorum*

DESCRIPTION: This is an evergreen tree with a straight trunk; a narrow, pointed crown; and slender branches of aromatic, gray-green foliage often drooping at the ends. Its leaves are opposite in four rows, forming slender, four-angled twigs; they are 1/16" (1.5 mm) long, scalelike, pointed, and gray-green. The bark is reddish-brown, thin, fibrous, and shreddy.

FRUIT: Cones: 1/4" (6 mm) in diameter; berrylike, bright blue with a whitish coat; they are soft, juicy, sweetish, resinous, and usually two-seeded; maturing in the second year. Male or pollen cones occur on separate trees.

HABITAT: Rocky soils, especially on limestone and lava outcrops; in open woodlands at lower border of trees to the north; in foothills with pinyons to the south.

RANGE: Generally in mountains, from central British Columbia, east to western North Dakota, and south to Trans-Pecos Texas; also in North Mexico. It occurs at 5000–9000' (1524–2743 m) to the south, and almost to sea level in the north.

USES: This species is commonly planted as an ornamental tree and the wood is used for cedar chests, fence posts, and lumber.

SIMILAR SPECIES: Eastern Red Cedar, but the ranges of the two species do not overlap.

CONSERVATION STATUS: LC

This species is widespread and the population is increasing with no identified threats. Rocky Mountain Juniper is protected in Nevada.

Diameter: 1½' (0.5 m)
Height: 20–50' (6–15 m)

EASTERN RED CEDAR *Juniperus virginiana*

■ DENSE POPULATION
■ LIGHT POPULATION
■ NATURALIZED

Eastern Red Cedar is native in 37 states and eastern Canadian provinces; it is locally abundant in the East. This tree is resistant to extremes of drought, heat, and cold. Colonists prized this widespread species for its wood, used for building furniture, rail fences, and log cabins. Many kinds of wildlife consume the juicy "berries," including the Cedar Waxwing (*Bombycilla cedrorum*), a bird named for its relationship with this tree. Eastern Red Cedar can be harmful to nearby apple orchards because it is an alternate host for a fungal disease called cedar-apple rust.

DESCRIPTION: This is an evergreen, aromatic tree with a trunk that is often angled and buttressed at the base and a narrow, compact, columnar crown; sometimes becoming broad and irregular. Its leaves are evergreen, opposite in four rows forming slender four-angled twigs; they are ¹⁄₁₆" (1.5 mm) long, to ³⁄₈" (10 mm) long on leaders, scalelike, not toothed, and dark green with a gland-dot. The bark is reddish-brown, thin, fibrous, and shreddy.

FRUIT: Cones: ¼–³⁄₈" (6–10 mm) in diameter, berrylike and dark blue with a bloom; they are soft, juicy, sweetish, and resinous, with one or two seeds. Pollen cones occur on separate trees.

CAUTION: The berries and leaves of this tree are mildly toxic. If accidentally ingested, contact your doctor or poison control center. Some people are allergic to the pollen, leaves, and/or wood.

HABITAT: From dry uplands, especially limestone, to flood plains and swamps; also abandoned fields and fence rows; often in scattered pure stands.

RANGE: Southern Ontario and widespread in the eastern half of the United States from Maine south to northern Florida, west to Texas, and north to North Dakota.

USES: The fragrant wood from this species is used for fence posts, cedar chests, cabinetwork, and carvings. The heartwood was once almost exclusively the source of wood for pencils; Incense Cedar is now used instead. It is also grown for Christmas trees, shelterbelts, and ornamental purposes.

SIMILAR SPECIES: Rocky Mountain Juniper, but the ranges of the two species do not overlap.

CONSERVATION STATUS: 🅛🅒
This species is widespread and the population is increasing due to better fire control practices.

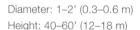

Diameter: 1–2' (0.3–0.6 m)
Height: 40–60' (12–18 m)

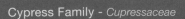

■ DENSE POPULATION
■ LIGHT POPULATION
■ NATURALIZED

This southeastern coastal variety of Eastern Red Cedar is distinguished by its often drooping foliage and smaller "berries." The Latin name for this variety, *silicicola,* means "growing in sand." This long-lived, slow-growing tree is often grown as an ornamental and is regarded as a low-maintenance planting.

DESCRIPTION: An evergreen aromatic tree with a narrow or spreading crown, the lower branches drooping; sometimes forming thickets. The leaves are evergreen; opposite in four rows forming slender, drooping, four-angled twigs; they are 1/16" (1.5 mm) long, to 3/8" (10 mm) long on leaders, scalelike, and dark green with a gland-dot. The bark is reddish-brown, thin, fibrous and shreddy.

FRUIT: Cones: 3/16" (5 mm) in diameter, berrylike, dark blue with a bloom; they are soft, juicy, sweetish, and resinous. It has one- or sometimes two-pointed, ridged seeds. Pollen cones occur on separate trees.

HABITAT: From dry uplands, especially on limestone, to wet soils of riverbanks and swamps and sandy soils near beaches; often in old fields and fence rows.

RANGE: Chiefly near the coast, from North Carolina south to Florida and west to southeastern Texas; near sea level.

USES: This species is planted as an ornamental and the wood is used for fence posts, cedar chests, cabinetwork, and carvings.

SIMILAR SPECIES: Eastern Red Cedar.

CONSERVATION STATUS: LC
The population of Southern Red Cedar is thought to be stable, with no known threats.

Diameter: 2' (0.6 m)
Height: 50' (15 m)

Dawn Redwood was once known only from 5-million-year-old fossils until 1947, when a forestry worker in China discovered several live groves. Although the wild population is severely restricted to a small area in south-central China, this "living fossil" is now intensely cultivated worldwide as an ornamental tree.

DESCRIPTION: This is a medium-sized to large deciduous tree with a symmetrical, conical crown, and a markedly tapering trunk that becomes deeply fluted with age. Dawn Redwood resembles Bald Cypress and also has soft, thin needles that are flat in cross section, but its branchlets and needles are arranged in opposite pairs in featherlike sprays (alternate in Bald Cypress). Young bark is reddish and flaky. Its needles are ¾–1½" (1.9–2.5 cm) long, bright green above and gray-green beneath; they turn brick red before shedding in fall.

FRUIT: Cones: ¾–1" (1.9–2.5 cm) long, dark brown, with blunt scales. The cones hang on a long stalk.

HABITAT: Cultivated in North America.

RANGE: Planted worldwide. Native to China.

USES: Dawn Redwood is commonly cultivated as an ornamental tree all over the world.

SIMILAR SPECIES: Bald Cypress.

CONSERVATION STATUS:
The natural population of this tree totals between 5000 and 6000, and is decreasing and severely fragmented. The remaining mature trees have been declared protected in China, but urbanization and changes in land use have limited the chances for these forests to regrow.

Height: more than 200' (61 m)

The sole member of the genus *Platycladus*, this species goes by several names, including Chinese Arborvitae and Oriental Arborvitae. Small and slow-growing, it is a popular ornamental tree around the world.

DESCRIPTION: Small evergreen tree or a shrub, often planted as a hedge. This species is easily identified by the hooklike points on the cones. Foliage sprays are flattened and oriented vertically. Its leaves are scalelike, tiny, up to ⅛" (3.2 mm) long; they are glossy, yellow-green, thick, and short-pointed.

FRUIT: Cone: ⅝" (1.2 cm) long; bright blue-green, becoming dark brown; rounded; erect. It has six thick scales, each tipped with a strongly curved point. It matures in the fall of its first year.

HABITAT: Cultivated in North America.

RANGE: Native to China and North Korea; it is widely planted and is not known to widely escape cultivation.

USES: This species is commonly cultivated as an ornamental tree in many parts of the world.

SIMILAR SPECIES: This tree is not easily confused with any other species in North America.

CONSERVATION STATUS: **NT**
Large-scale harvesting of mature trees for timber has caused a decline in the wild-growing forests, but there is little data about the native population, and it is unclear how much its overall numbers have been affected by exploitation.

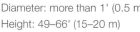
Diameter: more than 1' (0.5 m)
Height: 49–66' (15–20 m)

■ DENSE POPULATION
■ LIGHT POPULATION
■ NATURALIZED

Renowned for their majestic height and longevity, Redwoods are an iconic tree of the West. The world's tallest tree is a Redwood 368' (112 m) high. The age of these trees at maturity is 400–500 years; the highest recorded age counted in annual rings is 2200 years.

Redwoods maintain a pyramidal form and dark green foliage throughout the year. Circles of trees grow from sprouts around stumps and dead trunks. Redwoods grow around 3–5 feet per year, and are not affected significantly by any known pests or diseases. The bark of older trees turns a beautiful, bright orange.

Existing stands of Redwood occupy only a fraction of the large area in California and Oregon where they originally grew before the arrival of European settlers. Virgin forests remain in several state parks, as well as in the Redwoods National Park and along the Redwoods Highway. There is still some question concerning the status of the species outside of these parks. The Redwood industry maintains that selective logging, leaving seed trees, and planting in tree farms assure the future of this species. Conservationists feel that every effort should be made to maintain this magnificent tree at its present levels.

DESCRIPTION: The world's tallest tree, with a reddish-brown trunk much enlarged and buttressed at the base and often with rounded swellings or burls and slightly tapering; its crown is short, narrow, irregular and open with horizontal or drooping branches. The leaves are evergreen and of two kinds. Most are needlelike and unequal, ⅜–¾" (10–19 mm) long, flat and slightly curved, stiff and sharp-pointed, extending down the twig at the base; these are dark green above and whitish-green beneath, spreading in two rows. Leaves on leaders are scalelike, as short as ¼" (6 mm), keeled, concave, and spreading around the twig. The bark is reddish-brown, tough and fibrous, thick, and deeply furrowed into broad, scaly ridges; the inner bark is cinnamon-brown. The twigs are slender, dark green, and fork in one plane, ending in a scaly bud.

FRUIT: Cones: ½–1⅛" (1.2–3 cm) long; elliptical, reddish-brown, with many flat, short-pointed cone scales; hanging down at the end of a leafy twig; maturing in one season. It has two to five light brown, two-winged seeds under each cone scale.

HABITAT: Mostly alluvial soils on flats and benches or terraces. It forms pure stands in luxuriant dense forests; it is also found with Douglas Fir, Port Orford Cedar, and mixed conifers.

RANGE: Extreme southwestern Oregon south to central California in the fog belt, a coastal strip 5–35 miles (8–56 km) wide; from sea level to 3000' (914 m).

USES: The species is highly valuable to the lumber industry, having been logged extensively beginning in the 19th century. Redwood lumber is highly valued because it is attractive, light weight, and resistant to decay and fire.

SIMILAR SPECIES: Alaska Cedar, Arizona Cypress, Atlantic White Cedar.

CONSERVATION STATUS: (EN)
Endangered and decreasing. Conservation efforts are under way to protect the remaining native-growing stands in California and extreme southwestern Oregon. Redwoods have been cultivated horticulturally with limited success introducing it outside of its native range. Key to the species' survival is the preservation of remaining old-growth Redwood forests.

Diameter: 10–15' (3–4.6 m), sometimes larger
Height: 200–325' (61–99 m)

■ DENSE POPULATION
■ LIGHT POPULATION
■ NATURALIZED

Almost all Giant Sequoias are protected in Yosemite, Kings Canyon, and Sequoia national parks, in 4 national forests, and in state parks and forests. This rare species ranks among the world's oldest trees; with some having been aged at up to 3200 years old. Giant Sequoia has become a popular, large ornamental tree in moist, cool climates along the Pacific Coast and around the world. The lumber is no longer used; sadly, many mature trees were cut and wasted in the early logging days. This species is also known as Giant Redwood or Sierra Redwood.

DESCRIPTION: One of the world's largest trees, Giant Sequoia has a fibrous, reddish-brown trunk much enlarged and buttressed at the base, fluted into ridges, and conspicuously narrowed or tapered above; narrow, conical crown of short, stout, horizontal branches reaches nearly to base. Giant trees have tall, bare trunk and an irregular, open crown. Its leaves are evergreen; crowded and overlapping; ⅛–¼" (3–6 mm) long, to ½" (12 mm) on leaders; scalelike; ovate or lance-shaped, sharp-pointed; blue-green with two whitish lines. The bark is reddish-brown, fibrous, very thick, deeply furrowed into scaly ridges. Its twigs are much-branched, slender, drooping; blue-green turning brown.

FRUIT: Cones: 1¾–2¾" (4.5–7 cm) long; elliptical, reddish-brown; many flat, sharp-pointed cone scales; maturing in two seasons; hanging down at the end of a leafy twig and remaining attached; three to nine seeds under the cone scale, light brown, two-winged, falling gradually.

HABITAT: Granitic and other rocky soils in scattered groves in moist mountain sites, usually canyons or slopes; in coniferous forests.

Diameter: 20' (6 m), sometimes larger
Height: 150–250' (46–76 m)

RANGE: Western slope of Sierra Nevada, central California; at 4500–7500' (1372–2286 m); rarely at 3000–8900' (914–2713 m).

USES: Although highly resistant to decay, the wood from mature Giant Sequoias is quite brittle and seldom used for construction. The wood from young trees is sometimes used for lumber. The species is a popular ornamental tree and is sometimes used as a Christmas tree.

SIMILAR SPECIES: Redwood.

CONSERVATION STATUS: (EN)

Seedlings and saplings are vulnerable to forest fires, but the thick bark of mature trees offers resistance. Many of the species' population is protected in national forests, but these trees are vulnerable to competition from other species in the absence of periodic fires in many of these protected groves.

■ DENSE POPULATION
■ LIGHT POPULATION
■ NATURALIZED

Pond Cypress is native to the southeastern United States, found in shallow ponds and poorly drained areas from southeastern Virginia to southeastern Louisiana below 100' (30 m). It was formerly considered a subspecies of Bald Cypress, and some authorities still classify it as such. The two species differ in ecology, with Pond Cypress thriving in blackwater rivers, and Bald Cypress being a larger tree with longer needles. Both trees are deciduous, with the needles dropping off in fall, and both species depend on standing water for seed germination in the wild.

ALTERNATE NAMES: *Taxodium distichum* var. *imbricarium*, *Taxodium distichum* var. *nutans*

DESCRIPTION: This deciduous conifer is narrowly conical or columnar with spreading branches and upright branchlets. The crown is sometimes irregularly flat-topped, and the trunk is enlarged at the base, sometimes with short, rounded "knees." Its leaves are needles ½" (12 mm), overlapping, covering the twigs. The awl-shaped foliage is bright-green changing to a rich orange-brown in fall. The bark is brown to light gray, deeply furrowed into thick vertical plates.

FRUIT: Cones: 1½" (4 cm) long and round.

HABITAT: Shallow ponds and poorly drained areas, often in flatwoods.

RANGE: Coastal plains from southeastern Virginia to southeastern Louisiana and the tip of Florida, below 100' (30 m) elevation.

USES: This species is often used in home landscaping.

SIMILAR SPECIES: Bald Cypress.

CONSERVATION STATUS: LC
The population is stable with no threats identified.

Height: to 100' (30 m)

- ■ DENSE POPULATION
- ■ LIGHT POPULATION
- ■ NATURALIZED

Called "The Wood Eternal" because of the heartwood's resistance to decay, Bald Cypress is used for heavy construction, including docks, warehouses, boats, and bridges, as well as general millwork and interior trim. The trees are planted as ornamentals northward in colder climates and in drier soils. Pond Cypress, a closely related species of the Deep South with shorter scalelike leaves, was formerly considered a variety of *T. distichum*—some experts still consider these two trees conspecific. Bald Cypress is the state tree of Louisiana.

DESCRIPTION: This is a large, needle-leaf, aquatic, deciduous tree often with cone-shaped "knees" projecting from submerged roots, with trunks enlarged at the base and spreading into ridges or buttresses, and with a crown of widely spreading branches, flattened at the top. Its needles are deciduous; ⅜–¾" (10–19 mm) long; borne singly in two rows on slender green twigs, crowded and featherlike; they are flat, soft, flexible, and dull light green above and whitish beneath, turning brown and shedding with the twig in fall. The bark is brown or gray; with long fibrous or scaly ridges, peeling off in strips.

FRUIT: Cones: ¾–1" (2–2.5 cm) in diameter, round and gray, with one or two at the end of a twig; several flattened, four-angled, hard cone scales are shed at maturity in fall. It has two brown, three-angled seeds nearly ¼" (6 mm) long, under the cone scale. Tiny pollen cones occur in a narrow, drooping cluster 4" (10 cm) long.

HABITAT: Wet, swampy soils of riverbanks and floodplain lakes that are sometimes submerged. It is often found in pure stands.

RANGE: Southern Delaware to southern Florida, west to south Texas and north to southeastern Oklahoma and southwestern Indiana. It occurs below 500' (152 m) and locally in Texas to 1700' (518 m).

USES: Bald Cypress is a popular ornamental tree in North America as well as in Europe and Asia. Native Americans used this species to construct coffins, homes, drums, and canoes. It is also used commercially for heavy construction as well as for paneling, mulch, and shingles.

SIMILAR SPECIES: Pond Cypress, Dawn Redwood.

CONSERVATION STATUS: **LC**
The population is stable with no substantial threats.

Diameter: 3–5' (0.9–1.5 m),
rarely 10' (3 m) or more
Height: 100–120' (30–37 m) or more

NORTHERN WHITE CEDAR *Thuja occidentalis*

■ DENSE POPULATION
■ LIGHT POPULATION
■ NATURALIZED

Probably the first North American tree to be introduced to Europe, the Northern White Cedar was discovered by French explorers and grown in Paris about 1536. The year before, tea prepared from the foliage and bark, now known to be high in vitamin C, saved the crew of Jacques Cartier from scurvy. The trees grow slowly and reach an age of 400 years or more. The lightweight, easily split wood was preferred for canoe frames by Native Americans, who also used the shredded outer bark and the soft wood to start fires. Today, the wood is used principally for poles, cross-ties, posts, and lumber.

DESCRIPTION: This is a resinous and aromatic evergreen tree with an angled, buttressed, often branched trunk and a narrow, conical crown of short, spreading branches. Its leaves are evergreen, opposite in four rows, $\frac{1}{16}$–$\frac{1}{8}$" (1.5–3 mm) long, short-pointed, dull yellow-green above and paler blue-green beneath; the side pair is keeled, the flat pair has a gland-dot. The bark is light red-brown, thin, fibrous and shreddy; it is fissured into narrow connecting ridges. The twigs branch in a horizontal plane and are much flattened and jointed.

FRUIT: Cones: $\frac{3}{8}$" (10 mm) long; elliptical; light brown; upright from short curved stalk; with eight to ten paired, leathery, blunt-pointed cone scales, with four usually bearing two tiny, narrow-winged seeds each.

HABITAT: Adapted to swamps and to neutral or alkaline soils on limestone uplands; often found in pure stands.

RANGE: Southeastern Manitoba east to Nova Scotia and Maine, south to New York, and west to Illinois; south locally to North Carolina; to 3000' (914 m) in the south.

USES: The wood from the Northern White Cedar is used for utility poles, cross-ties, posts, and lumber. Cedar oil is distilled from the twigs and used for medicinal purposes and for perfumes.

SIMILAR SPECIES: Western Red Cedar.

CONSERVATION STATUS: LC
The population is increasing with no identified threats.

Diameter: 1–3' (0.3–0.9 m)
Height: 40–70' (12–21 m)

■ DENSE POPULATION
■ LIGHT POPULATION
■ NATURALIZED

In mixed-species and uneven-aged stands, Western Red Cedars are able to tolerate shady understory conditions and can grow slowly over long periods. The wood from Western Red Cedar is particularly resistant to rot and has been used extensively for outdoor construction and for boatbuilding. Native Americans of the Northwest carved totem poles and split lumber for lodges from this durable softwood, and made special war canoes from hollowed-out trunks of this species. Native Americans also used the wood for boxes, batons, and helmets, and the fibrous inner bark for rope, roof thatching, blankets, and cloaks.

DESCRIPTION: This is a large tree with a tapering trunk, buttressed at the base, and with a narrow, conical crown of short, spreading branches drooping at ends; foliage is resinous and aromatic. Its leaves are evergreen, opposite in four rows, ¹⁄₁₆–⅛" (1.5–3 mm) long, scalelike, sharp-pointed, and shiny dark green, usually with whitish marks beneath; the side pair keeled, and the flat pair is usually without a gland-dot. The bark is reddish-brown, thin, fibrous, and shreddy. The twigs are much-branched in a horizontal plane, slightly flattened in fanlike sprays, and jointed.

FRUIT: Cones: ½" (12 mm) long; clustered and upright from short, curved stalk; elliptical, brown; with 10–12 paired, thin, leathery, sharp-pointed cone scales; six usually bearing two to three seeds with two wings.

HABITAT: Moist, slightly acid soils, often forming widespread forests with Western Hemlock; also found with other conifers.

RANGE: Southeastern Alaska southeast along the coast to northwest California; also southeastern British Columbia south in the Rocky Mountains to western Montana; to 3000' (914 m) in the north; to 7000' (2134 m) in the south.

USES: The wood from this species is highly resistant to rot and used extensively for utility poles, patios, shingles, siding, and paneling. It is also used for boat construction and for guitar soundboards.

SIMILAR SPECIES: Northern White Cedar.

CONSERVATION STATUS: LC
The population is stable and notable specimens can be seen in famous national parks in the U.S. and Canada. This species is subject to large-scale harvesting but remains very common.

Diameter: 2–8' (0.6–2.4 m) or more
Height: 100–175' (30–53 m) or more

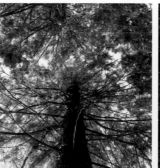

This tree is cultivated in North America as a landscaping evergreen. It is native to mountain forests in Japan.

DESCRIPTION: This is a small evergreen tree with leaves larger than those of similar trees. They are ¼" (6.4 mm) long, scalelike, slightly succulent and waxy, and dark green, with broad, white stripes beneath.

FRUIT: Cones: ½–¾" (1.3–1.9 cm) long, blue-green becoming blue-brown, rounded with six to eight scales.

HABITAT: In cultivation in North America; its native habitat is mountain forests.

RANGE: Native to Japan; widely cultivated in North America.

USES: This evergreen tree is cultivated in gardens and parks in North America.

SIMILAR SPECIES: Western Red Cedar, Northern White Cedar.

CONSERVATION STATUS: LC
The population is stable with no identified threats.

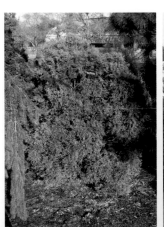

■ DENSE POPULATION
■ LIGHT POPULATION
■ NATURALIZED

The Pacific Yew is a conifer native to the Pacific Northwest. Its strong wood has been used for archery bows, poles, canoe paddles, and small cabinetwork; however, the limited population and small yield from this tree limits its value as timber. Its main commercial value is for the synthesis of paclitaxel, a chemotherapy medication derived from its bark. Although most parts of yew plants are deadly poisonous, the red, juicy cup around the seed of Pacific Yew is reported to be edible, provided the poisonous seed is not chewed or swallowed. Birds eat these cups and scatter the seeds.

DESCRIPTION: This is a poisonous, non-resinous, evergreen tree with an angled, often twisted or irregular trunk and with a broad crown of slender, horizontal branches; it is sometimes shrubby. Its needles are ½–¾" (12–19 mm) long, ¹⁄₁₆" (1.5 mm) or more wide, evergreen, spreading in two rows, flattened, short-pointed at both ends, soft and flexible, short-stalked, and deep yellow-green above and light green with two broad, whitish bands beneath. The bark is purplish-brown, thin, and smooth, with red-brown papery scales. Twigs are green, becoming light brown; they are slender and slightly drooping, with two lines below each leaf.

FRUIT: Seeds and male cones occur on separate trees. Its elliptical seeds are ¼" (6 mm) long, stalkless, blunt-pointed, two- to four-angled, and brown; nearly enclosed by a scarlet cup ⅜" (10 mm) in diameter, which is soft, juicy, and sweet; scattered and single on leafy twigs.

CAUTION: Poisonous: Do not eat or ingest any part of this species. If accidentally ingested, contact your doctor or poison control center.

HABITAT: Moist soil of stream banks and canyons; in understory of coniferous forests.

RANGE: Extreme southeast Alaska south along the coast to central California; also southeast British Columbia south in Rocky Mountains to central Idaho; from sea level in north to 7000' (2134 m) in south.

USES: The leaves, seeds, and twigs are poisonous to humans. Native Americans used the wood to make bows and canoe paddles. Pacific Yew contains the cancer-fighting compound paclitaxel.

SIMILAR SPECIES: California Torreya, Canada Yew.

CONSERVATION STATUS:
Until the 1990s, almost all paclitaxel used for chemotherapy treatments was derived from bark of the Pacific Yew, harvested in a process that killed the tree. The species became widely cultivated but the wild population declined as a result of heavy exploitation. Fires and logging have also led to population reduction. Exploitation of the species has decreased since the development of alternative methods for paclitaxel production.

Diameter: 2' (0.6 m)
Height: 50' (15 m)

■ DENSE POPULATION
■ LIGHT POPULATION
■ NATURALIZED

Canada Yew is native to central and eastern North America, where it occurs in swampy woods and ravines and along riverbanks and lake shores. This low-growing, shrubby plant is often cultivated as an evergreen hedge plant or as a foundation shrubbery. The colorful berries of this species are poisonous, but many birds eat the fleshy red cups and deposit the toxic seeds without being harmed.

DESCRIPTION: This is a low, straggling, evergreen shrub with short, straight, flat needles and spreading limbs that ascend at the tips. Its needles are ¾" (2 cm) long, pointed, flattened, in two rows. They are dark green above and pale green below, and take on a reddish-brown tint in winter. The bark is shreddy and reddish brown.

FRUIT: Small, pointed seed enclosed in fleshy, bright red, berry-like cup.

CAUTION: Poisonous: Do not eat or ingest any part of this species. If accidentally ingested, contact your doctor or poison control center.

HABITAT: Cool, moist, mixed woods.

RANGE: Newfoundland to southeastern Manitoba south through the northeastern United States west to Minnesota, south to northern Illinois and northeastern Iowa, and locally to Virginia and Tennessee.

USES: Native Americans used trace amounts of the leaves for medicinal purposes. It is also used today in cancer research and treatment.

SIMILAR SPECIES: Florida Yew, Pacific Yew.

CONSERVATION STATUS: LC
The population is protected but decreasing due to logging and forest fires. It is threatened in 12 states and one Canadian province in eastern North America.

Height: 3–6' (0.9–1.8 m) tall; may be twice as wide

■ DENSE POPULATION
■ LIGHT POPULATION
■ NATURALIZED

This rare species, along with Florida Torreya (*Torreya taxifolia*), is endemic in the wild to a small area along the Apalachicola River in Torreya State Park in northwestern Florida. It is also cultivated in botanical gardens. As with other yews, the seeds and needles are poisonous and can be fatal when consumed by people or livestock. The red juicy cup around the seed, however, is apparently harmless.

DESCRIPTION: Florida Yew is a faintly aromatic but non-resinous, poisonous shrub or small bushy tree with many stout, spreading branches. Its needles are ¾–1" (2–2.5 cm) long, 1⅛" (1.5 cm) or more wide, and evergreen, spreading in two rows; they are short-pointed at both ends, soft and flexible, flattened and slightly curved, dark green above and light green with two gray bands beneath; they have a very short leafstalk. The bark is dark purplish-brown, very thin, smooth, shedding in irregular plates. Twigs are yellow-green, turning red-brown; slender, with two lines below the base of each leaf.

FRUIT: Seeds and male cones occur on separate trees. The seeds are small and blunt-pointed; egg-shaped or elliptical; brown, nearly enclosed by a scarlet cup ⅜" (10 mm) in diameter; soft, juicy, and sweet; borne singly; stalkless. Male or pollen cones are ⅛" (3 mm) in diameter.

CAUTION: Poisonous: Do not eat or ingest any part of this species. If accidentally ingested, contact your doctor or poison control center.

HABITAT: Moist ravines in hardwood forests.

RANGE: Local in northwestern Florida in Gadsden and Liberty counties, mainly along eastern side of Apalachicola River; at less than 100' (30 m).

USES: The seeds and leaves of this species are poisonous to humans; like other yews, however, the bark contains the cancer-fighting compound paclitaxel.

SIMILAR SPECIES: Florida Torreya, Canada Yew.

CONSERVATION STATUS: **CR**

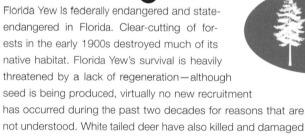

Florida Yew is federally endangered and state-endangered in Florida. Clear-cutting of forests in the early 1900s destroyed much of its native habitat. Florida Yew's survival is heavily threatened by a lack of regeneration—although seed is being produced, virtually no new recruitment has occurred during the past two decades for reasons that are not understood. White tailed deer have also killed and damaged individual stems by rubbing and browsing.

Diameter: 1' (0.3 m)
Height: 13–20' (4–6 m)

■ DENSE POPULATION
■ LIGHT POPULATION
■ NATURALIZED

This species, endemic to California, goes by several names, including "California Nutmeg Tree" and "Stinking Cedar." The former refers to the resemblance of the tree's fragrant seeds to those of the unrelated commercial spice nutmeg (*Myristica fragrans*). The latter alludes to the unpleasant odor resulting from crushing this tree's needles. Native Americans used to make bows from the California Torreya's strong wood.

DESCRIPTION: This is a strongly aromatic tree with a conical or rounded crown and rows of slender, spreading branches. Its needles are 1–2¾" (2.5–7 cm) long, less than ⅛" (3 mm) wide, evergreen, spreading in two rows; mostly paired; flattish and slightly curved; long, sharp point at tip, short-pointed and almost stalkless at base; stiff; shiny dark green above, green with two narrow, whitish lines beneath. The bark is gray-brown, thin, irregularly fissured into narrow scaly ridges. Its twigs are mostly paired, slender, with two lines below the base of each leaf; yellow-green turning reddish-brown.

FRUIT: Seeds and male cones occur on separate trees. The seeds are 1–1½" (2.5–4 cm) long; elliptical; fleshy outer layer green with purplish markings and shedding; inner layer

yellow-brown, thick-walled, stalkless; scattered and single on leafy twigs; maturing in two seasons.

HABITAT: Mixed evergreen forests along mountain streams, especially in shady canyon bottoms; also on exposed slopes.

RANGE: Mountains of central and northern California including Coast Ranges and western slope of Sierra Nevada; at 3000–6500' (914–1981 m), also down almost to sea level near the coast.

USES: Native Americans used the seeds as food, and the wood to construct bows. The wood is also sometimes used to make boards for the strategy game Go.

SIMILAR SPECIES: Pacific Yew, Florida Yew.

CONSERVATION STATUS: (LC)
This species saw a reduction in population due to logging in the early 19th and 20th centuries, but the population is slowly being restored. Deforestation in parts of California has contributed to population loss.

Diameter: 8"–2' (0.2–0.6 m), sometimes larger
Height: 16–70' (5–21 m)

■ DENSE POPULATION
■ LIGHT POPULATION
■ NATURALIZED

The only cycad native to the United States, primarily Florida, Coontie is becoming rare in the wild as its native habitats are destroyed. This tree is the host plant for the Atala Butterfly (*Eumaeus atala*), also native to Florida.

ALTERNATE NAME: *Zamia integrifolia*

DESCRIPTION: This is a fern-like or palm-like tree with a short, stout trunk. Its leaves are 3' (1 m), pinnately compound, with stiff 7" (17.8 cm) leaflets.

FRUIT: Cones: 8" (20.3 cm), cylindrical, rusty-brown to purplish, and velvety, found atop the plant.

CAUTION: Poisonous: Do not eat or ingest any part of this species. If accidentally ingested, contact your doctor or poison control center.

HABITAT: Hammocks, pine-oak woods, scrub, shell mounds, towns.

RANGE: Georgia south to peninsular Florida.

USES: This tree is often planted as an ornamental or for groundcover in landscaping.

SIMILAR SPECIES: This species is not easily confused with any other species in North America.

CONSERVATION STATUS:

Zamia is one of only a few primitive genera that are living remnants of plants that were abundant about 325 million years ago. Coontie is rare and decreasing in Florida as people dig up wild plants to put in their gardens or to sell; it may be extinct in Haiti and Puerto Rico.

Height: 3' (1 m)

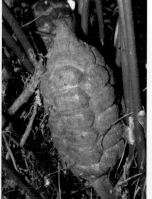

Ginkgo is known as a living fossil, the sole survivor of an ancient and formerly widespread order of plants. This tree has long been cultivated and possibly preserved from extinction by Buddhist monks on temple grounds in China, Japan, and Korea. The seeds, known as Ginkgo nuts, are popularly eaten in Japan and other Asian countries. This tree's alternate name, Maidenhair Tree, alludes to the resemblance of the leaves to that fern. Ginkgo is a hardy plant resistant to smoke, dust, wind, ice, insects, and disease.

DESCRIPTION: This deciduous tree, introduced in North America, has a straight trunk and an open, pyramid-shaped crown, becoming wide-spreading and irregular with age. Its leaves are 3–5 in cluster on spurs, or alternate; 1–2" (2.5–5 cm) long, 1½–3" (4–7.5 cm) wide; oddly fan-shaped, slightly thickened, slightly wavy on broad edge, often two-lobed, with fine forking parallel veins but no midvein; they are dull light green, turning yellow and shedding in autumn, and have a long leafstalk. The bark is gray, becoming rough and deeply furrowed. The twigs are light green to light brown, hairless, long and stout, with many spurs or short side twigs bearing crowded leaf scars.

FRUIT: Seeds and male cones occur on separate trees in early spring. Seeds are 1" (2.5 cm) long; elliptical, naked, and yellowish, with a thin, juicy, malodorous pulp and a large, thick-walled edible kernel. They grow one or two at the end of a long stalk, maturing and shedding in fall. Male or pollen cones are ¾" (19 cm) long.

Diameter: 2' (0.6 m)
Height: 50–70' (15–21 m)

CAUTION: Poisonous when eaten in large quantities or over a prolonged period: Do not eat or ingest any part of this species. If accidentally ingested, contact your doctor or poison control center. Some people are sensitive to the chemicals in the leaves and fleshy seedcoats.

HABITAT: Lawns and along streets in moist soil, in humid temperate regions.

RANGE: Apparently native in southeastern China. Planted in the eastern United States and on the Pacific Coast.

USES: This species is widely cultivated, including for use in growing bonsai trees. The seeds are a popular snack in several Asian countries and have long been used in traditional cooking and medicine.

SIMILAR SPECIES: The sole extant species in its order, Ginkgo is unlikely to be confused with any other trees.

CONSERVATION STATUS: **EN**

The species has been widely cultivated for centuries as both a food source and as traditional medicine. Ginkgo was thought to be extinct in the wild, but is now known to grow wild in at least two sites in eastern China. It is the only remaining species of the division Ginkgophyta, which was widespread prehistorically.

■ DENSE POPULATION
■ LIGHT POPULATION
■ NATURALIZED

This tree is native to the southeastern United States, in eastern Louisiana, southern Mississippi and Alabama, and northwestern Florida. It has been planted as a deciduous shrub north to Philadelphia. The leaves, when crushed, have a distinctive, sweet odor. Although Florida Anise-tree has no commercial uses, the fruit of a similar species in China, Star Anise (*Illicium verum*), is used as a culinary spice, and the oil distilled from the fruit and leaves is used in medicine and to flavor alcoholic drinks.

DESCRIPTION: This evergreen can be a shrub or a small tree with flowers that have a strong, unpleasant fishy odor. It has slender, crooked trunks, often leaning; a small, open, rounded crown; and curious, wheel-shaped fruit. Its leaves are evergreen, 2½–6" (6–15 cm) long and 1–2" (2.5–5 cm) wide; elliptical to lance-shaped, long-pointed at the ends, not toothed, leathery, hairless, crowded at twig ends, dark green above, and paler with gland-dots beneath. The bark is gray or brown and smooth, becoming fissured. The twigs are brown, slender, upright, and hairless.

FLOWERS: Spring. They are nearly 2" (5 cm) wide, with 20–30 dark red or purple very narrow sepals, and with a strong fishy odor. They are showy but often hidden, solitary on a long, curved-down stalk at the leaf base.

FRUIT: 1–1½" (2.5–3 cm) wide, half-round, and spreading like a wheel or star; composed of 11–15 individual dry fruits, narrow and pointed, opening on top, and one-seeded, maturing in summer.

CAUTION: This plant can be toxic and should not be confused with the culinary spice derived from the related Star Anise (*Illicium verum*).

HABITAT: Wet soils of streams, ravines, and swamps.

RANGE: Northwestern Florida to central Alabama and southeastern Louisiana; to 500' (152 m).

USES: Florida Anise-tree has no commercial uses but has been planted as a shrub in areas outside its native range.

SIMILAR SPECIES: Star Anise.

CONSERVATION STATUS: LC
This species is widely distributed. The large population is stable with no major threats.

Diameter: 3" (7.5 cm)
Height: 10–20' (3–6 m)

■ DENSE POPULATION
■ LIGHT POPULATION
■ NATURALIZED

Also known as Tuliptree, Yellow Poplar is native to the eastern United States and was introduced to Europe from Virginia by the earliest colonists. It is also grown on the Pacific Coast. Very tall trees with massive trunks once stood in pristine, old-growth forests, but most were cut for the valuable soft wood. Pioneers would hollow out a single log to make a long, lightweight canoe. Yellow Poplar is now a primary commercial hardwood, used for furniture, as well as for crates, toys, musical instruments, and pulpwood.

DESCRIPTION: One of the tallest and most beautiful eastern hardwoods, Yellow Poplar has a long, straight trunk, a narrow crown that spreads with age, and large showy flowers resembling tulips or lilies. Its leaves are 3–6" (7.5–15 cm) long and wide, and have blades of an unusual shape, with a broad tip and base nearly straight like a square, and with four or sometimes six short-pointed paired lobes; they are hairless and long-stalked, shiny dark green above and paler beneath, turning yellow in fall. The bark is dark gray, becoming thick and deeply furrowed. The twigs are brown, stout, and hairless, with ring scars at the nodes.

FLOWERS: Spring. 1½–2" (4–5 cm) long and wide, cup-shaped, with six rounded green petals (orange at base); solitary and upright at the end of a leafy twig.

FRUIT: 2½–3" (6–7.5 cm) long; conelike; light brown; composed of many overlapping one- or two-seeded nutlets 1–1½" (2.5–4 cm) long (including narrow wing); shedding from upright axis in autumn; the axis is persistent in winter.

HABITAT: Moist well-drained soils, especially valleys and slopes; often found in pure stands.

RANGE: Extreme southern Ontario east to Vermont and Rhode Island, south to northern Florida, west to Louisiana, and north to southern Michigan; to 1000' (305 m) in the north and to 4500' (1372 m) in the southern Appalachians.

USES: The wood from this species is used for furniture, as well as for crates, toys, musical instruments, and pulpwood.

SIMILAR SPECIES: Chinese Tuliptree, which is occasionally cultivated in North America.

CONSERVATION STATUS: LC
The species is widespread and stable across its range. There are no specific threats to the population. The bark of mature trees has proven very fire resistant, but saplings are more susceptible to fires.

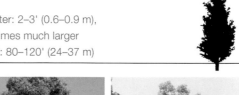

Diameter: 2–3' (0.6–0.9 m), sometimes much larger
Height: 80–120' (24–37 m)

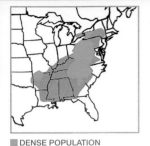

■ DENSE POPULATION
■ LIGHT POPULATION
■ NATURALIZED

One of the largest magnolias, and the only one native to Canada, is the Cucumbertree. This species is named for its long, oblong-shaped fruit. Cucumbertrees have grown as large as 8' (4.2 m) in diameter and 96' (29 m) tall. They can also live for more than 100 years.

DESCRIPTION: This large magnolia tree has a straight trunk and a narrow crown of short upright to spreading branches. Its leaves are 5–10" (13–25 cm) long, often larger, and 3–6" (7.5–15 cm) wide. They are elliptical or ovate, abruptly short-pointed, with straight or wavy edges, and green, becoming hairless above and paler and often with soft hairs beneath, and turning dull yellow or brown in fall. The bark is dark brown, furrowed into narrow, scaly, forking ridges. The twigs are stout with ring scars at nodes; young twigs and buds are densely hairy.

FLOWERS: Spring. 2½–3½" (6–9 cm) wide; bell-shaped, with six large greenish-yellow or bright yellow petals; solitary at the end of twig.

FRUIT: 2½–3" (6–7.5 cm) long, conelike, oblong, dark red, and composed of many pointed fruits that split open, each with two seeds that hang down on threads; maturing in late summer.

HABITAT: Moist soils of mountain slopes and valleys in mixed forests.

RANGE: Extreme southern Ontario and western New York south to northwestern Florida, west to Louisiana, and north to Missouri; at 100–4000' (30–1219 m).

USES: Cucumbertrees are planted as ornamental and shade trees.

SIMILAR SPECIES: Fraser Magnolia, Pyramid Magnolia.

CONSERVATION STATUS: (LC)
The population is stable with a widespread distribution in eastern North America. There are no identified threats.

Diameter: 2' (0.6 m)
Height: 60–80' (18–24 m)

■ DENSE POPULATION
■ LIGHT POPULATION
■ NATURALIZED

This rare and localized species in northwestern Florida is closely related to Bigleaf Magnolia, which is taller and has larger leaves and flowers. Ashe's Magnolia is named for its discoverer, William Willard Ashe (1872–1932), an early forester of the United States Forest Service.

DESCRIPTION: This magnolia can be a shrub or a small tree with a broadly conical crown of many slender, upright branches, and with very large leaves and flowers. Its leaves are 8–24" (20–61 cm) long, 4–11" (10–28 cm) wide; reverse ovate, blunt or short-pointed at the tip, broadest beyond the middle, tapering to a notched two-lobed base, with straight or wavy edges; they are thin; shiny light green above, whitish or silvery and often hairy beneath. The leafstalks are 2–4" (5–10 cm) long. The bark is gray-brown; smooth or slightly rough. The twigs are stout with silvery hairs when young, and with ring scars at the nodes.

FLOWERS: Late spring. 12" (30 cm) wide; cup-shaped corolla of six curved creamy white petals; fragrant.

FRUIT: 1¾–2¾" (4.5–7 cm) long; conelike; narrowly cylindrical; rose-red to brown; composed of many separate pointed two-seeded fruits that split open in fall.

HABITAT: Upland bluffs in hardwood forests.

RANGE: Only in northwestern Florida at 100–200' (30–61 m).

USES: Ashe's Magnolia is sometimes planted as an ornamental tree.

SIMILAR SPECIES: Bigleaf Magnolia.

CONSERVATION STATUS: VU
Ashe's Magnolia is restricted to scattered sites on the Florida panhandle. It doesn't respond well to disturbance or to competition with more aggressive plants. It is state-endangered in Florida.

Diameter: 4" (10 cm)
Height: 13–30' (4–9 m)

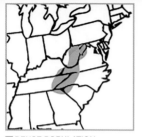

■ DENSE POPULATION
■ LIGHT POPULATION
■ NATURALIZED

This species, also known as Mountain Magnolia, is native to the Appalachian Mountains and is fairly common in Great Smoky Mountains National Park. This magnolia is named for John Fraser (1750–1811), a Scottish botanist who introduced many North American plants to Europe. It is planted as an ornamental for its large flowers and coarse foliage.

DESCRIPTION: This magnolia of eastern mountains is often branched near the base, and has an open crown of spreading branches, large leaves, and very large flowers. Its leaves are crowded, 8–18" (20–46 cm) long, 5–8" (13–20 cm) wide, usually reverse ovate (sometimes ear-shaped), broadest beyond the middle, and short-pointed at the tip, with two large, pointed lobes at the narrow base; they are hairless, not toothed, and bright green above and pale and whitish beneath. The bark is light gray, smooth or becoming scaly, and thin. The twigs are brown and stout, with ring scars at the nodes.

FLOWERS: Spring. 8–10" (20–25 cm) wide with six to nine cream-colored petals; they are fragrant and solitary at the end of the twig.

FRUIT: 4–5" (10–13 cm) long, conelike, oblong, rose-red, and composed of many long-pointed, hairless, two-seeded fruits that split open in early fall.

HABITAT: Moist soils of mountain valleys in hardwood forests.

RANGE: Virginia and West Virginia south to northern Georgia; at 800–5000' (244–1524 m).

USES: This species is planted as an ornamental, but is not otherwise used for commercial purposes.

SIMILAR SPECIES: Pyramid Magnolia, Cucumbertree.

CONSERVATION STATUS: LC
Fraser Magnolia is widely distributed with no significant threats.

Diameter: 1–2' (0.3–0.6 m)
Height: 30–70' (9–21 m)

■ DENSE POPULATION
■ LIGHT POPULATION
■ NATURALIZED

Planted around the world in warm temperate and subtropical regions, Southern Magnolia is a popular ornamental and shade tree. Several horticultural varieties have been developed, including some that are cold-hardy. The Southern Magnolia is named for its large range throughout all of the southern states from South Carolina to Texas. It matures quickly, producing its first seeds within a decade, and the seeds, which are readily distributed by birds and small mammals, have a high rate of germination. It can often be found growing with Sweetgum, Blackgum, Yellow Poplar, oaks, White Ash, hickories, Red Maple, Sweetbay Magnolia, and American and Winged elms. The tree is often found in parks and public places, including, famously, on the White House grounds, where one planted by Andrew Jackson thrived for nearly 200 years until damaged in the 1940s and finally removed in 2018. Other old and impressive examples grow in many southern cities from Houston to Charleston.

DESCRIPTION: One of the most beautiful native trees, Southern Magnolia is an evergreen with a straight trunk, conical crown, and large, fragrant white flowers. Its leaves are evergreen, 5–8" (13–20 cm) long, 2–3" (5–7.5 cm) wide, oblong or elliptical; they are thick and firm with edges slightly turned under, shiny bright green above and pale with rust-colored hairs beneath, on stout leafstalks with rust-colored hairs. The bark is dark gray and smooth, becoming furrowed and scaly. The twigs are covered with rust-colored hairs when young, with ring scars at the nodes, ending in buds also covered with rust-colored hairs.

FLOWERS: Late spring–summer. Flowers are 6–8" (15–20 cm) wide, cup-shaped, with three white sepals and six or more petals; they are quite fragrant and solitary at the end of the twig.

FRUIT: 3–4" (7.5–10 cm) long, conelike, oblong, pink to brown, covered with rust-colored hairs, and composed of many separate, short-pointed, two-seeded fruits that split open in early fall.

HABITAT: Moist soils of valleys and low uplands with various other hardwoods.

RANGE: Eastern North Carolina to central Florida and west to eastern Texas; to 400' (122 m).

USES: Principal uses of the wood are furniture, boxes, cabinetwork, and doors. The dried leaves are used by florists in decorations.

SIMILAR SPECIES: Ashe's Magnolia, Bigleaf Magnolia, Chinese Magnolia.

CONSERVATION STATUS: LC

This species is widespread and abundant in the Southeast and along the Gulf Coast. The bark of mature trees is fire resistant, and they can resprout even if the tops are burned. Droughts in the winter can cause die-back and mortality, and the fungus *Mycosphaerella milleri* is known to afflict seedlings.

Diameter: 2–3' (0.6–0.9 m)
Height: 60–80' (18–24 m)

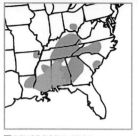

■ DENSE POPULATION
■ LIGHT POPULATION
■ NATURALIZED

This beautiful tree is native to the southeastern United States and planted throughout the East as an ornamental—although in windy places the giant leaves become torn and unsightly. Bigleaf Magnolia was named by the French naturalist and explorer André Michaux (1746–1802), who discovered the tree in North Carolina in 1789 and called it the "queenliest of all the deciduous magnolias."

DESCRIPTION: Sporting the largest leaves and flowers of any native North American plant, Bigleaf Magnolia has a broad, rounded crown of stout, spreading branches. Its leaves are 15–30" (38–76 cm) long and 6–10" (15–25 cm) wide, reverse ovate, broadest beyond the middle, mostly blunt at the tip; they are notched with two rounded lobes at the base, not toothed, bright green above with silvery hairs beneath. The stout, hairy leaf-stalks are 3–4" (7.5–10 cm) long. The bark is light gray; smooth, thin. The twigs are stout and hairy with large leaf-scars at the nodes and ending in large buds covered with white hairs.

FLOWERS: Late spring–early summer. 10–12" (25–30 cm) wide, cup-shaped with six white petals with a spot at the base; they are fragrant.

FRUIT: 2½–3" (6–7.5 cm) long, conelike, elliptical or nearly round, rose-red, and composed of many separate short-pointed, two-seeded, hairy fruits; maturing in autumn.

HABITAT: Moist soil of valleys, especially ravines; it is found in the understory of hardwood forests.

RANGE: Central North Carolina south to western Georgia and west to Louisiana; local in southern Ohio, northeastern Arkansas, and southeastern South Carolina.

USES: This species is planted as an ornamental north to Massachusetts.

SIMILAR SPECIES: Southern Magnolia, Ashe's Magnolia, Umbrella Magnolia.

CONSERVATION STATUS: LC
The population is stable with no identified threats.

Diameter: 1½' (0.5 m)
Height: 30–40' (9–12 m).

■ DENSE POPULATION
■ LIGHT POPULATION
■ NATURALIZED

This rare and localized species of the Coastal Plain is named for its pyramid-shaped crown. It is closely related to Fraser Magnolia, a mountain species that has larger leaves and larger flowers with pale yellow petals. Pyramid Magnolia was discovered by William Bartram (1739–1823), a naturalist from Pennsylvania who named the tree in his 1791 book, *Travels through North and South Carolina, Georgia, East and West Florida, the Cherokee Country, the Extensive Territories of the Muscogulges or Creek Confederacy, and the Country of the Chactaws*.

DESCRIPTION: This rare tree has large, showy leaves and flowers and a pyramid-shaped crown of upright branches. Its leaves are 3½–4½" (9–11 cm) wide, 6–9" (15–23 long, crowded, reverse ovate, broadest beyond the middle, and blunt at the tip, with two large, pointed lobes at the narrow base. They are short-stalked, not toothed, bright green above, and pale and whitish beneath. The bark is dark brown, smoothish or becoming scaly, and thin. The twigs are stout with ring scars at the nodes.

FLOWERS: Spring. 4" (10 cm) wide, with six to nine creamy white petals; solitary at the end of the twig.

FRUIT: 2–2½" (5–6 cm) long, conelike, oblong, rose-red, composed of many separate two-seeded fruits ending in short points curved outward; maturing in fall.

HABITAT: Moist valley soils.

RANGE: Scattered from South Carolina to eastern Texas; to 400' (122 m).

USES: This species is sometimes planted as an ornamental tree.

SIMILAR SPECIES: Fraser Magnolia, Ashe's Magnolia, Southern Magnolia, Sweet Bay.

CONSERVATION STATUS:

This species is much less abundant than Fraser Magnolia and occurs in a smaller range. It is threatened by habitat destruction in some areas.

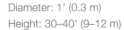
Diameter: 1' (0.3 m)
Height: 30–40' (9–12 m)

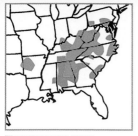

- ■ DENSE POPULATION
- ■ LIGHT POPULATION
- ■ NATURALIZED

This large-leaved deciduous tree is fairly common at low altitudes in Great Smoky Mountains National Park in North Carolina and Tennessee. The arrangement of spreading leaves somewhat resembles the ribs of an umbrella, hence the common name. This tree's scientific name means "three petals" and probably refers to the three sepals, which are longer than the more numerous petals.

DESCRIPTION: Umbrella Magnolia has large leaves and large flowers, and a broad, open crown of spreading branches; often with sprouts at base. Its leaves are 10–20" (25–51 cm) long, 5–10" (13–25 cm) wide; reverse ovate; broadest beyond middle; not toothed; crowded; short-stalked; green above, with silky hairs beneath when young. The bark is light gray; smooth, thin. The twigs are stout; with ring scars at nodes.

Diameter: 1' (0.3 m)
Height: 30–40' (9–12 m)

FLOWERS: Spring. 7–10" (18–25 cm) wide, with three cup-shaped, light green sepals and six or nine shorter white petals; these flowers have a disagreeable odor and occur at the end of the twig.

FRUIT: 2½–4" (6–10 cm) long, conelike, oblong, rose-red, and composed of many separate two-seeded, short-pointed fruits; maturing in fall.

HABITAT: Moist soils of mountain valleys; it is found in hardwood forests.

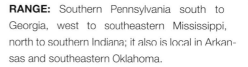

RANGE: Southern Pennsylvania south to Georgia, west to southeastern Mississippi, north to southern Indiana; it also is local in Arkansas and southeastern Oklahoma.

USES: This species is not commonly used for any commercial purposes, although it is occasionally planted as a shade tree.

SIMILAR SPECIES: Bigleaf Magnolia.

CONSERVATION STATUS: LC
The species is widespread with no identified threats.

■ DENSE POPULATION
■ LIGHT POPULATION
■ NATURALIZED

This native ornamental is popular for its fragrant flowers, showy conelike fruit, bicolored leaves, and smooth grayish bark. Its crushed leaves and twigs have a spicy scent and its flowers have a light lemon scent. It occurs as a shrub in the northern part of its range and is more treelike farther south. The first *Magnolia* species to be introduced to Europe, Sweet Bay was first sent to England in 1678 by the English naturalist and missionary John Banister (1654–1692).

DESCRIPTION: An attractive, fragrant tree native to eastern swamplands, Sweet Bay has a narrow, rounded crown that sheds its leaves in winter or is almost evergreen southward. Its leaves and twigs have a spicy scent. Its leaves are 3–6" (7.5–15 cm) long, 1¼–2½" (3–6 cm) wide, oblong, blunt at the tip, without teeth, slightly thickened, and short-stalked, becoming shiny green above, whitish and finely hairy beneath. The bark is gray, smooth, thin, and aromatic. Twigs with ring scars at the nodes end in buds covered with whitish hairs.

FLOWERS: Late spring–early summer. 2–2½" (5–6 cm) wide, cup-shaped with nine to twelve white petals; the flowers are fragrant.

FRUIT: 1½–2" (4–5 cm) long, cone-like, elliptical, dark red, and composed of many separate pointed fruits, each with two red seeds; maturing in early fall.

HABITAT: Wet soils of coastal swamps and borders of streams and ponds.

RANGE: Long Island south to southern Florida and west to southeastern Texas; it also is local in northeastern Massachusetts; to 500' (152 m).

USES: This species is not commonly harvested, but its wood is occasionally used for furniture, cabinets, and paneling. Sweet Bay is a popular ornamental.

SIMILAR SPECIES: Red Bay.

CONSERVATION STATUS: LC
Sweet Bay is abundant in parts of its range, and there are no major threats to this species known. Infection by the fungus *Mycosphaerella milleri* is fairly common in these trees and causes small leaf spots in early summer.

Diameter: 1½' (0.5 m)
Height: 20–60' (6–18 m)

Also known as Saucer Magnolia, this hybrid magnolia was first noted in a nurseryman's Paris garden in 1820. It is planted as an ornamental around the world for its gorgeous flowers, although its fragile limbs tend to break in ice storms and under heavy snow. It blooms a week earlier than many other magnolias and when only 2–3' (61–91 cm) tall.

DESCRIPTION: This popular magnolia is a small tree or large shrub with deciduous, simple leaves. Typically it has several stems and a broad, rounded crown of low branches; it can be as broad as it is tall. This cultivated tree produces an abundance of white to deep purple, large and showy flowers in the spring. Its leaves are 5–8" (13–20 cm) long, elliptic to obovate, with abruptly pointed tips, and untoothed margins; they are dark green above and paler and fuzzy beneath. The terminal leaf bud is fuzzy. Its fall color is yellow-brown.

FLOWERS: Spring–early summer. 5–10" (13–25 cm) wide, showy, white to pink to deep purple-pink; they can be shallow or deep, cup- to saucer-shaped, usually with nine elliptic tepals. The flowers are sometimes fragrant.

FRUIT: A 2¼–4" (6–10 cm) long, conelike cluster of follicles, narrowly ovoid and knobby, green at first turning pink at maturity. Late summer–early autumn; usually few fruits are produced.

HABITAT: Planted as an ornamental in North America.

RANGE: Introduced in North America, this species is a hybrid of *M. denudata* and *M. liliiflora*, both native to China.

USES: Its huge, fragrant flowers make this hybrid a popular ornamental and container plant.

SIMILAR SPECIES: Southern Magnolia, Sweet Bay.

CONSERVATION STATUS:
This species is a horticural hybrid not found in the wild.

Height: 20–25' (6–7.5 m)

POND APPLE   *Annona glabra*

■ DENSE POPULATION
■ LIGHT POPULATION
■ NATURALIZED

The Pond Apple, native to Florida, is a tropical fruit tree in the family Annonaceae. This family includes more than 2000 species of small to medium-sized flowering trees and shrubs, only a handful of which occur in North America. Pond Apple fruit is often eaten by people and a variety of animals, including American Alligators, giving this tree the alternate name Alligator Apple.

DESCRIPTION: This tree has a rounded, spreading crown and a short, buttressed trunk. Its leaves are 5" (12.5 cm) long, elliptical to ovate, leathery, and evergreen. The bark is dark reddish brown and shallowly fissured.

FLOWERS: April–June. 1" (2.5 cm), cup-shaped, creamy white to pale yellow, nodding.

FRUIT: 5" (12.5 cm) diameter, roundish, yellow. The fruit is edible and ripe September–November.

HABITAT: Swamps, wet hammocks, streamsides.

RANGE: Central and southern Florida.

USES: The fruit of this species is edible and sometimes made into a jam or used in drinks.

SIMILAR SPECIES: Florida Strangler Fig, Shortleaf Fig.

CONSERVATION STATUS: LC

This species is widely distributed in Tropical America and the coast of West Africa. It is common in swamps and on pond banks in southern Florida.

Height: 50' (15 m)

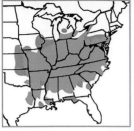

■ DENSE POPULATION
■ LIGHT POPULATION
■ NATURALIZED

This is the northernmost representative of a mostly tropical family of trees, Annonaceae, which includes the popular fruits sugar-apple and soursop. Common Pawpaw is known by numerous names, including Wild Banana, Appalachian Banana, Prairie Banana, and Hillbilly Mango. The fruit, which resembles a banana in taste and appearance, was once commonly harvested, but the supply has decreased greatly due to the clearing of forests. The fruit is still gathered and eaten locally, and is also readily consumed by opossums, squirrels, raccoons, some birds, and other wildlife. The name Pawpaw comes from the Spanish word *papaya*, an unrelated but superficially similar tropical fruit.

DESCRIPTION: Common Pawpaw is a shrub or small tree that forms colonies from root sprouts, with a straight trunk, spreading branches, and large leaves. Its leaves are 7–10" (18–25 cm) long, 3–5" (7.5–13 cm) wide, spreading in two rows or long twigs, reverse ovate, broadest beyond the middle, short-pointed at the tip, tapering to the base and short leafstalk. They are covered with rust-colored hairs when young, green above, and paler beneath; turning yellow in autumn. Bruised foliage has an unpleasant odor. The bark is dark brown, warty, and thin. The twigs are brown, often with rust-colored hairs, ending in small hairy buds.

FLOWERS: Early spring. 1½" (4 cm) wide; three triangular green to brown or purple outer petals, hairy with prominent veins, nodding singly on slender stalks.

FRUIT: 3–5" (7.5–13 cm) long, 1–1½" (2.5–4 cm) in diameter, berrylike, brownish, cylindrical and slightly curved, suggesting a small banana; the edible, soft, yellowish pulp has the flavor of custard. It has several shiny, brown, oblong seeds.

CAUTION: Poisonous parts: Although some parts of this species are edible, other parts are toxic to humans. Eat only if you are absolutely sure which parts are edible.

HABITAT: Moist soils; especially flood plains; in understory of hardwood forests.

RANGE: Ontario and western New York, south to Florida, west to Texas, and north to Nebraska; to 2600' (792 m) in the southern Appalachians.

USES: The fruit of the Common Pawpaw is gathered locally and eaten raw or used in a variety of desserts and other recipes. It is not more widely cultivated because the fruits ferment quickly after being picked, but it is easily grown from seed and has few pests.

SIMILAR SPECIES: Georgia Fevertree or Pinckneya, and magnolias.

CONSERVATION STATUS: LC

This tree species has a wide distribution and no threats have been identified. It is state-endangered in New Jersey and threatened in New York.

Diameter: 8" (20 cm)
Height: 30' (9 m)

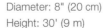

■ DENSE POPULATION
■ LIGHT POPULATION
■ NATURALIZED

Once popular as a street tree in the southeastern United States, Camphor-tree is now considered an invasive pest in that region. It produces dense stands and shades out all other growth, including our native plants. This species was originally introduced to North America for camphor production.

DESCRIPTION: Introduced to North America, this aromatic evergreen tree has a rounded, dense crown, wider than high, and an odor of camphor in crushed foliage. Its leaves are evergreen, partly opposite, 2½–4" (6–10 cm) long, 1–2¼" (2.5–6 cm) wide, elliptical, and pointed, with three main veins from the rounded base. The leaves are not toothed, slightly thickened, long-stalked, and pinkish when young, becoming shiny green above and dull whitish beneath. The bark is gray and smoothish, becoming rough, thick, and furrowed. The twigs are green, slender, and hairless.

FLOWERS: Spring. ⅛" (3 mm) long, yellowish, in clusters 1½–3" (4–7.5 cm) long.

FRUIT: A black, one-seeded berry, ⅜" (10 mm) in diameter, with a greenish cup and the spicy taste of camphor.

HABITAT: Moist soils, roadsides and waste places in humid subtropical regions.

RANGE: Native of tropical Asia from eastern China to Vietnam, Taiwan, and Japan. Extensively cultivated and naturalized locally from North Carolina to southern Texas.

USES: This tree is cultivated for camphor, a substance used in creams, ointments, and lotions, and for wood, which is insect-repellent and often used to make cabinets and chests.

SIMILAR SPECIES: Lancewood, Red Bay.

CONSERVATION STATUS:
This plant can be weedy and problematic in the Southeast and in Hawaii. The conservation status of this species in the wild has not been fully assessed.

Diameter: 2' (0.6 m)
Height: 40' (12 m)

This shrub is known as the "forsythia of the wild" because of its early spring flowering, which gives a subtle yellow tinge to many lowland woods where it is common. A tea can be made from the spicy leaves and twigs of this species.

DESCRIPTION: This is a deciduous shrub with dense clusters of tiny, pale yellow flowers that bloom before the leaves from globose buds along the twigs. Its leaves are 2–5½" (5–13.8 cm) long, dark green, oblong, smooth, and untoothed, and have an aromatic, spicy fragrance when crushed.

FLOWERS: March–April. ⅛" (3 mm) wide; six sepals and petals, all alike. Male and female flowers occur on separate plants.

FRUIT: Ovoid, shiny, red, berrylike drupes.

HABITAT: Swamps and wet woods.

RANGE: Maine south to Florida; west to Texas; north to Missouri, Iowa, and Ontario.

USES: Spicebush is often cultivated as an ornamental. Native Americans used this plant to treat various ailments, and a tea can be made from the spicy leaves and twigs.

SIMILAR SPECIES: Bog Spicebush, Eastern Sweetshrub.

CONSERVATION STATUS: **LC**
The population is widespread and stable with no major threats identified.

Height: 6–17' (1.8–5.1 m)

LANCEWOOD *Nectandra coriacea*

■ DENSE POPULATION
■ LIGHT POPULATION
■ NATURALIZED

This small tree is native to eastern and southern Florida. With its lovely white flowers, attractive fruit, and dark green leaves, Lancewood is a popular choice for native landscaping and gardening in Florida.

ALTERNATE NAME: *Ocotea coriacea*

DESCRIPTION: A narrow-trunked evergreen tree, Lancewood has dense, spreading branches and a dark reddish brown, rough, warty bark. Its leaves are up to 6" (15 cm) long, lanceolate, pointed at the tip.

FLOWERS: March–June. ⅓" (8 mm), star-shaped, creamy white, fuzzy, in branched clusters.

FRUIT: ½" (12 mm), berrylike, olive-shaped drupe; green to purple-black, in red cup.

HABITAT: Hammocks and other woodlands.

RANGE: Native of eastern and southern Florida and southward into the Caribbean and Mexico.

USES: Lancewood is often planted as an ornamental.

SIMILAR SPECIES: Camphor-tree, Red Bay.

CONSERVATION STATUS: (LC)
Although not overly common in Florida, this tropical species is widespread in the Caribbean and on the Yucatán Peninsula. There are no major threats to the population.

Height: to 33' (10 m).

■ DENSE POPULATION
■ LIGHT POPULATION
■ NATURALIZED

This handsome evergreen native to the southeastern United States is sometimes a shrub, but is usually a small tree. It is sometimes planted as an ornamental tree in parks and gardens. Swampbay (var. *pubescens*) is a variety found in coastal swamps and characterized by twigs and lower leaf surfaces covered with rust-colored hairs.

DESCRIPTION: This is a small evergreen tree with a dense crown. Its leaves are evergreen, 3–6" (7.5–15 cm) long, ¾–1½" (2–4 cm) wide, elliptical or lance-shaped, short-stalked, thick and leathery, with the edges slightly rolled under; they are shiny green above and pale with whitish or rust-colored hairs beneath. The bark is dark or reddish-brown and furrowed into broad, scaly ridges.

Diameter: 2' (0.6 m)
Height: 60' (18 m)

FLOWERS: Spring. ³⁄₁₆" (5 mm) long, light yellow, several in a long-stalked cluster at the leaf base.

FRUIT: ½–⅝" (12–15 mm) long, nearly round; it is shiny, dark blue-black. It has a six-lobed cup at the base, a thin pulp, and a rounded seed; maturing in fall.

HABITAT: Wet soils of valleys and swamps, also sandy uplands and dunes; it is found in mixed forests.

RANGE: Southern Delaware south to southern Florida and west to southern Texas; to 400' (122 m).

USES: The wood from this species is harvested locally and used for fine cabinetwork and boatbuilding. The spicy leaves are sometimes used to flavor soups and meats. This tree is also planted as an ornamental.

SIMILAR SPECIES: Sweet Bay.

CONSERVATION STATUS: LC
This species is widespread and the population is stable with no identified threats.

■ DENSE POPULATION
■ LIGHT POPULATION
■ NATURALIZED

This is the northernmost New World representative of a mostly tropical family of trees. Sassafras has been widely used by humans for a variety of purposes, including to brew sassafras tea and to flavor root beer. Early colonists had the roots and bark shipped back to Europe. The greenish twigs and leafstalks have a pleasant, spicy, slightly gummy taste. Many mammals, including American Black Bears (*Ursus americanus*), eat the berries, as do many important native bird species.

DESCRIPTION: Sassafras is an aromatic tree or thicket-forming shrub with variously shaped leaves and a narrow, spreading crown of short, stout branches. Its leaves are 3–5" (7.5–13 cm) long, 1½–4" (4–10 cm) wide; they are elliptical, often with two mitten-shaped lobes or three broad and blunt lobes. The leaves are not toothed, the base is short-pointed, and they have long, slender leafstalks; they are shiny green above and paler and often hairy beneath; turning yellow, orange, or red in fall. The bark is gray-brown, becoming thick and deeply furrowed. The twigs are greenish, slender, and sometimes hairy.

FLOWERS: Early spring. ⅜" (10 mm) long, yellow-green, with several clustered at the end of leafless twigs; male and female flowers usually occur on separate trees.

FRUIT: Elliptical, ⅜" (10 mm) long, shiny bluish-black berries; each is held in a red cup on a long, red stalk containing one shiny brown seed; maturing in autumn.

Diameter: 1½' (0.5 m), sometimes larger
Height: 30–60' (9–18 m)

HABITAT: Moist, particularly sandy, soils of uplands and valleys, often in old fields, clearings, and forest openings.

RANGE: Extreme southern Ontario east to southwestern Maine, south to central Florida, west to eastern Texas, and north to central Michigan; to 5000' (1524 m) in the southern Appalachians.

USES: Sassafras has long been used for culinary and medicinal purposes. It is the key ingredient in traditional root beer, and the ground leaves are used in traditional Creole cuisine, particularly gumbo. Native Americans and early colonists used the roots to treat a variety of ailments and diseases. Sassafras is also a common ornamental tree.

SIMILAR SPECIES: This tree is unlikely to be confused with any other species in its native range.

CONSERVATION STATUS: (LC)

Sassafras is common throughout its wide range. Laurel wilt disease, a fungal infection transmitted by an introduced ambrosia beetle, is known to afflict other *Sassafras* species and has appeared in the southeastern portion of this tree's range.

■ DENSE POPULATION
■ LIGHT POPULATION
■ NATURALIZED

A handsome ornamental and street tree on the West Coast, this species is also known as California Bay. When crushed, the leaves, twigs, and other parts are strongly fragrant. The attractive light brown wood with darker streaks takes a beautiful finish and is used for veneer in furniture and paneling, cabinet-work, and interior trim. Prized for novelties and woodenware, it is often marketed as "Oregon Myrtle," although it is a member of the laurel family.

DESCRIPTION: An evergreen tree with a short trunk usually forked into several large, spreading branches, forming a broad, rounded, dense crown of aromatic, peppery foliage; in exposed situations it can be a low, thicket-forming shrub. Its leaves are evergreen, 2–5" (5–13 cm) long, ½–1½" (1.2–4 cm) wide, ellip-tical or lance-shaped, short-pointed or rounded at the ends; they are thick and leathery, with edges slightly turned under, and shiny dark green above and dull and paler beneath with a promi-nent network of veins; turning yellow or orange before shedding gradually after the second year. The bark is dark reddish-brown and thin with flat scales. The twigs are stout, hairy, and yellow-green when young, becoming yellow-brown.

FLOWERS: Late winter–early spring. Flowers are ¼" (6 mm) long, pale yellow, and numerous, clustered on a stalk at the leaf bases.

FRUIT: An elliptical or nearly round berry, ¾–1" (2–2.5 cm) long, greenish to purple, with a thin pulp and large brown seed; maturing in late autumn.

HABITAT: Moist soils, especially in mountain canyons and valleys; it is found in mixed forests.

RANGE: Southwest Oregon south in the Coast Ranges and Sierra Nevada to southern California; to 4000' (1219 m); at the southern limit, 2000–6000' (610–1829 m).

USES: Native Americans used parts of this tree for food and for medicinal purposes. Some people still eat the fruit. The wood is used for furniture and other purposes.

SIMILAR SPECIES: Bay Laurel.

CONSERVATION STATUS: LC

California Laurel is widespread and common, and often can be dominant within Californian forests. There are no major threats to the species and the population is stable.

Diameter: 1½–2½' (0.5–0.8 m)
Height: 40–80' (12–24 m)

EVERGLADES PALM *Acoelorrhaphe wrightii*

■ DENSE POPULATION
■ LIGHT POPULATION
■ NATURALIZED

This rare palm, also known as Paurotis Palm, is restricted in the wild to Everglades National Park. It is the only species in its genus.

DESCRIPTION: This is a small to medium-sized tree. The trunk is slender, 6" (15 cm) in diameter, clustered, and covered with reddish-brown leafstalk bases. Its leaves are 3' (1 m) long, fan-shaped, and deeply palmately divided into folded, linear segments. The leafstalk is spiny-edged and orange-tinged.

FLOWERS: In hanging clusters, to 7" (22 cm) long.

FRUIT: Small round berrylike fruits, in clusters, turning from green to orange to black.

HABITAT: Swamps, wet areas, towns.

RANGE: Cultivated in southern Florida; in North America it grows wild only in Everglades National Park.

USES: This species has been commonly cultivated as an ornamental and nursery plant.

SIMILAR SPECIES: Senegal Date Palm.

CONSERVATION STATUS:
Everglades Palm is considered threatened in Florida, and habitat for the species is threatened by heavy development, wetland drainage, and other forms of degradation. The population is scattered in the Caribbean, Florida, eastern Mexico, and central America, although its abundance is unclear.

Diameter: 6" (15 cm)
Height: 35' (10.76 m)

■ DENSE POPULATION
■ LIGHT POPULATION
■ NATURALIZED

Fishtail Palms can reach heights of 25–30 feet (10 m), and are named for the extremely large leaves that resemble the shape of a fish's tail. The tree prefers moist soil, growing densely and usually sending up multiple stems or trunks. Fishtail Palm is sometimes planted outdoors in Florida, and can be planted in atriums indoors. The tree parts contain oxalic acid and can cause severe skin irritation; the fruits are poisonous. When the fruit of the lowest node matures, the individual stem dies, leaving the remaining stems unaffected. The regular production of new stems allows a single clump of Fishtail Palms to remain vital for many years.

DESCRIPTION: This is a small, clump-forming, multi-stemmed tree with a crown of arching leaves. The trunk is light brown and smooth, with horizontal rings. The fruits and flowers hang from the leaf nodes in long, flowing clusters. Its leaves are 6–10' (1.8–3 m) long, bipinnately compound, and slender, without spines, with an erect central axis. Leaf segments are 4–8" (10–20 cm) long, obovate (inversely triangular), with coarsely veined, ragged margins on the outer end, resembling the tail of a fish.

FLOWERS: Year-round. White flowers occur in dense, branched, hanging clusters. The first flower cluster of a mature stem emerges from the base of the uppermost leaves. Succeeding flower clusters emerge from successively lower leaf nodes until the lowest node is reached.

FRUIT: A ½" (1.3 cm) wide, nearly round drupe, dark red, purplish, or black. The fruits are poisonous. They are abundant, and occur year-round on the plant in hanging clusters.

CAUTION: Poisonous: Do not eat or ingest any part of this species. If accidentally ingested, contact your doctor or poison control center.

HABITAT: Disturbed hammocks.

RANGE: Introduced in North America; naturalized in disturbed hammocks in southern Florida. Native from India to the Philippines.

USES: Grown for ornament for their attractive foliage. It can be grown indoors as a large houseplant.

SIMILAR SPECIES: Cabbage Palmetto, California Fan Palm, Canary Island Date.

CONSERVATION STATUS: LC
The Fishtail Palm population is stable. This species can be invasive in Florida and should not be planted outdoors casually.

Diameter: to 6" (15 cm)
Height: 15–25' (4.6–7.6 m)

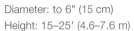

■ DENSE POPULATION
■ LIGHT POPULATION
■ NATURALIZED

This native, slow-growing palm is an endangered species in Florida. It is grown widely as an ornamental in residential and commercial landscapes.

DESCRIPTION: This small palm has a smooth, slender trunk, growing to about 6" (15 cm) in diameter, and has long-stalked, fan-shaped leaves. Its leaves are 30" (75 cm) across, deeply palmately divided into pointed, linear segments, and densely silvery-hairy below.

FLOWERS: Small, creamy-white clusters tucked in among the leaves.

FRUIT: Branched clusters of small, reddish- to purplish-black berries.

HABITAT: Rocky pinelands, coastal dunes, scrub.

RANGE: Extreme southern mainland Florida and Florida Keys; also found in the Bahamas.

USES: This native palm is cultivated as an ornamental.

SIMILAR SPECIES: Key Thatch Palm.

CONVERSATION STATUS:

CONSERVATION STATUS: NT

This species is threatened in Florida and on a few Bahamian islands. The global population is relatively stable and widespread but occurs with spotty distribution within its range. Many subpopulations in Florida consist of only a few individual plants.

Diameter: 6" (15 cm)
Height: to 25' (7.5 m)

■ DENSE POPULATION
■ LIGHT POPULATION
■ NATURALIZED

The Coconut Palm is native to southeastern Asia, but has long been cultivated throughout the world, including in North America. It has become naturalized in some areas of southern Florida.

DESCRIPTION: This is a tall, tropical palm that bears the familiar coconut. The trunk is smooth, grayish-brown, straight, leaning, or gently curving; it is swollen at the base with conspicuous rings and vertical cracks. Its leaves are feather-like, 18–23' (5.4–7 m) long, and pinnately divided into linear-lanceolate segments.

FRUIT: Coconut enclosed in large, green to brown husk, to 16" (40 cm) long; occurs year-round.

HABITAT: Beaches, disturbed areas, towns.

RANGE: Naturalized in coastal southern Florida.

USES: The Coconut Palm has been used for a myriad of purposes, from food to medicine to cosmetics. It is grown in more than 90 countries, with most production occurring in Indonesia, the Philippines, and India.

SIMILAR SPECIES: Queen Palm.

CONSERVATION STATUS:
The conservation status of this domesticated species has not been assessed. The evolutionary history of the Coconut Palm prior to human dispersal is ambiguous, with its origin unclear.

Height: 80–100' (24–30 m)

■ DENSE POPULATION
■ LIGHT POPULATION
■ NATURALIZED

This palm is the official symbol of the Canary Islands, where it is native, and is cultivated in the southern United States. It is sometimes called the Pineapple Palm because of the resemblance of its trunk to pineapple skin. The fruit of this species is less palatable than that of the Date Palm (*P. dactylifera*), which is cultivated as a fruit tree in southeastern California and other hot, dry regions.

DESCRIPTION: This ornamental palm sports a massive, unbranched trunk covered by bases of old leaves and with very large, pinnately compound leaves. Those leaves are evergreen, numerous, upright, and spreading around the top, and 12–20' (3.7–6.1 m) long. The leafstalks are stout, bearing long, green spines at the edges. Its many leaflets are 1½' (0.5 m) long, 1" (2.5 cm) wide, very narrow, long-pointed, and folded into a short stalk at the base, with edges turned upward; they are leathery, straight, and not drooping. The color is dull green; dead leaves are light brown, hanging down. The trunk is stout, often wider near the crown, with masses of air roots forming near an enlarged base; it has rough, light brown leaf bases often bearing air plants and ferns.

FLOWERS: Male and female on separate trees; small, whitish; numerous, along much-branched orange flower stalks 3–6' (0.9–1.8 m) long; among leaves.

FRUIT: ¾" (2 cm) long, egg-shaped, yellow, edible dates that are one-seeded; they occur in heavy clusters.

HABITAT: Subtropical regions.

RANGE: Native of the Canary Islands. Introduced along the southern border of the United States in California, Arizona, and the Gulf states east to Florida.

USES: This species is cultivated in the southern United States as an ornamental palm. The sap is used to make syrup in the Canary Islands.

SIMILAR SPECIES: Date Palm.

CONSERVATION STATUS: (LC)

The native population of this tree is small but appears to be increasing. One major threat to this species in the wild is extensive hybridization with the Date Palm (*P. dactylifera*), which has been introduced in the tree's native range. Another threat is beetle infestation by various weevil species, which mostly affects cultivated populations. Fire and habitat degradation are local threats to the native population on the Canary Islands, where the species is listed as Near Threatened. The species is legally protected in many areas and efforts to promote cultivation are under way.

Diameter: 2–3' (0.6–0.9 m)
Height: 50' (15 m)

■ DENSE POPULATION
■ LIGHT POPULATION
■ NATURALIZED

This species has been cultivated for its sweet fruit in southern California and other hot, dry regions. It is showing some invasive tendencies in some areas and is included on many invasive species lists in the United States and Canada. Fossils show that the Date Palm has existed for at least 50 million years; it has been widely domesticated since ancient times.

DESCRIPTION: This introduced palm has a rounded crown and a stout, gray trunk often covered with woody bases of old leaves. The leaves can be up to 20' (6 m) long; they are feather-like, erect, gray-green, and evergreen.

FLOWERS: February–April. Tiny and white, they occur in dense clusters on branched, orange stalks.

FRUIT: Blackish, edible dates; ripe September–December.

HABITAT: Gardens, other cultivated sites.

RANGE: Native to northern Africa or southwestern Asia; it is cultivated in the southwestern United States.

USES: This palm is cultivated for its fruit in parts of Africa, Asia, and North America.

SIMILAR SPECIES: Canary Island Date Palm.

CONSERVATION STATUS:
This domesticated tree species has been cultivated in North Africa and the Middle East for thousands of years.

Height: 23–50' (7–15 m)

■ DENSE POPULATION
■ LIGHT POPULATION
■ NATURALIZED

Also known as the Cuban Royal Palm, this large, elegant tree is endangered in the wild in Florida. It is widely cultivated and is a beautiful plant for a native southern Florida landscape. It requires full sun and moist, rich soil. It is the only native North American representative of its genus.

DESCRIPTION: The trunk of this stately palm towers at more than 100' (30 m) tall, and the large, pinnate leaves are 10' (3 m) long or more. The long, smooth, green, sheathing leaf bases and the straight, smooth, pale-gray trunk that usually bulges at the base are distinctive. Its leaves are 9–15' (2.7–4.5 m) long, feather-like, pinnately divided into linear segments in four rows.

FLOWERS: White, in clusters of two males and one female: male flowers are ⅓" (8 mm) long with three spreading petals; female flowers are tiny and budlike.

HABITAT: Moist, rich hammocks, marshes, and towns and gardens.

RANGE: Extreme southern mainland Florida and Florida Keys; found in Everglades National Park.

USES: This palm tree is widely cultivated throughout the world as an ornamental. It is commonly used in Palm Sunday observances.

SIMILAR SPECIES: Florida Cherry Palm.

CONSERVATION STATUS:
Florida Royal Palm is considered endangered in Florida. This species is widely planted, but it is threatened by urban development and horticultural collection in Florida.

Height: 130' (39 m)

■ DENSE POPULATION
■ LIGHT POPULATION
■ NATURALIZED

There are more than 2500 plants in the palm family world-wide, but only about a dozen are native to the United States. The Dwarf Palmetto is one of the most cold-tolerant of these, thriving in diverse habitats across the southeastern and south-central United States and northeastern Mexico. It is known to survive temperatures near 0°F (−18°C), but needs hot, humid summer climates to grow well. This shrub-like palm generally tops out around 6 feet in height. Its stem is sometimes not visible because it resides underground and is quite short. Dwarf Palmetto is simple to grow and tolerates a wide variety of soils.

DESCRIPTION: Dwarf Palmetto is a clump-forming shrub. Its leaves are 3' (1 m) long, fan-shaped, and divided into approximately 30 segments.

FLOWERS: May–July. The flowers are tiny and whitish, clustered on long, arching stalks.

FRUIT: ½" (1.3 cm) black, shiny drupes.

CAUTION: Poisonous—do not eat or ingest any part of this species. If accidentally ingested, contact your doctor or poison control center.

HABITAT: Lowland, rich woods, swamps, watersides.

RANGE: Southeastern coastal plain from North Carolina to Texas, Oklahoma, and Arkansas.

USES: Dwarf Palmetto is widely cultivated and is a popular landscape palm. Palmetto leaves are used to weave baskets and were once used to construct thatched homes. This species is also an important source of food for wildlife, notably for birds, raccoons, and squirrels.

SIMILAR SPECIES: Cabbage Palmetto, California Fan Palm, Canary Island Date.

CONSERVATION STATUS: LC
The population is stable with few threats.

Height: 6' (2 m)

CABBAGE PALMETTO *Sabal palmetto*

■ DENSE POPULATION
■ LIGHT POPULATION
■ NATURALIZED

The Cabbage Palmetto is native to the southern United States and is the official state tree of both Florida and South Carolina, even appearing on the latter state's official flag. This tree is also a popular ornamental and street tree, the northernmost New World palm and one of the most hardy. Many of these trees had been killed in the past in order to eat the large leaf buds as a cabbagelike salad. The name comes from the Spanish word *palmito*, meaning "little palm."

DESCRIPTION: Cabbage Palmetto is a medium-sized, spineless, evergreen palm with a stout, unbranched trunk and extremely large, fan-shaped leaves spreading around the top. Its leaves are 4–7' (1.2–2.1 m) long and nearly as broad, folded into many long, narrow segments; they are long-pointed and drooping, coarse, stiff, and leathery, and splitting apart nearly to the stout midrib, with threadlike fibers separating at the edges. They are shiny dark green. It has stout, stiff leafstalks that are 5–8' (1.5–2.4 m) long, green, and ridged above, with a long, fibrous, shiny brown sheath at a wedge-shaped base, which splits and hangs down with age. The trunk is gray-brown and rough or ridged.

FLOWERS: Early summer. The flowers are ³⁄₁₆" (5 mm) long with a deeply six-lobed whitish corolla; they are fragrant, and nearly stalkless, occurring in curved or drooping, much-branched clusters arising from leaf bases.

FRUIT: Berries, ⅜" (10 mm) in diameter, nearly round, shiny black with thin, sweet, dry flesh; they are one-seeded; maturing in autumn.

HABITAT: Sandy shores, crowded in groves; it is found inland in hammocks.

RANGE: Near the coast from southeastern North Carolina to southern and northwestern Florida, including the Florida Keys.

USES: This species is a popular ornamental tree in the southern United States. The trunks are sometimes used for wharf pilings, docks, and poles. Brushes and whisk brooms are made from young leafstalk fibers, and baskets and hats from the leaf blades.

SIMILAR SPECIES: Washington Fan Palm, Mexican Palmetto.

CONSERVATION STATUS: LC
The population is stable with few threats.

Diameter: 1½' (0.5 m)
Height: 30–50' (9–15 m) or more

■ DENSE POPULATION
■ LIGHT POPULATION
■ NATURALIZED

This native palm is endemic to the southeastern United States, found throughout Florida and in the southernmost parts of Mississippi, Alabama, Georgia, and South Carolina. With its large, fanned-out leaves and horizontal branching, Saw Palmetto makes an interesting addition to any southeastern garden.

DESCRIPTION: This is a clump-forming shrub that occasionally becomes a small tree. The trunks are prostrate and sprawling (sometimes erect), covered with reddish leafstalk bases and brown fibers. Its leaves are evergreen, 3' (0.9 m) or more across, fan-shaped, deeply palmately divided into folded linear segments; they have harsh, saw-like "teeth" or spines along the petiole.

FLOWERS: Tiny, white, fragrant flowers occurring on plume-like, branched stalks from leaf axils.

FRUIT: Berries: the fruit turns from orange to black when mature.

HABITAT: Pinelands, hammocks, coastal dunes, sandhills.

RANGE: Coastal plain from South Carolina to Florida and Texas.

USES: An extract made from Saw Palmetto fruit has been used in medical research and treatment for urinary and reproductive conditions including prostate cancer. This species is also planted as an ornamental.

SIMILAR SPECIES: Paurotis Palm.

CONSERVATION STATUS: LC
There are no major threats to the population.

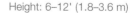

Height: 6–12' (1.8–3.6 m)

This palm is native to South America but cultivated through-out the world as a popular ornamental tree, including in central and southern Florida where it has naturalized and is considered an invasive species. Birds and other wildlife are known to eat the fallen fruit.

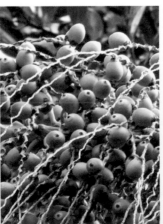

DESCRIPTION: This small to medium-sized tree has a gray, smooth, trunk that is ringed with widely spaced old leaf scars, with an 8–10' (2–3 m) tall crownshaft and a crown of huge, loosely spread, upright, and arching leaves. Its leaves are pinnately compound, their segments glossy, bright green, and flexi-ble; they do not lie flat. Just below the leaves, the trunk is surrounded by dead leafstalks and persistent leafstalk bases that form a conspicu-ous, cuplike structure called a "boot." The leaves have 150–250 pairs of leaf segments per leaf; they lack spines. The leaf segments are 1½–3' (46–91 cm) long, soft, fine, textured, and flexible, and glossy, bright green.

FLOWERS: Summer. Small, yellow or creamy white flowers in dense, large, 3–6' (1–2 m) long, brushlike clusters at leaf bases.

FRUIT: Winter. This tree produces a 1" long, round drupe; it is yellow to bright orange and datelike.

HABITAT: Introduced in North America.

RANGE: Native to Brazil. This species is introduced in North America and has naturalized in central and southern Florida.

USES: This palm is a popular street and landscaping tree. It is also grown as an indoor houseplant.

SIMILAR SPECIES: Coconut Palm.

CONSERVATION STATUS:
The conservation status of this species has not been fully assessed.

Diameter: 1' (30.48 cm)
Height: 25–50' (7.6–15 m)

■ DENSE POPULATION
■ LIGHT POPULATION
■ NATURALIZED

California Fan Palm is the largest native palm of the continental United States, as well as the only western species. It is also known as Desert Fan Palm, and as Petticoat Palm for the shaggy mass of dead leaves that hangs against the trunk. Groves of these trees grow in Palm Canyon near Palm Springs and in Joshua Tree National Park. California Fan Palm is widely cultivated as an ornamental along streets and avenues in southern California, southern Arizona, the Gulf States east to Florida, and in subtropical regions around the world. This genus is named for the first president of the United States.

DESCRIPTION: A large palm of the far southwestern United States, California Fan Palm sports a massive, unbranched trunk and very large, fan-shaped leaves. Its leaves are evergreen, numerous, spreading around the top; if not burned or cut, old dead leaves hang down against the trunk in a thick thatch. The leafstalks are 3–5' (0.9–1.5 m) long and stout with hooked spines along edges. Leaf blades are 3–5' (0.9–1.5 m) in diameter, gray-green, and split into many narrow, folded, leathery segments, with the edges frayed into many threadlike fibers. The trunk is gray and smooth with horizontal lines and vertical fissures.

FLOWERS: ⅜" (10 mm) long with a funnel-shaped, deeply three-lobed white corolla; they are short-stalked and slightly fragrant. Many occur together in much-branched clusters 6–12" (1.8–3.7 m) long, drooping from leaf bases.

FRUIT: An elliptical black berry, ⅜" (10 mm) in diameter, with thin, sweetish, edible pulp and one elliptical brown seed.

HABITAT: Moist soils along alkaline streams and in canyons of the mountains in the Colorado and Mojave deserts.

RANGE: Southeastern California (San Bernardino County to San Diego County), southwestern Arizona (Kofa Mountains, Yuma County, and also southern Yavapai County, where it is perhaps introduced), and northern Baja California; at 500–3000' (152–914 m).

USES: Native Americans ate the sweet berries and ground the seeds into a flour. They also cut out and roasted the terminal bud at the top of the tree, the source of all new growth, thereby killing the tree. The leaves were used for clothing, sandals, thatching, and basketry. Contemporary uses include trunks as wharf pilings, docks, and poles. Brushes and whisk brooms are made from young leafstalk fibers, and baskets and hats from the leaf blades.

SIMILAR SPECIES: Cabbage Palmetto, Mexican Fan Palm.

CONSERVATION STATUS: NT
Grazing, urbanization, and habitat degradation all affect the wild population. This species is protected in Arizona, where the plants are subject to damage and vandalism.

Diameter: 2–3' (0.6–0.9 m)
Height: 20–60' (6–18 m)

■ DENSE POPULATION
■ LIGHT POPULATION
■ NATURALIZED

The Latin species name, meaning "robust," is more appropriate to the other species of this genus, California Fan Palm, which has a stout, massive trunk. Mexican Fan Palm has a narrower trunk and tends to grow slightly taller and faster. Also known as the Washington Fan Palm, this tree is better adapted to planting near the coast but is less cold-hardy than its native relative.

ALTERNATE NAME: Mexican Washingtonia

DESCRIPTION: Tall, introduced, ornamental palm has a slender, unbranched trunk and very large, fan-shaped leaves. Those leaves are evergreen and numerous, spreading around the top; old, dead, light brown leaves hang down against the trunk in a thick thatch, if not burned or cut. The leafstalks are 2½–4' (0.8–1.2 m) long, reddish-brown, and stout, with hooked spines along the edges. The leaf blades are 3–5' (0.9–1.5 m) in diameter, bright green, and split into many narrow, folded, leathery segments; edges with threadlike fibers when young. The trunk is light brown, nearly smooth or becoming finely fissured, with horizontal rings, and enlarged at the base.

FLOWERS: ⅜" (10 mm) long with funnel-shaped, deeply 3-lobed, white corolla; short-stalked, slightly fragrant, and numerous, in much-branched 6–12' (1.8–3.7 m) long clusters, spreading or drooping from leaf bases.

FRUIT: An elliptical black berry, ⅜" (10 mm) in diameter, with thin, sweetish, edible pulp and one seed.

HABITAT: Subtropical regions.

RANGE: Native of canyons in northern Baja California and Sonora, Mexico. Introduced along the southern border of the United States in California, southern Arizona, and the Gulf states east to Florida.

USES: This species is grown as an ornamental tree.

SIMILAR SPECIES: Cabbage Palmetto, Mexican Palmetto.

CONSERVATION STATUS:
The conservation status of this species has not been fully assessed.

Diameter: 1–1½' (0.3–0.5 m)
Height: 50–70' (15–21 m)

■ DENSE POPULATION
■ LIGHT POPULATION
■ NATURALIZED

This small shrub is native to California and Baja California. It may become quite common several years after a fire in chaparral. There is only one other species in the genus, Channel Islands Tree Poppy (*D. harfordii*), found on islands off the coast of southern California.

DESCRIPTION: Tree Poppy is a stiff, roundish shrub with brilliant yellow, cup-shaped flowers at the ends of short branches or on long stalks in leaf axils. Its leaves are 1–4" (2.5–10 cm) long, leathery, lanceolate, bluish green, and twisted so that the flat sides of the blades face sideways.

FLOWERS: April–June. The flowers are 1–2½" (2.5–6.5 cm) wide, yellow, and four-petaled with many short stamens.

FRUIT: 2–4" (5–10 cm) and cylindrical with many smooth, brown or black seeds.

HABITAT: Dry slopes in chaparral.

RANGE: Northern California south to Mexico.

USES: This species is cultivated as an ornamental plant.

SIMILAR SPECIES: Channel Islands Tree Poppy.

CONSERVATION STATUS: (NT)
More research needs to be done to better understand the conservation status of this species.

Height: 4–20' (1.2–6 m)

This large shade tree is a hybrid of the American Sycamore of the eastern United States and the Oriental Planetree (*Platanus orientalis*) of southeastern Europe and Asia Minor. It is also popular as a street tree in Europe, where it originated probably before 1700. The plants can be clipped into screens and arbors. The genus name *Platanus* is the classical Latin and Greek name of Oriental Planetree, from the Greek word for "broad," which describes the leaves.

ALTERNATE NAME: *Platanus acerifola*

DESCRIPTION: This large, introduced shade tree has a straight, stout trunk and a broad, open crown of spreading to slightly drooping branches and coarse foliage. Its leaves are shiny green above and pale beneath, 5–10" (13–25 cm) long and wide and palmately three- or five-lobed; they have shallow, short-pointed lobes, with few large teeth or none, and three or five main veins from a notched base, becoming hairless or nearly so. The leafstalk is long and stout, covering the side bud at an enlarged base. The bark is smooth with patches of brown, green, and gray, peeling off in large flakes. The twigs are greenish, slender, zigzagged, and hairy, with ring scars at the nodes.

FLOWERS: Spring. The flowers are tiny and in greenish, ball-like, drooping clusters with males and females on separate twigs.

FRUIT: 1" (2.5 cm) in diameter, usually two bristly, brown balls hanging on a long stalk, composed of many narrow nutlets with hair tufts; maturing in autumn, separating in winter.

HABITAT: Moist soils in humid temperate regions, hardy and tolerant of city conditions.

RANGE: Planted across the United States.

USES: This hybrid is cultivated as an ornamental and street tree.

SIMILAR SPECIES: American Sycamore.

CONSERVATION STATUS:
This species is a horticultural hybrid.

Diameter: 2' (0.6 m)
Height: 70' (21 m)

Platanus occidentalis **AMERICAN SYCAMORE**

■ DENSE POPULATION
■ LIGHT POPULATION
■ NATURALIZED

The American Sycamore is a large, stately tree, familiar from southern New England to Georgia and west to eastern Texas. It closely resembles its close relative, the London Planetree, with a distinctive mottled bark with scale-like patches of light and dark brown or gray on a cream-colored surface. It can grow to great sizes, with some early records at more than 160' (48.8 m) high with trunks reaching as much as 15' (4.6 m) in diameter. The present record trunk is about 11' (3.4 m) in diameter. The Sycamore grows to a larger trunk diameter than any other native hardwood. The largest trunks are often hollow, and there are records of their being used by early settlers as dwelling places. History records a Virginia family that lived in the hollow bole of a huge Sycamore.

DESCRIPTION: One of the largest eastern hardwoods, American Sycamore has an enlarged base; a massive, straight trunk; and large, spreading, often crooked branches forming a broad open crown. Its leaves are 4–8" (10–20 cm) long and wide (larger on shoots), broadly ovate, bright green above and paler beneath, and becoming hairless except on the veins, turning brown in fall. They have three or five shallow, broad, short-pointed lobes, wavy edges with scattered large teeth, and three or five main veins from a notched base. The leafstalk is long and stout, covering the side bud at the enlarged base. The bark is smooth, whitish, and mottled, peeling off in large thin flakes, exposing patches of brown, green, and gray; the base of large trunks is dark brown and deeply furrowed into broad, scaly ridges. The twigs are greenish, slender, and zigzagging, with ring scars at the nodes.

FLOWERS: Spring. Tiny, greenish, occurring in one or two ball-like, drooping clusters, with male and female clusters on separate twigs.

FRUIT: 1" (2.5 cm) in diameter, usually one brown ball hanging on a long stalk, composed of many narrow nutlets with hair tufts; maturing in fall, separating in winter.

HABITAT: Wet soils of stream banks, flood plains, and the edges of lakes and swamps; occasionally dominant in mixed forests.

RANGE: Southwestern Maine, south to Florida, west to Texas, north to eastern Nebraska; also found in Mexico; to 3200' (975 m).

USES: Sycamore wood is used for furniture parts, millwork, flooring, and specialty products such as butcher blocks, as well as pulpwood, particleboard, and fiberboard.

SIMILAR SPECIES: California Sycamore, London Planetree (a hybrid of the American Sycamore).

CONSERVATION STATUS: ⓛⓒ
American Sycamore is rarely dominant but is common over a wide range, and its large population is stable. This species is preyed upon by the sycamore leaf beetle (*Neochlamisus platani*), and subject to a fungus called anthracnose that causes cankers on the trees and spreads as wind-borne spores. The London Planetree is largely immune from anthracnose infection and is thus often planted instead of the American Sycamore.

Diameter: 2–4' (0.6–1.2 m), sometimes much larger
Height: 60–100' (18–30 m)

■ DENSE POPULATION
■ LIGHT POPULATION
■ NATURALIZED

Also known as the Western Sycamore, this large tree is common in the valleys of California, where the barrel-shaped trunks often lean and fork into picturesque shapes. It is also a common ornamental and shade tree. The mottled bark and coarse, light green foliage are distinctive. Individual specimens have measured 116' (35.4 m) high, 27' (8.2 m) in trunk circumference, and 158' (48.2 m) in crown spread.

DESCRIPTION: This tree has an enlarged base, a stout trunk often branched near the base, and a broad, irregular, open crown of thick, spreading branches. Its leaves are 6–9" (15–23 cm) long and wide, slightly star-shaped, and deeply divided about halfway to the base into five (sometimes three) narrow, long-pointed lobes. The leaves have wavy edges with few large teeth, with five (sometimes three) main veins from the notched or blunt base. They are light green above and paler and hairy beneath. The leafstalk is long and stout, covering the side bud at the enlarged base. The bark is whitish, smooth, and thin on the branches; peeling in brownish flakes and mottled on the trunk, becoming dark gray or brown; and rough, thick, deeply furrowed at the base. The twigs are slender and light brown with ring scars at the nodes; they are hairy when young.

FLOWERS: In spring with leaves. Tiny male and female flowers are borne on separate, ball-like clusters of two to seven.

FRUIT: ⅞" (22 m) in diameter; two to seven balls or heads hanging on long stalks, composed of many narrow nutlets with tufts of hairs at the base; maturing in fall, and separating in winter.

HABITAT: Wet soils of stream banks in valleys, foothills, and mountains.

RANGE: California and northern Baja California; to 4000' (1219 m).

USES: This attractive species is commonly cultivated as an ornamental or shade tree in residential and commercial landscapes.

SIMILAR SPECIES: American Sycamore.

CONSERVATION STATUS:
This species has a stable population with no major threats.

Diameter: 2–4' (0.6–1.2 m), sometimes much larger
Height: 40–80' (12–24 m)

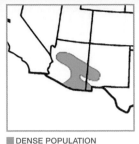

■ DENSE POPULATION
■ LIGHT POPULATION
■ NATURALIZED

This is one of the largest deciduous trees in the southwestern United States and is valuable in preventing erosion along stream banks. These large trees with their spreading whitish branches and huge, mottled trunks are a common and conspicuous sight along desert valleys and canyons. This species is especially common in Sycamore Canyon near Williams in northern Arizona. Woodpeckers and other desert birds nest in the hollow trunks of old Arizona Sycamore trees.

ALTERNATE NAME: *Platanus racemosa wrightii*

DESCRIPTION: This large, handsome tree has a stout trunk, often branched near the base, and a broad, irregular, open crown of thick, spreading branches. Its leaves are 6–9" (15–23 cm) long and wide, slightly star-shaped, deeply divided into three or five (sometimes seven) narrow, long-pointed lobes, usually without teeth, generally five main veins from deeply notched base; they are light green above, paler and hairy beneath. The bark is whitish and smooth, thin on the branches, peeling in brownish flakes and mottled on the trunk, and becoming dark gray, rough, thick, and deeply and irregularly furrowed at the base. The twigs are slender and light brown with ring scars at the nodes; they are hairy when young.

FLOWERS: Spring. Tiny male and female flowers in (usually) two to four separate, ball-like clusters.

FRUIT: ¾–1" (2–2.5 cm) in diameter, two to four balls or heads hanging on long stalks, composed of many narrow nutlets with tufts of hairs at the base; maturing in fall, and separating in winter.

HABITAT: Wet soils along streams and canyons in foothills and mountains, deserts, desert grasslands, and with oaks.

RANGE: Southwest New Mexico, Arizona, and northwestern Mexico; at 2000–6000' (610–1829 m).

USES: This species is planted as an ornamental and shade tree in urban parks and residential landscaping.

SIMILAR SPECIES: California Sycamore, American Sycamore.

CONSERVATION STATUS: LC
The population is stable with no major threats.

Diameter: 2–4' (0.6–1.2 m)
Height: 40–80' (12–24 m)

SWEETGUM *Liquidambar styraciflua*

■ DENSE POPULATION
■ LIGHT POPULATION
■ NATURALIZED

This large, fragrant tree is native to the southeastern United States. It is a popular ornamental tree as well as an important timber tree, its reddish-brown wood used for furniture, cabinet-work, and other items. In the past, a gum was obtained from the trunks by peeling the bark and scraping off the resinlike solid. This gum was used medicinally as well as for chewing gum. Commercial storax, a fragrant resin used in perfumes and medicines, is from the related Oriental Sweetgum (*Liquidambar orientalis*) of western Asia.

DESCRIPTION: Sweetgum is a large tree with a straight trunk and a conical crown that becomes round and spreading. Its star-shaped leaves are 3–6" (7.5–15 cm) long and wide and maple-like, with five (sometimes seven) long-pointed, finely saw-toothed lobes and five main veins from the notched base. They have a resinous odor when crushed. The leafstalks are slender, nearly as long as the blades, and shiny dark green above, turning reddish in fall. The bark is gray and deeply furrowed into narrow scaly ridges. The twigs are green to brown and stout, often forming corky wings.

FLOWERS: The flowers are tiny and occur in greenish, ball-like clusters in spring, with male flowers in several clusters along a stalk and female flowers in a drooping cluster on the same tree.

FRUIT: 1–1¼" (2.5–3 cm) in diameter; a long-stalked, drooping brown ball composed of many individual fruits, each ending in two long, curved, prickly points and each with one or two long-winged seeds; maturing in fall and persistent into winter.

HABITAT: Moist soils of valleys and lower slopes; it is found in mixed woodlands. This tree is often a pioneer after logging or clearing, and in old fields left to regenerate.

RANGE: Extreme southwestern Connecticut south to central Florida, west to eastern Texas, and north to southern Illinois; also a variety is found in eastern Mexico. It occurs to 3000' (914 m) in the southern Appalachians.

USES: This species is an important timber tree, second in production only to oaks among hardwoods. Its wood is used for furniture, cabinetwork, plywood, barrels, and boxes. It is also a popular ornamental tree.

SIMILAR SPECIES: Japanese Maple, American Sycamore.

CONSERVATION STATUS:
The population is widespread with no major threats.

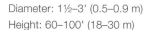
Diameter: 1½–3' (0.5–0.9 m)
Height: 60–100' (18–30 m)

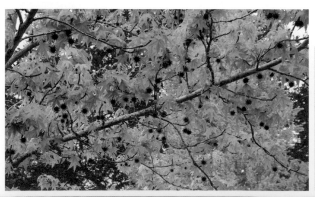

- ■ DENSE POPULATION
- ■ LIGHT POPULATION
- ■ NATURALIZED

This shrub is native to eastern North America and one of the few that flowers in late autumn and winter. The fragrant extract of Witch-hazel leaves, twigs, and bark is used in mildly astringent lotions and toilet water. It was formerly a common belief, still held by some, that a forked branch of this shrub could be used to locate underground water.

DESCRIPTION: A slightly aromatic shrub or small tree, Witch-hazel has a broad, open crown of spreading branches and small yellow flowers that are present in fall or winter. Its leaves are 3–5" (7.5–13 cm) long, 2–3" (5–7.5 cm) wide, broadly elliptical, pointed or rounded at the tip, blunt to notched and unequal at the base, broadest and wavy-lobed beyond the middle, with five to seven straight veins on each side. The leaves are hairy when young. They are dull dark green above and paler below, turning yellow in fall. The bark is light brown and smooth or scaly. Twigs are slender and zigzagging, with gray or rust-colored hairs.

Diameter: 4–8" (10–20 cm)
Height: 20–30' (6–9 m)

FLOWERS: Fall–winter. The flowers are threadlike and twisted with four bright yellow petals, 1" (2.5 cm) long, and produced in clusters along the leafless twigs.

FRUIT: ½" (12 mm) long; a hard elliptical capsule ending in four sharp curved points, light brown, opening in two parts, maturing in fall, with one or two shiny blackish seeds ¼" (6 mm) long, and ejected with force by contracting capsule walls.

HABITAT: Moist soil in understory of hardwood forests.

RANGE: Southern Ontario east to Nova Scotia, south to central Florida, west to eastern Texas, and north to central Wisconsin; local in northeastern Mexico. It occurs to 5000' (1524 m), sometimes higher in the southern Appalachians.

USES: Native Americans, and later European settlers, boiled leaves, twigs, and bark from this plant to produce an extract used to treat a variety of ailments. Forked twigs were, and still are, cut and used as divining rods.

SIMILAR SPECIES: Ozark Witch-hazel (*Hamamelis vernalis*), a species native to the Ozark Plateau in central North America, which is usually a large shrub.

CONSERVATION STATUS: 🄻🄲
The species is widespread and abundant throughout its range. Overharvesting for use of the species in cosmetics and for medicinal purposes poses a minor threat.

This species, native to China and Japan, is introduced in North America. The symmetrical canopy, heart-shaped leaves, and showy, redolent autumn foliage make Katsura Tree an attractive ornamental. The strong, sweet smell given off by dying leaves in fall and during droughts can dominate the tree's surroundings.

DESCRIPTION: This is a small to medium-sized tree with deciduous, simple leaves. It has a straight trunk, sometimes multi-stemmed, and a pyramidal or ovoid crown of densely spaced, upright branches spreading 20–30' (6–9 m) wide. The new leaves are reddish purple; the shaggy brown bark peels and curls in thin strips. In fall the brilliant amber-colored, heart-shaped leaves have a spicy, burnt-sugar aroma. The leaves are 2–4" (5–10 cm) long and wide and heart-shaped with bluntly toothed margins. They are dark green above and beneath. New leaves are reddish purple.

FLOWERS: Spring. ½" (1.3 cm) wide, green, tuftlike flowers that lack petals.

FRUIT: Fall. A ½–¾" (1.3–2 cm) long, podlike follicle; it is green, ripening to brown, appearing in clusters of two to four.

HABITAT: Introduced in North America; planted in parks and residential properties.

RANGE: Native to China and Japan; the species is widely introduced in North America.

USES: This attractive tree is cultivated as an ornamental. It makes an excellent shade tree.

SIMILAR SPECIES: This species, native to Asia and one of only two in its genus, is not likely to be confused with any other species in North America.

CONSERVATION STATUS:
The conservation status of this species has not been fully assessed.

Height: 40–60' (12–18 m)

■ DENSE POPULATION
■ LIGHT POPULATION
■ NATURALIZED

This small, flowering tree (or shrub) is native to southwestern Texas and northern Mexico. The roots have been used to treat rheumatism and other conditions. This plant is also known as Texas Guaiacum and Soapbush. The genus name is sometimes called *Guajacum*.

DESCRIPTION: A small tree or shrub with evergreen, pinnately compound leaves. It has a short trunk and a clumplike, dense crown. In spring, this plant has distinctive, purple-blue, star-shaped flowers set against dark green, feathery leaves. The exposed, large, red seeds are conspicuous in fall. It has three leaflets, each 2–4" (5–10 cm) long, with finely toothed margins. They are bright green above and hairy beneath.

FLOWERS: Blue to purple flowers, about ½–¾" (1–2 cm) in diameter, with five sepals, five petals, and ten stamens. Blooms appear after it rains; the blooming period lasts from March to September.

FRUIT: A 1–2" (2.5–5 cm) long, inflated, papery capsule with three lobes; it is green to brown. Fruit appears in late summer and persists into winter.

HABITAT: Moist soils along waterways, usually in hardwood forests.

RANGE: Central and southern Texas at elevations to 2000' (610 m).

USES: This plant is cultivated as an ornamental and the hard, dense wood is used for fence posts, tool handles, and firewood. The roots are used in soaps and for medicinal purposes.

SIMILAR SPECIES: Creosote Bush.

CONSERVATION STATUS: **LC**
The wild population of this plant is thought to be secure, but it has not been fully assessed.

Height: 23' (7 m)

■ DENSE POPULATION
■ LIGHT POPULATION
■ NATURALIZED

Creosote Bush is a prominent plant in western North America, and a species characteristic of hot deserts. Its strong fragrance fills the air following rains. Many "bunches" of plants that occur together are actually clones formed from the old crowns of an older plant. The foliage of Creosote Bush hides species of grasshoppers, praying mantids, and crickets that occur only on this plant. A fly, the Creosote Gall Midge (*Asphondylia* spp.), may cause numerous leafy galls on the plant.

DESCRIPTION: This is a medium to large evergreen shrub with numerous flexible stems usually arising from the base at an angle. It has two opposite leaflets, united at the base, pointed, yellowish-green, resinous, and pungent especially when wet; they are ¼–½" (6–13 mm) long.

FLOWERS: February–August, occasionally year-round. The flowers are yellow and 1" (2.5 cm) wide, with five sepals, twisted like the blades of a fan.

FRUIT: Globose, densely woolly, white to reddish flowers exhibiting a persistent style.

HABITAT: On well-drained plains and slopes, especially those with a caliche layer; to 4000' (1220 m). It is often the most abundant shrub, frequently forming pure stands.

RANGE: Southern Nevada, extreme southwest Utah, southeastern California, southern Arizona, southern New Mexico north along major river valleys to Albuquerque, western Texas, and south into Mexico.

USES: Native Americans used this species to treat a variety of ailments, and it is still used as a traditional herbal medicine, although its use is not endorsed by the mainstream medical community due to potential toxicity to the liver and kidneys.

SIMILAR SPECIES: Texas Lignum-vitae. Other similar, closely related species exist in South America.

CONSERVATION STATUS: 🅛🅒

The population of this species has increased in recent years due to better land management techniques. Overharvesting of this plant, collected for use in traditional herbal medicine, as well as wildfires, drought, and urbanization place pressure on this species.

Height: usually less than 4' (120 cm) but can reach 12' (360 cm) in well-watered situations

■ DENSE POPULATION
■ LIGHT POPULATION
■ NATURALIZED

Sweet Acacia occurs as a spiny shrub or small tree and is found in many parts of the world. In North America this species is found in the southwestern and southeastern United States. This species is extensively planted in southern Europe for its flowers, which are used in making perfume. After drying in the shade, the pleasantly fragrant flowers can be used as sachets to keep clothes smelling fresh. Livestock may eat the tender leaves and pods; it is also a honey plant.

ALTERNATE NAME: *Acacia smallii*

DESCRIPTION: This is a spiny, much-branched shrub or small tree with a widely spreading, flattened crown and fragrant yellow balls of tiny flowers. Its leaves are alternate or clustered, bipinnately compound 2–4" (5–10 cm) long, usually with three to five pairs of side axes. It has 10–20 pairs of leaflets ⅛–¼" (3–6 mm) long, oblong, mostly hairless, stalkless, and gray-green. The bark is grayish-brown, thin, and smooth or scaly. The twigs are slightly zigzagged, slender, and covered with fine hairs when young, with straight, slender, paired white spines at the nodes.

FLOWERS: The ³⁄₁₆" (5 mm) long, yellow or orange flowers are fragrant and include many tiny stamens clustered in stalked balls ½" (12 mm) in diameter; flowering mainly occurs in late winter and early spring.

FRUIT: The fruit is a dark brown or black, hard, cylindrical pod 1½–3" (4–7.5 cm) long, ⅜–½" (10–12 mm) in diameter, short-pointed at the ends; maturing in summer and remaining attached, often opening late, with many elliptical, flattened, shiny, brown seeds.

HABITAT: Sandy and clay soils, especially in open areas and along borders of woodlands, and roadsides.

RANGE: Southern Texas and local in southern Arizona; it is also found in Mexico. It is cultivated and naturalized from Florida west to Texas and southern California. It occurs to 5000' (1524 m).

USES: The flowers of this plant are used as a perfume ingredient. Mucilage is produced from the gum of the trunk, and tannin and dye from the pods and bark.

SIMILAR SPECIES: Catclaw Acacia.

CONSERVATION STATUS: LC
The population is secure. This species is widely distributed in tropical America and spread by cultivation and naturalization.

Diameter: 4" (10 cm)
Height: 16' (5 m)

■ DENSE POPULATION
■ LIGHT POPULATION
■ NATURALIZED

Catclaw Acacia is so named because its sharp, stout, hooked spines grab onto passersby and can tear through clothing and flesh. Its colloquial name "Wait-a-Minute Bush" suggests the need for travelers to wait a minute to remove the menacing spines. Catclaw honey (also known as Uvalde honey, from the Texas county of that name) is from the flowers of this and related species. The Latin name honors Josiah Gregg, an explorer and amateur naturalist of the American Southwest.

ALTERNATE NAME: *Senegalia greggii*

DESCRIPTION: This is a spiny, much-branched, thicket-forming shrub, or occasionally a small tree with a broad crown. Its leaves are clustered, bipinnately compound, and 1–3" (2.5–7.5 cm) long; the slender axis usually has two or three pairs of side axes. Three to seven pairs of leaflets are ⅛–¼" (3–6 mm) long, oblong, rounded at the ends, thick, usually hairy, almost stalkless, and dull green. The bark is gray and thin, becoming deeply furrowed. The twigs are brown, slender, angled, and covered with fine hairs, with many scattered, stout spines ¼" (6 mm) long, hooked or curved backward.

FLOWERS: Early spring–summer. ¼" (6 mm) long, light yellow, fragrant, and stalkless, including many tiny stamens in long, narrow clusters 1–2" (2.5–5 cm) long.

FRUIT: The fruit is a 2½–5" (6–13 cm) long and ½–¾" (12–19 mm) wide, thin, flat, ribbonlike, oblong pod that is brown, curved, twisted, and often narrowed between the seeds, maturing in summer and shedding in winter, remaining closed, with several beanlike, nearly round, flat, brown seeds.

CAUTION: Poisonous: Do not eat or ingest any part of this species. If accidentally ingested, contact your doctor or poison control center.

HABITAT: Along streams, in canyons, and on dry, rocky slopes of plains and foothills.

RANGE: Central, southern, and Trans-Pecos Texas and northeastern Mexico, west to southern California and southern Nevada; to 2000' (610 m).

USES: The wood from this species is used locally for souvenirs, tool handles, and fuel. Honey is made from the flowers and Native Americans made meal from the seeds.

SIMILAR SPECIES: Sweet Acacia.

CONSERVATION STATUS: LC
The population is thought to be secure, but the status of the wild population has not been fully assessed.

Diameter: 6" (15 cm)
Height: 20' (6 m)

■ DENSE POPULATION
■ LIGHT POPULATION
■ NATURALIZED

Silktree is a native of southwestern and eastern Asia, where it is known by many common names including Persian Silktree or Pink Siris. It is sometimes called Mimosa Tree because the flowers are similar to those of the herbaceous sensitive-plants in the genus *Mimosa*. The hardiest tree of its genus, Silktree has an unusually long flowering period. Silktree leaflets slowly fold downward at night or during heavy rain, resulting in the colloquial nickname "Sleeping Tree."

DESCRIPTION: This small ornamental tree has a short trunk or several trunks, and a broad, flattened crown of spreading branches with showy pink flower clusters. Its leaves are bipinnately compound, 6–15" (15–38 cm) long, and fernlike, with five to twelve pairs of side axes covered with fine hairs. Each axis has 15–30 pairs of oblong pale green leaflets, ⅜–⅝" (10–15 mm) long. The bark is blackish or gray and nearly smooth. The twigs are brown or gray and are often angled.

Diameter: 8" (20 cm)
Height: 20' (6 m)

FLOWERS: Throughout summer. Flowers are more than 1" (2.5 cm) long with long, threadlike pink stamens that are whitish toward the base, crowded in long-stalked ball-like clusters 1½–2" (4–5 cm) wide, and grouped at ends of twigs.

FRUIT: The fruit is a 5–8" (13–29 cm) long, flat-pointed, oblong, yellow-brown pod; maturing in summer, remaining closed, with several beanlike, flattened, shiny, brown seeds.

HABITAT: Open areas including wasteland and dry gravelly soils.

RANGE: Native from Iran to China; it is naturalized from Maryland to southern Florida, west to eastern Texas, and north to Indiana; to 2000' (610 m).

USES: This species was once widely planted in the United States as an ornamental tree, however, it is no longer recommended due to its invasive tendencies and susceptibility to disease.

SIMILAR SPECIES: Woman's Tongue (*Albizia lebbeck*), a related cultivated species native to Southeast Asia and Australia that has naturalized in many areas.

CONSERVATION STATUS:
The conservation status of this species in the wild has not been fully assessed. It is exotic in North America and often considered an invasive pest.

EASTERN REDBUD *Cercis canadensis*

- ■ DENSE POPULATION
- ■ LIGHT POPULATION
- ■ NATURALIZED

This tree brightens eastern landscapes in early spring with its showy pink flowers. Eastern Redbud is found in moist valleys and woodlands and is also planted as an ornamental. Although most widespread in the East, it can thrive as far west as California. It is the state tree of Oklahoma. The flowers can be eaten raw as a salad, boiled, or fried. Green twigs from the tree are sometimes used as seasoning for wild game.

DESCRIPTION: Eastern Redbud is a tree with a short trunk, a rounded crown of spreading branches, and pink flowers that cover the twigs in spring. Its leaves are 2½–4½" (6–11 cm) long and broad, heart-shaped with broad short points and without teeth, with five to nine main veins. The leaves are dull green above, paler and sometimes hairy beneath, turning yellow in autumn. The bark is dark gray or brown and smooth, becoming furrowed into scaly plates. The twigs are brown, slender, and angled.

FLOWERS: Early spring before leaves. ½" (12 mm) long and pea-shaped with five slightly unequal purplish-pink petals, rarely white. Four to eight flowers appear in a cluster on slender stalks.

FRUIT: 2½–3¼" (6–8 cm) long, flat, narrowly oblong pods that are pointed at the ends. It is pink, turning blackish, splitting open on one edge, falling in late autumn or winter, with several beanlike, flat, elliptical, dark brown seeds.

HABITAT: Moist soils of valleys and slopes and in hardwood forests.

RANGE: New Jersey south to central Florida, west to southern Texas, and north to southeastern Nebraska; also Mexico; to 2200' (671 m).

USES: Native Americans ate the flowers and seeds of this tree, and used an astringent from the bark to make medicine for treating a variety of ailments. It is planted as an ornamental for its showy flowers.

SIMILAR SPECIES: Judas Tree (*Cercis siliquastrum*), a native of western Asia and southern Europe.

CONSERVATION STATUS: LC

The global population of this tree is thought to be secure, although its status has not been fully assessed in several U.S. states. It may be imperiled at the state level in some Atlantic Coast states, and it is presumed extirpated in Ontario.

Diameter: 8" (20 cm)
Height: 40' (12 m)

■ DENSE POPULATION
■ LIGHT POPULATION
■ NATURALIZED

Also called Western Redbud, this large, flowering shrub is common in Grand Canyon National Park and is a handsome ornamental with showy flowers that cover the twigs in early spring. Native Americans used to make bows from the wood.

ALTERNATE NAME: *Cercis orbiculata*

DESCRIPTION: California Redbud is a large shrub or small tree with a rounded crown of many spreading branches. Its leaves are 1½–3½" (4–9 cm) long and broad, nearly round, and long-stalked, with seven to nine veins from a notched base, thickened, mostly hairless. They are dark green above and paler beneath. The bark is gray, smooth, and becoming fissured. The twigs are reddish-brown when young, becoming dark gray, and hairless.

FLOWERS: Early spring, before leaves. ½" (12 mm) long and pealike, with five slightly unequal purplish-pink petals, rarely white. They appear in clusters on slender stalks, scattered along twigs.

FRUIT: 2–3½" (5–9 cm) long, narrowly oblong, flat, thin pods that are brown or purplish, maturing in late summer and splitting open on one edge. Many hang in clusters along twigs, with several beanlike seeds.

HABITAT: Canyons and slopes of foothills and mountains.

RANGE: Northern California east to southern Utah and south to southern Arizona; at 500–6000' (152–1829 m).

USES: This species is planted as an ornamental and for wildlife habitat. It is useful as a soil stabilizer along watercourses and can withstand periodic flooding.

SIMILAR SPECIES: Eastern Redbud, Texas Redbud.

CONSERVATION STATUS:
The conservation status of this species in the wild has not been fully assessed. The species is susceptible to a few fungal pathogens that can cause foliage or flower blight and root rot, but these infections are usually not enough to kill mature trees. Periodic fires control these pathogens and encourage new growth in the tree, which resprouts vigorously after a fire.

Diameter: 4" (10 cm)
Height: 16' (5 m)

■ DENSE POPULATION
■ LIGHT POPULATION
■ NATURALIZED

This handsome ornamental tree is native to the southeastern United States, where it occurs in scattered patches in Great Smoky Mountains National Park and in several of the surrounding states. Charles Sprague Sargent (1841–1927), author of the classic 14-volume *Silva of North America*, called this rare species one of the most beautiful flowering trees of the American forests. The clear yellow heartwood, which turns light brown on exposure, has been used as a source of a yellow dye. The genus name is from Greek words meaning "branch" and "brittle."

ALTERNATE NAME: *Cladrastis lutea*

DESCRIPTION: This is a medium-sized tree with a short trunk and a broad, rounded crown of spreading branches. Its leaves are pinnately compound and 8–12" (20–30 cm) long, with the base of the leafstalk enclosing the bud. Its 7–11 leaflets are 2½–4" (6–10 cm) long, 1¼–2" (3–5 cm) wide, elliptical, nearly in pairs except at end, short-stalked, not toothed, becoming hairless. They are shiny green above and paler beneath, turning yellow in autumn. The bark is gray; smooth, and thin. The twigs are brown, slender, zigzag, and hairless.

FLOWERS: Late spring. 1–1¼" (2.5–3 cm) long, pea-shaped, fragrant flowers with five white petals in drooping clusters about 1' (30 cm) long.

FRUIT: 2–3¼" (5–8 cm) long, flat, narrowly oblong pod hanging in clusters, maturing and falling in early autumn before splitting open, with two to six beanlike seeds.

HABITAT: Moist soils, especially limestone cliffs, stream banks, and rich rocky coves (or deep valleys of mountains); in hardwood forests.

RANGE: Southwestern Virginia, western North Carolina, and northeastern Georgia west to eastern Oklahoma; it also is local north to southern Indiana. It occurs at 300–3500' (91–1067 m).

USES: Yellowwood is commonly grown as an ornamental tree and its wood has been used to make furniture and gunstocks.

SIMILAR SPECIES: This species is the only one in its genus to occur in North America and is unlikely to be confused with any other native tree.

CONSERVATION STATUS:
The global population is probably secure, but unsustainable forest management practices and various diseases and pests may threaten the species. It is state-listed as threatened in parts of its area of occupancy; its protected status varies by state.

Diameter: 1½' (0.5 m)
Height: 50' (15 m)

■ DENSE POPULATION
■ LIGHT POPULATION
■ NATURALIZED

Western Coral Bean is found on rocky hillsides in the southwestern United States and northwestern Mexico. Although handsome in flower, this plant is leafless and unattractive most of the year.

DESCRIPTION: This is a shrub or small tree, leafless much of the year, with prickly stems and leafstalks, and long, bright red flowers in racemes near ends of the branches. The leaves are compound, with three leaflets, each to 3" (7.5 cm) long.

FLOWERS: March–May, also September. The flowers are up to 3" (7.5 cm) long and basically "pea-like," but modified so that the upper petal (banner) is straight and beak-like, nearly hiding four lower petals. The calyx is waxy white, nearly without lobes.

FRUIT: Pods with several large, scarlet, bean-like seeds.

CAUTION: Poisonous: Do not eat or ingest any part of this species. If accidentally ingested, contact your doctor or poison control center.

HABITAT: Rocky hillsides among oaks, juniper, and brush.

RANGE: Southern Arizona and southwestern New Mexico to northwestern Mexico.

USES: The red, bean-like seeds, although poisonous if ingested, have been used to make jewelry, particularly necklaces.

SIMILAR SPECIES: Coral Bean.

CONSERVATION STATUS: LC
This species has a widespread, stable population. No major threats have been identified.

Height: to 15' (4.5 m), but usually much shorter

■ DENSE POPULATION
■ LIGHT POPULATION
■ NATURALIZED

Coral Bean is found throughout the southeastern United States and is planted for its showy flowers and seeds, although the brittle branches are subject to damage by windstorms. This unusual tropical tree extends its range northward as a shrub or perennial herb, but is killed back to the ground each winter.

DESCRIPTION: This is a spiny shrub with many slender stems, sometimes a small tree with a crooked trunk, spreading brittle branches, and a rounded crown. Its leaves are bipinnately compound and 6–8" (15–20 cm) long with slender stalks, sometimes prickly, with three leaflets, 1½–3" (4–7.5 cm) long and almost as wide, sometimes larger. They are triangular or slightly three-lobed, not toothed. The midvein is often prickly beneath. They are light green. The bark is light gray and smooth, becoming thick and furrowed, and sometimes spiny. The twigs are light green, stout, and brittle, with scattered short, curved spines or prickles.

FLOWERS: 2" (5 cm) long, narrow, showy flowers with a dark red tubular calyx and five narrow unequal red or scarlet petals, with many arranged in upright long-stalked clusters 8–12" (20–30 cm) long.

FRUIT: 4–8" (10–20 cm) long, dark brown or black cylindrical pod, long-pointed at the ends, narrowed between seeds, maturing in late summer and opening along one edge, with several beanlike, shiny red, poisonous seeds.

CAUTION: Poisonous: Do not eat or ingest any part of this species. If accidentally ingested, contact your doctor or poison control center.

HABITAT: Moist, sandy soils.

RANGE: North Carolina through Florida, including the Florida Keys, west to Texas; to 500' (152 m).

USES: This species is occasionally grown as an ornamental. It is a good shrub to plant to attract hummingbirds. The beanlike seeds are used to make necklaces. The toxic seeds have been used for poisoning rats and fish.

SIMILAR SPECIES: Western Coral Bean.

CONSERVATION STATUS:
This species has a widespread, stable population. No major threats have been identified.

Diameter: 8" (20 cm)
Height: 20' (6 m)

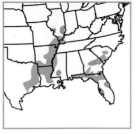

■ DENSE POPULATION
■ LIGHT POPULATION
■ NATURALIZED

This swamp tree is found in wet areas of the southeastern United States. Both the common and scientific names of this tree refer to its association with wet habitats. The genus name honors Johann Gottlieb Gleditsch (1714–1786), director of the Berlin Botanical Gardens and a pioneer in researching plant reproduction. The species is also called Swamp Locust.

DESCRIPTION: This is a medium-sized, spiny tree with a short trunk and a broad, flattened crown of spreading branches. Its leaves are pinnately and bipinnately compound, 4–8" (10–20 cm) long, the axis often with three or four pairs of side axes or forks. Numerous paired leaflets are ½–¼" (1.2–3.2 cm) long and oblong, with finely wavy edges, nearly hairless, stalkless, shiny dark green above, and dull yellow-green beneath. The bark is gray or brown, thin, smooth, and fissured into small scaly plates, with large branched spines. The twigs are brown, usually with stout, shiny brown spines to 4" (10 cm) long, often slightly flattened and curved, and mostly unbranched.

FLOWERS: Late spring. ¼" (6 mm) wide, greenish-yellow, bell-shaped, with five spreading petals, covered with fine hairs, occurring in short narrow clusters at the leaf bases. Male and female flowers are usually on separate twigs or trees.

FRUIT: The fruit is a 1–2" (2.5–5 cm) long, ¾" (19 mm) wide, flat, elliptical, shiny brown pod with thin papery walls and without

pulp, appearing in drooping clusters, maturing in late summer, opening late, shedding in fall, and usually with one rounded, flattened, brown seed.

HABITAT: Wet soils of riverbanks, flood plains, and swamps, especially where submerged for long periods; it is found in flood-plain forests.

RANGE: South Carolina to central Florida, west to eastern Texas, and north to southern Illinois and extreme southwestern Indiana; to 400' (122 m).

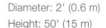

USES: The durable wood from this species is often used for making fence posts.

SIMILAR SPECIES: Honey Locust.

CONSERVATION STATUS: LC
The population is widespread and stable. No major threats have been identified.

Diameter: 2' (0.6 m)
Height: 50' (15 m)

- ■ DENSE POPULATION
- ■ LIGHT POPULATION
- ■ NATURALIZED

Native to the moist river valleys of the central United States, this tree has been introduced throughout the world, and in many regions, it has become an aggressive, invasive pest. Honey Locust is easily recognized by the large, branched spines on the trunk; thornless forms, however, are common in cultivation and are sometimes found in the wild. This hardy tree is popular for shade, hedges, and attracting wildlife. Livestock and many forms of wildlife consume the honeylike, sweet pulp of the tree's pods.

DESCRIPTION: This large, spiny tree has an open, flattened crown of spreading branches. Its leaves are pinnately and bipinnately compound, 4–8" (10–20 cm) long, the axis often with three to six pairs of side axes or forks. It has many oblong leaflets ⅜–1¼" (1–3 cm) long, paired and stalkless, with finely wavy edges. They are shiny dark green above, dull yellow-green, and nearly hairless beneath, turning yellow in fall. The bark is gray-brown or black and fissured in long, narrow, scaly ridges, with stout brown spines, usually branched, sometimes 8" (20 cm) long, with three or more points. The twigs are shiny brown, stout, and zigzag, with long spines.

FLOWERS: Late spring. ⅜" (10 mm) wide and bell-shaped, with five greenish-yellow petals, and covered with fine hairs, appearing in short, narrow clusters at the leaf bases; usually males and females are on separate twigs or trees.

FRUIT: 6–16" (15–41 cm) long, 1¼" (3 cm) wide, flat, dark brown pod, hairy, slightly curved and twisted, and thick-walled, shedding unopened in late autumn, with many beanlike, flattened, dark brown seeds in sweetish, edible pulp.

HABITAT: Moist soils of river flood plains in mixed forests, sometimes on dry upland limestone hills; it is also found in waste places.

RANGE: Extreme southern Ontario to central Pennsylvania, south to northwestern Florida, west to southeastern Texas, and north to southeastern South Dakota; naturalized farther eastward. It occurs to 2000' (610 m).

USES: This hardy species is a popular ornamental tree. It produces quality, durable wood that is sometimes used to make furniture and fence posts. The thorns have been used as pins.

SIMILAR SPECIES: Water Locust, Kentucky Coffeetree.

CONSERVATION STATUS: ⓛⓒ
This species has a widespread and abundant population with no major threats.

Diameter: 2½' (0.8 m)
Height: 80' (24 m)

■ DENSE POPULATION
■ LIGHT POPULATION
■ NATURALIZED

Although rare and scattered in the wild, Kentucky Coffeetree is popularly planted as an ornamental for its large leaves and stout twigs, which are bare except in summer. As the leaves develop late in spring and shed early, the leafless trees often appear to be dead. The genus name, from Greek, means "naked branch." The roasted seeds from this species were once used as a coffee substitute; raw seeds, however, are poisonous.

DESCRIPTION: Kentucky Coffeetree is usually a short-trunked tree with a narrow, open crown of coarse branches and very large, twice-compound leaves. Those leaves are bipinnately compound, 12–30" (30–76 cm) long, and the axis has three to eight pairs of side axes or forks. The upper axis has six to fourteen mostly paired leaflets (sometimes one at the end) 1–3" (2.5–7.5 cm) long, ¾–2" (2–5 cm) wide, ovate, without teeth, pink when unfolding, becoming nearly hairless, dull green above, paler beneath, turning yellow in fall. The bark is gray, thick, deeply furrowed into narrow scaly ridges often projecting toward one side. The twigs are brown, few, stout, and hairy when young, with thick brown pith.

FLOWERS: ⅝–¾" (15–19 cm) long and 1½–2" (4–5 cm) wide.

FRUIT: A dark red-brown pod, thick-walled and hanging on stout stalks, and falling unopened in winter. It contains several beanlike seeds that are ¾" (19 mm) in diameter, rounded, shiny dark brown, and thick-walled.

CAUTION: The seeds and pods are toxic unless properly cooked. Do not eat or ingest any part of this species. If accidentally ingested, contact your doctor or poison control center.

HABITAT: Moist soils of valleys with other hardwoods.

RANGE: Extreme southern Ontario east to central New York, southwest to Oklahoma, and north to southern Minnesota; naturalized eastward; at 300–2000' (91–610 m).

USES: This attractive tree is cultivated as an ornamental. It makes a great shade tree. The roasted seeds were once used as a coffee substitute and the wood is sometimes used for cabinetwork.

SIMILAR SPECIES: Common Honey Locust.

CONSERVATION STATUS: ⓛⓒ
Kentucky Coffeetree is widely distributed and its population is stable, but it is uncommon to rare in the wild. The species is threatened in Canada, where a small, scattered population is primarily found in southwestern Ontario.

Diameter: 2' (0.6 m)
Height: 70' (21 m)

■ DENSE POPULATION
■ LIGHT POPULATION
■ NATURALIZED

Lead Tree is a small, fast-growing tree that has been widely cultivated around the world. It can be highly invasive and has become considered an invasive weed in many parts of the world, but it continues to be widely planted for forage, reforestation, and a number of other uses. It forms dense thickets that can exclude native plants; once established it is very difficult to remove.

DESCRIPTION: This is a broad, multi-stemmed, thornless shrub or small tree with a spreading crown. Its leaves are 12" (30 cm) long and feather-like, with many oblong ⅝" (15 mm) leaflets. The bark is dark brown, ridged, and scaly.

FLOWERS: Year-round. Minute, white to yellowish flowers in pincushion-like ¾" (2 cm) heads.

FRUIT: 6" (15 cm) long, flat, brown pods.

HABITAT: Hammocks, both planted and natural forests, coastal strands, scrub, croplands, disturbed areas.

RANGE: A native of Central America; it is introduced in the United States and naturalized in Georgia, Florida, Texas, and Arizona.

USES: This introduced species is widely cultivated for lumber, pulpwood, livestock feed, and reforestation. It is effective for nitrogen fixation and may be useful as biomass for biofuel production.

SIMILAR SPECIES: Bahama Lysiloma (Wild Tamarind).

CONSERVATION STATUS:
The conservation status of this species in the wild has not been fully assessed. This fast-growing tree can be highly invasive; it is introduced and has escaped in the southern tier of the U.S., especially in and near urban areas.

Height: to 36' (10.8 m)

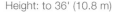

- ■ DENSE POPULATION
- ■ LIGHT POPULATION
- ■ NATURALIZED

This native of far southern Florida is commonly grown as an ornamental tree in its natural range for its delicate foliage and fragrant flowers. It prefers a sunny location, average to dry conditions, and can tolerate saline soils. It is often used as a street tree.

ALTERNATE NAMES: False Tamarind, Bahama Lysiloma

DESCRIPTION: This tree has an umbrella-like crown, zig-zagging branches, and feather-like leaves. Its leaves are 7" (17.5 cm) long, semi-evergreen, bipinnately compound, with many ovate to oblong, ½" (12 mm) leaflets. The bark is gray to brown and smooth to scaly.

FLOWERS: March–August. Minute flowers, in ¾" (2 cm), pincushion-like, greenish-white heads.

FRUIT: Flat, black pods to 8" (20 cm) long.

HABITAT: Found in hammocks.

RANGE: The southern tip of the Florida peninsula and the Florida Keys; also found in the Caribbean and Central America.

USES: It is cultivated as an ornamental tree. The wood is harvested and used locally.

SIMILAR SPECIES: Horseflesh Mahogany.

CONSERVATION STATUS: LC
The population is stable. The tree is locally common, especially along lake banks.

Height: 30' (9 m)

■ DENSE POPULATION
■ LIGHT POPULATION
■ NATURALIZED

This spiny tree is native to deserts of the southwestern United States and northwestern Mexico. It is known locally as Ironwood and in Spanish as *palo de hierro*. The hard, dark brown wood with thin, yellow sapwood is easily polished but dulls tools used to work with it. It is one of the heaviest native woods; only Leadwood (*Krugiodendron ferrum*), a small tropical tree of southern Florida, is heavier. The Desert Ironwood is the single species of a genus named for Stephen Thayer Olney (1812–1878), a businessman and botanist of Rhode Island.

DESCRIPTION: This is a spiny evergreen tree with a short trunk and a widely spreading, rounded, dense crown often broader than high with numerous purplish, pea-shaped flowers in late spring. Its leaves are evergreen or nearly so, densely clustered, pinnately compound, and 1–2¼" (2.5–6 cm) long. Two to ten pairs of leaflets are ¼–¾" (6–9 mm) long, oblong, generally rounded at the tip and short-pointed at the base, without teeth, thick, short-stalked, and blue-green, with fine pressed hairs. The bark is gray, smooth, and thin, becoming much fissured, scaly, and shreddy. The twigs are greenish, slender, and covered with gray hairs when young, with short, slender, straight spines paired at the nodes.

FLOWERS: Late spring with new leaves. ½" (12 mm) long and pea-shaped, with five unequal purple petals; few, in short clusters along twigs; fragrant, abundant and showy.

FRUIT: The fruit is a 2–2½" (5–6 cm) long cylindrical pod; it is short-pointed, slightly narrowed between seeds, light brown, covered with sticky hairs, and thick-walled, with one to five beanlike, shiny brown seeds; maturing in late summer, splitting in two parts.

HABITAT: Sandy and gravelly washes in rocky foothills in deserts.

RANGE: Arizona, California, and Mexico; to 2500' (762 m).

USES: The hard wood from this species is used to make knife handles, bowls, and small boxes, and also makes excellent firewood. The seeds are edible after roasting.

SIMILAR SPECIES: Catclaw Acacia, Blue Paloverde, Jerusalem Thorn.

CONSERVATION STATUS: **NT**

This species is declining despite protections. A parasitic mistletoe occurs on the branches—the mistletoe's reddish, juicy berries attract birds such as the Phainopepla (*Phainopepla nitens*); the birds in turn spread the sticky seeds of the parasite to other trees. Overharvesting for fuel and ironwood carvings—a popular tourist item—has caused additional population decline. Urbanization also threatens this species as habitat is cleared for development or degraded by recreational and tourist use. This valuable tree is known to be illegally extracted from protected reserves in both the U.S. and Mexico.

Diameter: 2' (0.6 m)
Height: 30' (9 m)

■ DENSE POPULATION
■ LIGHT POPULATION
■ NATURALIZED

Also commonly called Paloverde or Palo Verde, this species is native to the southwestern United States as well as northern Mexico. This flowering, fast-growing tree is widely used as an ornamental and hedge plant in warm regions and has been spread by humans to many places around the world. The foliage and pods have been used as emergency forage for livestock, as well as by wildlife. Bees produce fragrant honey from the flowers. The word "Jerusalem" in this tree's common name is derived from a Spanish and Portuguese word *girasol*, meaning "turning toward the sun."

DESCRIPTION: This is a spiny tree with an open, spreading crown of drooping twigs and narrow evergreen "streamers." It appears leafless most of the year. The leaves are bipinnately compound but appear pinnately compound, with a short, spine-tipped axis and one to three pairs of wiry, flattened, narrow, ever-green, drooping axes or "streamers" that are 8–20" (20–51 cm) long. It has 25–30 pairs of leaflets ¼" (6 mm) or less in length, narrowly oblong, and light green, remaining on the tree only a short time before falling. The bark of the trunk and branches is yellow-green and smooth, becoming scaly at the base of large trunks. The twigs are yellow-green, smooth, slightly zig-zag, and slender with paired, short spines at the nodes border-ing a third, larger brownish spine of the leaf axis. The twigs are finely hairy when young.

FLOWERS: ¾" (19 mm) wide with five rounded, golden-yellow petals, the largest red-spotted and turning red in withering. They appear in loose, upright clusters to 8" (20 cm) long in spring and summer or nearly continuously in tropical climates.

FRUIT: The fruit is a narrowly cylindrical, dark brown pod 2–4" (5–10 cm) long, long-pointed at the ends, narrowed between the seeds, hanging down, with one to eight beanlike seeds, maturing in summer and fall and remaining closed.

HABITAT: Moist valley soils; rare in foothills and mountain can-yons of desert and desert grassland.

RANGE: South to Trans-Pecos Texas and local in southern Ari-zona; to 4500' (1372 m). It is planted and becoming naturalized along the southern border of the United States, sometimes act-ing as a weed. It is widely distributed in tropical America.

USES: This tree is widely cultivated as an ornamental. It is an invasive species in several regions around the world, especially in Australia.

SIMILAR SPECIES: Blue Paloverde and Foothill Paloverde in southern California and Arizona.

CONSERVATION STATUS: LC
This species has a wide distribution and large population. There are no major threats.

Diameter: 1' (0.3 m)
Height: 40' (12 m)

■ DENSE POPULATION
■ LIGHT POPULATION
■ NATURALIZED

This small, spiny tree is native to the deserts of the southwestern United States and northwestern Mexico. It spends most of the year leafless; photosynthesis is performed by the blue-green branches and twigs. Blue Paloverde is important for native wildlife. The twigs, pods, and seeds are eaten by rodents and birds, and the flowers are a source of honey. The common name is derived from the Spanish *paloverde*, meaning "green tree" or "green pole."

ALTERNATE NAME: *Cercidium floridum*

DESCRIPTION: This is a small tree that is leafless most of the year, with a short blue-green trunk and a widely spreading, open crown. The leaves are few and scattered, bipinnately compound, 1" (2.5 cm) long, with a short axis forking into two side axes, with two or three pairs of leaflets on each side axis, ¼" (6 mm) long, oblong, pale blue-green, appearing in spring but soon shedding. The bark of the trunk and branches is blue-green and smooth, with the base of large trunks becoming brown and scaly. The twigs are blue-green, smooth, slightly zigzag, and hairless, with a straight, slender spine less than ¼" (6 mm) long at each node.

FLOWERS: Spring, sometimes again in late summer. ¾" (19 mm) wide with five bright yellow petals; the largest has a few red spots. Four or five flowers appear in a cluster less than 2" (5 cm) long.

FRUIT: 1½–3¼" (4–8 cm) long, narrowly oblong, flat, thin pods, short-pointed at the ends. They are yellowish-brown, maturing and falling in summer, with two to eight beanlike seeds.

HABITAT: Along washes and valleys and sometimes on lower slopes of deserts and desert grasslands.

RANGE: Central and southern Arizona, southeastern California, and northwestern Mexico; 4000' (1219 m).

USES: Native Americans used the seeds as a food source. This tree is now cultivated as an ornamental plant and for wildlife habitat.

SIMILAR SPECIES: Desert Ironwood, Foothill Paloverde, Jerusalem Thorn.

CONSERVATION STATUS: LC
The population is considered to be secure; its numbers appear to be stable.

Diameter: 1½' (0.5 m)
Height: 30' (9 m)

■ DENSE POPULATION
■ LIGHT POPULATION
■ NATURALIZED

The Spanish common name *paloverde*, meaning "green tree" or "green pole," describes the smooth, green branches and twigs of this species and its relatives. During most of the year, in the absence of leaves, the branches and twigs manufacture food. This adaptation to a desert habitat exposes less surface to the sun, aiding moisture retention. Blue Paloverde (*P. florida*), which has blue-green branches and foliage, occurs chiefly along drainages, blooms earlier, and has larger, deeper yellow flowers. Native Americans ground and ate the beanlike seeds of both species.

ALTERNATE NAME: *Cercidium microphyllum*

DESCRIPTION: This is a small, spiny tree that is leafless most of the year, with a yellow-green trunk and a wide, much-branched, open crown. Its leaves are few; they are bipinnately compound but appear as if a pinnately compound pair; they are ¾–1" (2–2.5 cm) long; consisting of a very short axis with 2 forks, each with 3–7 pairs of minute leaflets, elliptical, slightly hairy, and yellow-green, appearing in spring but soon shedding. The bark is yellow-green and smooth. The twigs are short and stiff, ending in long, straight spines about 2" (5 cm) long.

FLOWERS: Spring. About ½" (12 mm) wide with 5 pale yellow petals, the largest petals being white or cream. They appear in clusters to 1" (2.5 cm) long.

FRUIT: 2–3" (5–7.5 cm) long cylindrical pods, ending in long, narrow points, constricted between seeds, and remaining attached.

HABITAT: Associated with Saguaro on desert plains and rocky slopes of foothills and mountains.

RANGE: Arizona, southeastern California, and northwestern Mexico; at 500–4000' (152–1219 m).

USES: Native Americans used the seeds as a food source. This tree is now cultivated as an ornamental plant.

SIMILAR SPECIES: Blue Paloverde, Jerusalem Thorn.

CONSERVATION STATUS: LC

This species is one of the more common trees in the Sonoran Desert. Its population appears stable, but the limited population in southeastern California is considered vulnerable.

Diameter: 1' (0.3 m)
Height: 25' (7.6 m)

■ DENSE POPULATION
■ LIGHT POPULATION
■ NATURALIZED

This small, tropical tree is native to the Caribbean. Its range in the United States is limited to coastal and southern Florida. In the West Indies, powder made from the roots, bark, and leaves was used to stun fish, hence its common name.

DESCRIPTION: This is a small tree with twisted branches, an irregularly rounded crown, and gray, mottled, scaly bark. Its leaves can be up to 10" (25 cm) long, and they are grayish-green, leathery, evergreen, and pinnately compound, with five to eleven ovate leaflets up to 2¾" (7 cm) long.

FLOWERS: April–May. ⅓" (8 mm), pea-like, lavender-white, in 4" (10 cm) clusters.

FRUIT: Pods, 4" (10 cm) long, rectangular, light brown, with four papery, wavy-edged wings.

HABITAT: Shell mounds, coastal hammocks.

RANGE: Mainly coastal southern Florida; also Caribbean.

USES: This tree is cultivated as an ornamental, and its decay-resistant wood is used for boats, fence posts, and poles.

SIMILAR SPECIES: This species is not likely to be confused with other native trees.

CONSERVATION STATUS: **LC**
The population is stable with no major threats. It is abundant in the Florida Keys.

Height: 50' (15 m)

- DENSE POPULATION
- LIGHT POPULATION
- NATURALIZED

Native to the southwestern United States and northern Mexico, Honey Mesquite is a thorny shrub found in a wide range of habitats. Its seeds are eaten by a variety of mammals and birds, including jackrabbits and quail, as well as livestock. These shrubs often invade grasslands where ranchers work to eradicate them. As the common name suggests, this species is also a honey plant. The word "mesquite" is a Spanish adaptation of the Aztec name *mizquitl*.

DESCRIPTION: This is a shiny, large, thicket-forming shrub or small tree with a short trunk; an open, spreading crown of crooked branches; and narrow, beanlike pods. Its bipinnately compound leaves are 3–8" (7.5–20 cm) long, and the short axis bears one pair of side axes or forks, each fork with 7–17 pairs of stalkless leaflets ⅜–1¼" (1–3 cm) long, ⅛" (3 mm) wide, narrowly oblong, hairless or nearly so, and yellow-green. The bark is dark brown, rough, thick, and becoming shreddy. The twigs are slightly zigzag, with stout, yellowish, mostly paired spines ¼–1" (0.6–2.5 cm) long at enlarged nodes, which afterwards bear short spurs.

FLOWERS: Spring–summer. ¼" (6 mm) long, nearly stalkless, light yellow, fragrant flowers, crowded in narrow clusters 2–3" (5–7.5 cm) long.

FRUIT: A narrow pod 3½–8" (9–20 cm) long, less than ⅜" (10 mm) wide, ending in a long narrow point, slightly flattened, and wavy-margined between the seeds. The fruit has sweet pulp, maturing in summer, remaining closed, with several beanlike seeds within a four-sided case.

HABITAT: Sandy plains and sandhills and along valleys and washes; it is found in short grass, desert grasslands, and deserts.

RANGE: Eastern Texas and southwestern Oklahoma west to extreme southwestern Utah and southern California. It is also found in northern Mexico, and naturalized north to Kansas and Colorado. It occurs to 4500' (1372 m).

USES: The deep taproots, often larger than the trunks, are used as firewood. Southwestern Native Americans prepared meal and cakes from the pods.

SIMILAR SPECIES: Velvet Mesquite.

CONSERVATION STATUS: LC

This species may become weedy or invasive in some habitats and may displace other vegetation if not properly managed. It is considered a noxious weed in Florida.

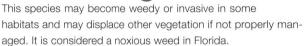

Diameter: 1' (0.3 m)
Height: 20' (6 m)

■ DENSE POPULATION
■ LIGHT POPULATION
■ NATURALIZED

Native to the southwestern United States, this spiny shrub, sometimes a small tree, is easily recognized by its unusual pods, which are the basis of both its English and Spanish common names. These sweetish, nutritious pods are edible and are often fed to domestic livestock. This is an important tree to wildlife; many small mammals and birds eat the seeds and foliage, and the thicket-forming trees host nests for several species of birds.

DESCRIPTION: This is a spiny shrub or small tree with long, slender branches and odd, screwlike pods. The leaves are clustered, bipinnately compound, and 2–3" (5–7.5 cm) long, with a stalk of ½" (12 mm) and one pair (sometimes two) of side axes. It has five to eight pairs of leaflets ¼–⅜" (6–10 mm) long, ⅛" (3 mm) wide, oblong, short-pointed, finely covered with gray hairs, dull green, and stalkless. The bark is light brown, smooth, thick, separating in long, fibrous strips and becoming shaggy. The twigs are slender, covered with gray hairs when young, with slender, whitish, paired spines about ⅜" (10 mm) long, united with base of leafstalk at nodes.

FLOWERS: Spring–summer. ³⁄₁₆" (5 mm) long, light yellow, crowded in narrow clusters about 2" (5 cm) long.

FRUIT: 1–2" (2.5–5 cm) long, a pod tightly coiled into a narrow spiral like a large screw, pale yellow or light brown, hard, with sweetish pulp and many tiny, beanlike seeds; several to many are crowded on a stalk, often abundant, maturing in summer, and not splitting open.

HABITAT: Along streams and valleys in deserts; often forming thickets.

RANGE: Trans-Pecos Texas west to extreme southwestern Utah and southeastern California; it is also found in adjacent northern Mexico. It occurs to 5500' (1676 m).

USES: Native Americans made meal, cakes, and syrup from the pods and used the root bark to treat wounds. The durable, hard wood is used for fence posts, tool handles, and firewood.

SIMILAR SPECIES: Honey Mesquite.

CONSERVATION STATUS: LC
This species has a wide distribution and large population. No major threats have been identified.

Diameter: 8" (20 cm)
Height: 20' (6 m)

■ DENSE POPULATION
■ LIGHT POPULATION
■ NATURALIZED

The common name of this tree describes the hazy appearance of these gray, leafless plants when seen from a distance. These plants have leaves for only a few weeks each year; most of the photosynthesis for its food production occurs in the smoky-gray twigs. Quite showy when covered with flowers, this species is sometimes grown as an ornamental in frost-free regions. Smoke Tree is common in Joshua Tree National Park. It also goes by the common name Smokethorn.

ALTERNATE NAME: *Dalea spinosa*

DESCRIPTION: This is a spiny, much-branched shrub or small tree with a short, crooked trunk; a compact or irregular crown of smoky-gray branches; and small leaves. It is leafless most of the year. When it does have leaves, they are ⅜–1" (1–2.5 cm) long, ⅛–½" (3–12 mm) wide, reverse lance-shaped, rounded at the tip, long-pointed at the base, with wavy edges. They are stalkless or nearly so, gray, densely hairy, and gland-dotted. They are few in early spring, shedding after a few weeks and before flowering. The bark is dark gray-brown, furrowed, and scaly. The twigs are smoky-gray, with dense, pressed hairs and brown gland-dots. They are zigzagged and slender, ending in slender, sharp spines.

FLOWERS: Late spring–early summer when leafless. ½" (12 mm) long, pea-shaped, with five unequal, dark purple or violet petals; they are fragrant and gland-dotted, and generally few, occuring in clusters to 1¼" (3 cm) long, along twigs.

FRUIT: A small, egg-shaped pod ⅜" (10 mm) long, ending in a curved point; it is hairy and gland-dotted, containing usually one brown, beanlike seed; maturing in late summer and not opening.

HABITAT: Sandy and gravelly washes in desert; in subtropical regions.

RANGE: Western Arizona, extreme southern Nevada, south-eastern California, and northwestern Mexico; from below sea level to 1500' (457 m).

USES: This species is sometimes grown as an ornamental.

SIMILAR SPECIES: Frémont Indigo Bush (*P. fremontii*), Dye Weed.

CONSERVATION STATUS: **NT**
The overall population is apparently secure, but this species is state-listed as vulnerable in Arizona and endangered in Nevada.

Diameter: 1' (0.3 m)
Height: 20' (6 m)

■ DENSE POPULATION
■ LIGHT POPULATION
■ NATURALIZED

This species, native to the southeastern United States, is often cultivated as an ornamental shrub and is sometimes planted for erosion control. In the north it is planted along highways, where it acts as a snow fence. It is also called Moss Locust.

DESCRIPTION: This is a shrub or small tree with bristly twigs. Its leaves are 10" (25.4 cm) long, pinnately compound, with leaflets 1½" (4.25 cm) long and ovate or elliptical.

FLOWERS: April–July. The tiny flowers are pinkish, purplish to reddish-purplish, or rose-colored, and arranged along a 4" (10 cm) elongated axis.

FRUIT: 2½" (6.25 cm), hairy pods.

HABITAT: Slopes, ridges, sandhills, open woods.

RANGE: Ontario to Nova Scotia and throughout East south to Florida, west to New Mexico, Utah, and Minnesota; also in Washington.

USES: This species is cultivated as an ornamental. It is used for erosion control and, in the north, as a snow fence along highways.

SIMILAR SPECIES: Black Locust, Clammy Locust.

CONSERVATION STATUS: (LC)
Bristly Locust has a wide distribution and occurs in protected areas. There are no major threats to this species.

Height: 10' (3 m)

■ DENSE POPULATION
■ LIGHT POPULATION
■ NATURALIZED

A native of the southwestern United States, a notable population of New Mexican Locust at the north rim of Grand Canyon National Park puts on a spectacular flower display in early summer. This shrub or small tree is sometimes planted as an ornamental for the attractive flowers. Livestock and a variety of wildlife, especially deer, browse the foliage, and cattle eat the flowers.

DESCRIPTION: This is a spiny shrub or small tree with an open crown and showy, fragrant, purplish-pink, pea-shaped flowers. It often forms thickets. Its leaves are pinnately compound, 4–10" (10–25 cm) long, with 13–21 leaflets, paired except at the end, ½–1½" (1.2–4 cm) long, ¼–1" (0.6–2.5 cm) wide, elliptical, rounded at the ends, with a tiny bristle tip, not toothed, finely hairy when young, pale blue-green, and nearly stalkless. The bark is light gray, thick, and furrowed into scaly ridges. The twigs are brown with rust-colored gland hairs when young, with stout, brown, paired spines ¼–½" (6–12 mm) long at the nodes.

FLOWERS: Late spring–early summer. ¾" (19 mm) long and pea-shaped, with five unequal purplish-pink petals; on stalks with sticky hairs; many in large, unbranched, drooping clusters at the base of the leaves.

FRUIT: The fruit is a narrowly oblong, flat, thin, brown pod 2½–4½" (6–11 cm) long with bristly and often glandular hairs, splitting open, maturing in early autumn, with three to eight beanlike, flattened, dark brown seeds.

HABITAT: Canyons and moist slopes; with Gambel's Oak, Ponderosa Pine, and pinyons.

RANGE: Mountains from southeastern Nevada east to southern and central Colorado, south and east to Trans-Pecos Texas, and west to southeastern Arizona; at 4000–8500' (1219–2591 m). It is also found in northern Mexico.

USES: This species is sometimes planted as an ornamental for the handsome flowers and is also valuable for erosion control, sprouting from roots and stumps and rapidly forming thickets.

SIMILAR SPECIES: Black Locust.

CONSERVATION STATUS: LC
The population is stable and there are no major threats to this species.

Diameter: 8" (20 cm)
Height: 25' (7.6 m)

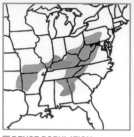

■ DENSE POPULATION
■ LIGHT POPULATION
■ NATURALIZED

This species is commonly used for erosion control, particularly on strip-mined areas and other low-quality disturbed sites. It grows rapidly and spreads by sprouts like a weed, and thus has spread far beyond its native range. In many areas of North America it is considered an alien invasive plant, while in others it is recommended as a native species for the garden. Although it does have fragrant flowers and gives a pretty, dappled shade, the tree is short-lived and brittle, and the pods are messy. Once established, it spreads profusely and is extremely difficult to eradicate. Gardeners would do better to select less aggressive native trees.

DESCRIPTION: This is a medium-sized, spiny tree with a forking, often crooked, and angled trunk and an irregular, open crown of upright branches. Its leaves are pinnately compound, 6–12" (15–30 cm) long, with 7–19 leaflets 1–1¾" (2.5–4.5 cm) long, ½–¾" (12–19 mm) wide, paired except at end, elliptical, with tiny bristle tip, without teeth, hairy when young, drooping and folding at night, dark blue-green above, pale and usually hairless beneath. The bark is light gray, thick, deeply furrowed into long rough forking ridges. The twigs are dark brown, with stout paired spines ¼–½" (6–12 mm) long at nodes.

FLOWERS: Late spring. ¾" (19 mm) long and pea-shaped, with five unequal white petals, the largest being yellow near the base. They are fragrant and are arranged in showy, drooping clusters 4–8" (10–20 cm) long at the base of the leaves.

FRUIT: 2–4" (5–10 cm) long, a dark brown, narrowly oblong flat pod, maturing in autumn, remaining attached into winter, splitting open, with 3–14 dark brown, flattened, beanlike seeds.

CAUTION: Poisonous: Do not eat or ingest any part of this species. If accidentally ingested, contact your doctor or poison control center.

HABITAT: Moist to dry sandy and rocky soils, especially in old fields and other open areas, and in woodlands.

RANGE: This plant's native range stretches from central Pennsylvania and southern Ohio south to northeastern Alabama, and from southern Missouri to eastern Oklahoma; it is naturalized from Maine to California and in southern Canada. It occurs from 500' (152 m) to above 5000' (1524 m) in the southern Appalachians.

USES: Black Locust has been widely planted for ornament and shelterbelts, and is also used for erosion control, particularly on strip-mined areas. Native American tribes in Virginia made bows from the wood and apparently planted the trees eastward. The durable wood was also used as corner posts for the colonists' first homes.

SIMILAR SPECIES: Clammy Locust, Bristly Locust.

CONSERVATION STATUS: **LC**

This species is common, widespread, and increasing. Because of its tendency toward invasiveness, this species should not be planted outdoors casually.

Diameter: 1–2' (0.3–0.6 m)
Height: 40–80' (12–24 m)

■ DENSE POPULATION
■ LIGHT POPULATION
■ NATURALIZED

This small tree or shrub, native to the southeastern United States, was discovered in 1776 by the early American botanist William Bartram (1739–1823) in South Carolina. It is easily recognized by the sticky or clammy secretion of the gland-hairs on the twigs, leafstalks, flower stalks, and pods.

DESCRIPTION: Clammy Locust is a small tree with a spreading crown of slender branches, or it may be a shrub. It is often spiny, with showy pink flowers and sticky or clammy gland-hairs on twigs, leafstalks, flower stalks, and pods. Its leaves are pinnately compound, 5–10" (13–25 cm) long, with 13–21 leaflets ¾–1½" (2–4 cm) long, ½–¾" (12–19 mm) wide, paired (except at end), elliptical, with a tiny bristle tip, not toothed, and hairy when young. They are dark green above and paler and finely hairy beneath. The bark is dark brown, and smooth, becoming furrowed. Twigs are dark brown with small, paired spines at some nodes.

FLOWERS: Late spring. ¾" (19 cm) long and pea-shaped, with five unequal pale pink petals, the largest being pale yellow near the base. They are crowded in drooping clusters to 3" (7.5 cm) long at the base of leaves.

FRUIT: 2–3¼" (5–8 cm) long, narrowly oblong, flat brown pod, sticky with gland-hairs, splitting open, with several mottled-brown, flattened, beanlike seeds.

HABITAT: Open forests, clearings, and waste places; forming thickets.

RANGE: West Virginia southwest to central Alabama; it is naturalized beyond this range in the eastern United States and southeastern Canada. It occurs at 500–4000' (152–1219 m).

USES: This species is cultivated as an ornamental tree or shrub.

SIMILAR SPECIES: Bristly Locust, Black Locust.

CONSERVATION STATUS:
The population of this species is declining in its native range for unknown reasons. Forest succession remains the primary threat to this tree. It is widely established outside of its native range as an exotic plant.

Diameter: 8" (20 cm)
Height: 30' (9 m)

■ DENSE POPULATION
■ LIGHT POPULATION
■ NATURALIZED

Also known as Texas Sophora, this deciduous tree is native to the south-central United States, found in Texas and parts of Oklahoma, Arkansas, and Louisiana. It is recognized by its pink-tinged white flowers.

DESCRIPTION: This is a shrub or small tree with a rounded crown of stout, spreading branches, pinkish flowers, and pods resembling a string of beads. Its leaves are pinnately compound and 6–9" (15–23 cm) long. It usually has 13–19 leaflets ¾–1½" (2–4 cm) long; they are paired (except at end), elliptical, short-pointed without teeth, and covered with gray hairs when young. They are yellow-green above and paler and slightly hairy beneath. The bark is gray-brown, thin, and scaly, becoming furrowed. The twigs are greenish, slightly zigzagged, slender, and covered with fine hairs.

FLOWERS: Early spring. ½" (12 mm) long and pea-shaped with five unequal white petals tinged with pink and covered with fine hairs, appearing in lateral clusters 2–4½" (5–11 cm) long.

FRUIT: The fruit is a black, leathery pod, 1–5" (2.5–13 cm) long and ⅜" (10 mm) in diameter, slightly hairy; it is long-pointed at the ends and narrow between the seeds, with thin, fleshy, sweetish walls; maturing in summer, and remaining closed and attached in winter.

CAUTION: Poisonous: Do not eat or ingest any part of this species. If accidentally ingested, contact your doctor or poison control center.

HABITAT: Moist soils along streams and on limestone hills; it is often found in groves.

RANGE: Southwestern Arkansas and northwestern Louisiana west to southern Oklahoma and central Texas; at 300–1800' (91–549 m).

USES: This species is cultivated as an ornamental, often in parking lots and along streets and highways.

SIMILAR SPECIES: Japanese Pagoda Tree.

CONSERVATION STATUS: NT
The population is apparently secure but has a limited native range. More study is needed to assess the status of this species in the wild.

Diameter: 8" (20 cm)
Height: 30' (9 m)

■ DENSE POPULATION
■ LIGHT POPULATION
■ NATURALIZED

This evergreen shrub goes by several common names, including Texas Mountain Laurel, Frijolito, and Mescalbean. It is often cultivated in warm regions for its shiny, green foliage and showy, purple flowers. Native Americans made necklaces as well as a narcotic powder from the seeds.

DESCRIPTION: Mescalbean is an evergreen shrub or sometimes a small, much-branched tree with a narrow crown and bright red, beanlike, poisonous seeds and showy, purple flowers. Its leaves are evergreen, pinnately compound, 3–6" (7.5–15 cm) long, usually with 5–11 leaflets ¾–2" (2–5 cm) long, ⅜–1" (1–2.5 cm) wide. They are paired (except at the end), elliptical, rounded or slightly notched at the tip, without teeth, thick and leathery, and shiny dark green, becoming hairless above and paler and hairless (or nearly so) beneath. The bark is dark gray and fissured into narrow, flattened, scaly ridges; becoming rough and shaggy. The twigs are densely covered with white hairs when young, becoming light brown.

FLOWERS: Early spring with new leaves, sometimes also in fall. The flowers are ¾–1" (2–2.5 cm) long and pea-shaped, with five unequal bluish-purple or violet petals, fragrant, and crowded on one side of the stalk in clusters 2–4½" (5–11 cm) long consisting of many flowers each.

FRUIT: The fruit is a cylindrical pod 1–5" (2.5–13 cm) long, about ⅝" (15 mm) in diameter, densely covered with brown hairs and pointed at the ends. It is slightly narrowed between the seeds, hard and thick-walled, maturing in late summer and not opening.

CAUTION: Poisonous: Do not eat or ingest any part of this species. If accidentally ingested, contact your doctor or poison control center.

HABITAT: Moist soils of streams, canyons, and hillsides, mainly on limestone; it often forms thickets.

RANGE: Central to southwest and Trans-Pecos Texas and southeastern New Mexico; it is also found south to central Mexico. It occurs to 6500' (1981 m).

USES: This is a popular ornamental plant, valued especially for its showy, strongly fragrant flowers.

SIMILAR SPECIES: Texas Sophora, Japanese Pagoda Tree.

CONSERVATION STATUS: LC
The population is widespread and stable with no major threats.

Diameter: 6" (15 cm)
Height: usually less than 5' (1.5 m), sometimes 20' (6 m)

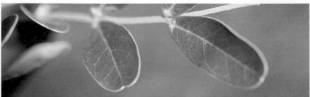

This fairly small tree is readily identified by its large, branched flower clusters and seed pods with deep, often stringlike constrictions between the seeds. The native Eve's Necklacepod (*S. affine*) has a blunt or notched leaflet tip and smaller, less-branching clusters of pink-tinged white flowers. Both species were formerly in the genus *Sophora*.

DESCRIPTION: This is a small to medium-sized tree with deciduous, pinnately compound leaves and a rounded crown of spreading branches. Young branches are green and hairless or nearly so. Its leaves are 5–11" (13–28 cm) long, dark green above and paler to whitish beneath, turning yellow in fall. Its seven to seventeen leaflets are 1–2" (2.5–5 cm) long, usually ovate, with a pointed tip and a rounded base. The margins are untoothed.

FLOWERS: Late summer. 1–2" (2.5–5 cm) long, whitish yellow with five petals. They appear in large, branched clusters up to 12" (30 cm) long.

FRUIT: Fall, persisting through winter. 2–3" (5–7.6 cm) long, hairless green pod with distinct constrictions between the two to eight dark brown seeds.

HABITAT: Introduced in North America; planted in parks and as a street tree in cities.

RANGE: Native to Asia; it is introduced in North America, especially in Ohio and Pennsylvania south to North Carolina.

USES: This species is cultivated as an ornamental tree in parks and urban areas.

SIMILAR SPECIES: Eve's Necklacepod.

CONSERVATION STATUS:
The conservation status of this species in the wild has not been fully assessed. It is an exotic in North America.

Diameter: 5' (1.5 m)
Height: 65' (19.5 m)

■ DENSE POPULATION
■ LIGHT POPULATION
■ NATURALIZED

Tamarind, native to Africa and the Arabian Peninsula, has been introduced in many parts of the world. It is established in southern Florida. Tamarind is a member of the pea family and has the telltale pinnately compound leaves and elongated pod that splits along the seam to reveal the seeds.

DESCRIPTION: The crown of this tree is dense and rounded, and the trunk is massive and often twisted. The bark is dark brown to blackish, and rough. Its leaves are 5" (12.5 cm) long, evergreen, and pinnately compound, with 20–30 elliptical, ¾" (2 cm) leaflets.

FLOWERS: April–June. 1" (2.5 cm), star-shaped, pale yellow, in long clusters.

FRUIT: The fruit is a brown, beanlike pod 6" (15 cm) long with edible pulp. It is ripe May–July.

HABITAT: Hammocks and yards.

RANGE: Native of Africa and the Arabian Peninsula; introduced from central Florida southward along the coast.

USES: This tree has been cultivated for commercial use in many regions of the world, particularly in India and the United States. The fruit pulp has many culinary uses, including flavoring for curries and as an ingredient in Worcestershire sauce. The wood is used to make furniture and other items.

SIMILAR SPECIES: This is the only species in genus; it is not likely to be confused with any native plant in North America.

CONSERVATION STATUS: 🔵LC
Tamarind has a wide geographical distribution. There are no major threats and the population in its native range is large and stable.

Diameter: 6" (15 cm)
Height: usually less than 5' (1.5 m), sometimes 20' (6 m)

■ DENSE POPULATION
■ LIGHT POPULATION
■ NATURALIZED

This flowering shrub or small tree is found on coasts in tropical areas around the world. In North America it is also planted as an ornamental.

DESCRIPTION: This is a clump-forming shrub, or sometimes, a small tree, with a conical crown and a short trunk. Its leaves are 1½" (4.3 cm) long, linear-oblanceolate, grayish green, hairy, fleshy, and evergreen. The bark is dark gray or brown, cracked or shaggy, peeling in ragged strips.

FLOWERS: Year round. The yellow flowers have a diameter of ½" (1.25 cm).

HABITAT: Dunes, beaches, coastal scrub, sandy thickets.

RANGE: Coastal, from central Florida south.

USES: Bay Cedar is planted as an ornamental along the coast. The strong, durable wood is sometimes used to make small items.

SIMILAR SPECIES: As the only species in its genus, this plant is unlikely to be confused with anything else in its range.

CONSERVATION STATUS: (LC)
This species has a wide distribution and large population with no threats.

Height: 30' (10 m)

- ■ DENSE POPULATION
- ■ LIGHT POPULATION
- ■ NATURALIZED

One of six serviceberry species native to North America, this one, also known as Saskatoon Serviceberry, is found throughout western Canada and into the northwestern United States. The fruit of this and related species are eaten fresh, prepared in puddings, pies, and muffins, and dried like raisins and currants. They are also an important food for wildlife from songbirds to squirrels and bears. Deer and livestock browse the foliage, although in concentrated doses the twigs and leaves—which contain cyanide compounds—can fatally poison these animals.

DESCRIPTION: This is a shrub or small tree, usually with several trunks, and with star-shaped white flowers. Its leaves are ¾–2" (2–5 cm) long and almost as broad. They are broadly elliptical to nearly round, rounded at both ends, coarsely toothed above the middle, usually with 7–9 straight veins on each side, dark green and becoming hairless above, and paler and hairy when young beneath. The bark is gray or brown, thin, and smooth or slightly fissured. The twigs are red-brown, slender, and hairless.

FLOWERS: Spring. ¾–1¼" (2–3 cm) wide with 5 narrow white petals, arranged in small, terminal clusters.

FRUIT: ½" (12 mm) in diameter, like a small apple, purple or blackish, edible, juicy and sweet with several seeds; in early summer.

HABITAT: Moist soils in forests and openings.

RANGE: Central Alaska southeast to Manitoba, western Minnesota, Colorado, and western to northern California; local east to southeastern Quebec; to 6000' (1829 m).

USES: The fruit of this and similar species are eaten by people, fresh or dried. This species is also cultivated.

SIMILAR SPECIES: Utah Serviceberry.

CONSERVATION STATUS: LC
The Western Serviceberry is considered stable with no known threats.

Diameter: 8" (20 cm)
Height: 30' (9 m)

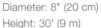

■ DENSE POPULATION
■ LIGHT POPULATION
■ NATURALIZED

This plant occurs as a flowering shrub or a small tree, and grows in upland hardwood forests throughout much of eastern North America. It is also called Shadbush or Shadblow, names that allude to the fact that this plant's showy masses of white flowers tend to occur at the same time that shad ascend the rivers in early spring to spawn. An older name for this species is Sarvis.

DESCRIPTION: This species can be a tree with a narrow, rounded crown, or an irregularly branched shrub, with star-shaped, white flowers. Its leaves are 1½–4" (4–10 cm) long, 1–2" (2.5–5 cm) wide, ovate or elliptical, pointed at the tip, notched at the base, and finely saw-toothed, with soft hairs when young, especially beneath. They have 11–17 straight veins on each side and are dull green above, paler beneath, turning yellow to red in fall. The bark is light gray and smooth, becoming furrowed into narrow ridges. The twigs are red-brown and slender, often covered with white hairs when young.

FLOWERS: Spring. 1¼" (3 cm) wide with five narrow white petals, arranged on slender stalks in terminal clusters.

FRUIT: ¼–⅜" (6–10 mm) in diameter, like a small apple, purple, edible, nearly dry or juicy and sweet, with several seeds; in early summer.

HABITAT: Moist soils in hardwood forests.

RANGE: Southern Newfoundland and Nova Scotia south to northwest Florida, west to Louisiana and eastern Oklahoma, and north to Minnesota; to 6000' (1829 m) in the southern Appalachians.

USES: This species is sometimes planted as an ornamental for its showy clusters of flowers.

SIMILAR SPECIES: Canadian Serviceberry, Allegheny Serviceberry.

CONSERVATION STATUS: LC

Downy Serviceberry is susceptible to occasional problems with diseases and pests. With a large geographic range, however, the species faces no major conservation threats.

Diameter: 1' (0.3 m)
Height: 40' (12 m)

■ DENSE POPULATION
■ LIGHT POPULATION
■ NATURALIZED

Serviceberries are subject to many disease and insect problems, but damage to the plant from these problems is usually cosmetic rather than life threatening. The fruit is a valuable food source for a variety of birds, including thrushes, waxwings, robins, and woodpeckers, as well as squirrels and chipmunks. The plant is also important for the larvae of several butterfly species.

DESCRIPTION: Canadian Serviceberry is a small, understory tree or large, multi-trunked shrub usually growing in clumps with many upright branches. Its crown is delicate and open. Enduring fall foliage is orange to rusty-red. The deciduous plant grows 6–20' (1.8–6 m) high, sometimes taller.

FLOWERS: Long-petaled white blossoms, opening before leaves emerge, are followed by fruits.

FRUIT: Small, crimson-colored, edible, apple-like fruit.

HABITAT: Wood borders; moist, upland woods.

RANGE: Maine to Georgia, west to central New York.

USES: This species is often cultivated as an ornamental plant and for wildlife. The wood is sometimes used for handles, walking sticks, fishing rods, and other similar items.

SIMILAR SPECIES: Downy Serviceberry.

CONSERVATION STATUS: LC
The population is considered stable with no known threats.

Height: 6–20' (1.8–6 m),
sometimes taller

■ DENSE POPULATION
■ LIGHT POPULATION
■ NATURALIZED

This widespread native of eastern North America is easy to grow as an ornamental plant and provides year-round interest. The berries are edible and juicier than those of the similar *A. arborea*. Serviceberries are subject to many disease and insect problems, but damage from these problems is usually cosmetic rather than life-threatening to the plant.

DESCRIPTION: This serviceberry is a multiple-trunked tree or shrub, 15–25' (4.5–7.6 m) tall, with dense, fine-textured branching. The blue-green summer foliage can become orange or red in fall. The bark is smooth and slate-gray with white, longitudinal stripes.

FLOWERS: White flowers occur in terminal clusters before the leaves appear and are followed by summer berries.

FRUIT: Edible summer berries turning from red to purple or nearly black.

HABITAT: Cool, rich woods, moist to drier thickets, swamp margins, and clearings.

RANGE: Newfoundland to southern Ontario, south to Delaware, Kentucky, and Iowa; it is also found in the mountains to Georgia and Tennessee.

USES: This species is cultivated as an ornamental plant. The sweet, edible fruit can be eaten raw or cooked.

SIMILAR SPECIES: Downy Serviceberry.

CONSERVATION STATUS:  LC

Allegheny Serviceberry is considered stable with no major threats to the species.

Height: 15–25' (4.5–7.6 m)

■ DENSE POPULATION
■ LIGHT POPULATION
■ NATURALIZED

Native to eastern and central North America, this plant is most common in eastern Canada, New England, and the Great Lakes region. This species is recognized by the nearly round leaves, as the common name suggests. The Latin species name, meaning "blood red," refers to the color of the young twigs. It is also known as the Red-twigged Shadbush.

DESCRIPTION: This is a shrub or small tree with one or several slender trunks, upright branches, and star-shaped, white flowers. Its leaves are 1¼–2¾" (3–7 cm) long, elliptical to nearly round, rounded or blunt at the tip, coarsely toothed, with 12–15 straight veins on each side, becoming hairless or nearly so. They are dull green above and paler beneath. The bark is gray and smooth, becoming slightly fissured. The twigs are rounded when young, and slender.

FLOWERS: Spring, before leaves. Nearly 1" (2.5 cm) wide with five narrow white petals on slender stalks in terminal clusters.

FRUIT: ⁵⁄₁₆" (8 mm) in diameter, like a small apple, blackish, edible, juicy and sweet, with several seeds; in summer.

HABITAT: Rocky slopes and stream banks in open forests.

RANGE: Southwest Quebec and Maine south to northern New Jersey and west to northern Iowa and northern Minnesota; also south in the mountains to western North Carolina and eastern Tennessee; to 2000' (610 m).

USES: This species is cultivated as an ornamental plant and for wildlife.

SIMILAR SPECIES: Downy Serviceberry.

CONSERVATION STATUS: (LC)
The overall population is secure, but this species is considered state-endangered or of special concern in several U.S. states.

Diameter: 4" (10 cm)
Height: 20' (6 m)

UTAH SERVICEBERRY *Amelanchier utahensis*

■ DENSE POPULATION
■ LIGHT POPULATION
■ NATURALIZED

This widespread plant of the mountainous western United States occurs in a variety of habitats. Often occurring as a low shrub, it is browsed by Mule Deer (*Odocoileus hemionus*), Desert Bighorn Sheep (*Ovis canadensis*), and domestic livestock. A variety of birds, including the Sage Grouse (*Centrocercus urophasianus*), commonly eat the berries of this plant.

DESCRIPTION: This is a low shrub or small tree. Its leaves are simple and stalked. The blades are 1" (2.5 cm), ovate to round, with serrated margins above the middle half.

FLOWERS: April–May. White, ½" (1.3 cm) petals; the flowers are showy and occur in clusters of three to six.

FRUIT: ½" (1.3 cm) diameter berries, purplish, globose, juicy but usually dry and brownish at maturity; persistent.

HABITAT: Streamsides, arid mountain slopes and canyons, sagebrush, mixed shrub, pinyon-juniper, and Ponderosa Pine communities.

RANGE: Western United States from Washington west to Montana and south to California and Texas.

USES: This species is cultivated as an ornamental plant and for wildlife.

SIMILAR SPECIES: Western Serviceberry.

CONSERVATION STATUS: LC
This species is common and its population is considered stable, with no major threats identified.

Height: 10' (3 m)

Cercocarpus ledifolius CURL-LEAF MOUNTAIN MAHOGANY

■ DENSE POPULATION
■ LIGHT POPULATION
■ NATURALIZED

This small tree of the western United States is characteristic of low mountains and slopes throughout the Great Basin, generally in areas with low annual precipitation. Deer browse the evergreen foliage year-round. The name "Mountain Mahogany" applied to this genus is misleading; these shrubby trees are not related to true mahogany (genus *Swietenia*) of tropical America. The dark reddish-brown, mahogany-colored heartwood may have led to this name.

DESCRIPTION: This slightly resinous and aromatic evergreen shrub or small tree sports a compact, rounded crown of widely spreading, curved, and twisted branches and many stiff twigs. Its leaves are evergreen, usually clustered, ½–1¼" (1.2–3 cm) long, and less than ⅜" (10 mm) wide. They are narrowly lance-shaped or elliptical, thick and leathery, with edges rolled under, slightly resinous and aromatic, almost stalkless, shiny dark green with a grooved midvein and obscure side veins above, and pale and with fine hairs beneath. The bark is reddish-brown, thick, and deeply furrowed into scaly ridges. The twigs are reddish-brown and hairy when young.

FLOWERS: Early spring. Flowers are ⅜" (10 mm) long, funnel-shaped, slightly five-lobed, yellowish, hairy, without petals, and stalkless, with one to three occurring at the leaf bases.

FRUIT: ¼" (6 mm) long; narrowly cylindrical, hairy, with a twisted tail 1½–3" (4–7.5 cm) long, covered with whitish hairs; maturing in summer.

HABITAT: Dry, rocky mountain slopes; in grassland, with sagebrush, pinyons, and oaks, and in coniferous forests.

Diameter: ½–1½' (0.15–0.5 m)
Height: 15–30' (4.6–9 m)

RANGE: Extreme southwest Washington east to southern Montana, south to northern Arizona, and west to southern California. It occurs at 4000–10,500' (1219–3200 m).

USES: The hard, heavy wood of this species is an important source of fuel in local mining operations. The wood is also used for novelties, as it takes a high polish. Various Native American groups used parts of this plant for medicinal purposes, and the wood was used for making bows. The Navajo made a red dye from the roots by grinding and then mixing them with juniper ashes and powdered alder bark.

SIMILAR SPECIES: Birchleaf Mountain Mahogany.

CONSERVATION STATUS: LC
The population of this common species is secure, with no significant threats identified.

BIRCHLEAF MOUNTAIN MAHOGANY *Cercocarpus montanus*

- ■ DENSE POPULATION
- ■ LIGHT POPULATION
- ■ NATURALIZED

Birchleaf Mountain Mahogany commonly occurs as a shrub in chaparral vegetation in the western United States and northern Mexico, where it often sprouts after a fire. It is an important browse plant for deer and domesticated cattle and sheep. The common and scientific names both refer to the resemblance of the leaves to those of shrubby birches. It is sometimes called Hardtack, perhaps from its ability to withstand cutting, fire, drought, and heavy browsing. *Cercocarpus*, from the Greek words for "tail" and "fruit," describes the hairy tails or plumes from the elongated flower style. These hairy fruits are carried long distances by the wind; they also become entangled in the fur of various animals, further aiding dispersal. The plumes are straight when moist but twist like a corkscrew upon drying, helping to bury the seed in the soil.

ALTERNATE NAME: *Cercocarpus betuloides*

DESCRIPTION: This is a large evergreen shrub or small tree with a single trunk and spreading crown. Its leaves are evergreen, 1–1¼" (2.5–3 cm) long, ⅜–½" (10–12 mm) wide, elliptical, rounded at the tip, broadest and finely toothed beyond the middle, tapering toward the short-stalked base, slightly leathery, with five to eight straight, sunken veins on each side. They are dark green above and pale green or grayish and slightly hairy beneath. The bark is dark brown, smooth, and becoming scaly and reddish-brown. The twigs are reddish-brown and often hairy.

FLOWERS: Early spring to summer. ⅜" (10 mm) long, funnel-shaped, slightly five-lobed, yellowish, hairy, without petals, nearly stalkless, with one to three at the leaf base.

FRUIT: ⅜" (10 mm) long and narrowly cylindrical with a twisted tail 2–3¼" (5–8 cm) long. It is one-seeded and covered with whitish hairs, maturing in late summer and autumn.

HABITAT: Dry, rocky soils of coastal areas, foothills, and mountain slopes; in chaparral and oak woodland.

RANGE: Southwestern Oregon, Montana, and South Dakota south to Baja California and Trans-Pecos, Texas; at 3500–8000' (1067–2438 m).

USES: This species is cultivated as an ornamental shrub and for wildlife habitat. Native Americans used the hard wood to make fishing spears and other tools.

SIMILAR SPECIES: Curl-leaf Mountain Mahogany.

CONSERVATION STATUS: LC
This species is infrequent but locally common throughout the western United States. Although the global population is secure, limited populations are considered threatened in some states and have been extirpated from some localities.

Diameter: 6" (15 cm)
Height: 20' (6 m)

■ DENSE POPULATION
■ LIGHT POPULATION
■ NATURALIZED

This is one of the many native hawthorns of eastern North America. With its brilliant fall color and wildlife-attracting berries, it is a good choice for a backyard wildlife habitat. Once established, the tree becomes drought-resistant and wind tolerant, making it a popular tree for shelterbelts and urban planting. Both the Latin species name and common name describe the pear-shaped fruit, which is longer than it is broad.

DESCRIPTION: This is a shrub or small tree with a broad, flattened crown of slender, spreading, nearly horizontal branches, pear-shaped fruit, and brilliant autumn foliage, mostly without spines. Its leaves are 2–4½" (5–11 cm) long and 1–3" (2.5–7.5 cm) wide. They are elliptical to ovate, short- or long-pointed, usually doubly saw-toothed with gland-tipped teeth, often with shallow lobes and dotlike glands on the leafstalks. They are dull yellow-green with sunken veins (and hairy when young) above, and finely hairy beneath, turning orange and scarlet in fall. The bark is dark gray and scaly, becoming thick and furrowed. The twigs are densely hairy when young; they are mostly without spines, but sometimes have a few short spines.

FLOWERS: Spring. More than ½" (12 mm) wide, with five white petals, 15–20 pink stamens, and two or three styles, with many flowers arranged in broad, hairy clusters.

FRUIT: Elliptical or pear-shaped fruits are ⅜" (10 mm) in diameter and orange-red or red. They have a thin, sweet, juicy pulp and two or three nutlets, with many appearing in upright clusters; maturing in fall and remaining attached.

HABITAT: Moist soils of valleys; especially rocky stream banks and uplands.

RANGE: Southern Ontario and New York, south to Georgia, west to eastern Texas, north to Minnesota; to 2000' (610 m) or above.

USES: This species is commonly cultivated as an ornamental plant and for wildlife habitat. The hard, strong wood is occasionally used for tool handles.

SIMILAR SPECIES: Fleshy Hawthorn, Green Hawthorn, and other hawthorn species.

CONSERVATION STATUS: LC
The population of Pear Hawthorns is stable, with no global threats identified.

Diameter: 6" (15 cm)
Height: 20' (6 m)

■ DENSE POPULATION
■ LIGHT POPULATION
■ NATURALIZED

This small, flowering tree is found in Kansas and other parts of the midwestern United States, as well as in eastern Canada. The Latin species name refers to this tree's resemblance to Scarlet Hawthorn (*C. coccinea*), which has smaller flowers and smaller fruit with thin, dry pulp and calyx lobes that are not enlarged. Kansas Hawthorn provides important cover and nesting sites for a variety of small bird species.

DESCRIPTION: This is a small tree with a broad, rounded, dense crown of spreading branches; large flowers; large, shiny, dark red fruit; and showy autumn foliage. It can also be a large thicket-forming shrub. Its leaves are 2½–3" (6–7.5 cm) long and 2–2½" (5–6 cm) wide. They are broadly ovate, short-pointed at the tip, rounded or straight at the base, sharply and often doubly saw-toothed with several shallow lobes beyond the middle. They are reddish-tinged when young, turning dull dark green above, paler and slightly hairy beneath, becoming orange and red in fall. The leafstalks are often reddish with red gland-dots. The bark is dark brown and scaly. The twigs are hairless with many stout spines.

FLOWERS: Spring. ¾–1" (2–2.5 cm) wide with five white petals, 20 pink or red stamens, and five styles. Four to seven flowers are arranged in compact, usually hairless clusters. Blooms in spring.

FRUIT: To ¾" (19 mm) in diameter, rounded but flattened at the ends, often angled. The fruit is shiny, dark red with many pale dots, and has a reddish gland-toothed calyx with enlarged lobes. They have a thick, juicy pulp and five nutlets, with many in spreading clusters; maturing in autumn.

HABITAT: Dry upland slopes, especially limestone hills, at the edges of woods.

RANGE: Southern Illinois south to northeastern Oklahoma and northern Arkansas and west to southeastern Kansas; at 400–1500' (122–457 m).

USES: This species is planted as an ornamental tree. The wood is occasionally used for tool handles.

SIMILAR SPECIES: Scarlet Hawthorn.

CONSERVATION STATUS:

The conservation status of this species has not been formally evaluated. Although widely cultivated, its limited population in the wild may be threatened. This species is susceptible to a disease called cedar-hawthorn rust caused by the fungus *Gymnosporangium globosum*. The pathogen requires both a cedar or juniper host and a hawthorn host to complete its life cycle, which may destroy the hawthorn species through severe defoliation. Many of the possible host plants are common ornamental landscaping shrubs and trees; care should be taken not to plant juniper and hawthorn species in the same landscape planting.

Diameter: 8" (20 cm)
Height: 20' (6 m)

■ DENSE POPULATION
■ LIGHT POPULATION
■ NATURALIZED

This common hawthorn is a small tree native to eastern North America. Common and widespread, this species has been planted as an ornamental tree and as a hedge since colonial times. The long spines and shiny, dark green, spoon-shaped leaves make this one of the most easily recognized hawthorns. The common and Latin species names of this species both describe the numerous and extremely long spines, which are used locally as pins. Some cultivated varieties have reduced spines.

DESCRIPTION: This is a small, spiny, thicket-forming tree with a short, stout trunk, and a broad, dense crown of spreading and horizontal branches; hairless throughout. Its leaves are 1–4" (2.5–10 cm) long and ⅜–2" (1–5 cm) wide. They are spoon-shaped or narrowly elliptical, short-pointed or rounded at the tip, widest beyond the middle, tapering to a narrow base, sharply saw-toothed beyond the middle with gland-tipped teeth, and slightly thick and leathery. They are shiny dark green above and pale with a prominent network of veins beneath, turning orange and scarlet in fall. The bark is dark gray or brown and scaly, with branched spines. The twigs are stout, usually with many long, slender, brown spines.

FLOWERS: Spring–summer. ½–⅝" (12–15 mm) wide with five white petals and 10 to sometimes 20 pink or pale yellow stamens, with many flowers arranged in large clusters.

FRUIT: Rounded pomes ⅜–½" (10–12 mm) in diameter, greenish or dull dark red, with thin, hard pulp, and usually two or three nutlets. Several fruits appear in drooping clusters, maturing in late fall and persisting until spring. Some popular cultivated varieties have yellow fruits.

HABITAT: Moist soils of valleys and low upland slopes.

RANGE: Southern Ontario and Quebec south to northern Florida, west to eastern Texas, and north to Iowa; to 2000' (610 m).

USES: This species is commonly cultivated as an ornamental tree.

SIMILAR SPECIES: This species is more easily identified than other hawthorn species, but it is most similar to Barberry Hawthorn.

CONSERVATION STATUS: LC
The population is stable and without known threats.

Diameter: 1' (0.3 m)
Height: 30' (9 m)

■ DENSE POPULATION
■ LIGHT POPULATION
■ NATURALIZED

This small tree has showy white flowers, glossy foliage, and shiny black fruits. Although primarily concentrated in the Pacific Northwest, this species is the most widespread western member of its genus, and the only *Crataegus* species found north to southeastern Alaska. Livestock often browse the foliage; pheasants, partridges, quail, and other birds consume the berries. It is also an important host plant to a number of western butterfly species. It is named for the Scottish botanical explorer David Douglas (1798–1834) and is sometimes known by the common names Douglas' Hawthorn or Douglas' Thornapple.

DESCRIPTION: This handsome small tree has a compact, rounded crown of stout, spreading branches. It is also often a thicket-forming shrub. Its leaves are 1–3" (2.5–7.5 cm) long and ⅝–2" (1.5–5 cm) wide. They are obovate to ovate, broadest toward the short-pointed tip, sharply saw-toothed, and often slightly lobed. They are shiny dark green becoming nearly hairless above, and paler beneath. The bark is gray or brown, and smooth or becoming scaly. The twigs are shiny red, slender, and hairless, often with straight or slightly curved spines to 1" (2.5 cm) long.

FLOWERS: Spring. ½" (12 mm) wide with five white petals, 10–20 pink stamens, and three to five styles on long, slender stalks in broad, leafy clusters.

FRUIT: Pomes ½" (12 mm) in diameter, turning shiny black, with thick, light yellow pulp and three to five nutlets. They are sweetish and mealy but somewhat insipid-tasting, with several on long stalks in drooping clusters, maturing in late summer.

HABITAT: Moist soils of mountain streams and valleys; with sagebrush and conifers.

RANGE: Local in southern Alaska and from British Columbia south to central California, east to New Mexico, and north to southern Saskatchewan; it is also local near Lake Superior. It occurs near sea level in the north, and to 6000' (1829 m) in the south.

USES: This species is planted as an ornamental tree. Native Americans ate the fruit. The wood is sometimes used for tool handles and other small items.

SIMILAR SPECIES: Columbia Hawthorn and other hawthorn species.

CONSERVATION STATUS: LC
The population is widespread and stable. There are no major conservation concerns.

Diameter: 1' (0.3 m)
Height: 30' (9 m)

■ DENSE POPULATION
■ LIGHT POPULATION
■ NATURALIZED

Parsley Hawthorn is found in moist valleys of the southeastern United States. This distinctive hawthorn species is easy to recognize by its small, divided leaves and small, bright red, oblong fruit. It typically grows in moist clay or sandy loams in full to partial sun, and it also adapts well to garden soils. The Latin species name honors Humphry Marshall (1722–1801), a U.S. botanist.

DESCRIPTION: This is a small tree with wide-spreading, slender branches and a broad, irregular, open crown of parsley-like foliage. It can also be a low much-branched shrub. Its leaves are ¾–2" (2–5 cm) long and ¾–1½" (2–4 cm) wide. They are broadly ovate and deeply divided nearly to the midvein into five to seven narrow, short-pointed, saw-toothed lobes, with the veins running to the notches as well as to the points of the lobes. The leaves are hairless or nearly so with long, slender leafstalks, and are shiny green above and paler beneath. The bark is gray, thin, and smooth, peeling off in patches and mottled with brown. The twigs are light brown and hairy when young, with straight spines.

FLOWERS: Spring. ⅝" (15 mm) wide, with five white petals, about 10 red stamens, and one to three styles. Three to twelve flowers are clustered on long, slender, hairy stalks.

FRUIT: ⅜" (10 mm) long and half as wide, oblong, and bright red, with thin, juicy pulp, and one to three (usually two) nutlets; maturing in fall and persisting until winter.

HABITAT: Moist valley soils.

RANGE: Southeast Virginia south to central Florida, west to eastern Texas, and north to southeastern Missouri; to 600' (183 m).

USES: This species is grown as an ornamental plant.

SIMILAR SPECIES: Oneseed Hawthorn.

CONSERVATION STATUS: LC
The population is considered stable, and there are no conservation threats identified for this species.

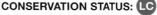

Diameter: 4" (10 cm)
Height: 20' (6 m)

■ DENSE POPULATION
■ LIGHT POPULATION
■ NATURALIZED

One of the largest trees of its genus, Downy Hawthorn is wide-spread in North America, most heavily concentrated in the Midwest. This tree is early to bloom and early to fruit, often dropping many of its leaves by late summer or early in the fall. The common and Latin species names both refer to this tree's soft, hairy foliage. It was originally called White Thorn, likely a reference to the tree's white flowers and sharp spines.

DESCRIPTION: This handsome tree sports a tall trunk and a compact, rounded crown of spreading branches, large broad hairy leaves, many large flowers, and large scarlet fruit. Its leaves are 3–4" (7.5–10 cm) long and wide, broadly ovate, short-pointed at the tip, rounded or slightly notched at the base, doubly saw-toothed, with four or five veins on each side ending in shallow pointed lobes. They are densely covered with white hairs when young and are dark yellow-green above and pale and slightly hairy beneath. The bark is brown to gray and fissured into scaly plates, becoming thick. The twigs are covered with white hairs when young. They typically have stout spines, although sometimes they are nearly thornless.

FLOWERS: Spring. To 1" (2.5 cm) wide with five white petals, 20 light yellow stamens, and four or five styles, arranged in broad clusters.

FRUIT: ¾" (19 mm) in diameter, nearly round or short oblong fruit with scarlet or crimson with dark dots. It is slightly hairy with thick, juicy, edible pulp and four or five nutlets. A few fruits appear in drooping clusters, maturing in late summer or fall.

HABITAT: Moist soil of valleys and hillsides in open woods.

RANGE: Quebec and Nova Scotia, south to West Virginia and Alabama, west to south central Texas, and north to North Dakota; to 1500' (457 m) or higher.

USES: This species is not typically cultivated. It is a food source for birds and other wildlife.

SIMILAR SPECIES: Green Hawthorn and other hawthorn species.

CONSERVATION STATUS: (LC)
The global population of this widespread species is considered stable. It is state-endangered in some areas, especially at the edges of its natural range.

Diameter: 1' (0.3 m)
Height: 40' (12 m)

Crataegus monogyna **ONESEED HAWTHORN**

■ DENSE POPULATION
■ LIGHT POPULATION
■ NATURALIZED

Also called English Hawthorn or Common Hawthorn, Oneseed Hawthorn is the most common small tree over much of Great Britain because of its usefulness to farmers as a hedge. During the 17th to 19th centuries, farmers planted countless miles of these small trees, which grew into tough, stout hedges with sharp thorns that successfully contain sheep and cattle. The species was introduced and has naturalized in North America and elsewhere around the world, where it can sometimes become weedy and invasive. This hawthorn species is distinguishable by the single nutlet, or seed, instead of the two to five seen in other hawthorns. Cultivated varieties have double flowers, pink or red petals, many or no spines, variegated leaves, and a narrow, upright, drooping, or dwarf habit.

DESCRIPTION: This introduced shrub or small tree has a dense, rounded crown of spreading branches. Its leaves are 1–2' (2.5–5 cm) long, ½–1" (1.2–2.5 cm) wide, ovate or reverse ovate, blunt at the tip, broadest beyond the middle, tapering to the base, deeply three- to seven-lobed, the lobes often slightly toothed, with side veins ending in both notches and lobes, and hairless. The leafstalks are long and slender. The leaves are shiny green above and paler beneath, shedding late in fall. The bark is gray and scaly. The twigs are brown or gray, slender, and hairless (or nearly so), with slender spines.

FLOWERS: Spring. ⅜–½" (10–12 mm) wide with five white petals, 20 red stamens, and one style, with many flowers arranged on slender stalks in broad, flat clusters.

FRUIT: ⁵⁄₁₆–⅜" (8–10 mm) in diameter, elliptical or rounded, bright red, with one nutlet, maturing in early autumn.

HABITAT: Scattered in moist soils, along roadsides and edges of forests, and in open areas and waste places.

RANGE: Native of Europe, northern Africa, and western Asia. Planted, escaped from cultivation, and naturalized locally from southern British Columbia to California, east to Nova Scotia, south to the Great Lakes and West Virginia, and scattered localities southward.

USES: This species is commonly used for medicinal purposes and is also cultivated as a hedge plant. The fruit can be eaten raw or in jellies or jams.

SIMILAR SPECIES: English Hawthorn, Parsley Hawthorn.

CONSERVATION STATUS: LC

This species is abundant throughout its native range in Europe. It is widely introduced and naturalized locally in parts of North America.

Diameter: 8" (20 cm)
Height: 25' (7.6 m)

WASHINGTON HAWTHORN *Crataegus phaenopyrum*

■ DENSE POPULATION
■ LIGHT POPULATION
■ NATURALIZED

This species is one of the showiest and most popular hawthorns grown for ornamental purposes. In the early 19th century, it was introduced into Pennsylvania from Washington, D.C., as a hedge plant and is thus called Washington Hawthorn. The Latin species name refers to the pearlike foliage. The shiny, red fruits are both showy and edible, possessing a mild, pleasant flavor.

DESCRIPTION: This is a shrub or small tree with a short trunk and a regular, rounded crown of upright branches, abundant small flowers in spring, many small, round, red fruits, and brilliant autumn foliage. It is hairless throughout. Its leaves are 1½–2½" (4–6 cm) long and 1–1¾" (2.5–4.5 cm) wide. They are broadly ovate to triangular or three-lobed, short-pointed at the tip, nearly straight to slightly notched at the base, coarsely saw-toothed, often with five shallow lobes, and slightly hairy when young. They are tinged with red, becoming shiny dark brown above, and paler beneath, turning scarlet and orange in fall. The bark is light brown, smooth, and thin, becoming scaly. The twigs are shiny brown with slender spines.

FLOWERS: Late spring. They are more than ½" (12 mm) wide with five white petals, 20 pale yellow stamens, and three to five styles. Many flowers appear in compact, hairless clusters.

FRUIT: Shiny red or scarlet berries ¼" (6 mm) in diameter, with a ring scar from the shed calyx. They have a thin, dry pulp and three to five nutlets exposed at the ends. They mature in fall and persist until spring. The fruits are edible, but the seeds are poisonous.

CAUTION: The seeds of this species are poisonous. If ingested in quantities, contact your doctor or poison control center.

HABITAT: Moist valley soils.

RANGE: Virginia south to northern Florida, west to Arkansas, and north to southern Missouri; naturalized locally northeast to Massachusetts; to 2000' (610 m).

USES: This species is commonly grown as an ornamental or hedge plant. The hard wood is sometimes used to make tools and other small items.

SIMILAR SPECIES: Red Maple.

CONVERSATION STATUS: NT

The native range of this species is not fully determined. Washington Hawthorn is susceptible to diseases including apple scab, powdery mildew, and fire blight, as well as some fungi and pests. The species does, however, have a good resistance to the fungus that causes cedar-apple rust (*Gymnosporangium juniperi-virginianae*).

Diameter: 1' (0.3 m)
Height: 30' (9 m)

■ DENSE POPULATION
■ LIGHT POPULATION
■ NATURALIZED

This small deciduous tree is native to much of eastern North America. The common and Latin species names refer to the small, whitish dots on the colorful fruits. It is also called White Haw, in reference to this tree's pale gray bark.

DESCRIPTION: This is a small, thicket-forming tree with a round or flattened, dense crown of stout, spreading branches, often broader than it is high. Its leaves are 2–3" (5–7.5 cm) long, ¾–2" (2–5 cm) wide, obovate or elliptical, blunt at the tip, broadest beyond the middle and tapering to the base, sharply and often doubly saw-toothed and lobed, and slightly thickened. They are dull green with sunken veins and becoming hairless above, paler and hairy beneath, and turning orange or scarlet in fall. The bark is gray or brown, fissured, and scaly. The twigs are hairy when young, usually with many spines.

FLOWERS: Late spring. ½–¾" (12–19 mm) wide with five white petals, 20 pink or yellow stamens, and two to five styles. Numerous flowers are arranged in compact hairy clusters.

FRUIT: ½–¾" (12–19 mm) in diameter, rounded or slightly elliptical, dull red or sometimes yellow, with whitish dots. They have thick, mellow pulp and two to five nutlets. Many appear in drooping clusters, maturing and falling in autumn.

HABITAT: Moist soils of valleys and on rocky upland slopes, especially on limestone; it is found in forest openings and borders.

RANGE: Southern Quebec and Newfoundland south to Georgia, west to eastern Oklahoma, and northern Minnesota; to nearly 6000' (1829 m) in the southern Appalachians.

USES: This species is cultivated as an ornamental tree.

SIMILAR SPECIES: Downy Hawthorn and other hawthorn species.

CONSERVATION STATUS: LC
The population is widespread and stable, but this species is highly susceptible to cedar-hawthorn rust. Care should be taken not to plant cedar/juniper and hawthorn species in the same landscape planting to prevent the spread of this pathogen.

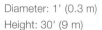
Diameter: 1' (0.3 m)
Height: 30' (9 m)

■ DENSE POPULATION
■ LIGHT POPULATION
■ NATURALIZED

This widespread hawthorn sometimes flowers when it is a low shrub. Its Latin name refers to the succulent, soft fruit. This species goes by many other common and colloquial names, including Succulent Hawthorn and Round-fruited Cockspurthorn; many variant forms of this plant have been given species names, with perhaps two dozen synonyms recognized.

DESCRIPTION: This is a shrub or small tree with a short trunk and a broad, irregular, dense crown of stout branches. Its leaves are 2–2½" (5–6 cm) long and 1–1½" (2.5–4 cm) wide. They are elliptical, gradually narrowed from the middle to the base, doubly saw-toothed, shallow-lobed beyond the middle, with four to seven fine sunken veins on each side, slightly thickened, and hairy when young. They are shiny dark green above and pale yellow-green with fine hairs on the midvein beneath. The bark is dark red-brown and scaly. The twigs are stout and hairless, with many long, stout, slightly curved spines.

FLOWERS: Late spring. ⅝–¾" (15–19 mm) wide with five white petals, 10–20 white or pink stamens, and two to four styles in broad, hairy clusters.

FRUIT: ½–⅝" (12–15 mm) in diameter, bright red, with a gland-toothed calyx at the tip. The fruits have thick, juicy, sweet pulp and two to four nutlets; they are arranged in drooping clusters on long stalks, maturing in early fall.

HABITAT: Moist soils of valleys and open upland areas.

RANGE: Southern Manitoba east to Nova Scotia, south to western North Carolina, and west to Kansas and Nebraska; to 3000' (914 m); locally to Colorado at 5000–7000' (1524–2134 m).

USES: This species is popular as an ornamental plant.

SIMILAR SPECIES: Pear Hawthorn, Green Hawthorn.

CONSERVATION STATUS: LC

The population of this widespread species is large and stable. Small, isolated subpopulations may be vulnerable to localized disturbance or habitat loss; the species is listed as state-endangered in Massachusetts, New Jersey, and Vermont.

Diameter: 6" (15 cm)
Height: 20' (6 m)

■ DENSE POPULATION
■ LIGHT POPULATION
■ NATURALIZED

This species, native to the southeastern United States, occurs in tree form only in northern Florida. Otherwise, it is a small shrub. It is sometimes planted as a novelty in botanical gardens because of its dwarf size. It is commonly known as One-flowered Hawthorn—a reference to its solitary flowers—and also as Dwarf Hawthorn because of its size.

DESCRIPTION: A low spreading shrub or sometimes a small, bushy tree, this plant has short, stout trunks and a dense, rounded crown of stout, crooked branches. Its leaves are ¾–1½" (2–4 cm) long, ½–1¼" (1.2–3.2 cm) wide, spoon-shaped, broadest beyond the middle and tapering to an almost stalkless base, coarsely and wavy saw-toothed, sometimes lobed, slightly thick, and with sunken veins. They are shiny dark green, rough, and becoming hairless above, and paler and hairy beneath. The bark is gray-brown and smooth, becoming fissured into scaly plates near the base. The twigs are stiff and hairy, with many slender spines.

FLOWERS: Late spring. Nearly ⅝" (15 mm) wide with five white petals, about 20 pale yellow or white stamens, and three to five styles. They are usually solitary (sometimes with two or three) on short stalks.

FRUIT: To ½" (12 mm) in diameter, rounded, yellow to dull red to brown, with a large, gland-toothed calyx at the tip. They have hard, dry, mealy pulp, and three to five nutlets, maturing in late summer or fall.

HABITAT: Dry sandy and rocky uplands in open woods, forest borders, and old fields.

RANGE: New York and New Jersey south to northern Florida, west to eastern Texas, and northeast to southern Missouri; to 2000' (610 m) or above.

USES: This species is sometimes cultivated as an ornamental plant.

SIMILAR SPECIES: This is a distinct hawthorn not likely to be confused with other species.

CONSERVATION STATUS: 🔵LC
The global population is considered secure with no major threats identified. Small native populations are state-endangered in Ohio and New York.

Diameter: 6" (15 cm)
Height: 2–5' (0.6–1.5 m),
sometimes to 16' (5 m)

■ DENSE POPULATION
■ LIGHT POPULATION
■ NATURALIZED

This species, also known as Southern Thorn, is native to the southeastern United States. It is commonly cultivated as an ornamental, with many variable forms. Carl Linnaeus gave this species its Latin name, meaning "green," in 1753 from a specimen with shiny green foliage sent from Virginia.

DESCRIPTION: This is a thicket-forming tree with a straight, often fluted trunk; a rounded, dense crown of spreading branches; shiny foliage; showy flowers; and small red to yellow fruit. Its leaves are 1–2½" (2.5–6 cm) long, ½–1½" (1.2–4 cm) wide, elliptical or nearly 4-angled, short-pointed at the tip, gradually narrowed to a long-pointed base, finely saw-toothed, sometimes slightly 3-lobed, shiny dark green and becoming hairless above, and pale with tufts of hairs in vein angles beneath, often turning scarlet in autumn. The bark is pale gray with orange-brown inner bark, and scaly. The twigs are gray and hairless, usually without spines.

FLOWERS: Spring. ⅝" (15 mm) wide with 5 white petals, about 20 pale yellow stamens, and usually 5 (2–5) styles. Many flowers (mostly 8–20) appear on long, slender stalks in branching clusters.

FRUIT: ¼" (6 mm) in diameter and bright red, orange-red, or yellow. It has thin juicy pulp and usually 5 nutlets. Many appear in drooping clusters, maturing in autumn and persisting into winter.

HABITAT: Wet or moist soils of valleys and low upland slopes.

RANGE: Delaware south to northern Florida, west to eastern Texas, and north to southwestern Indiana; to about 500' (152 m).

USES: This species is popularly cultivated as an ornamental tree.

SIMILAR SPECIES: Pear Hawthorn, Fleshy Hawthorn.

CONSERVATION STATUS: LC
The global population is considered secure with no widespread conservation threats. Populations along the northern tier of its natural range may be more vulnerable.

Diameter: 1½' (0.5 m)
Height: 40' (12 m)

■ DENSE POPULATION
■ LIGHT POPULATION
■ NATURALIZED

Native to China, Loquat is a tropical-looking tree with large, leathery, green foliage and bright yellow-orange fruit. It is planted as an ornamental and cultivated for its edible fruit. It may be invasive in some areas, such as southern Florida.

DESCRIPTION: This is a small tree or large shrub with evergreen, simple leaves. The trees can sometimes grow to 30' (9 m) tall. It has a short trunk, and the erect branches form a compact, symmetrical crown. Its leaves are 5–12" (12.7–30 cm) long, narrowly elliptic to lance-shaped, with toothed margins. They are glossy, dark green above and whitish beneath. The twigs are woolly when new.

FLOWERS: Late fall–winter. ½–¾" (1.3–1.9 cm) wide, appearing in white clusters, and fragrant.

FRUIT: 1–2" (2.5–8 cm) long, round pome, yellow-orange, arranged in clusters. Spring.

HABITAT: This species is native to Asia and cultivated in North America.

RANGE: Introduced in North America; planted in California and the Gulf states; naturalized in Florida. Native to Asia.

USES: This species is cultivated for its citrusy fruit and as an ornamental plant.

SIMILAR SPECIES: This species is unlikely to be confused with other trees.

CONSERVATION STATUS:
The conservation status of this species has not been fully assessed. It is introduced in North America.

Diameter: 6–8" (15–20 cm)
Height: 15–20' (4.5–6 m)

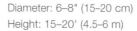

■ DENSE POPULATION
■ LIGHT POPULATION
■ NATURALIZED

The only species in genus *Heteromeles*, Toyon is showy in winter with its evergreen leaves and abundant red fruit. A pioneer plant on eroded soil, it sprouts vigorously after fire or cutting. It can be a prominent member of coastal sage scrub and chaparral plant communities as well as California oak woodlands, and is an important food plant for wildlife. The common name Toyon is of Native American origin. This plant is popular for Christmas decorations; it is also known as Christmas Berry or California Holly.

DESCRIPTION: One of the most beautiful native shrubs or small trees, this species is evergreen with a short trunk, many branches, and a rounded crown. Its leaves are evergreen, 2–4" (5–10 cm) long, ¾–1½" (2–4 cm) wide, oblong lance-shaped, sharply saw-toothed, thick, and short-stalked. They are shiny dark green above and paler beneath. The bark is light gray, smooth, and aromatic. The twigs are dark red and slender, and hairy when young.

FLOWERS: Early summer. ¼" (6 mm) wide with five white petals. Many are arranged in upright clusters 4–6" (10–15 cm) wide.

FRUIT: Like small apples, ¼–⅜" (6–10 mm) long, red (sometimes yellow), mealy and sour, usually two-seeded, maturing in fall and remaining attached in winter.

HABITAT: Along streams and on dry slopes, often on sea cliffs; it is found in chaparral and woodland zones.

RANGE: California in the Coast Ranges and Sierra Nevada foothills and Channel Islands; it is also found in Baja California. It occurs to 4000' (1219 m).

USES: This native evergreen is cultivated as an ornamental plant. Native Americans used the fruit and leaves for food and for medicinal purposes.

SIMILAR SPECIES: As the only species in its genus, this plant is not commonly confused with other trees or shrubs.

CONSERVATION STATUS: LC
This species is adapted to withstand drought, and the population is considered stable with no known threats. Some cultivars are susceptible to fireblight.

Diameter: 1' (0.3 m)
Height: 30' (9 m)

■ DENSE POPULATION
■ LIGHT POPULATION
■ NATURALIZED

This species—the only member of its genus—is endemic to the Channel Islands of California. Lyontree occurs in two forms: the tree with mostly simple leaves is confined to Santa Catalina Island (ssp. *floribundus*), and the shrub (or sometimes tree) with pinnately compound leaves is found on the other islands (ssp. *aspleniifolius*). Both are grown as ornamentals in California. The genus name, meaning "Lyon's shrub," honors William Scrugham Lyon (1852–1916), the U.S. horticulturist and forester who first collected this species in 1884.

ALTERNATE NAME: Catalina Ironwood

DESCRIPTION: This is an evergreen shrub with several crooked stems, or occasionally, a small tree with a straight trunk. The leaves are evergreen, opposite, 4–7" (10–18 cm) long, ½–1" (1.2–2.5 cm) wide, thick, shiny dark green above, and yellow-green and densely hairy when young beneath. They are lance-shaped or oblong, or pinnately compound with three to seven narrowly lance-shaped leaflets deeply lobed on the edges. The bark is dark reddish-brown, thin, papery and shedding in narrow strips, becoming gray. The twigs are orange and hairy when young, becoming shiny red.

FLOWERS: Early summer. ½" (12 mm) wide, with five rounded, white petals, with many appearing in flattened clusters 4–8" (10–20 cm) wide.

FRUIT: ³⁄₁₆" (5 mm) long, narrow, long-pointed, hairy, hard capsules, each usually four-seeded, and maturing in late summer.

HABITAT: Dry, rocky soils of canyons and chaparral.

RANGE: Found only on Santa Rosa, Santa Cruz, Santa Catalina, and San Clemente islands of California; it occurs at 500–2000' (152–610 m).

USES: Lyontree is cultivated as an ornamental plant. Native Americans used the tough wood, known locally as Ironwood, to make spear handles and shafts, and European settlers used it for making fishing poles and canes.

SIMILAR SPECIES: This species is not likely to be confused with other plants.

CONSERVATION STATUS:
This species has experienced declines due to severe weather and habitat alteration resulting from climate change. Exotic grazing animals and woodpeckers damage these trees; many feral herbivores have been removed from San Clemente Island. Recent assessment has shown resprouting from stumps as well as seedlings, which may signal a stabilization of this plant's population.

Diameter: 1' (0.3 m)
Height: 40' (12 m)

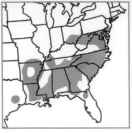

■ DENSE POPULATION
■ LIGHT POPULATION
■ NATURALIZED

This crabapple grows at low altitudes in the southeastern and south-central United States, often forming thickets. The fruit is a popular food source for wildlife, including bobwhites, grouse, rabbits, squirrels, opossums, raccoons, skunks, and foxes.

DESCRIPTION: This is a small tree with a short trunk, spreading branches, and a broad, open crown. Its leaves are 1–2¾" (2.5–7 cm) long, ½–¾" (12–19 mm) wide, elliptical or oblong, usually blunt at the tip, wavy saw-toothed, and hairy when young. They are dull green above and paler beneath, turning brown in autumn. The leaf-stalks are slender, ½–¾" (12–19 mm) long, and hairy when young. The bark is gray or brown and furrowed into narrow scaly ridges. The twigs are brown, and densely covered with hairs when young.

FLOWERS: Spring. 1–1½" (2.5–4 cm) wide with five fragrant, rounded, pink petals; they are arranged in clusters on long stalks.

FRUIT: Resembles small apples, ¾–1" (2–2.5 cm) in diameter, yellow-green, sour, long-stalked, and maturing in late summer.

HABITAT: Moist soils of valleys and lower slopes, borders of forests, fence rows, and old fields.

RANGE: Southern Virginia south to northern Florida, west to Louisiana, and north to Arkansas; local from southern New Jersey to southern Ohio, and in southeastern Texas. It occurs to 2000' (610 m).

USES: This species is often grown as an ornamental plant. The hard, heavy wood has been used to make tool handles.

SIMILAR SPECIES: This tree is sometimes confused with some hawthorn species.

CONSERVATION STATUS: LC
Considered stable, this species has a relatively large geographic range. There are no known wide-ranging threats, but small subpopulations are considered threatened or of special concern in several states.

Diameter: 1' (0.3 m)
Height: 30' (9 m)

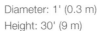

■ DENSE POPULATION
■ LIGHT POPULATION
■ NATURALIZED

This is the most common crabapple species of the Ohio Valley and the Great Lakes region, with smaller numbers stretching across a broad range in the East. Sweet Crabapple is sometimes planted as an ornamental, with double-flowered varieties that have a greater number of larger, deeper pink flowers. The small, sour, often fragrant fruits can be made into preserves and cider.

ALTERNATE NAME: *Pyrus coronaria*

DESCRIPTION: This is a small tree with a short trunk and several stout branches forming a broad, open crown. Its leaves are 2–4" (5–10 cm) long, 1½" (4 cm) wide, ovate, coarsely sawtoothed beyond the middle, and slightly lobed on young twigs; both blades and leafstalks have fine reddish hairs when young. The leaves are yellow-green above and pale beneath, turning yellow in fall. The bark is red-brown, fissured, and scaly. The twigs are red-brown, and covered with gray hairs when young.

FLOWERS: Spring. Nearly 1½" (4 cm) wide with five rounded white or pink petals, arranged in clusters on long stalks.

FRUIT: Resembles a small apple, 1–1¼" (2.5–3 cm) in diameter, yellow-green, long-stalked, and maturing in late summer. The fruit is often fragrant and sour.

HABITAT: Moist soils in openings and borders of forests.

RANGE: Southern Ontario east to New York, south to extreme northern Georgia, west to northeastern Arkansas, and north to northern Illinois; to 3300' (1006 m) in the southern Appalachians.

USES: This crabapple is sometimes planted as an ornamental, and the fruit can be made into preserves, cider, or vinegar.

SIMILAR SPECIES: Prairie Crabapple, and some species of hawthorn.

CONVERSATION STATUS: **LC**
The population is considered stable with no known conservation threats.

Diameter: 1' (0.3 m)
Height: 30' (9 m)

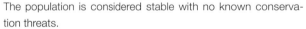

■ DENSE POPULATION
■ LIGHT POPULATION
■ NATURALIZED

This is the only western species of crabapple; it grows abundantly in the temperate coniferous forests in its range. This tree bears oblong fruit, whereas the three eastern crabapple species have round fruit. The fruit is used for jellies and preserves and can be eaten raw, but is sour. Grouse and other birds readily consume the crabapples as well. The tree's strong, durable wood can be made into excellent tool handles.

ALTERNATE NAMES: Oregon Crabapple, *Pyrus fusca*, *Malus diversifolia*

DESCRIPTION: This is a small tree, often with several trunks and many branches, or a thicket-forming shrub. It is sometimes spiny. Its leaves are 1½–3½" (4–9 cm) long, ¾–1½" (2–4 cm) wide, ovate, elliptical, or lance-shaped, sharply saw-toothed, and sometimes slightly three-lobed toward the tip, with slender, hairy leafstalks. They are shiny green and becoming hairless above, and pale and usually slightly hairy beneath, turning orange and red in fall. The bark is gray, smooth to slightly scaly, and thin. The twigs are reddish, slender, and sometimes spiny, and hairy when young.

FLOWERS: Early summer. ¾–1" (2–2.5 cm) wide with five rounded white or pink petals, with several arranged in long-stalked, upright clusters.

FRUIT: Resembles an apple, ½–¾" (12–19 mm) long, oblong, yellow or red, sour, and edible, with few seeds, maturing in late summer.

HABITAT: Along streams and valleys; also on low slopes and in coniferous forests.

RANGE: Southern Alaska south near the Pacific Coast to northwest California; to 1000' (305 m).

USES: The fruit of this species is made into jellies and preserves, and the strong wood is used to make tool handles and other small items. Native Americans valued this tree as a food source and a medicinal plant.

SIMILAR SPECIES: Other crabapple species and hawthorn species.

CONSERVATION STATUS: LC

The global population is considered abundant, stable, and genetically diverse. Overall, *Malus fusca* faces no major population threats. However, subpopulations in southern Alaska are known to be threatened by habitat changes from timber harvest runoff, agriculture, habitat loss, invasive species, and potentially climate change.

Diameter: 1' (0.3 m)
Height: 30' (9 m)

■ DENSE POPULATION
■ LIGHT POPULATION
■ NATURALIZED

This crabapple species is native to the central United States and is concentrated mainly in the eastern prairie region in the upper Mississippi Valley. Numerous species of birds, including bobwhites and pheasants, as well as squirrels, rabbits, and other mammals readily consume the fruit. A variety known as Texas Crabapple (var. *texana*) is found in a small area in central Texas. A handsome double-flowered variety is grown as an ornamental.

ALTERNATE NAME: Iowa Crabapple, *Pyrus ioensis*

DESCRIPTION: This species can be a spiny shrub or a small tree, with spreading branches and a broad, open crown. Its leaves are 2½–4" (6–10 cm) long, 1–1½" (2.5–4 cm) wide. They are elliptical, wavy saw-toothed and often slightly lobed, with prominent veins. They are slightly thickened, hairy when young, with slender, hairy leafstalks. They are shiny dark green above, and paler and hairy beneath, turning yellow in fall. The bark is reddish-brown and scaly. The twigs are reddish-brown, and hairy when young.

FLOWERS: Spring. 1½–2" (4–5 cm) wide with five rounded pink or white petals, arranged in clusters on long stalks.

FRUIT: Like a small apple, 1–1¼" (2.5–3 cm) in diameter, yellow-green, and long-stalked, maturing in late summer.

HABITAT: Moist soils along streams in prairie and forest borders.

RANGE: Northern Indiana south to Arkansas and Oklahoma, and north to extreme southeastern South Dakota; it is local in central Texas and Louisiana. It occurs at 500–1500' (152–457 m).

USES: This species is sometimes grown as an ornamental.

SIMILAR SPECIES: Sweet Crabapple.

CONSERVATION STATUS: NT

Although this species of crabapple is apparently secure, it faces threats from a variety of diseases, pests, and habitat loss, all of which may be exacerbated by climate change.

Diameter: 1' (0.3 m)
Height: 10–30' (3–9 m)

■ DENSE POPULATION
■ LIGHT POPULATION
■ NATURALIZED

This is the familiar apple tree, a native of Europe and West Asia, which has been cultivated since ancient times. Although the species is well known, it is sometimes not recognized when growing wild. For nearly 50 years American pioneer Jonathan Chapman (1774–1845)—better known as Johnny Appleseed—traveled North America mostly on foot, distributing apple seeds to everybody he met. With seeds from cider presses, he helped to establish orchards from Pennsylvania to Illinois. Numerous improved varieties have been developed from this species and from hybrids with related species. Various wildlife species will eat the fallen fruit after harvest.

ALTERNATE NAME: *Malus domestica*

DESCRIPTION: This familiar fruit tree has a short trunk, a spreading rounded crown, showy pink-tinged blossoms, and delicious red fruit. Its leaves are 2–3½" (5–9 cm) long, 1¼–2¼" (3–6 cm) wide, ovate or elliptical, wavy saw-toothed, with hairy leafstalks. They are green above and densely covered with gray hairs beneath. The bark is gray, and fissured and scaly. The twigs are greenish, turning brown, and densely covered with white hairs when young.

FLOWERS: Early spring. 1¼" (3 cm) wide with 5 rounded petals, and white tinged with pink.

FRUIT: The familiar edible apple, it is 2–3½" (5–9 cm) in diameter, shiny red or yellow, sunken at the ends, with thick, sweet pulp and a star-shaped core containing up to 10 seeds, maturing in late summer.

HABITAT: Moist soils near houses, fences, roadsides, and clearings.

RANGE: Native of Europe and West Asia; naturalized locally across southern Canada, in the eastern United States, and in the Pacific states.

USES: This is the apple of commerce, widely grown for its familiar fruit.

SIMILAR SPECIES: May be confused with other apple or crab-apple trees.

CONSERVATION STATUS: 🅛🅒
This stable species does not have any known conservation threats.

Diameter: 1–2' (0.3–0.6 m)
Height: 30–40' (9–12 m)

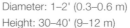

■ DENSE POPULATION
■ LIGHT POPULATION
■ NATURALIZED

This woodland native of the Pacific Coast and westernmost mountain ranges of North America is the only member of genus *Oemleria*. It is often shrubby with multiple trunks. It is a recommended plant for California and northwestern wildlife gardens; numerous bird species relish its fruits. The foliage is said to smell like cucumbers when fresh. This plant grows best in a moist, shady spot.

DESCRIPTION: This is a shrub or small tree with aromatic leaves and smooth, purplish bark. Its leaves are 2–5" (5–12.5 cm) long and elliptical, with the edges sometimes rolled. The bark is smooth and reddish brown to dark gray. The twigs are slender and green, turning reddish brown.

FLOWERS: February–April. ½" (12 mm), greenish white, funnel-shaped, in hanging clusters. Male and female flowers occur on different plants.

FRUIT: A plumlike drupe, ½" (12 mm) long, white turning blue-black with a pale bloom, and edible but bitter; ripe June–July.

HABITAT: Shady coniferous forests, open lowland forests, and chaparral.

RANGE: British Columbia; west of the Cascades in Washington and Oregon; south along the California coast to Ventura County, and the western slopes of the Cascades and Sierra Nevada.

USES: The strong wood from this species is used to make various small tools. Native Americans used the plant for food and for medicinal purposes.

SIMILAR SPECIES: The only species in its genus, it is not commonly confused with other plants.

CONSERVATION STATUS: **LC**
This species has a widespread range and is considered stable, with no known conservation concerns.

Height: to about 17' (5.1 m)

■ DENSE POPULATION
■ LIGHT POPULATION
■ NATURALIZED

The American Plum, also known as the Wild Plum, is native to central and eastern North America. It is a handsome ornamental with large flowers and relatively large fruit, which is eaten fresh and also used in jellies and preserves. Many kinds of birds and mammals eat the fruit. Numerous cultivated varieties with improved fruit have been developed.

DESCRIPTION: This is a thicket-forming shrub or small tree with a short trunk, many spreading branches, a broad crown, showy large white flowers, and red plums. Its leaves are 2½–4" (6–10 cm) long, 1¼–1¾" (3–4.5 cm) wide, elliptical, long-pointed at the tip, sharply and often doubly saw-toothed, slightly thickened, dull green with slightly sunken veins above, and paler and often slightly hairy on the veins beneath. The bark is dark brown and scaly. The twigs are light brown, slender, and hairless, with short twigs ending in a spine.

FLOWERS: Early spring before leaves. ¾–1" (2–2.5 cm) wide with five rounded white petals in clusters of two to five on slender equal stalks, with a slightly unpleasant odor.

FRUIT: Plum, ¾–1" (2–2.5 cm) in diameter, with thick, red skin and juicy, sour, edible pulp and a large stone; and maturing in summer.

CAUTION: Poisonous parts: Although some parts of this species are edible, wilted leaves, twigs, and seeds are toxic to humans and livestock. Eat only if you are absolutely sure which parts are edible.

HABITAT: Moist soils of valleys and low upland slopes.

RANGE: Saskatchewan east to New Hampshire, south to Florida, west to Oklahoma, and north to Montana; it occurs to 3000' (914 m) in the South and to 6000' (1829 m) in the Southwest.

USES: This species is often cultivated as an ornamental and for its fruit. It is also grown for erosion control, spreading by its roots.

SIMILAR SPECIES: Canada Plum and other plum species.

CONVERSATION STATUS: 🅛🅒

CONSERVATION STATUS: Because of its large population size and range, the American Plum is considered globally secure; however, several states and provinces classify the American Plum as either vulnerable or threatened in the wild. The species is susceptible to a number of diseases, including branch cankers, stem decay, and plum pocket (bladder plum), as well as the fungal disease black knot. American Plums can also be damaged by overgrazing, browsing, and trampling by herbivores including deer and livestock.

Diameter: 1' (0.3 m)
Height: 30' (9 m)

■ DENSE POPULATION
■ LIGHT POPULATION
■ NATURALIZED

This small, evergreen cherry is native to the southeastern United States. It may also be found growing wild in parts of California, where it has escaped cultivation. It is a handsome ornamental and hedge plant, and an important host and nectar plant for several species of butterflies. The fruits are eaten by many songbirds, turkeys and quail, and raccoons and other mammals.

DESCRIPTION: This evergreen tree has a narrow or broad crown of spreading branches, glossy foliage, and inedible black fruit. Its leaves are evergreen, 2–4" (5–10 cm) long, ¾–1½" (2–4 cm) wide, elliptical or narrowly elliptical, thick and slightly turned under at the edges, sometimes with scattered teeth, hairless, and aromatic when crushed. They are shiny dark green with obscure veins above, and paler beneath. The bark is gray, thin, and smooth or fissured. The twigs are greenish, slender, and hairless.

Diameter: 10" (25 cm),
larger in cultivation
Height: 40' (12 m)

FLOWERS: Early spring. ³⁄₁₆" (5 mm) wide with five small rounded cream-colored petals, arranged in dense clusters along an axis at the leaf base.

FRUIT: A plum ⅜–½" (10–12 mm) long, elliptical, and short-pointed. It has shiny black, thick skin; thin, dry, inedible pulp; and a large stone, maturing in fall and remaining attached until spring.

CAUTION: Poisonous: Do not eat or ingest any part of this species. If accidentally ingested, contact your doctor or poison control center.

HABITAT: Moist soils of valleys and lowlands, forests, and forest borders, often forming dense thickets.

RANGE: Southeastern North Carolina south to central Florida and west to eastern Texas; to 500' (152 m).

USES: This species is often grown as an ornamental tree or hedge plant.

SIMILAR SPECIES: Sometimes confused with other cherry species.

CONSERVATION STATUS: 🆗
The population is secure with no known conservation threats.

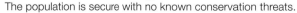

■ DENSE POPULATION
■ LIGHT POPULATION
■ NATURALIZED

This is the common domesticated plum, long cultivated for its edible fruit. Prunes—large plums with firm flesh that is readily dried—are derived from this species as well. It is also commonly known as the European Plum. The native range of this species is unknown, but it has been domesticated since ancient times, and was introduced to North America by early British and French colonists. Numerous improved varieties have been developed for fruit; others, including a double-flowered form, are grown as ornamentals.

DESCRIPTION: This small, introduced fruit tree has a spreading, rounded, open crown. Its leaves are 2–4" (5–10 cm) long and 1–2" (2.5–5 cm) wide. They are elliptical or reverse ovate, coarsely wavy saw-toothed, thickened, dull green (and hairy when young) above, and hairy with a prominent network of veins beneath. The bark is gray, smooth or fissured, and thick. The twigs are reddish, stout, and often hairy, with a few short twigs ending in spines.

FLOWERS: Early spring. ¾–1" (2–2.5 cm) wide with five rounded white petals, with one or two on hairy stalks in clusters, blooming in early spring.

FRUIT: A large plum, 1–2" (2.5–5 cm) in diameter, elliptical, and bluish-black with thick, juicy, sweetish, edible pulp; it has a smooth stone, maturing in summer.

CAUTION: Poisonous parts: Although some parts of this species are edible, other parts are toxic to humans. Eat only if you are absolutely sure which parts are edible.

HABITAT: Introduced in North America; it is often found along roadsides and fence rows.

RANGE: Naturalized locally in southeastern Canada and the northeastern and northwestern United States.

USES: This species is sometimes cultivated as an ornamental, and improved varieties have been developed for fruit.

SIMILAR SPECIES: American Plum.

CONSERVATION STATUS:
This species is widely cultivated; its native range is unknown and has not been evaluated. In some areas, efforts are under way to eradicate this plant from the landscape where it has escaped cultivation.

Diameter: 1' (0.3 m)
Height: 30' (9 m)

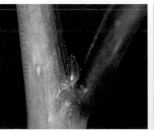

■ DENSE POPULATION
■ LIGHT POPULATION
■ NATURALIZED

This is the most common western cherry, often found in recently disturbed areas or open woods. It can form thickets, spreading not only by seed but also by sending out underground stems that sprout nearby. The scientific name describes its notched petals. As the common name suggests, this plant's small fruits are intensely bitter and inedible to humans. However, many birds and mammals consume the fruit, and deer and livestock browse the leaves. The plant is also a larval host for several moth and butterfly species.

DESCRIPTION: This is a thicket-forming shrub or a small tree with a rounded crown; slender, upright branches; bitter foliage; and small, bitter cherries. Its leaves are 1–2½" (2.5–6 cm) long and ⅜–1¼" (1–3 cm) wide. They are oblong to elliptical, rounded or blunt at the tip, short-pointed at the base with one or two dotlike glands, finely saw-toothed, with blunt, gland-tipped teeth. They are dark green above, and paler and sometimes hairy beneath. The bark is dark brown, smooth, and very bitter. The twigs are shiny red, slender, and hairy when young.

Diameter: 8" (20 cm)
Height: 20' (6 m)

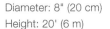

FLOWERS: In spring with leaves. ½" (12 mm) wide with five rounded, notched, white petals, with three to ten arranged on slender stalks.

FRUIT: A round cherry ⁵⁄₁₆–⅜" (8–10 mm) in diameter with thick, red to black skin; thin, juicy, and bitter pulp; and a pointed stone, maturing in summer.

CAUTION: Poisonous: Do not eat or ingest any part of this species. If accidentally ingested, contact your doctor or poison control center.

HABITAT: Moist soils of valleys and on mountain slopes; in chaparral and coniferous forests.

RANGE: British Columbia, Washington, and western Montana south to southern California and southwestern New Mexico; to 9000' (2743 m) in the south.

USES: Some Native Americans used parts of this plant for medicinal purposes.

SIMILAR SPECIES: Hollyleaf Cherry, Common Chokecherry.

CONSERVATION STATUS: 🔵 **LC**
Although the population is large and stable overall, the species is considered critically imperiled in Utah. Bitter Cherry is susceptible to numerous pests, including tent caterpillars, borers, and aphids, as well as to fungi that cause trunk and root rot. The species is also threatened by overgrazing from herbivores such as deer.

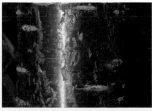

■ DENSE POPULATION
■ LIGHT POPULATION
■ NATURALIZED

This small, evergreen cherry is native to the chaparral areas of coastal California and Baja California, and has long been planted as an ornamental and a hedge plant. It is also a highly valuable wildlife plant. It is a host plant for a number of butterflies, and the flowers attract bees and other pollinators. Although sweetish and edible, the fruits are mostly stone and are consumed mainly by birds and small mammals. The common and scientific names both refer to the holly-like leaves, which are used as Christmas decorations.

DESCRIPTION: This is a small evergreen tree with a short trunk, a dense crown of stout, spreading branches, spiny-toothed leaves, and red cherries. It is hairless throughout, and is often a shrub. The leaves are evergreen, 1–2" (2.5–5 cm) long, ¾–1¼" (2–3 cm) wide, ovate to rounded, short-pointed or rounded at the ends, coarsely spiny-toothed, thick and leathery, and crisp, with a fragrance resembling almonds when crushed. They are shiny green above and paler beneath. The bark is dark reddish-brown and fissured into small, square plates. The twigs are gray or reddish-brown and slender.

FLOWERS: Early spring. ¼" (6 mm) wide with five rounded white petals, arranged in upright, narrow, unbranched clusters less than 2½" (6 cm) long at the leaf bases.

FRUIT: A rounded cherry ½–⅝" (12–15 mm) in diameter, usually red but sometimes purple or yellow, with thin, juicy, sweetish pulp and a smooth stone marked by branching lines. It matures in late fall.

CAUTION: Poisonous: Do not eat or ingest any part of this species. The cherry pits may be toxic if ingested without proper preparation. If accidentally ingested, contact your doctor or poison control center.

HABITAT: Dry slopes and in moist soils along streams; in chaparral and foothill woodland.

RANGE: The Pacific Coast from central California south to Baja California; to 5000' (1524 m).

USES: This species is often grown as an ornamental or a hedge plant. Native Americans used to crack the dried fruit and prepare meal from the ground seeds after leaching them of poisons.

SIMILAR SPECIES: Bitter Cherry.

CONSERVATION STATUS: 🔵 LC
The species is considered stable with no known threats. It is specially adapted to survive fires by resprouting from the root crown.

Diameter: 1' (0.3 m)
Height: 25' (7.6 m)

■ DENSE POPULATION
■ LIGHT POPULATION
■ NATURALIZED

This is a mostly coastal species native to the eastern United States. Dense stands of this shrub may be found between sand hills in secondary dunes behind a beach. In early summer, the dunes are bright with multitudes of Beach Plum blossoms, their flowers forming clusters toward the tips of the stems. The fruit it produces in late summer and early fall is eaten by many birds and small mammals. The plums are also edible to humans and may be eaten fresh or cooked, and are often used to make preserves. This species has also been successfully planted at sites farther inland, where it tends to grow much taller.

DESCRIPTION: This is a straggling, sometimes treelike, fruiting shrub. The branches are brown with tan dots. The leaves are alternate, elliptical, and finely toothed, to 2¾" (7 cm) long.

FLOWERS: June. Snowy white, becoming pinkish before dropping.

FRUIT: ½" (15 mm) in diameter, ranging in color from dark red to bluish purple, and maturing September–October.

CAUTION: Poisonous parts: Although some parts of this species are edible, other parts are toxic to humans. Eat only if you are absolutely sure which parts are edible.

HABITAT: Secondary coastal sand dunes.

RANGE: New Brunswick to New Jersey and Delaware.

USES: This species is cultivated for its fruit, which is used to make jam and sometimes wine.

SIMILAR SPECIES: American Plum.

CONSERVATION STATUS: Although grown commercially, the population in the wild exists in a restricted range that is often coveted for tourism and beachfront development. About half of the species' rangewide habitat has been lost, and much of the rest has become fragmented, although many stands remain intact in managed areas. The species is considered endangered in several states along the East Coast.

Diameter: to 4–8" (10–20 cm) at the root collar
Height: to 6' (1.8 m); occasionally to 13' (3.9 m)

■ DENSE POPULATION
■ LIGHT POPULATION
■ NATURALIZED

This species, native to North America and fairly widespread across southern Canada, is often called Fire Cherry because its seedlings come up after forest fires. This plant matures rapidly and is fairly short-lived, usually with a lifespan of 20–40 years after maturing. During that time it serves as a "nurse" tree, providing cover and shade for the establishment of seedlings of the next generation of larger hardwoods. The cherries are made into jelly and are also an important source of food for wildlife.

DESCRIPTION: This is a small tree or shrub with horizontal branches; a narrow, rounded, open crown; shiny red twigs; bitter, aromatic bark and foliage; and tiny red cherries. Its leaves are 2½–4½" (6–11 cm) long, ¾–1¼" (2–3 cm) wide, broadly lance-shaped, long-pointed, finely and sharply saw-toothed, and becoming hairless. They are shiny green above and paler beneath, turning bright yellow in autumn. Its slender leafstalks often have two gland-dots near the tip. The bark is reddish-gray, smooth, and thin, becoming gray and fissured into scaly plates.

FLOWERS: Spring. ½" (12 mm) wide with five rounded white petals, with three to five flowers on long, equal stalks.

FRUIT: Cherry ¼" (6 mm) in diameter with red skin; thin, sour pulp; and a large stone; maturing in summer.

CAUTION: Poisonous parts: Although some parts of this species are edible, other parts are toxic to humans. Eat only if you are absolutely sure which parts are edible.

HABITAT: Moist soil, often in pure stands on burned areas and clearings; it is often found with aspens, Paper Birch, and Eastern White Pine.

RANGE: British Columbia and South Mackenzie east across Canada to Newfoundland, south to northernmost Georgia, west to Colorado; to 6000' (1829 m) in the southern Appalachians.

USES: The edible fruit is sometimes made into jams and jellies. The light, soft wood is sometimes used for firewood and pulpwood but does not have significant commercial value.

SIMILAR SPECIES: Black Cherry, Common Chokecherry.

CONSERVATION STATUS: (LC)
The population is widespread and globally secure. Isolated populations are listed as vulnerable in some states and provinces.

Diameter: 1' (0.3 m)
Height: 30' (9 m)

■ DENSE POPULATION
■ LIGHT POPULATION
■ NATURALIZED

The familiar Peach, native to China, has been grown as a fruit tree since ancient times. Spanish colonists introduced the Peach into Florida, and Native Americans then planted it widely. It has naturalized locally across the continent and seems to do especially well in the Southeast. Numerous cultivated varieties include freestone peaches, with pulp separating from the stone; clingstones, with pulp adhering to the stone; and smaller, hairless fruits known as nectarines. Other varieties are grown as ornamentals, some with double flowers and with white or red petals.

DESCRIPTION: This is a well-known small fruit tree with a short trunk, a spreading rounded crown, showy pink blossoms, long narrow leaves, and yellow to pink juicy fruit. Its leaves are 3½–6" (9–15 cm) long, ¾–1" (2–2.5 cm) wide, lance-shaped or narrowly oblong, finely saw-toothed. The sides often curve up from the midvein, and the leafstalks are short with glands near the tip. The leaves are shiny green above and paler beneath. Crushed foliage has a strong odor and bitter taste. The bark is dark reddish-brown and smooth, becoming rough. The twigs are greenish turning reddish-brown, long, slender, and hairless.

FLOWERS: Early spring, before leaves. 1–1¼" (2.5–3 cm) wide with five rounded pink petals, usually single and nearly stalkless.

FRUIT: A nearly round peach 2–3" (5–7.5 cm) in diameter with fine velvety hairs covering the yellow-to-pink skin, thick and sweet edible pulp, a large elliptical pitted stone, and maturing in summer.

CAUTION: Poisonous parts: Although some parts of this species are edible, other parts are toxic to humans. Eat only if you are absolutely sure which parts are edible.

HABITAT: Introduced in North America, it is often found along roadsides and fence rows.

RANGE: Native to northwestern China. Naturalized locally in eastern United States, southern Ontario, and California; mostly in the Southeast.

USES: This species has been widely cultivated for its fruit and, in some cases, as an ornamental tree.

SIMILAR SPECIES: This familiar, distinctive tree is not likely to be confused with other species.

CONVERSATION STATUS:
The population of this widely cultivated species is secure. Although it is introduced in North America, no unfavorable ecological impacts have been reported, and the plant's spread is relatively easy to manage.

Diameter: 1' (0.3 m)
Height: 30' (9 m)

■ DENSE POPULATION
■ LIGHT POPULATION
■ NATURALIZED

This widespread species is the largest and most important native cherry. It is common in both North and South America. A pioneer species, it is fast-growing and often spreads into roadsides, waste spaces, and forest margins. About a half dozen geographical varieties have been distinguished; varieties in the southwestern United States tend to be smaller. Its fruits are important food for many types of wildlife, including passerine birds, game birds, and mammals. Pitted fruits are edible and eaten raw, but the pits and the tree's foliage (especially wilted foliage) contain concentrations of toxic cyanide compounds as a deterrent to browsing by herbivores.

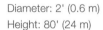

DESCRIPTION: This aromatic tree has a tall trunk, an oblong crown, abundant small white flowers, and small black cherries. Crushed foliage and bark have a distinctive cherry-like odor. Its leaves are 2–5" (5–13 cm) long, 1¼–2" (3–5 cm) wide, elliptical, with one or two dark red glands at the base, finely saw-toothed with curved or blunt teeth, and slightly thickened. They are shiny dark green above, and light green and often hairy along the midvein beneath, turning yellow or reddish in autumn. The bark is dark gray and smooth with horizontal lines, becoming irregularly fissured and scaly, exposing reddish-brown inner bark. The bark is aromatic and bitter. The twigs are red-brown, slender, and hairless.

FLOWERS: Late spring. ⅜" (10 mm) wide with five rounded, white petals, with many flowers arranged along a spreading or drooping axis of 4–6" (10–15 cm) at the end of a leafy twig.

FRUIT: A cherry ⅜" (10 mm) in diameter, with dark red skin turning blackish; slightly bitter, juicy, edible pulp; and an elliptical stone, maturing in late summer.

CAUTION: Poisonous parts: Although the pitted fruits of this species are edible, other parts are toxic to humans. Eat only if you are absolutely sure which parts are edible.

HABITAT: On many sites except for extremely wet or extremely dry soil; it is sometimes found in pure stands.

RANGE: Southern Quebec to Nova Scotia, south to central Florida, west to eastern Texas, and north to Minnesota; varieties occur from central Texas west to Arizona and south to Mexico. It occurs to 5000' (1524 m) in the southern Appalachians and at 4500–7500' (1372–2286 m) in the Southwest.

USES: This is our most important native cherry species. Its wood is used for cabinetry, furniture, and a variety of other items, and even as a spice in food. Wild cherry syrup, a cough medicine, is obtained from the bark. The fruit is used for preserves and wine. This tree is also sometimes planted as an ornamental.

SIMILAR SPECIES: Pin Cherry, Common Chokecherry, Alabama Cherry.

CONSERVATION STATUS: LC
This widespread and common species is stable and faces no known or anticipated conservation threats.

Diameter: 2' (0.6 m)
Height: 80' (24 m)

■ DENSE POPULATION
■ LIGHT POPULATION
■ NATURALIZED

This is the only wild plum in the Pacific states and is easily identified when in fruit. The plant occurs in forests in California and western and southern Oregon, both near the coasts and in the Sierra Nevada and Cascade Ranges. The fruit, often borne in large quantities, can be eaten fresh or dried, or in preserves and jellies. Deer and livestock often consume the leaves.

DESCRIPTION: This is a thicket-forming shrub or a small tree with a short trunk; stiff, crooked, nearly horizontal branches; and reddish-purple plums. Its leaves are 1–2½" (2.5–6 cm) long, ½–1¾" (1.2–4.5 cm) wide, elliptical to rounded, with a blunt or rounded tip, and a blunt to notched base. They are sharply saw-toothed, becoming nearly hairless, slightly thickened, with prominent side veins, dark green above, pale beneath, and turning red to yellow in fall. The bark is gray-brown and furrowed into long, scaly plates. The twigs are bright red and short, often ending in spines.

FLOWERS: Early spring. ⅝" (15 mm) wide with five rounded white petals, turning pink. Two to four flowers appear on slender, equal stalks.

FRUIT: A rounded plum ⅝–1" (1.5–2.5 cm) in diameter, reddish-purple or sometimes yellow, with sour, edible pulp and a slightly flattened, smooth stone, maturing in late summer.

CAUTION: Poisonous parts: Although some parts of this species are edible, other parts are toxic to humans. Eat only if you are absolutely sure which parts are edible.

HABITAT: Dry, rocky slopes in moist valleys.

RANGE: Western and southern Oregon south to central California in the Coast Ranges and Sierra Nevada; to 6000' (1829 m).

USES: The tart fruit from this species is eaten fresh or dried and is often made into jellies.

SIMILAR SPECIES: Beach Plum, Bitter Cherry.

CONSERVATION STATUS: **LC**
Despite a somewhat restricted range, the wild population is considered secure. The species is found in numerous protected areas. Like other *Prunus* species, Klamath Plum is susceptible to rust mites (*Aculus cornutus* and *A. fockeui*).

Diameter: 8" (20 cm)
Height: 20' (6 m)

■ DENSE POPULATION
■ LIGHT POPULATION
■ NATURALIZED

Also commonly known as Hog Plum, this native species is found in a variety of habitats in the southeastern United States. The fruit—which matures late in the summer compared to other plums—is often gathered for making jellies and jams. The scientific name describes the plant's flower clusters, known as umbels, which are rounded and have equal stalks like an umbrella.

DESCRIPTION: This is a small tree with a short trunk and a broad, compact, flattened crown of slender branches. It is sometimes a thicket-forming shrub. Its leaves are 1½–2¾" (4–7 cm) long, ¾–1½" (2–4 cm) wide, elliptical, and finely saw-toothed, usually with two dark glands at base. They are dark green above, and paler with a hairy midvein beneath. The bark is dark brown and scaly. The twigs are shiny red and slender, and hairy when young.

FLOWERS: Early spring before leaves. ⅝" (15 mm) wide with five rounded white petals. Two to five flowers are clustered on slender, equal stalks.

FRUIT: A plum ⅜–⅝" (10–15 mm) in diameter with thick black, red, or yellow skin with a whitish bloom, thick sour pulp, and an elliptical stone, maturing quite late in summer.

CAUTION: Poisonous: Do not eat or ingest any part of this species. If accidentally ingested, contact your doctor or poison control center.

HABITAT: Along streams and in sandy or rocky slopes, uplands, edges of forests, and open areas.

RANGE: Southern North Carolina south to central Florida, west to southern Arkansas and eastern Texas; to 1000' (305 m).

USES: The fruits of this species are often made into jellies and jams.

SIMILAR SPECIES: Chickasaw Plum.

CONSERVATION STATUS: LC
Although its native global population is apparently secure and stable, the species is listed as imperiled in North Carolina, at the edge of its natural area of occupancy.

Diameter: 6" (15 cm)
Height: 20' (6 m)

Prunus virginiana **COMMON CHOKECHERRY**

■ DENSE POPULATION
■ LIGHT POPULATION
■ NATURALIZED

This species is widespread across southern Canada and the northern United States. Chokecherry fruits are astringent or puckery, especially when immature or raw, but they can be made into preserves and jelly. This species is sometimes divided into three geographic varieties based on minor differences in the leaves and fruits. Tent caterpillars (*Malacosoma* spp.) often construct their webs on the branches of this plant. It is sometimes known by the common names Bitter-berry or Virginia Bird Cherry.

DESCRIPTION: This is a shrub or small tree, often forming dense thickets, with dark red or blackish chokecherries. Its leaves are 1½–3¼" (4–8 cm) long, ⅝–1½" (1.5–4 cm) wide, elliptical, finely and sharply saw-toothed, and slightly thickened. They are shiny dark green above, light green and sometimes slightly hairy beneath, and turning yellow in fall. The leafstalks are slender, usually with two gland-dots. The bark is brown or gray, and smooth or becoming scaly. The twigs are brown and slender, with a disagreeable odor.

FLOWERS: Late spring. ½" (12 mm) wide with five rounded white petals, arranged in unbranched clusters to 4" (10 cm) long.

FRUIT: A chokecherry ¼–⅜" (6–10 mm) in diameter with shiny dark red or blackish skin; juicy, astringent or bitter pulp; and a large stone, maturing in summer.

CAUTION: Poisonous parts: Although some parts of this species are edible, other parts are toxic to humans. Eat only if you are absolutely sure which parts are edible.

HABITAT: Moist soils, especially along streams in mountains, forest borders, clearings and roadsides.

RANGE: British Columbia east to Newfoundland, south to North Carolina, west to California; to 8000' (2438 m) in the Southwest.

USES: The fruits from this species were an important food source for many Native Americans. Other parts of the trees were used for medicinal and ceremonial purposes. Now the fruits are used for preserves, jellies, and sometimes wine, particularly in the western United States.

SIMILAR SPECIES: Bitter Cherry, Black Cherry, Pin Cherry, Desert Almond.

CONVERSATION STATUS: (LC)
The global population is thought to be stable and secure, with no specific conservation threats identified. The plant can be weedy or invasive in parts of the Northeast and the Great Plains. Several southeastern states, however, list the species as state-threatened.

Diameter: 6" (15 cm)
Height: 20' (6 m)

- ■ DENSE POPULATION
- ■ LIGHT POPULATION
- ■ NATURALIZED

This native to the southwestern United States and northern Mexico is notably abundant on the south rim at Grand Canyon National Park, Arizona. This species is also an attractive ornamental and is often planted for erosion control. It is an important browse plant for deer and livestock, especially in winter. It is sometimes called "Quininebush" because of the bitter-tasting foliage, or Stansbury's Cliffrose for American explorer Howard Stansbury (1806–1863).

ALTERNATE NAME: *Cowania stansburiana, Cowania mexicana* var. *stansburiana*

DESCRIPTION: This is a small-leaf, evergreen, resinous, spreading shrub or small tree with a crooked trunk; an irregular, open crown of many short, stiff, erect branches; and showy white flowers in spring and summer. Its leaves are evergreen, crowded, and ¼–⅝" (6–15 mm) long. They are wedge-shaped, divided into three to seven narrow lobes, thick and leathery, with the edges rolled under, with white sticky resin-dots, and bitter tasting. They are dark green and often loosely hairy above, and densely covered with white, woolly hairs beneath. The bark is reddish-brown or gray and shreddy, splitting into long, narrow strips. The twigs are reddish-brown, hairy, and gland-dotted.

FLOWERS: Spring–summer. ¾–1" (2–2.5 cm) wide with five broad white or pale yellow, spreading petals borne singly at the end of side twigs. They are fragrant.

FRUIT: ¼" (6 mm) long, each with a long, whitish, feathery tail to 2" (5 cm) long. Five to ten appear in a cluster, developed from a single flower; maturing in autumn.

HABITAT: Dry, rocky hills and plateaus, especially limestone; it is found in deserts and with oaks, pinyons, and junipers.

RANGE: Southeastern Colorado to northern Utah, eastern California, and New Mexico; also south to Mexico. It occurs at 3500–8000' (1067–2438 m).

USES: This species is often planted as an ornamental and for erosion control. Native Americans used to make rope, sandals, and clothing from the shreddy bark and arrow shafts from the stems.

SIMILAR SPECIES: Bitterbrush.

CONSERVATION STATUS: LC
The population is stable with no known threats to the species.

Diameter: 6" (15 cm)
Height: 20' (6 m)

- ■ DENSE POPULATION
- ■ LIGHT POPULATION
- ■ NATURALIZED

Bitterbrush, or Antelope Brush, is among the West's most important native plants for wildlife. Several species of rodents gather and cache the seeds, and the plant provides cover for numerous small animals and birds. It provides high-quality spring and winter browse for large herbivores such as mule deer, elk, moose, bighorn sheep, and pronghorns. Domestic livestock also find the plant highly palatable and are prone to overgraze, leaving the plant denuded and unavailable to wildlife. Some Bitterbrush populations are able to sprout after a fire and quickly reinvade a burned area; other populations do not sprout and may thereby be eliminated from extensive areas following a range fire. Where this occurs on lands already overgrazed, the resulting damage can be disastrous.

DESCRIPTION: Individual Bitterbrush plants exhibit considerable variation in growth form; usually it is an erect, much-branched silvery shrub. It has fragrant yellow flowers and many small, wedge-shaped three-lobed leaves. Its leaves are evergreen or sometimes deciduous. They are alternate but crowded and apparently in bundles, about ½" (12 mm) long, with margins turned under slightly. They are gray-green and downy above, and white and covered with feltlike hairs beneath. The bark is gray or brown, with young branches more or less sticky and covered with fine hairs.

FLOWERS: May–July. About ½" (12 mm) across, fragrant, creamy yellow, usually solitary at ends of short branchlets. They have five petals, five sepals, and about 25 stamens.

FRUIT: A leathery, hairy, teardrop-shaped achene tapering in a persistent style.

HABITAT: Dry, well-drained slopes, usually with sagebrush but ranging upslope in pinyon-juniper woodlands, and dry, open conifer forests nearly to timberline.

RANGE: British Columbia and Montana south to central California and New Mexico.

USES: This species is planted on reclaimed surface mines and sometimes for ornamental purposes. It is valuable to livestock and wildlife.

SIMILAR SPECIES: Cliffrose.

CONSERVATION STATUS: LC
The global population is thought to be secure but has not been widely evaluated. Overgrazing is the biggest threat to communities of this plant.

Height: 3–10' (0.9–3 m)

■ DENSE POPULATION
■ LIGHT POPULATION
■ NATURALIZED

Also called Callery Pear, these attractive trees native to China and Vietnam were introduced to North America as ornamental plantings beginning in the 1960s. Bradford Pear became widely used in commercial plantings and as an urban street tree because of its ability to tolerate a variety of growing conditions and withstand pollution and poor soil. It is fast-growing but short-lived and prone to storm damage. In many places it is spreading into natural habitats and is considered by many to be an invasive weed.

DESCRIPTION: This teardrop-shaped deciduous tree grows 30–50' (9–15 m) in height and 20–30' (6–9 m) in width. It produces masses of white flowers in spring and brilliant fall foliage. Its leaves are alternate, simple, 1–3" (2.5–7.5 cm) long, and wavy, with slightly-toothed margins.

FLOWERS: April–May. White, five-petaled, terminal clusters occur early in the spring before leaves appear.

FRUIT: The inedible, small, round fruits, are green to brown and hard, almost woody.

HABITAT: Roadsides, fields, disturbed areas, urban areas.

RANGE: This native of China has naturalized throughout the eastern, central and southern parts of the United States.

USES: This species is popular as an ornamental tree in both residential and commercial landscapes.

SIMILAR SPECIES: Common Pear and Chinese Pear, although neither is common in North America.

CONSERVATION STATUS:
The conservation status of this species in the wild has not been fully evaluated. This species can be invasive in the eastern and midwestern United States, crowding out native plants and forming nearly pure stands in some disturbed areas.

Height: 30–50' (9–15 m)

■ **DENSE POPULATION**
■ **LIGHT POPULATION**
■ **NATURALIZED**

American Mountain Ash is a handsome tree native to eastern North America. It is often grown as an ornamental. The showy red fruit of this tree persists into winter. The berries are readily eaten by birds, especially grouse, grosbeaks, and waxwings. Moose browse the leaves, winter twigs, and fragrant inner bark. According to one story, Roan Mountain at the North Carolina-Tennessee border was named for the Rowan Trees (another name for this species) along its summit.

DESCRIPTION: This is a small tree with a spreading crown or a shrub with many stems, and with showy white flowers, and bright red berries. Its leaves are pinnately compound, 6–8" (15–20 cm) long, 11–17 stalkless, lance-shaped leaflets 1½–4" (4–10 cm) long, ½–1" (1.2–2.5 cm) wide, long-pointed, saw-toothed, becoming hairless, yellow-green above, paler beneath, turning yellow in autumn. The bark is light gray, smooth or scaly, and thin. The twigs are reddish-brown, stout, and hairy when young.

FLOWERS: Late spring. ¼" (6 mm) wide with five white rounded petals, numerous flowers, and crowded in upright clusters 3–5" (7.5–13 cm) wide.

FRUIT: Resembles small apples, ¼" (6 mm) diameter, with bright red skin, bitter pulp, and few seeds; many occur in clusters, maturing in fall.

HABITAT: Moist soils of valleys and slopes; in coniferous forests.

RANGE: Ontario to Newfoundland, south to Georgia, and northwest to Illinois; to 5000–6000' (1524–1829 m) in the southern Appalachians.

USES: This species is planted as an ornamental tree, especially in parks and gardens.

SIMILAR SPECIES: Northern Mountain Ash, Staghorn Sumac, Smooth Sumac.

CONSERVATION STATUS:
American Mountain Ash has a large, widely distributed, and stable population. There are no current or anticipated threats to this species.

Diameter: 8" (20 cm)
Height: 30' (9 m)

■ DENSE POPULATION
■ LIGHT POPULATION
■ NATURALIZED

This shrubby species takes the form of a small tree in southeastern Alaska. Numerous birds and mammals eat the small, applelike fruit. The common name for the species honors Edward Lee Greene (1843–1915), the American botanist who first described it. It also is commonly called Cascade Mountain Ash.

DESCRIPTION: This species is a shrub that forms dense clumps, or is rarely a small tree, with many small, white flowers and tiny, applelike fruit. Its leaves are pinnately compound and 4–9" (10–23 cm) long, with 11–15 stalkless leaflets 1¼–2½" (3–6 cm) long. They are lance-shaped, pointed at the tip, sharply saw-toothed almost to an unequal, rounded base, and becoming hairless. They are shiny dark green above and slightly paler beneath. The bark is gray or reddish and smooth. The twigs are light brown and slender, with whitish hairs when young.

FLOWERS: Early summer. ⅜" (10 mm) wide with five white rounded petals, fragrant, and many in upright, rounded clusters 1¼–3" (3–7.5 cm) wide.

FRUIT: Resembles a tiny apple, less than ⅜" (10 mm) in diameter, shiny red, bitter, with several seeds, and with fewer than 25 in a cluster, maturing in summer and persisting into winter.

HABITAT: Moist soils, in openings and clearings; in coniferous forests.

RANGE: Alaska east to Southwest Mackenzie and south in the mountains to central California and southern New Mexico, locally beyond. It occurs down to sea level in the north, and at 4000–9000' (1219–2743 m) in the south.

USES: This species does not have any significant commercial or popular use. The fruit is a valuable food source for wildlife.

SIMILAR SPECIES: Sitka Mountain Ash.

CONSERVATION STATUS: LC
The population is secure and there are no known threats.

Diameter: 4" (10 cm)
Height: 20' (6 m)

■ DENSE POPULATION
■ LIGHT POPULATION
■ NATURALIZED

This species is named for Sitka, Alaska, where it was first discovered, although it is widespread throughout far-western North America. It is somewhat similar to, and often confused with, Greene Mountain Ash. One key difference is that the leaves of Sitka Mountain Ash are sharply saw-toothed only above the middle rather than over most of their length. Grosbeaks, waxwings, grouse, and other birds consume the fruit and spread the seeds. The species also is commonly called Western Mountain Ash. California Mountain Ash is an aggregate species of closely related western mountain ashes that is sometimes classified as a subspecies of Sitka Mountain Ash and sometimes treated as a distinct species (*S. californica*).

DESCRIPTION: This is a shrub or a small tree with a rounded crown and with many fragrant, small white flowers in early summer, followed by small, applelike fruit. Its leaves are pinnately compound and 4–8" (10–20 cm) long, usually with 9–11 (sometimes 7–13) stalkless leaflets 1¼–2½" (3–6 cm) long. They are elliptical or oblong, rounded or blunt at the ends, and sharply saw-toothed above the middle. They are dull blue-green above, and pale and hairless or nearly so beneath. The bark is gray and smooth, and the twigs have rust-colored hairs when young.

FLOWERS: Early summer. ¼" (6 mm) wide with five rounded, white petals, with many appearing in upright, rounded clusters 2–4" (5–10 cm) wide.

FRUIT: Resembles a small apple, ⅜–½" (10–12 mm) in diameter, red, turning orange or purple, bitter, with few seeds, with several in a cluster, maturing in late summer.

HABITAT: Moist soils in coniferous forests.

RANGE: Southwest Alaska southeast along the coast to Washington and in the mountains to central California and northern Idaho. It occurs to the timberline northward and at 5000–10,000' (1524–3048 m) in the south.

USES: This species does not have significant commercial or popular use. The fruit is a valuable food source for wildlife.

SIMILAR SPECIES: Greene Mountain Ash.

CONSERVATION STATUS: 🔵 LC
The population is stable, rather large, and considered secure due to its wide range. The species faces no current or anticipated conservation threats.

Diameter: 6" (15 cm)
Height: 20' (6 m)

■ DENSE POPULATION
■ LIGHT POPULATION
■ NATURALIZED

This shrubby evergreen has a limited range, found only in parts of southern Arizona and Baja California. The beautiful, dark brown wood streaked with red is hard and very heavy; the tree's scarcity, small size, and slow growth have limited its commercial value. It is also called Arizona Rosewood. This genus is named for Louis Nicolas Vauquelin (1763–1827), a renowned French chemist who worked extensively with plants.

DESCRIPTION: This is an evergreen, much-branched shrub or a small tree with stiff, twisted branches. Its leaves are evergreen, 1½–4" (4–10 cm) long and ¼–⅝" (6–15 mm) wide. They are narrowly lance-shaped, long-pointed at the tip, short-pointed at the base, saw-toothed, thick and leathery, and short-stalked. They are bright yellow-green with a sunken midvein and many fine side veins above, covered with fine white hairs beneath. The bark is dark reddish-brown, thin, and broken into small square scales or shaggy. The twigs are reddish-brown, slender, and densely hairy.

FLOWERS: Late spring. ⅜" (10 mm) wide with five spreading, rounded white petals, turning red, with many crowded in terminal, upright, hairy clusters 2–3" (5–7.5 cm) wide.

FRUIT: A hard, hairy capsule ¼" (6 mm) long, splitting into five parts, each with two winged seeds, and maturing in summer, remaining attached during winter.

HABITAT: Canyons and mountains in high-elevation deserts and oak woodlands.

RANGE: Mountains of southern Arizona; also Baja California; at 2500–5000' (762–1524 m).

USES: This species is often cultivated as an ornamental or hedge plant.

SIMILAR SPECIES: This species is fairly distinctive and not commonly confused with other trees or shrubs.

CONSERVATION STATUS: NT

The wild population of this species is apparently secure but occupies a limited range.

Diameter: 8" (20 cm)
Height: 20' (6 m)

■ DENSE POPULATION
■ LIGHT POPULATION
■ NATURALIZED

Russian Olive became a popular ornamental and roadside tree because it is tolerant of cold, drought, poor soil, and pollution, but is now considered a noxious or invasive weed in North America. It has been most problematic in the West, where it sometimes takes over native riparian habitats. The plants sprout and spread quickly from an extensive underground network of roots that can be extremely difficult to eradicate, and their seeds also are dispersed by wildlife. Many birds consume the fruit, including waxwings, robins, grosbeaks, pheasants, and quail. This species is not related to the Olive (*Olea europaea*), which also has narrow, gray leaves.

DESCRIPTION: This introduced shrub, or small tree, often has a crooked or leaning trunk, a dense crown of low branches, silvery foliage, and sometimes spiny twigs. Its leaves are 1½–3¼" (4–8 cm) long and ⅜–¾" (10–19 mm) wide. They are lance-shaped or oblong and without teeth, and short-stalked. They are dull gray-green with obscure veins above, and silvery, scaly, and brown-dotted beneath. The bark is gray-brown, thin, fissured and shedding in long strips. The twigs are silvery, scaly when young, becoming reddish-brown, long and slender, and often ending in a short spine.

FLOWERS: Late spring–early summer. ⅜" (10 mm) long and bell-shaped with four calyx lobes, yellow inside, silvery outside (petals absent), fragrant, short-stalked, and scattered along twigs at leaf bases.

FRUIT: ⅜–½" (10–12 mm) long, berrylike, elliptical, yellow to brown with silvery scales, becoming shiny. It has thin, yellow, mealy, sweet edible pulp; a large brown stone; and scattered along the twig. It matures in late summer and fall.

HABITAT: Moist soils, from salty to alkaline; spreading in valleys.

RANGE: Native to southern Europe and Asia; it is planted and naturalized from British Columbia east to Ontario and from New England west to Texas and California. It occurs to 5000' (1524 m) or above.

USES: This shrubby tree has long been cultivated as an ornamental plant but is now considered to be an invasive, problematic species in most of North America.

SIMILAR SPECIES: Autumn Olive.

CONSERVATION STATUS: 🅛🅒
There is little data about the wild population of this species, but it is widespread and presumed to be secure. In North America this species tends to be invasive; it thrives in poor soil and outcompetes many native plant species, and it spreads quickly and can be extremely difficult to remove once established.

Diameter: 4" (10 cm)
Height: 20' (6 m)

■ DENSE POPULATION
■ LIGHT POPULATION
■ NATURALIZED

In the United States, Autumn Olive was once heavily planted in disturbed habitats for revegetation purposes and along roadsides and other areas as an ornamental. Its berries are readily consumed by wildlife, which increased its popularity in gardens and planting projects. However, it is now widely considered a noxious or invasive weed, and many agencies and organizations have implemented programs to eradicate it in native habitats and to prevent further plantings. The birds and mammals that consume the berries deposit them into native habitats, contributing to the spread of this plant throughout eastern North America.

DESCRIPTION: This is a deciduous shrub that bears masses of berries in the fall. Its leaves are 1–3½" (2.5–9 cm) long, alternate, ovate to lanceolate, untoothed, and shiny green above, silvery below.

FLOWERS: ¾" (2 cm) long and tubular, in small drooping clusters. They are fragrant and creamy white.

FRUIT: Abundant brownish to red berries ¼" (6 mm) in diameter.

HABITAT: Disturbed habitats, such as roadsides and fields; open woods and woodland edges.

RANGE: Native of China, Japan, and Korea; it is naturalized in Ontario and in the eastern United States from Maine to Florida and west to Nebraska and Louisiana.

USES: This species has been heavily planted on reclaimed surface mines and other areas requiring revegetation. It is now considered an invasive pest in the United States.

SIMILAR SPECIES: Russian Olive.

CONSERVATION STATUS: LC

The native population in China is stable and unthreatened. This fast-spreading species is considered invasive in much of North America. It rapidly invades prairies, savannas, and wetlands and can be difficult to contain or eradicate. The plant alters natural plant community structure and composition by creating dense thickets and shading out native species.

Height: to 20' (6 m)

■ DENSE POPULATION
■ LIGHT POPULATION
■ NATURALIZED

This silvery, shrubby tree, also known as Bull Berry or Thorny Buffaloberry, is native to central and western North America. Its berries are an important food source for the Sharp-tailed Grouse (*Tympanuchus phasianellus*). They are also sometimes eaten by people, although they have a sour taste.

DESCRIPTION: This deciduous, thorny shrub or small tree has silvery leaves, young twigs, and berries. Its branches are opposite, and the twigs are often spine-tipped. Its leaves are 1½" (3.8 cm) long, opposite, and oblong. The stems are thorny, silvery-scurfy when young, becoming brownish with age. The roots are shallow and much-branched; they sprout readily.

FLOWERS: April–May. Tiny, yellowish, four-lobed flowers in clusters.

FRUIT: Tiny red or yellow, egg-shaped berries, often crowded and showy.

HABITAT: River bottoms, ditches, and meadows.

RANGE: Most common in northern Great Plains; also found from British Columbia to Manitoba, south throughout the western United States and to the Great Lakes.

USES: The sour berries are edible and sometimes eaten raw or dried, or used for flavoring other food.

SIMILAR SPECIES: Canada Buffaloberry (*S. canadensis*) is a shrub that resembles the shrubby form of this species.

CONSERVATION STATUS: LC
The population is stable with no major threats to the species identified.

Height: 18' (5.5 m)

FELTLEAF CEANOTHUS   *Ceanothus arboreus*

■ DENSE POPULATION
■ LIGHT POPULATION
■ NATURALIZED

This distinctive species is endemic to coastal southern California, growing wild on three of the Channel Islands. It is also sometimes called Island Ceanothus or Island Mountain Lilac. The species has been planted along the coast as an ornamental and along roadsides. Wildlife, especially butterflies and quail, use this species as a food plant. One of the largest species of its genus, it has an appropriate species name that means "treelike."

DESCRIPTION: This is an evergreen large shrub or small tree with a short, straight trunk, a rounded crown of many stout, spreading branches, and abundant, pale blue flowers. Its leaves are evergreen, 1–3" (2.5–7.5 cm) long, and ¾–1½" (2–4 cm) wide. They are broadly ovate to elliptical, with three prominent, sunken main veins. They are finely wavy saw-toothed, slightly thickened, dull green and finely hairy above, and paler and densely covered with white hairs beneath. The bark is gray, smooth, and thin, becoming dark brown and fissured into small, square, scaly plates. The twigs are light brown, slender, slightly angled, and covered with soft hairs.

FLOWERS: Early spring. ⅛" (3 mm) wide with five pale blue petals, fragrant, in branched clusters 2–6" (5–15 cm) long, and crowded on slender, hairy stalks at the ends of leafy twigs.

FRUIT: ¼" (6 mm) in diameter, rough, black capsules, three-lobed, splitting into three one-seeded nutlets, maturing in summer.

HABITAT: Chaparral on dry slopes and in canyons.

RANGE: Restricted to Santa Rosa, Santa Cruz, and Santa Catalina islands off California; near sea level.

USES: This species is used as an ornamental plant and commonly planted along roadside.

SIMILAR SPECIES: Greenbark Ceanothus, Blue Blossom.

CONSERVATION STATUS: VU

A species endemic to three of the Channel Islands, the species is considered vulnerable due to its limited range and because of the threat of overgrazing and trampling by wild and domesticated herbivores on the islands. It is common where it occurs outside of its native range.

Diameter: 1' (0.3 m)
Height: 25' (7.6 m)

■ DENSE POPULATION
■ LIGHT POPULATION
■ NATURALIZED

This species has a limited natural range in southern California and northern Baja California. These plants—often occurring as shrubs—are drought-resistant and sprout from stumps, but are killed back in cold winters. Also known as Greenbark Whitethorn, the species also is sometimes called Redheart because of its dark red wood, which makes good fuel. Deer and elk commonly browse the leaves of this and related species.

DESCRIPTION: This is an evergreen, spiny shrub or small tree with an irregular, spreading, open crown and many showy flowers resembling lilacs. The leaves are evergreen, ½–1¼" (1.2–3 cm) long, and ⅝–¾" (15–19 mm) wide. They are elliptical to oblong, generally not toothed, with one midvein, thick and stiff, and shiny green on both surfaces. On young shoots and seedlings, the leaves are larger, toothed, and three-veined. The bark is olive green and smooth, becoming dark red-brown, rough, and scaly. The twigs are bright green, slender, and widely forking, with some ending in short, sharp spines.

FLOWERS: Early spring. ⅛" (3 mm) wide with five pale blue to whitish petals, fragrant, crowded on slender stalks in branched clusters 2–6" (5–15 cm) long, at the ends of new leafy twigs.

FRUIT: Sticky black capsules ¼" (6 mm) in diameter, slightly three-lobed, splitting into three one-seeded nutlets.

HABITAT: Dry hillsides and mountain canyons usually near coast; in chaparral and sage shrub.

RANGE: Pacific Coast Ranges of southwest California and Baja California, to 3000' (914 m).

USES: This species is often grown as an ornamental and screen along roadsides and is also useful for erosion control. The wood is sometimes used as firewood.

SIMILAR SPECIES: Feltleaf Ceanothus, Blue Blossom.

CONSERVATION STATUS:
The wild population is apparently secure but occupies a limited range. There are no specific threats to the species identified.

Diameter: 6" (15 cm)
Height: 20' (6 m)

■ DENSE POPULATION
■ LIGHT POPULATION
■ NATURALIZED

This is the hardiest and largest *Ceanothus* species. Each spring the highways of the West Coast display masses of Blue Blossom flowers. Plants can be grown in screens, in hedges, and against walls. Elk and deer consume the leaves. Dense thickets of these plants can form after fires and logging. The species is sometimes known colloquially as the "California Lilac" because of its flowers, but this plant is not closely related to true lilacs (genus *Syringa*).

DESCRIPTION: This is a large evergreen shrub or a small tree with a short trunk, many spreading branches, and showy blooms resembling lilacs. Its leaves are evergreen, ¾–2" (2–5 cm) long, and ½–¾" (12–19 mm) wide. They are oblong or elliptical, rounded to short-pointed at both ends, finely wavy saw-toothed, with three main veins, and slightly thickened. They are shiny green above, and paler and slightly hairy on raised veins beneath. The bark is red-brown and fissured into narrow scales. The twigs are pale yellow-green, angled, and slightly hairy when young.

FLOWERS: Spring. ³⁄₁₆" (5 mm) wide with five light to deep blue petals (rarely almost white). The flowers are fragrant and crowded on slender stalks in branched clusters 1–3" (2.5–7.5 cm) at the base of the upper leaves.

FRUIT: Smooth, sticky, black capsules ³⁄₁₆" (5 mm) in diameter, slightly three-lobed, splitting into three one-seeded nutlets, and maturing in summer.

HABITAT: Mountain slopes and canyons; it is found in chaparral, and in Redwood and mixed evergreen forests.

RANGE: Southwest Oregon south to southern California in the outer Pacific Coast Range; to 2000' (610 m).

USES: This species is grown as an ornamental along roadsides and as a hedge plant.

SIMILAR SPECIES: Greenbark Ceanothus, Feltleaf Ceanothus.

CONSERVATION STATUS: LC
This species is robust and prevalent; its population is stable and faces no known conservation threats.

Diameter: 8" (20 cm)
Height: 20' (6 m)

■ DENSE POPULATION
■ LIGHT POPULATION
■ NATURALIZED

Also commonly known as European Buckthorn and Alder Buckthorn, this is a handsome ornamental with shiny, alderlike leaves, turning yellow in autumn. It has long been planted both as a tree and as a tall hedge, especially in Europe, although it does spread rapidly and can become a pest. Although it was introduced to North America in the 1800s it did not widely naturalize until the early 1900s. SInce then, it has invaded wetlands and wet forests in the eastern half of North America, forming dense stands that crowd out native plants. Its main means of dispersal is via the many birds that eat its fruit from the summer through the fall. The Latin name *Frangula*, meaning "to break," refers to the brittle wood.

ALTERNATE NAME: *Rhamnus frangula*

DESCRIPTION: This is an introduced, invasive shrub or small tree with glossy foliage and red-to-black berries. Its leaves are 1½–2¾" (4–7 cm) long and ¾–2" (2–5 cm) wide. They are elliptical, usually widest above the middle, not toothed, with several almost straight parallel side veins, and nearly hairless. They are shiny dark green above and paler beneath, turning clear yellow in autumn. The bark is grayish, thin, and slightly fissured and warty. Twigs: slender; covered with fine hairs; thornless, ending in naked bud of tiny hairy leaves.

FLOWERS: Late spring and early fall. The flowers are ⅛" (3 mm) wide and bell-shaped, with five pointed, greenish-yellow sepals; they are arranged in short clusters at the leaf bases.

FRUIT: ⁵⁄₁₆" (8 mm) in diameter, berrylike; turning from red to black, clustered, with two or three seeds, maturing in late summer and fall.

CAUTION: Poisonous: Do not eat or ingest any part of this species. If accidentally ingested, contact your doctor or poison control center.

HABITAT: Hardy in various soils, escaping especially in wet areas along fences and in bogs.

RANGE: Native to Europe, western Asia, and northern Africa. Naturalized from southern Manitoba east to Nova Scotia, south to Maryland, Tennessee, and Iowa, and west to Wyoming and Colorado.

USES: This species is cultivated as an ornamental tree or shrub. The bark and berries were once used as a laxative, and the bark yields a yellow dye.

SIMILAR SPECIES: Carolina Buckthorn.

CONSERVATION STATUS:
The wild population of this species has not been formally evaluated. It is an invasive species of medium to high concern in parts of North America, especially in the North and East.

Diameter: 4" (10 cm)
Height: 20' (6 m)

■ DENSE POPULATION
■ LIGHT POPULATION
■ NATURALIZED

California Coffeeberry, also called California Buckthorn, is one of the most common shrubs of canyons and hillsides in California, where it occurs in redwood forest, Douglas-fir forest, coastal pine forest, oak woodland, mixed evergreen woodland, and chaparral. It is variable in form across subspecies and can grow as a tree to around 15' (4.5 m) in some conditions. The fruits and foliage of this long-lived plant are an important food source for wildlife, including various birds and mammals. About a half dozen subspecies are recognized.

ALTERNATE NAMES: *Rhamnus californica, Rhamnus tomentella*

DESCRIPTION: This is usually a tall shrub with leathery, elliptical leaves and black berries. The leaves are evergreen, alternate, and 1–3" (2.5–7.5 cm) long. They are usually more or less elliptical, with margins without teeth or sharply to bluntly toothed. They are dark green and hairless or sparsely downy beneath. In hot, dry places the leaves are smaller, thicker, and often hairy on one or both surfaces. In shady places the leaves are larger, thinner, less leathery, and mostly hairless. The bark is gray-brown or reddish.

FLOWERS: April–June. Inconspicuous and greenish, with several in stalked umbels.

FRUIT: A two- or three-part, berrylike drupe about ¼" (6 mm) long, green, turning red and then black at maturity.

CAUTION: Poisonous: Do not eat or ingest any part of this species. If accidentally ingested, contact your doctor or poison control center.

HABITAT: Chaparral, woodland, and lower montane forests.

RANGE: Southern Oregon to Baja California, east to the western slope of the Sierra Nevada and the mountains of the Mojave Desert.

USES: California Coffeeberry is cultivated as an ornamental plant and is often used in wildlife gardens and native landscaping.

SIMILAR SPECIES: Cascara.

CONSERVATION STATUS: **LC**
This species is relatively common in the wild in a variety of habitats; its population is considered secure.

Height: usually 4–6' (1.2–1.8 m), less often 10–20' (3–6 m)

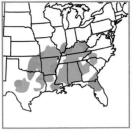

- ■ DENSE POPULATION
- ■ LIGHT POPULATION
- ■ NATURALIZED

This deciduous plant is found in a wide variety of habitats, and can occur as either a shrub or small tree. Birds and other wildlife consume the Carolina Buckthorn's berries, which are reported to have medicinal properties but can affect the liver. Although commonly called a buckthorn, this species typically has no spines. It was first described in South Carolina, hence the common and Latin species names, but it occurs across a wide range in the southeastern, south-central, and midwestern United States.

ALTERNATE NAME: *Rhamnus caroliniana*

DESCRIPTION: This is a shrub or small tree with a spreading crown of many slender branches. Its leaves are 2–5" (5–13 cm) long and ¾–1½" (2–4 cm) wide. They are elliptical, finely wavy-toothed, with many nearly straight side veins, covered with rust-colored hairs when young, becoming dark green above, and paler and often hairy beneath, turning yellow in autumn. The bark is gray, often with blackish patches, thin, and slightly fissured. The twigs are green or reddish, slender, becoming hairless, and ending in hairy naked buds of tiny brown leaves. They are foul-smelling when crushed.

FLOWERS: Late spring–early summer. ³⁄₁₆" (5 mm) wide and bell-shaped, with five pointed greenish-yellow sepals, arranged in clusters of several flowers at the leaf bases.

FRUIT: ⅜" (10 mm) in diameter, berrylike, ripening from red to shiny black, with thin juicy pulp, usually three seeds, and maturing in late summer and fall, often remaining attached.

CAUTION: This plant and other species in the genus can affect the liver if ingested. Eating or ingesting any part of this species is not recommended. If accidentally ingested, contact your doctor or poison control center.

HABITAT: Moist soils of stream valleys and upland limestone ridges.

RANGE: Extreme southern Ohio and West Virginia south to central Florida, west to central Texas, and north to central Missouri; to 2000' (610 m); it is also found in northeastern Mexico.

USES: This species is not widely used commercially but is known to have medicinal uses as a laxative. It is an important food source for wildlife.

SIMILAR SPECIES: Glossy Buckthorn.

CONSERVATION STATUS: 🄻🄲
This species is stable and does not currently have any conservation concerns.

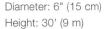

Diameter: 6" (15 cm)
Height: 30' (9 m)

■ DENSE POPULATION
■ LIGHT POPULATION
■ NATURALIZED

This shade-tolerant, slow-growing plant is primarily an understory species native to western North America. The bark is the source of the traditional laxative drug cascara sagrada, meaning "sacred bark" in Spanish. It is harvested commercially in Washington and Oregon by stripping bark from wild trees. When a tree is cut down, several sprouts grow from the stump. The berries are eaten by many birds and mammals, including bears—hence the common name Bearberry that is sometimes associated with this species. It is also sometimes called Cascara Buckthorn.

ALTERNATE NAME: *Rhamnus purshiana*

DESCRIPTION: This is a large shrub or small tree with a short trunk and a crown of many stout, upright branches. Its leaves are often clustered near the ends of twigs. They are 2–6" (5–15 cm) long, 1–2½" (2.5–6 cm) wide, broadly elliptical, and finely wavy-toothed, with many nearly straight side veins. They are dull green and nearly hairless above, and paler and slightly hairy beneath, turning pale yellow in late autumn. The bark is gray or brown, thin, and fissured into short, thin scales, and bitter, with yellow inner bark, turning brown upon exposure. The twigs are gray, slender, and hairy when young, ending in a bud of tiny brown leaves covered with rust-colored hairs.

FLOWERS: Spring–early summer. ³⁄₁₆" (5 mm) wide and bell-shaped, with five pointed, greenish-yellow sepals, arranged in clusters at leaf bases.

FRUIT: ⅜" (10 mm) in diameter, berrylike, red, turning to purplish-black, with thin, juicy, sweetish pulp and two or three seeds, maturing in late summer or autumn.

CAUTION: Eating or ingesting any part of this species can cause abdominal pain and diarrhea. The U.S. Food and Drug Administration (FDA) banned the use of this plant as an over-the-counter laxative in 2002 after studies investigating it as a possible carcinogen were inconclusive.

HABITAT: Moist soils in open areas, along roadsides, and in understory of coniferous and mixed evergreen forests.

RANGE: British Columbia south to northern California; it is also found in the Rocky Mountain region south to northern Idaho and western Montana. It occurs to 5000' (1524 m).

USES: This species is used for medicinal purposes, as a laxative. The fruit is edible, but also has a laxative effect. It is sometimes used as flavoring in drinks and ice cream. The wood is used locally for fence posts and firewood.

SIMILAR SPECIES: California Buckthorn.

CONSERVATION STATUS: **LC**
This is a relatively widespread and sometimes common species, but recent assessment shows a decline in the population of mature trees. The plant is subject to heavy exploitation using harvesting methods that are not sustainable—often mature trees are cut down to harvest the bark, or bark is stripped from live trees without leaving enough on the tree to ensure its survival.

Diameter: ½–1' (0.15–0.3 m)
Height: 30' (9 m)

■ DENSE POPULATION
■ LIGHT POPULATION
■ NATURALIZED

This medium-sized tree, quite similar to Northern Hackberry (*Celtis occidentalis*), grows in moist soils throughout the southeastern United States. Robins, mockingbirds, and many other birds relish the sweet fruits. The wood from this tree is often used to make furniture.

DESCRIPTION: This tree has a broad, rounded, open crown of spreading or slightly drooping branches. Its leaves are arranged in two rows and they are 2½–4" (6–10 cm) long and ¾–1¼" (2–3 cm) wide. They are broadly lance-shaped, long-pointed, often curved, and thin, with the two sides unequal, usually without teeth but sometimes with a few, and three main veins from the base. They are dark green and typically smooth above, and paler and usually hairless beneath. The bark is light gray, thin, and smooth, with prominent corky warts. The twigs are greenish, slender, and mostly hairless.

FLOWERS: Early spring. ⅛" (3 mm) wide, greenish, with both male and female at the base of young leaves.

FRUIT: Drupes are ¼" (6 mm) in diameter, orange-red or purple, one-seeded, dry and sweet, and slender-stalked at leaf bases.

HABITAT: Moist soils, especially clay, on river floodplains; it is sometimes found in pure stands but usually occurs with other hardwoods.

RANGE: Southeastern Virginia south to southern Florida, west to central and southwestern Texas, and north to central Illinois; it is also found in northeastern Mexico. It occurs to 2000' (610 m).

USES: This species is often planted as an ornamental or shade tree. The wood is used for furniture, sporting goods, and plywood.

SIMILAR SPECIES: Northern Hackberry, Lindheimer Hackberry.

CONSERVATION STATUS: LC
Although considered stable and without serious threats, the species can be susceptible to damage from various pests and diseases. Subpopulations found in Mexico are also at greater risk of habitat loss and possible future impacts from climate change.

Diameter: 1½' (0.5 m)
Height: 80' (24 m)

NETLEAF HACKBERRY *Celtis reticulata*

■ DENSE POPULATION
■ LIGHT POPULATION
■ NATURALIZED

Also known as Western Hackberry, this is the native hackberry of the western United States, mainly in the Southwest, but extending eastward into the prairie states. The sweet fruit is eaten by wildlife and was a food source for Native Americans. This hackberry is mostly confined to areas with a constant water supply. It is subject to cosmetic damage from various diseases and pests: The branches often have deformed bushy growths produced by mites and fungi, and the leaves may bear rounded, swollen galls caused by plant lice. Netleaf Hackberry is sometimes considered a variety of Sugarberry but is often elevated to a full species.

ALTERNATE NAME: *Celtis laevigata* var. *reticulata*

DESCRIPTION: This shrub or small tree has a short trunk and an open, spreading crown. Its leaves are arranged in two rows, and they are 1–2½" (2.5–6 cm) long and ¾–1½" (2–4 cm) wide. Their shape is highly variable, but they are mostly ovate, short- or long-pointed; with three main veins from an unequal-sided, rounded, or slightly notched base; without teeth or sometimes coarsely saw-toothed, and usually thick. They are dark green and rough above, and yellow-green with a prominent network of raised veins and slightly hairy beneath, shedding in late fall or winter. The bark is gray, smooth or becoming rough and fissured, with large, corky warts. The twigs are light brown, slender, slightly zigzag, and hairy.

FLOWERS: Early spring. ⅛" (3 mm) wide, greenish, with male and female at the base of young leaves.

FRUIT: Orange-red, one-seeded, sweet berries ¼–⅜" (6–10 mm) in diameter, short-stalked at the leaf bases and maturing in autumn.

HABITAT: Moist soils usually along streams, in canyons, and on hillsides in desert, grassland, and woodland zones.

RANGE: Central Kansas south to Texas, west to southern California, and north to eastern Washington; it is also found in northern and central Mexico. It usually occurs at 1500–6000' (457–1829 m).

USES: This species is cultivated as an ornamental plant. The berries were a significant food source for Native Americans and are still eaten by many people.

SIMILAR SPECIES: Sugarberry, Northern Hackberry.

CONSERVATION STATUS: LC
The population is secure with no conservation threats identified.

Diameter: 1' (0.3 m)
Height: 20–30' (6–9 m)

■ DENSE POPULATION
■ LIGHT POPULATION
■ NATURALIZED

This hardwood tree is widespread in the eastern and mid-western United States. Many birds, including quail, pheasants, woodpeckers, and waxwings, consume the sweet fruits. Branches of this and other hackberries may become deformed by bushy growths called witches'-brooms, produced by mites and fungi. The leaves often bear rounded galls caused by tiny, jumping plant lice. Also called Common Hackberry, this tree's common name was probably derived from "hagberry," a name used in Scotland for a cherry.

DESCRIPTION: This tree has a rounded crown of spreading or slightly drooping branches, often deformed as bushy growths called witches'-brooms. Its leaves are arranged in two rows, and they are 2–5" (5–13 cm) long, 1½–2½" (4–6 cm) wide, ovate, and long-pointed, usually sharply toothed except toward an unequally sided, rounded base, with three main veins. They are shiny green and smooth (sometimes rough) above, and paler and often hairy on veins beneath, turning yellow in fall. The bark is gray or light brown, smooth with corky warts or ridges, becoming scaly. The twigs are light brown, slender, mostly hairy, and slightly zigzag.

FLOWERS: Early spring. ⅛" (3 mm) wide and greenish, with male and female at the base of young leaves.

FRUIT: ¼–⅜" (6–10 mm) in diameter, orange-red to dark purple one-seeded drupes, dry and sweet; slender-stalked at leaf bases, and maturing in fall.

HABITAT: Mainly in river valleys, also on upland slopes and bluffs in mixed hardwood forests.

RANGE: Extreme southern Ontario east to New England, south to northern Georgia, west to northwest Oklahoma, north to North Dakota; local in southern Quebec and southern Manitoba; to 5000' (1524 m).

USES: This species is sometimes, but not commonly, planted as a landscape tree. The wood is used for furniture, sporting goods, boxes and crates, and plywood.

SIMILAR SPECIES: Sugarberry.

CONSERVATION STATUS: LC
Northern Hackberry has a large population that is spread across a large geographic region. There are no major threats to the species.

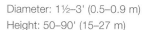

Diameter: 1½–3' (0.5–0.9 m)
Height: 50–90' (15–27 m)

WINGED ELM *Ulmus alata*

■ DENSE POPULATION
■ LIGHT POPULATION
■ NATURALIZED

This elm is found in both dry and moist woodlands of the southeastern and south-central United States. In the 18th and 19th centuries, this tree's fibrous inner bark was made into rope for fastening covers of cotton bales. Both the common and Latin species names refer to the distinctive broad, corky wings that form along the twigs. The Muscogee Native Americans referred to this tree as "Wahoo."

DESCRIPTION: This tree has a short trunk and an open, rounded crown. Its leaves are arranged in two rows, and they are 1¼–2½" (3–6 cm) long, elliptical, often slightly curved with sides unequal, doubly saw-toothed, with yellow midvein and many straight side veins, and thick and firm. They are dark green and hairless above, with soft hairs beneath, turning yellow in autumn. The bark is light brown, thin, and irregularly furrowed. The twigs are brownish, slender, and often with two broad corky wings.

FLOWERS: Early spring. ⅛" (3 mm) wide and greenish, clustered along twigs.

FRUIT: ⅜" (10 mm) long, elliptical, reddish, flat, with one-seeded keys (samaras), hairy, with a narrow wing having two curved points at the tip, maturing in early spring.

HABITAT: Dry uplands including abandoned fields, also in moist valleys; it is also found in hardwood forests.

RANGE: Southern Virginia, south to central Florida, west to central Texas, and north to central Missouri; to 2000' (610 m).

USES: This species is not commonly cultivated and does not have any significant commercial use.

SIMILAR SPECIES: Cedar Elm.

CONSERVATION STATUS: LC

The Winged Elm has a stable and abundant native population. The species faces no major threats and only minor pest concerns. Although mildly susceptible to Dutch elm disease, Winged Elm is not as strongly affected as is the population of American Elm.

Diameter: 1½' (0.5 m)
Height: 40–80' (12–24 m)

■ DENSE POPULATION
■ LIGHT POPULATION
■ NATURALIZED

American Elm is a widespread deciduous tree that occurs in a variety of habitats, especially rich, wet bottomlands, floodplains, stream banks, and swamplands, but also on well-drained soils on high or hilly areas. This well-known species was once abundant and familiar on lawns and city streets, but it has been ravaged by Dutch elm disease, caused by a fungus introduced accidentally around 1930 and spread by European and native elm bark beetles. Its natural range extends across the eastern United States into southeastern Canada, preferring moist soils.

DESCRIPTION: American Elm is a large, handsome, graceful tree, often with enlarged buttresses at the base, usually forked into many spreading branches, drooping at the ends, forming a broad, rounded, flat-topped or vaselike crown, often wider than high. Its leaves are arranged in two rows, and they are 3–6" (7.5–15 cm) long, 1–3" (2.5–7.5 cm) wide, elliptical, abruptly long-pointed, the base rounded with sides unequal, doubly saw-toothed, with many straight parallel side veins, usually hairless or slightly rough above, and paler and usually with soft hairs beneath, turning bright yellow in fall. The bark is light gray and deeply furrowed into broad, forking, scaly ridges. The twigs are brownish, slender, and hairless.

FLOWERS: Early spring. ⅛" (3 mm) wide and greenish, clustered along twigs.

FRUIT: ⅜–½" (10–12 mm) long, elliptical flat one-seeded keys (samaras), with wing hairy on edges, deeply notched with points curved inward, long-stalked, maturing in early spring.

HABITAT: Moist soils, especially valleys and floodplains; in mixed hardwood forests.

RANGE: Saskatchewan east to Cape Breton Island, south to central Florida, and west to central Texas; to 2500' (762 m).

USES: This graceful tree has been widely planted along streets and in parks. The hard, tough wood has been used to make wooden automobiles, and is also used for containers, furniture, and paneling.

SIMILAR SPECIES: Slippery Elm.

CONSERVATION STATUS: **EN**

The American Elm has seen the greatest historical decline of all elms due to the introduced and invasive Dutch elm disease. The disease poses the greatest risk to mature trees. Because American Elms are fast-growing and can set seed at a young age, and mortality does not occur from the disease until maturity, there are still large populations of smaller trees throughout the species' native range. However, the growth rate remains below the mortality rate, and the American Elm population is expected to decline between 50 and 80 percent if the current trend continues.

Diameter: about 4' (1.2 m)
Height: 100' (30 m) if able to achieve maturity

■ DENSE POPULATION
■ LIGHT POPULATION
■ NATURALIZED

Restricted in range to the south-central United States, this is the common native elm in eastern Texas, where it is commonly planted as a shade tree. It was named Cedar Elm because it is often found together with Ashe Juniper trees, which are locally known as cedars. The Latin species name means "thick leaf," a reference to the tree's thick, somewhat leathery foliage.

DESCRIPTION: This tree has a rounded crown of drooping branches and the smallest leaves of any native elm. Its leaves are arranged in two rows, and they are 1–2' (2.5–5 cm) long and ½–1" (1.2–2.5 cm) wide. They are elliptical or lance-shaped, blunt or short-pointed at the tip, the base rounded with unequal sides, coarsely saw-toothed with rounded teeth, thick and slightly leathery, with a yellow midvein and straight side veins. They are shiny dark green and rough above, with soft hairs beneath, turning bright yellow in late autumn. The bark is light brown, furrowed into broad, scaly ridges. The twigs are brownish, slender, and hairy, often with corky wings.

FLOWERS: Late summer. ⅛" (3 mm) wide and greenish, short-stalked, forming at the leaf bases.

FRUIT: ⅜–½" (10–12 mm) long, elliptical, flat, green, one-seeded keys (samaras) with narrow wings deeply notched at the tip, covered with soft white hairs, short-stalked at leaf bases, and maturing in fall.

HABITAT: Moist soils along streams and also upland limestone hills; with various hardwoods.

RANGE: Extreme southwestern Tennessee south to Mississippi, west to southern Texas and extreme northeastern Mexico, and north to southern Oklahoma; it is local in northern Florida. It occurs to 1500' (457 m).

USES: Cedar Elm is planted as an ornamental or shade tree.

SIMILAR SPECIES: Winged Elm, Chinese Elm, September Elm.

CONSERVATION STATUS: LC

The biggest known threat to the Cedar Elm is Dutch elm disease, but the species seems less susceptible to the disease than other elms and has not shown a clear population decline. Climate change may reduce the available habitat for this species by up to half in the future, based on some climate and habitat models, but the magnitude of the threat is unknown. The species also faces minor threats from various pests.

Diameter: 2' (0.6 m)
Height: 80' (24 m)

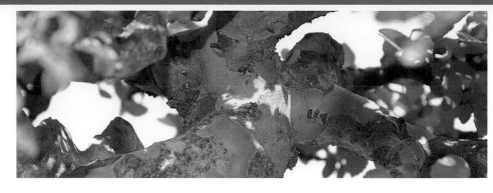

■ DENSE POPULATION
■ LIGHT POPULATION
■ NATURALIZED

Native to eastern Asia, Chinese Elm has been widely introduced around the world. Fast-growing and hardy, Chinese Elm is a handsome ornamental with showy bark and a compact crown. This tree is cultivated for shade and shelterbelts, and is highly resistant to Dutch elm disease. This species should not be confused with Siberian Elm (*Ulmus pumila*), a short-lived and disease-prone tree with weak wood sometimes erroneously called "Chinese Elm."

DESCRIPTION: This introduced tree has a dense, broad, rounded crown of spreading branches and small leaves. Its leaves are arranged in two rows, and they are ¾–2" (2–5 cm) long, ⅜–¾" (10–19 mm) wide, elliptical, saw-toothed, slightly thickened. They are shiny dark green above, and paler and hairy when young and in vein angles beneath, turning reddish or purplish in autumn or remaining nearly evergreen in warm climates. The bark is mottled brown, smooth, and shedding in irregular, thin flakes, exposing reddish-brown inner bark. The twigs are slender, slightly zigzag, and hairy.

FLOWERS: Autumn. ⅛" (3 mm) wide and greenish, clustered at the bases of the leaves.

FRUIT: ⅜" (10 mm) long, elliptical, flat, one-seeded keys (samaras), with broad, pale yellow wings, maturing in autumn.

HABITAT: Moist soils in humid temperate regions.

RANGE: Native of China, Korea, and Japan. It is widely planted across North America, especially in the Gulf and Pacific regions; it is naturalized in many areas.

USES: This species is a popular landscape tree, often found in parking lots, along streets, and in plazas. It is planted all over the world.

SIMILAR SPECIES: Cedar Elm.

CONSERVATION STATUS: LC
The species has no current threats assessed and is considered stable in its native range.

Diameter: 1½' (0.5 m)
Height: 50' (15 m)

■ DENSE POPULATION
■ LIGHT POPULATION
■ NATURALIZED

Slippery Elm is a medium-sized tree native to the eastern half of North America. It is named for the thick, slightly fragrant, glue-like, and edible inner bark, which is found by chewing through the outer bark of a twig. This inner bark can be dried and then moistened for use as a cough medicine or as a poultice. The Latin species name refers to the large, brown buds covered with rust-colored hairs.

DESCRIPTION: This tree has a broad, open, flat-topped crown of spreading branches and large, rough leaves. Its leaves are arranged in two rows. They are 4–7" (10–18 cm) long, 2–3" (5–7.5 cm) wide, elliptical, abruptly long-pointed, the base rounded with sides very unequal, doubly saw-toothed with many straight parallel side veins, and thick. They are green to dark green and very rough above, densely covered with soft hairs beneath, turning dull yellow in autumn. The bark is dark brown and deeply furrowed, the inner bark mucilaginous. The twigs are brownish, stout, and hairy.

FLOWERS: Early spring. ⅛" (3 mm) wide, greenish, numerous, and short-stalked along twigs.

FRUIT: ½–¾" (12–19 mm) long, nearly round, flat, one-seeded keys (samaras) with a light green, broad and hairless wing, slightly notched at the tip, maturing in spring.

HABITAT: Moist soils, especially lower slopes and flood plains, but often on dry uplands; in hardwood forests.

RANGE: Southern Ontario east to extreme southern Quebec and southwest Maine, south to northwestern Florida, west to central Texas, and north to southeastern North Dakota; to 2000' (610 m).

USES: This species has various medicinal uses, primarily as a demulcent. The wood is not widely used—however, the yoke of the Liberty Bell was made from Slippery Elm wood.

SIMILAR SPECIES: American Elm.

CONSERVATION STATUS: LC
The current population is extremely large and widespread, and its numbers are stable. This species is primarily susceptible to mortality from widespread pathogens including Dutch elm disease and elm yellows. Saplings and smaller trees can be susceptible to damage from bark stripping caused by browsing herbivores such as deer.

Diameter: 2–3' (0.6–0.9 m)
Height: 70' (21 m)

Japanese Zelkova is a medium-sized deciduous tree native to Japan, Korea, eastern China and Taiwan. This elm relative is an exquisite ornamental planted for its colorful fall foliage and unique, attractive bark. It is resistant to Dutch elm disease.

DESCRIPTION: This small to medium-sized tree sports a short, straight trunk with a domed or broad, rounded crown of wide-spreading, slender branches, starting low on the trunk. Its leaves are 2–3¾" (5–9.5 cm) long, narrowly elliptic, with a long-pointed tip, an asymmetrical base, coarsely toothed margins, and straight side veins. They are dark green and slightly rough above, and paler, slightly hairy on veins beneath. The bark is smooth with pink, brown, and orange horizontal stripes (most apparent in young bark), and peels to reveal orange inner bark. The twigs are brown, hairy when new, becoming smooth.

FLOWERS: Early spring. ⅛" (3.2 mm) wide, yellow-green, arranged in small clusters on short stalks.

FRUIT: ⅛–¼" (3.2–6.4 mm) long, ovoid drupe, brownish. They are borne singly, at the leaf base, maturing in fall.

HABITAT: Cultivated as an ornamental in North America.

RANGE: Native to Asia, it is introduced in North America.

USES: This species is cultivated as an ornamental tree.

SIMILAR SPECIES: This introduced tree is not likely to be confused with other native species.

CONSERVATION STATUS: NT

Although extensively planted in forests and populated areas in Japan, Europe, and North America, wild populations of this tree are extremely low in the Korean Peninsula and China, where the species is considered threatened. The major threat to this species is habitat loss as a result of urbanization, agricultural expansion, and unsustainable logging.

Diameter: 1–2½' (30–76 cm)
Height: 40–70' (12–21 cm)

■ DENSE POPULATION
■ LIGHT POPULATION
■ NATURALIZED

Native to central and southern Florida, this tree begins as a vine-like epiphyte, using another tree—often a Cabbage Palmetto—for support and sending out aerial roots. It eventually grows over and engulfs the host tree, effectively "strangling" it.

DESCRIPTION: In mature trees, the crown is broadly rounded, the trunks multiple, and the branches stout, with hanging roots. Its leaves are up to 5" (12.5 cm) long, elliptical to ovate, thick, and evergreen. The bark is broken into small, ash-gray to blackish plates.

FRUIT: ¾" (2 cm), round, red or yellow capsule.

CAUTION: Poisonous: Do not eat or ingest any part of this species. If accidentally ingested, contact your doctor or poison control center.

HABITAT: Hammocks, coastal islands, mangrove swamps.

RANGE: Central Florida and southward.

USES: Native Americans and early settlers used the edible fruit as a food source. This tree is sometimes planted as an ornamental tree and as a bonsai.

SIMILAR SPECIES: Shortleaf Fig.

CONSERVATION STATUS:
The Strangler Fig is fairly common in its range and its population is considered secure, with no identified threats to the species.

Height: to 65' (19.5 m)

■ DENSE POPULATION
■ LIGHT POPULATION
■ NATURALIZED

This small evergreen tree, native to Asia, escaped from cultivation and is now naturalized in southern Florida. Other common names for this species include Ficus Tree and Benjamin Fig. It is the official tree of Bangkok.

DESCRIPTION: This is a small tree with evergreen, simple leaves. It has a straight trunk and a symmetrical crown of weeping branches and dense foliage. Its leaves are 2–4" (5–10 cm) long. They are elliptic, with an abruptly long-pointed tip, eight to fifteen or more side veins angling away from the midvein at regular intervals and angles. They are shiny dark green.

FRUIT: ½" (12.7 mm) wide, nearly round, and yellowish or reddish.

HABITAT: Cultivated in North America, often as a potted plant. Naturalized in southern Florida and the Florida Keys.

RANGE: Introduced in North America; escaped from cultivation in southern Florida and the Florida Keys.

USES: This species is popular as a houseplant.

SIMILAR SPECIES: Other *Ficus* species.

CONSERVATION STATUS:  **LC**

The species is stable in its native range with no identified threats. In North America the species can escape from cultivation and spread to nearby areas thanks to roots that grow rapidly, sometimes growing under and lifting pavement, causing damage to sidewalks, patios, and driveways.

Height: to 100' (30 m)

■ DENSE POPULATION
■ LIGHT POPULATION
■ NATURALIZED

This tree, found growing in the river valleys of the central and eastern United States, and in other scattered areas, is commonly known by many names. One of those names is Bodark, from the French *bois d'arc*, meaning "bow wood," referring to Native Americans' use of the wood for archery bows. It is also called Horse Apple, because the fruit is sometimes eaten by livestock. Other names include Hedge Apple, Monkey Ball, and Monkey Brains. Rows of these spiny plants served as fences in the grassland plains before the introduction of barbed wire. Early settlers extracted a yellow dye for cloth from the root bark.

DESCRIPTION: This is a medium-sized, spiny tree with a short, often crooked trunk; a broad, rounded or irregular crown of spreading branches; single, straight, stout spines at the base of some leaves; and milky sap. Its leaves are 2½–5" (6–13 cm) long, 1½–3" (4–7.5 cm) wide, narrowly ovate, long-pointed, not toothed, and hairless. They are shiny dark green above and paler beneath, turning yellow in fall. The bark is gray or brown, thick, and deeply furrowed into narrow, forking ridges. The inner bark of the roots is orange, separating into thin, papery scales. The twigs are brown and stout, with a single spine ¼–1" (0.6–2.5 cm) long at some nodes and short twigs or spurs.

FLOWERS: Early spring. Tiny, greenish, and crowded in rounded clusters less than 1" (2.5 cm) in diameter, with male and female flowers on separate trees.

FRUIT: A heavy, yellow-green, hard and fleshy ball 3½–5" (9–13 cm) in diameter containing many light brown nutlets, maturing in fall and soon dropping.

CAUTION: Contact irritant: This species produces chemicals that irritate the skin. Do not touch. If you accidentally touch this species, thoroughly wash the affected area as soon as possible.

HABITAT: Moist soils of river valleys.

RANGE: The native range is uncertain. Southwest Arkansas to eastern Oklahoma and Texas; widely planted and naturalized in the eastern and some northwestern states.

USES: This species is planted as a hedge or a windbreak, and the heavy, dense wood is often used for fence posts, tool handles, and other items.

SIMILAR SPECIES: This tree is not likely to be confused with other species.

CONSERVATION STATUS: LC
The native range of this species is difficult to determine; the species is widely planted and commonly escapes cultivation. Its population is stable and there are no major threats affecting the species.

Diameter: 2' (0.6 m)
Height: 50' (15 m)

■ DENSE POPULATION
■ LIGHT POPULATION
■ NATURALIZED

White Mulberry has been cultivated for centuries, the leaves serving as the main food of silkworms that are used to produce silk commercially. Introduced in the southeastern United States, this tree grows rapidly and produces abundant, edible berries that are enjoyed by birds and many people alike. The trees spread like weeds in disturbed and urban areas, where the berries often litter the ground. Varieties include a hardy one for shelterbelts, another with drooping or weeping foliage, and a fruitless form.

DESCRIPTION: This naturalized small tree has a rounded crown of spreading branches, milky sap, and edible mulberries. Its leaves are in two rows. They are 2½–7" (6–18 cm) long, 2–5" (5–13 cm) wide, broadly ovate but variable in shape, with three main veins from a rounded or notched base, coarsely toothed, often divided into three or five lobes, and long-stalked. They are shiny green above and paler and slightly hairy beneath. The bark is light brown and smoothish, becoming furrowed into scaly ridges. The twigs are light brown and slender.

FLOWERS: Spring. Tiny, greenish, and crowded in short clusters with male and female on same or separate trees.

FRUIT: ⅜–¾" (10–19 mm) long, a cylindrical purplish, pinkish, or white mulberry composed of many tiny bead-like one-seeded fruits. They are sweet and juicy and edible, appearing in late spring.

HABITAT: Hardy in cities, drought-resistant, and adapted to dry, warm areas.

RANGE: Native of China. It is widely cultivated across the United States; it is naturalized in the East and in the Pacific states.

USES: This species is widely cultivated in the United States, Mexico, Australia, and many other countries to feed silkworms, used in the commercial production of silk. The fruit is sometimes eaten by people, dried, or made into wine.

SIMILAR SPECIES: Red Mulberry.

CONSERVATION STATUS: LC
White Mulberry is considered stable in its native range. In North America, this species has become widely naturalized in disturbed areas. It readily hybridizes with the native Red Mulberry, locally jeopardizing the viability of that species.

Diameter: 1' (0.3 m)
Height: 40' (12 m)

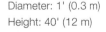

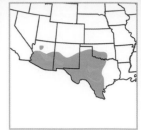

■ DENSE POPULATION
■ LIGHT POPULATION
■ NATURALIZED

This species is found in the southwestern United States, where it is also known as Texas Mulberry. In the U.S., it is found in limited areas of Arizona, New Mexico, and western Texas. It is well adapted to desert habitats. Similar to Red Mulberry, this species has a smaller leaf and a smaller, rounded fruit cluster.

ALTERNATE NAME: *Morus celtidifolia*

DESCRIPTION: This is a medium-sized tree with a short trunk and a rounded crown. The leaves are ½" (1.3 cm) long, elliptic or ovate, or two- or three-lobed.

FLOWERS: Tiny flowers crowded in short, drooping clusters, blooming in spring.

FRUIT: Small, red to black, edible mulberries occur in a small cluster, ripening in late spring.

HABITAT: Mountain canyons, dry limestone hills, and along streams. It is found at 650–7200' (200–2200 m).

RANGE: Texas, New Mexico and Arizona, and rarely in Oklahoma.

USES: This plant is not widely used due to its scarcity, but the berries are edible and sometimes eaten by people and wildlife. Native Americans are believed to have cultivated this plant for its edible fruit.

SIMILAR SPECIES: Red Mulberry.

CONSERVATION STATUS: LC
The global population of this plant is considered secure.

Diameter: 1' (30 cm)
Height: 10–20' (3–6 m)

■ DENSE POPULATION
■ LIGHT POPULATION
■ NATURALIZED

This mulberry is known for its tasty, edible black fruits. Its exact natural range is unknown due to its cultivation since ancient times, but the plant is believed to have originated in southwestern Asia, possibly in the mountains of ancient Mesopotamia and Persia. It has long been cultivated as a food plant and is introduced and naturalized west across much of Europe and the Middle East.

DESCRIPTION: This wide-spreading plant is quite similar to Red Mulberry, but the leaves are rarely lobed. The leaves are heart-shaped, 3–5" (7.5–12.7 cm) long, and dark green turning yellow in fall.

FLOWERS: May–June. The flowers are inconspicuous and greenish, and both sexes are found on the same plant.

FRUIT: Clusters of edible drupes, about 1" (2.5 cm) long and black when ripe; maturing August–September.

HABITAT: Mainly cultivated areas, such as parks and yards.

RANGE: Introduced and naturalized in pockets in the interior United States.

USES: Its black fruits are edible and are eaten raw or cooked, and used to make jams and various desserts.

SIMILAR SPECIES: White Mulberry, another widely cultivated and naturalized plant, sometimes also has black fruits.

CONSERVATION STATUS:
The conservation status of this species in the wild has not been fully assessed.

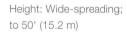

Height: Wide-spreading;
to 50' (15.2 m)

DENSE POPULATION
LIGHT POPULATION
NATURALIZED

This familiar and widespread native is found in moist forests throughout much of the eastern United States. People and wildlife, especially birds, eat the sweet berries. Choctaw Native Americans wove cloaks from the fibrous inner bark of young mulberry shoots.

DESCRIPTION: This is a medium-sized tree with a short trunk, a broad rounded crown, and milky sap. Its leaves are in 2 rows; they are 4–7" (10–18 cm) long, 2½–5" (6–13 cm) wide, ovate, abruptly long-pointed, with three main veins from an often unequal base, coarsely saw-toothed, and often with two or three lobes on young twigs. They are dull dark green and rough above, with soft hairs beneath, turning yellow in autumn. The bark is brown and fissured into scaly plates. The twigs are brown and slender.

FLOWERS: Spring. Tiny, about ⅛" (3 mm) long, crowded in narrow clusters, with male and female on same or separate trees.

FRUIT: 1–1¼" (2.5–3 cm) long, a cylindrical mulberry, red to dark purple, composed of many tiny, beadlike, one-seeded fruits that are sweet, juicy, and edible, appearing in late spring.

HABITAT: Moist soils in hardwood forests.

RANGE: Southern Ontario east to Massachusetts, south to southern Florida, west to central Texas and north to southeastern Minnesota; to 2000' (610 m).

USES: The sweet berries are eaten by people in many ways, and the wood is used locally for fence posts, furniture, interior finish, and agricultural implements.

SIMILAR SPECIES: White Mulberry, Mountain Mulberry.

CONSERVATION STATUS: LC
This common species covers a large geographic range and faces no current or anticipated threats.

Diameter: 2' (0.6 m)
Height: 60' (18 m)

■ DENSE POPULATION
■ LIGHT POPULATION
■ NATURALIZED

Until the first decades of the 20th century American Chestnut was North America's largest broadleaf tree. It is now gone from the forests, a victim of the chestnut blight caused by an introduced fungus that spreads in the breeze. This disease, thought to have originated in China, spread from imported trees planted on Long Island, NY, and discovered at the New York Botanical Garden in 1904. Within 40 years it had virtually wiped out this once abundant species.

DESCRIPTION: Formerly a large tree with a massive trunk and a broad, rounded, dense crown, American Chestnut now sprouts from the base of long-dead trees. The leaves are 5–9" (13–23 cm) long, 1½–3" (4–7.5 cm) wide, narrowly oblong, long-pointed with many straight parallel side veins, each ending in a curved tooth, and short-stalked. They are shiny yellow-green above and paler green below with a few hairs along the midvein, turning yellow in autumn. The bark is dark gray-brown and furrowed into flat ridges, on smooth sprouts. The twigs are green, slender, and hairless.

FLOWERS: Early summer. Many whitish male flowers ³⁄₁₆" (5 mm) long, in upright catkins 6–8" (15–20 cm) long at the base of the leaf. Few female flowers ⅜" (10 mm) long, bordered by narrow greenish scales, at the base of shorter catkins.

FRUIT: 2–2½" (5–6 cm) in diameter; short-stalked burs covered with many stout branched spines about ½" (12 mm) long, maturing in autumn and splitting open along 3–4 lines, 2–3 chestnuts ½–¾" (12–19 mm) long, broadly egg-shaped, becoming shiny dark brown, flattened and pointed; edible.

HABITAT: Moist upland soils in mixed forests.

RANGE: Extreme southern Ontario east to Maine, south to southwestern Georgia, west to Mississippi, north to Indiana; to 4000' (1219 m).

USES: For 300 years chestnut beams provided the structure for millions of houses and barns built in eastern North America. The wood of this species was also once the main domestic source of tannin, the edible chestnuts were a commercial crop, and the leaves were used in home medicines.

SIMILAR SPECIES: European Chestnut, Allegheny Chinkapin, Ozark Chinkapin, Chinese Chestnut.

CONSERVATION STATUS: CR

The American Chestnut suffered a population collapse in the 20th century due to chestnut blight, a disease caused by an Asian parasitic bark fungus (*Cryphonectria parasitica*) that wiped out billions of these once abundant trees. Today, sprouts continue from roots until killed back by the blight, and cultivated trees grow in western states and other areas where the parasite is absent. Blight-resistant chestnuts such as hybrids between American and Chinese species have been developed, a notable example being the Dunstan Chestnut in North Carolina.

Diameter: 4" (10 cm);
formerly 2–4' (0.6–1.2 m)
Height: 20' (6 m);
formerly 60–100' (18–30 m)

CHINESE CHESTNUT *Castanea mollissima*

This small tree, native to China, Taiwan, and Korea, has long been cultivated in eastern Asia for its edible nuts, with hundreds of recognized cultivars and hybrids. It was introduced to North America from China. The nuts are an important food source for various wildlife species.

DESCRIPTION: Chinese Chestnut is a small tree with deciduous, simple leaves. It has a straight trunk and a rounded crown of broad, spreading branches—nearly as wide as it is tall. Its leaves are 4–8" (10–20 cm) long, elliptic, with many side veins, each ending in a tooth at the margin. They have less strongly curved teeth than in American Chestnut. The bark is dark grayish brown and furrowed.

FLOWERS: Produced in clusters 1.5–7" (4–20 cm) long.

FRUIT: 2–3" (5–7.6 cm) wide; a prickly husk containing two or three nuts, each 1" (2.5 cm) wide, and brown.

HABITAT: Cultivated in North America.

RANGE: Native to Asia, it is introduced in North America, where it occasionally naturalizes.

USES: This species is widely cultivated for its sweet, edible nuts.

SIMILAR SPECIES: American Chestnut, Japanese Chestnut.

CONSERVATION STATUS: LC

This species has a widespread, stable population and is not considered threatened in its native range.

Height: to 30' (9 m)

■ DENSE POPULATION
■ LIGHT POPULATION
■ NATURALIZED

This native chestnut, also known as the American Chinquapin or Dwarf Chestnut, is found in dry, well-drained habitats across the southeastern United States. The famous explorer Captain John Smith published the first record of this nut-producing tree in 1612, noting its importance to the Native Americans of modern-day Virginia.

DESCRIPTION: This is a thicket-forming shrub or tree with a rounded crown, bearing spiny burs. Its leaves are 3–6" (7.5–15 cm) long, 1¼–2" (3–5 cm) wide, oblong or elliptical, short-pointed, with many straight parallel side veins, each ending in a short tooth. The leafstalks are short. The leaves are hairy and yellow-green above, with velvety white hairs beneath. The bark is reddish-brown and furrowed into scaly plates. The twigs are gray and hairy.

FLOWERS: Early summer. Many tiny whitish male flowers occur in upright catkins 4–6" (10–15 cm) long at the base of the leaf. Few female flowers, ⅛" (3 mm) long, occur at the base of smaller catkins.

FRUIT: ¾–1¼" (2–3 cm) in diameter, burs with many branched hairy spines, maturing in fall and splitting open. It has a single, edible, egg-shaped nut, shiny dark brown with whitish hairs.

HABITAT: Dry sandy and rocky uplands; in oak and hickory forests.

RANGE: New Jersey and southern Pennsylvania south to central Florida, west to eastern Texas, and north to southeastern Oklahoma, local in Ohio; to 4500' (1372 m).

USES: Native Americans used this plant for medicinal purposes. It is often planted to attract wildlife, as a variety of small mammals eat the nuts.

SIMILAR SPECIES: American Chestnut, Ozark Chinkapin, Sawtooth Oak.

CONSERVATION STATUS: LC
The population is widespread and stable with no major threats.

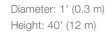

Diameter: 1' (0.3 m)
Height: 40' (12 m)

■ DENSE POPULATION
■ LIGHT POPULATION
■ NATURALIZED

Golden Chinkapin is endemic to the Pacific Coast of the United States. This is a handsome tree with a massive trunk, sometimes attaining an impressive size. Its showy, whitish blooms have a strong odor. The edible nuts, which resemble small chestnuts, are usually eaten by chipmunks and ground squirrels.

ALTERNATE NAME: *Castanopsis chrysophylla*

DESCRIPTION: This is an evergreen tree with a straight trunk, becoming grooved or fluted, and with stout, spreading branches and a broad, rounded crown. There is also a shrub variety. Its leaves are evergreen, 2–5" (5–13 cm) long, and ⅝–1½" (1.5–4 cm) wide. They are lance-shaped or oblong, thick and leathery, with edges slightly turned under, and without teeth. They are shiny dark green with scattered scales above, covered with tiny golden-yellow scales beneath, and turning yellow before falling. The shrub variety has leaves folded upward. The bark is gray and smooth when young, becoming reddish-brown, thick, and deeply furrowed into plates; the inner bark is bright red. The twigs are stiff and scurfy with tiny, golden-yellow scales when young, turning dark reddish-brown.

FLOWERS: Early summer. About ⅛" (3 mm) long, whitish, and stalkless, with many flowers arranged in catkins 2–2½" (5–6 cm) long. They are upright near the ends of twigs, and mostly male, with a few females at the base or in separate clusters.

FRUIT: A nearly stalkless, spiny bur 1–1½" (2.5–4 cm) in diameter; maturing in the autumn of the second year, splitting irregularly into four parts, with one or two nuts 5⁄16–½" (8–12 mm) long that are broadly egg-shaped or rounded, light brown, hard-shelled, and edible.

HABITAT: Gravelly and rocky soils in mountain slopes and canyons in Redwood and evergreen forests. The shrub variety occurs on dry ridges in chaparral and Knobcone Pine forests.

RANGE: The Pacific Coast region from Washington south in the Coast Ranges to central California; it is also local in the Sierra Nevada of central California. It occurs to 1500' (457 m), with the shrub variety occurring to 6000' (1829 m).

USES: This tree's edible nuts, commonly consumed by wildlife, are also sometimes eaten by people.

SIMILAR SPECIES: The golden underside of its leaves helps to separate this tree from other species.

CONSERVATION STATUS: LC

The species has an abundant population in California and Oregon but it is considered threatened in Washington State due to its small population.

Diameter: 1–3' (0.3–0.9 m)
Height: 40–80' (12–24 m)

- DENSE POPULATION
- LIGHT POPULATION
- NATURALIZED

Native to eastern North America, this species was recognized by the early colonists, who already knew the famous, closely related European Beech. The American Beech is a handsome shade tree and bears edible beechnuts, which are readily consumed by a variety of wildlife, including squirrels, raccoons, bears, and many birds. It provided an important food source for the now-extinct Passenger Pigeon, and it is likely that the clearing of beech forests was a major cause of the bird's extinction. Unlike most trees, beeches retain smooth bark in age, making them favorites for carving initials or dates that remain preserved indefinitely.

DESCRIPTION: This large tree has a rounded crown of many long, spreading and horizontal branches, producing edible beechnuts. Its leaves, spread in two rows, are 2½–5" (6–13 cm) long, 1–3" (2.5–7.5 cm) wide, elliptical or ovate, long-pointed at the tip, with many straight parallel slightly sunken side veins and coarsely saw-toothed edges, short-stalked, dull dark blue-green above, light green beneath, becoming hairless or nearly so, turning yellow and brown in fall. The bark is light gray, smooth, and thin. The twigs are slender and end in long, narrow, scaly buds, with short side twigs or spurs.

FLOWERS: With new leaves in spring. Male flowers are small and yellowish with many stamens, crowded in a ball ¾–1" (2–2.5 cm) in diameter, hanging on a slender, hairy stalk to 2" (5 cm) long. Female flowers are about ¼" (6 mm) long and bordered by narrow, hairy, reddish scales.

FRUIT: Short-stalked, light brown, prickly burs ½–¾" (12–19 mm) long, maturing in fall and splitting into four parts, usually with two nuts about ⅝" (15 mm) long; they are three-angled, shiny brown, and known as beechnuts.

HABITAT: Moist rich soils of uplands and well-drained lowlands; often in pure stands.

RANGE: Ontario east to Cape Breton Island, south to Florida, west to eastern Texas and north to Michigan; a variety is found in the mountains of Mexico. It occurs to 3000' (914 m) in the north and to 6000' (1829 m) in the southern Appalachians.

USES: The hard wood from this species is widely used for furniture, firewood, and a variety of other purposes. This tree is sometimes planted as an ornamental or shade tree.

SIMILAR SPECIES: European Beech, American Elm.

CONSERVATION STATUS: LC
The population is secure with no major threats.

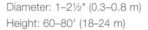

Diameter: 1–2½" (0.3–0.8 m)
Height: 60–80' (18–24 m)

EUROPEAN BEECH *Fagus sylvatica*

Widespread in Europe, European Beech was introduced to North America in the late 1600s. It is a popular ornamental and street tree and massive specimens can be found in parks and estates in North America. There are about 20 horticultural varieties of European Beech, some of which developed naturally and others made by horticultural selection. These include 'Copper Beech,' which has reddish purple foliage that turns blackish purple by summer, and 'Weeping Beech,' with a graceful form and pendant branches that reach from the crown to the ground.

DESCRIPTION: This is a medium-sized tree with deciduous, simple leaves; a short, stocky trunk; a dense, broad, rounded or pyramidal crown; and low-hanging branches. Older trees have knots and burls on the trunk, and a domed crown with heavy, twisted boughs descending to the ground. The leaves are similar to those of American Beech, but the margin teeth are bluntly rounded rather than pointed. They are 2–4" (5–10 cm) long, broadly elliptic, and with a pointed tip. The margins bear short, blunt, widely spaced teeth, and sometimes appear wavy, with five to nine pairs of side veins. They are glossy, dark green above and paler, often with long hairs beneath (silky-hairy when young). The autumn color is golden yellow, reddish bronze, rich orange-brown. The leaves persist through the winter. The leafstalk is short and hairy. The bark is silver-gray to dark gray-blue (usually darker than in American Beech). The twigs are light brown, turning darker. The winter bud is slender, glossy, and sharp-pointed.

FLOWERS: Female clusters are very small; male clusters hang on shorter stalks than in American Beech.

FRUIT: 1" (2.5 cm) long, ovoid, burlike husk, covered with short, soft bristles; similar to American Beech. Each nut is ⅝" (1.5 cm) long, angular, often with concave sides, oily, and deep brown, maturing in fall.

HABITAT: Introduced in North America, where it is planted in parks and landscapes.

RANGE: Native to Europe; it is introduced in North America.

USES: The hard wood from this species is used for furniture and a variety of other carpentry work. The tree is often planted as an ornamental, and the edible nuts are eaten by people and wildlife.

SIMILAR SPECIES: American Beech.

CONVSERVATION STATUS:

European Beech is widespread across Europe and its population is presumed to be quite large. It is vulnerable to various pests and diseases as well as to herbivory from squirrels and deer. The fungus *Ganoderma applanatum* causes mature trees to rot. Climate change also allows the pathogen *Phythophtora ramorum*, which causes sudden oak death, to enter European Beech's range.

Diameter: 2–4' (61–122 cm)
Height: 50–70' (15.2–21.3 m)

■ DENSE POPULATION
■ LIGHT POPULATION
■ NATURALIZED

Tanoak is placed in a separate genus with more than 100 species native to southeast Asia and the Indo-Malayian ecoregion. While the acorns resemble those of true oaks, the flowers are like those of chinkapins and chestnuts. Tanoaks are suited to well-drained, deep, fertile soils on coastal mountain slopes and ridges, but can also grow in stony or shallow soil. They are vulnerable to extreme temperatures. Tanoaks' form and shape is highly variable depending on the environment. Shady forests tend to include the taller Tanoak forms, while open areas with abundant sunlight tend to host shorter trees.

DESCRIPTION: This evergreen tree has a great central trunk and a crown varying from narrow and conical to broad and rounded. Sometimes this species is a shrub. Its leaves are evergreen, 2½–5" (6–13 cm) long, and ¾–2¼" (2–6 cm) wide. They are oblong, thick and leathery, with many straight parallel sunken side veins, with wavy-toothed border sometimes turning under, and stout, hairy leafstalks. They are shiny light green and becoming hairless or nearly so above, with whitish or yellowish hairs, woolly when young, beneath. The bark is brown, thick, and deeply furrowed into ridges and plates. The twigs are stout, with dull yellow hairs.

FLOWERS: Early spring, sometimes also in fall. Numerous, tiny, stalkless, whitish flowers form in catkins 2–4" (5–10 cm) long with an unpleasant odor, upright from the base of the leaf. They are usually all male, sometimes also with one or two tiny, greenish female flowers at the base.

FRUIT: Acorns: ¾–1¼" (2–3 cm) long, egg-shaped, with one or two on a stout, long stalk. They are yellow-brown, with a shallow, saucer-shaped cup covered by long, slender, spreading scales, and maturing in the second year.

HABITAT: Moist valleys and mountain slopes; in oak forests and sometimes in nearly pure stands.

RANGE: Pacific Coast from southwest Oregon south to southern California, and in the Sierra Nevada to central California; to 5000' (1524 m).

USES: Tanoak bark was once the main commercial western source of tannin. The tree's large acorns are a staple food for wildlife and were widely used by Native American groups in the coastal ranges of California. Native Americans ground flour from the acorns after removing the shells and washing the seeds in hot water to remove the bitter taste.

SIMILAR SPECIES: Allegheny Chinkapin, American Beech, American Chestnut.

CONSERVATION STATUS: LC
The population is stable. Tanoak is among the several species in northern and central California that have been affected by sudden oak death caused by the fungus *Phytophthora ramorum*. The disease is easily spread by beetles attracted to the sap of the infected trees. To keep trees healthy, apply a thick layer of mulch to the root zone area beneath the crown. Do not garden or disturb this area in any way, avoid frequent irrigation, and prune only from June to September when the fungus and insects are less active.

Diameter: 1–2½' (0.3–0.8 m)
Height: 50–80' (15–24 m)

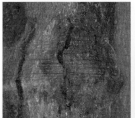

Fagales

SAWTOOTH OAK   *Quercus acutissima*

■ DENSE POPULATION
■ LIGHT POPULATION
■ NATURALIZED

Native to Japan, China, and Korea, this tree was widely planted in North America because of its rapid establishment and heavy fruit production at an early age, serving as a source of food in late summer and throughout fall for wildlife. The abundant crops of acorns, which are borne heavily every other year or sometimes annually, support numerous large birds, squirrels, deer, raccoons, opossums, and other mammals. The species slowly spreads from plantings, usually into disturbed areas, if its growth is left unchecked.

DESCRIPTION: This is a pyramidal, medium-sized tree that may reach 60' (18.2 m) tall and 60' (18.2 m) wide at maturity. Its leaves are alternate, glossy, oblong, simple, 3–7" (8–17.8 cm) long, and serrated. They are retained in the winter. The bark is ridged and furrowed, striated when young. The twigs are slender, red to gray-brown, with multiple terminal gray-brown buds, pubescent on the bud scale edges and somewhat pyramidal.

FLOWERS: Spring. Pendulous, yellow male catkins. Female catkins are borne on spikes, appearing with the leaves.

FRUIT: Acorns are oval; the cap covers half of the nut.

HABITAT: Introduced in North America; it is found in reclaimed surface-mined areas and other disturbed sites.

RANGE: Native to Asia. It is introduced and naturalized from northern Florida to eastern Texas and Oklahoma, north through Missouri and New York and into southern New England.

USES: This tree is widely planted in eastern North America and is becoming an invasive pest in some areas. The acorns provide food for a variety of wildlife species.

SIMILAR SPECIES: Allegheny Chinkapin.

CONSERVATION STATUS: LC

The population is widespread. Increased use of this species for timber and fuel may pose threats in the future. In North America it can become an invasive pest, but it is relatively easy to manage.

Diameter: 5' (1.5 m)
Height: 32–60' (10–18.2 m)

■ DENSE POPULATION
■ LIGHT POPULATION
■ NATURALIZED

This is the common oak of the California coast and foothills, often forming groves and found in a number of natural communities. The acorns were among those preferred by some Native American groups, who ground up shelled seeds into meal that was then washed to remove the bitter taste, and boiled into mush or baked in ashes as bread.

DESCRIPTION: This evergreen tree has a short, stout trunk; many large, crooked, spreading branches; and a broad, rounded crown. It is sometimes shrubby. Its leaves are evergreen, ¾–2½" (2–6 cm) long, and ½–1½" (1.2–4 cm) wide. They are oblong or elliptical, short-pointed or rounded at both ends, with edges turned under and bearing spiny teeth, and thick and leathery. They are shiny dark green above and yellow-green and often hairy beneath. The bark is dark brown, thick, and deeply furrowed.

FRUIT: Acorns: 1–1½" (2.5–4 cm) long, narrowly egg-shaped, with ⅓ enclosed by a deep, thin cup with many brownish, finely hairy scales outside and silky hairs inside, with one or a few together, stalkless, and maturing during the first year.

HABITAT: In valleys and on slopes, usually in open groves; it often is found with Canyon Live Oak and California Black Oak.

RANGE: Mostly found in the Coast Ranges, in central to southern California, including Santa Cruz and Santa Rosa Islands; it is also found in northern Baja California. It occurs to about 3000' (914 m).

USES: The acorns of this tree were an important food source for many Native Americans. The species is now commonly cultivated as a landscaping ornamental.

SIMILAR SPECIES: Canyon Live Oak.

CONSERVATION STATUS: **LC**
Although Coast Live Oak faces threats in certain localities, especially from human development and sudden oak death, its population is considered stable and has not experienced significant declines.

Diameter: 1–3' (0.3–0.9 m) or more
Height: 30–80' (9–24 m)

WHITE OAK   *Quercus alba*

■ DENSE POPULATION
■ LIGHT POPULATION
■ NATURALIZED

The spread of branches of a White Oak is among the largest of North American trees. (The Southern and Coastal Live Oaks can spread even farther.) The greatest example, a 460-year-old tree at Wye, Maryland, reached a span of 148' (45 m) before being toppled by wind in 2002. Several others attained a spread of more than 120' (36.5 m), with circumferences of up to 31' (9.5 m). The massive lower branches of large examples reach out horizontally in remarkable demonstration of the strength of the wood. Due to its strength and durability if properly seasoned, the White Oak has had an important history in shipbuilding, including in the construction of the frigate *Constitution*, whose gun deck, keel, and parallel framing is of White Oak from Massachusetts, New Jersey, and Maryland.

DESCRIPTION: This is the classic eastern oak, with wide-spreading branches and a rounded crown, and the trunk irregularly divided into spreading, often horizontal, stout branches. Its leaves are 4–9" (10–23 cm) long and 2–4" (5–10 cm) wide. They are elliptical, five- to nine-lobed, with the widest beyond the middle and tapering to the base, and hairless. They are bright green above and whitish or gray-green beneath, turning red or brown in fall, often remaining attached in winter. The bark is light gray; shallowly fissured into long broad scaly plates or ridges, often loose.

FRUIT: Acorns: ⅜–1¼" (1–3 cm) long; egg-shaped; about ¼ enclosed by a shallow cup; becoming light gray; with warty, finely hairy scales; maturing in the first year. The acorns germinate soon after falling, and any that survive harvesting by birds and rodents set roots before winter. Mast years occur at regular intervals of 4 to 10 years.

HABITAT: A range of habitats, favoring moist well-drained uplands and lowlands, often in pure stands.

RANGE: Southern Ontario and extreme southern Quebec east to Maine, south to northern Florida, west to eastern Texas, and north to east-central Minnesota; to 5500' (1676 m), or above in the southern Appalachians.

USES: The most important lumber tree of the White Oak group, its high-grade wood is useful for all purposes. It is sometimes called "Stave Oak" because the wood is outstanding in making tight barrels for whiskey and other liquids. In colonial times the wood was important in shipbuilding. The acorns are bitter when uncooked, but boiling sweetens them and they were an important food for Native Americans.

SIMILAR SPECIES: Northern White Oak.

CONSERVATION STATUS:
The White Oak population is decreasing, mostly because of fire management practices, land clearing, and deer browsing.

Diameter: 1–2' (0.3–0.6 m)
Height: 30–60' (9–18 m)

■ DENSE POPULATION
■ LIGHT POPULATION
■ NATURALIZED

One of the largest southwestern oaks, this handsome tree can be found in a variety of habitats, but it reaches its greatest size in canyons and other moist sites. It may grow up to 60' (18 m) in height with a trunk diameter of more than 3' (1 m). Although a good fuel, the hard, heavy wood of this species is difficult to cut and split.

DESCRIPTION: This is a medium-sized evergreen tree with an irregular, spreading crown of stout branches. Its leaves are evergreen, 1½–3" (4–7.5 cm) long, ¾–1½" (2–4 cm) wide, oblong or obovate, slightly wavy-lobed and toothed toward the tip, with the base notched or rounded; they are thick and stiff, dull blue-green and nearly hairless with sunken veins above, and paler and densely hairy with raised veins beneath, shedding gradually in spring as new leaves unfold. The bark is light gray and furrowed into narrow, scaly plates and ridges.

FRUIT: Acorns: ¾–1" (2–2.5 cm) long, oblong, about ⅓ enclosed by a deep cup of finely hairy scales, thickened at the base, with one or two on a short stalk or stalkless; maturing in the first year.

CAUTION: Poisonous: Do not eat or ingest any part of this species. If accidentally ingested, contact your doctor or poison control center.

HABITAT: Mountain slopes and canyons; it is found in oak woodlands with other evergreen oaks.

RANGE: Trans-Pecos Texas west to Arizona; it is also found in northern Mexico. It occurs at 5000–7500' (1524–2286 m).

USES: The hard wood of this tree is often used for fuel. The species is sometimes planted as an ornamental.

SIMILAR SPECIES: Emory Oak, Gray Oak, Net-leaf Oak.

CONSERVATION STATUS: LC

Subpopulations of Arizona White Oak are a conservation concern in Mexico, where the population is scattered. This species is susceptible to the wood decay fungus *Inonotus andersonii*. However, there are not believed to be any major threats that will adversely affect this species in the near future.

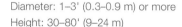

Diameter: 1–3' (0.3–0.9 m) or more
Height: 30–80' (9–24 m)

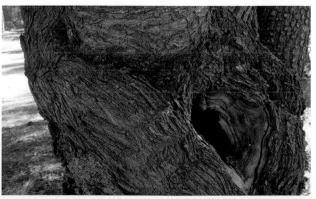

■ DENSE POPULATION
■ LIGHT POPULATION
■ NATURALIZED

This uncommon, shade-loving species occurs scattered with other oaks in a few southeastern states: eastern Texas, Arkansas, Louisiana, Mississippi, Alabama, Georgia, and Florida. It is believed to be an ancient species that once enjoyed wider distribution in the Coastal Plain. It was named in 1911 for the state of Arkansas, where it was discovered in a large stand.

DESCRIPTION: This is a medium-sized tree with a tall trunk and a narrow crown. Its leaves are 2–5" (5–13 cm) long, 1–2½" (2.5–6 cm) wide, and broadly obovate, broadest toward three-toothed or slightly three-lobed and rounded tips, and gradually narrowed to a short-pointed base. They are light yellow-green above, and paler with tufts of hairs in vein angles beneath. The bark is black, thick, and rough, and deeply furrowed into long narrow scaly ridges.

FRUIT: Acorns: ½" (12 mm) long, nearly round, with less than ¼ enclosed by a shallow cup; they are green becoming brown with one or two on a short stalk or stalkless, and maturing in the first year.

HABITAT: Well-drained sandy soils in hardwood forests.

RANGE: Southwest Georgia and northwest Florida west to eastern Texas and southwest Arkansas; to 400' (107 m).

USES: This tree is occasionally planted as an ornamental, and the wood is sometimes used for furniture, flooring, firewood, and other purposes.

SIMILAR SPECIES: Myrtle Oak, Chapman Oak, Blackjack Oak.

CONSERVATION STATUS: VU

The population is decreasing and severely fragmented, with most subpopulations too isolated to allow for seed dispersal. Threats from commercial logging, conversion of habitat to pine plantations, and poor land management practices have caused declines in habitat and population size. Arkansas Oak is also projected to lose more than 60% of its habitat due to climate change by 2050.

Diameter: 1' (0.3 m)
Height: 40–60' (12–18 m)

■ DENSE POPULATION
■ LIGHT POPULATION
■ NATURALIZED

This large oak is found in the north-central and northeastern United States, just crossing into Canada in southern Ontario and southern Quebec. It prefers moist habitats. The Latin species name, which means "two-colored," refers to the leaves, which are green above and whitish beneath.

DESCRIPTION: This is a large tree with a narrow, rounded, open crown of often-drooping branches. Its leaves are 4–7" (10–18 cm) long and 2–4½" (5–11 cm) wide. They are obovate, rounded or blunt at the tip, broadest beyond the middle, gradually narrowed to a pointed base; the edges are wavy with five to ten shallow, rounded lobes on each side. They are green and slightly shiny above, with soft whitish hairs beneath, turning brown to red in fall. The bark is light gray with large thin scales, becoming furrowed into plates.

FRUIT: Acorns: ¾–1¼" (2–3 cm) long and egg-shaped, with ⅓ or more enclosed by a deep cup of many distinct scales, becoming light brown; usually with two on a long, slender stalk, maturing in the first year.

Diameter: 2–3' (0.6–0.9 m)
Height: 60–70' (18–21 m)

HABITAT: Wet soils of lowlands, including stream borders, flood plains, and swamps subject to flooding; it is found in mixed forests.

RANGE: Extreme southern Ontario east to extreme southern Quebec and Maine, south to Virginia, west to Missouri, and north to southeastern Minnesota; local to southeastern Maine, North Carolina, and northeastern Kansas; to 1000' (305 m), locally to 2000' (610 m).

USES: Swamp White Oak is an important lumber species; its wood is used for a variety of purposes. It is also planted as a landscaping tree. Its acorns are an important food source for wildlife.

SIMILAR SPECIES: Bur Oak, Overcup Oak, Chinkapin Oak.

CONSERVATION STATUS: LC

The population of Swamp White Oak is secure. The species is susceptible to various pests, fungi, wilts, and canker. Severe fires can also top-kill mature trees, leaving the trees that survive fires susceptible to diseases and insects. Populations of Swamp White Oak are small in Maine and could be eliminated by logging. Climate change is also a concern and could cause this species to lose nearly 70% of its range by 2050.

■ DENSE POPULATION
■ LIGHT POPULATION
■ NATURALIZED

This small tree—or, sometimes, shrub—has a limited range in the southeastern United States, found in extreme southern Alabama, Florida, southeastern Georgia, and extreme southern South Carolina. It is named for Alvan Wentworth Chapman (1809–99), physician and botanist of Apalachicola, Florida, who first distinguished this oak species in his 1860 book, *Flora of the Southern United States*. Various wildlife species consume the acorns.

DESCRIPTION: This is a shrub or a small tree with a rounded crown of stout, spreading branches. It is often a low shrub. The leaves are 1½–3½" (4–9 cm) long and ¾–1½" (2–4 cm) wide. They are obovate or oblong, broadest beyond the middle and narrowed toward a blunt or rounded base, and the edges are straight or slightly wavy toward a rounded or three-lobed tip; they are slightly thickened. They are shiny dark green above and dull light green and often hairy on midvein beneath, turning yellow or reddish in fall and winter, and shedding gradually by early spring. The bark is light gray and shedding in scaly plates.

FRUIT: Acorns: ⅝–¾" (15–19 mm) long; egg-shaped; partly enclosed by deep half-round cup; becoming brown; stalkless or nearly so; maturing first year.

HABITAT: Sandy hills, ridges and coastal dunes; with Sand Pine and evergreen oaks.

RANGE: Extreme southern South Carolina and southeast Georgia to southern and northwestern Florida and southern Alabama; near sea level.

USES: This species has little commercial significance but is an important food source for wildlife.

SIMILAR SPECIES: Arkansas Oak, Myrtle Oak.

CONSERVATION STATUS: (LC)
This species has a high abundance within its range, most of which is located in Florida. Its population is presumed to be large and stable. Chapman Oak habitat faces threats from urbanization and development; off-road vehicles and foot traffic can also erode the soil it prefers.

Diameter: 8" (20 cm)
Height: 30' (9 m)

■ DENSE POPULATION
■ LIGHT POPULATION
■ NATURALIZED

This species is common in the mountainous regions of California, and many consider this to be the most beautiful of the California oaks. It also occurs less frequently in a variety of habitats throughout the Southwest. The species name, which means "golden-scale," refers to the yellowish acorn cups. Historically, this tree's hard, heavy wood was used locally for farm implements and wagon axles and wheels. Another common name for this species, Maul Oak, refers to the early use of the wood from this tree for the heads of mauls or wedges for splitting Redwood ties.

DESCRIPTION: This is an evergreen tree with a short trunk; large, spreading, horizontal branches; and a broad, rounded crown; it is sometimes shrubby. The leaves are evergreen, 1–3" (2.5–7.5 cm) long, ½–1½" (1.2–4 cm) wide, elliptical to oblong, short-pointed at the tip, rounded or blunt at the base, with the edges turned under and often with spiny teeth (especially on young twigs), and thick and leathery. They are shiny green above with yellow hairs or becoming gray and nearly hairless beneath. The bark is light gray, nearly smooth or scaly.

FRUIT: Acorns: ¾–2" (2–5 cm) long, variable in size and shape, egg-shaped, turban-like with a shallow, thick cup of scales densely covered with yellowish hairs, stalkless or short-stalked, and maturing in the second year.

HABITAT: In canyons and on sandy, gravelly, and rocky slopes; in pure stands and mixed forests.

RANGE: Southwestern Oregon south through the Coast Ranges and Sierra Nevada to southern California; it is local in western Nevada and in western and central Arizona. It occurs at 1000–6500' (305–1981 m) generally, and in Arizona at 5500–7500' (1676–2286 m).

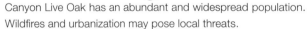

USES: Native Americans used the acorns of this tree as a food source. This species is also an important food source for a variety of birds and mammals.

SIMILAR SPECIES: Coast Live Oak.

CONSERVATION STATUS: LC
Canyon Live Oak has an abundant and widespread population. Wildfires and urbanization may pose local threats.

Diameter: 1–3' (0.3–0.9 m) or more
Height: 20–100' (6–30 m)

SCARLET OAK *Quercus coccinea*

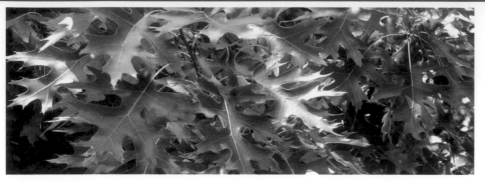

■ DENSE POPULATION
■ LIGHT POPULATION
■ NATURALIZED

This is a popular and handsome shade and street tree native to the eastern United States. The species is named for the bright scarlet color of the fall foliage, which helps to distinguish it from related trees. It is a member of the Red Oak group of related oak species, and the lumber is sometimes marketed as "Red Oak," which differs in its shallowly lobed, dull green leaves, and acorns with a shallow cup. Black Oak is also similar, but has yellow-green leaves with brown hairs beneath and acorns with a deep cup of loose hairy scales.

DESCRIPTION: This is a large tree with a rounded, open crown of glossy foliage, best known for its brilliant autumn color. Its leaves are 3–7" (7.5–18 cm) long and 2–5" (5–13 cm) wide. They are elliptical, deeply divided nearly to the midvein into 7 (rarely 9) lobes, broadest toward the tip, each lobe ending in several bristle-tipped teeth; the wide round sinuses between the lobes often form more than a half-circle; they have long, slender stalks. The leaves are shiny green above, and pale yellow-green, slightly shiny, and with tufts of hairs in vein angles along midvein beneath, turning scarlet in fall. The bark is dark gray and smooth, becoming blackish, thick, rough, and furrowed into scaly ridges or plates; the inner bark is reddish.

FRUIT: Acorns: ½–1" (1.2–2.5 cm) long, egg-shaped, becoming brown with two to four faint rings, with ⅓–½ of the acorn enclosed by a thick, deep, top-shaped cup of tightly pressed scales, tapering to a stalklike base, and maturing in the second year.

HABITAT: Various soils, especially poor and sandy, on upland ridges and slopes; it is found with other oaks and in mixed forests.

RANGE: Southwestern Maine south to Georgia, west to northeastern Mississippi, north to Missouri and Indiana; it is local in Michigan. It generally occurs to 3000' (914 m), locally to 5000' (1524 m).

USES: This species is sometimes planted as an ornamental tree, popular for its brilliant fall foliage color. Its wood is used for a variety of purposes.

SIMILAR SPECIES: Red Oak, Black Oak, Northern Pin Oak.

CONSERVATION STATUS:

Scarlet Oak is widely-distributed throughout the central-eastern United States, though it is locally endangered in Maine—likely because Maine is at the northernmost point of Scarlet Oak's range. Red Oak diseases, including oak wilt, can also cause population decline, but notable declines haven't been reported.

Diameter: 1–2½' (0.3–0.8 m)
Height: 60–80' (18–24 m)

■ DENSE POPULATION
■ LIGHT POPULATION
■ NATURALIZED

Recognizable at a distance by its bluish foliage, this handsome, drought-tolerant California oak was named for the famous Scottish botanical explorer David Douglas (1798–1834), who first described the species. It is used principally for firewood today, but Native Americans had many uses for this tree's acorns and wood. Livestock and wildlife enjoy the edible acorns, which are often produced in abundance.

DESCRIPTION: This tree has a short, leaning trunk; short, stout branches; a broad, rounded crown; and brittle, hairy twigs. It is sometimes shrubby. The leaves are 1¼–4" (3–10 cm) long and ¾–1¾" (2–4.5 cm) wide. They are oblong or elliptical, rounded or blunt at both ends, shallowly four- or five-lobed, coarsely toothed or without teeth, and thin but stiff. They are pale blue-green and nearly hairless above, and paler and slightly hairy beneath. The bark is light gray, thin, and scaly.

FRUIT: Acorns: ¾–1¼" (2–3 cm) long, elliptical, broad or narrow, with a shallow cup or warty scales; they are stalkless or nearly so, and mature in the first season.

HABITAT: Dry, loamy, gravelly, and rocky slopes; it often is found with other oaks and Digger Pine.

RANGE: This species is endemic to California, found mostly in foothills of the Coast Ranges and Sierra Nevada, at 300–3500' (91–1067 m).

USES: This species is commonly used as firewood. Its acorns have served as a food source for people and for a variety of wildlife and livestock.

SIMILAR SPECIES: Oregon White Oak, Valley Oak.

CONSERVATION STATUS: LC

The Blue Oak population is stable but seedling and sapling recruitment is low, raising questions about whether juvenile replacements will sustain current populations when mature Blue Oaks die. Low recruitment rates are exacerbated by several factors including fire suppression, livestock grazing, non-native annual grasses, and habitat fragmentation. Blue Oak is being monitored by the Integrated Hardwood Range Management Program, and recommendations for future legal protection may be imminent.

Diameter: 1' (0.3 m)
Height: 20–60' (6–18 m)

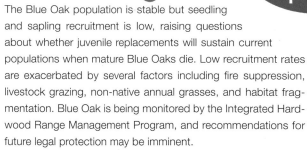

CALIFORNIA SCRUB OAK *Quercus dumosa*

■ DENSE POPULATION
■ LIGHT POPULATION
■ NATURALIZED

This variable, shrubby oak is found mainly in chaparral communities in California and Baja California. It readily hybridizes with tree species found nearby; some plants classified as trees may be hybrids, with their size inherited from the larger parent. The scientific name of this species means "bushy" or "shrubby." Other common names include Coastal Sage Scrub Oak and Nuttall's Scrub Oak. Various forms of wildlife, including birds and even some reptiles, feed on the acorns.

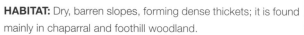

DESCRIPTION: California Scrub Oak is an evergreen, much-branched, thicket-forming shrub or small tree with a rounded crown. Its leaves are evergreen and ⅝–1" (1.5–2.5 cm) long; they are oblong or elliptical, short-pointed at the ends, often three- to nine-lobed, usually with short, sharp or spiny teeth, and thick and stiff. They are shiny green above, and paler and hairy beneath. The bark is gray or brown and scaly. The twigs are stiff, slender, and brown and hairy when young, becoming stout and gray.

FRUIT: Acorns: mostly ½–1" (1.2–2.5 cm) long, egg-shaped, ⅓ enclosed by a light brown, half round, thick-walled cup covered with many hairy, warty, overlapping scales; they are stalkless or nearly so and mature in the first season.

HABITAT: Dry, barren slopes, forming dense thickets; it is found mainly in chaparral and foothill woodland.

RANGE: The Coast Ranges and western base of the Sierra Nevada in northern California south to Baja California; to 5000' (1524 m).

USES: This species is sometimes planted as an ornamental.

SIMILAR SPECIES: Blue Oak.

CONSERVATION STATUS:

Extensive development in California has caused a population decline for California Scrub Oak, making this species rare. The species also takes a long time—up to 20 years—to regenerate.

Diameter: 6" (15 cm)
Height: 3–10' (0.9–3 m),
sometimes to 25' (7.6 m)

■ DENSE POPULATION
■ LIGHT POPULATION
■ NATURALIZED

This tree, native to the Great Lakes region, resembles Pin Oak and Scarlet Oak, both of which have more southern ranges. Its acorns are an important food source for squirrels, deer, and various birds, especially Blue Jays (*Cyanocitta cristata*). This tree also provides nesting cavities for Wood Ducks (*Aix sponsa*), Eastern Kingbirds (*Tyrannus tyrannus*), and several other bird species.

DESCRIPTION: This tree has a short trunk, many small branches, and a narrow crown. Its leaves are 3–5" (7.5–13 cm) long, 2½–4" (6–10 cm) wide, elliptical, deeply divided into five to seven lobes with a few bristle-tipped teeth and wide rounded sinuses. They are shiny green above and paler, often with tufts of hairs along midvein beneath, turning yellow, brown, or purple in fall, and often remaining attached in winter. The bark is gray or dark brown, smooth, becoming shallowly fissured or furrowed into narrow plates; the inner bark is light yellow.

FRUIT: Acorns: ½–¾" (12–19 mm) long, elliptical or nearly round; they are ⅓–½ enclosed by a deep cup tapering to a stalk-like base; becoming brown and maturing in the second year.

HABITAT: Well-drained dry or moist sandy and clay soils; it is found in pure stands, with other oaks, or in mixed forests.

RANGE: Southwestern Ontario, southeast to extreme northwestern Ohio, southwest to extreme northern Missouri, and north to extreme southeastern North Dakota; at 600–1300' (183–396 m).

USES: This species is commonly planted as an ornamental tree, popular for its brilliant fall color. Its heavy wood is used for furniture, flooring, fence posts, and many other purposes.

SIMILAR SPECIES: Pin Oak, Scarlet Oak.

CONSERVATION STATUS: LC

The Northern Pin Oak population is widespread and abundant but climate change could be a threat in the future.

Diameter: 1–2½' (0.3–0.8 m)
Height: 50–70' (15–21 m)

ENGELMANN OAK   *Quercus engelmannii*

■ DENSE POPULATION
■ LIGHT POPULATION
■ NATURALIZED

This member of the White Oak group was named for George Engelmann (1809–1884), a 19th century German-American botanist. A species of limited distribution, it is found only in a few counties in southwestern California. It is also known as Pasadena Oak.

DESCRIPTION: This tree has stout, spreading branches and a broad, irregular crown. Its leaves are evergreen or nearly so, 1–2¾" (2.5–7 cm) long, ½–1¼" (1.2–3.2 cm) wide, oblong or obovate, blunt or rounded at the ends, often wavy-toothed, and thick and leathery. They are blue-green or gray-green and nearly hairless above, and light yellow-green and usually hairy beneath, shedding in spring when new leaves appear. The bark is light gray, thin, and scaly. The twigs are stiff, brown, and densely hairy.

FRUIT: Acorns: ⅝–1" (1.5–2.5 cm) long, elliptical, with nearly ½ enclosed by a deep, scaly cup; they are stalkless or on slender stalks, maturing in the first year.

HABITAT: Dry slopes in foothills; with other oaks.

RANGE: Southwestern California and Santa Catalina Island; to 4000' (1219 m).

USES: This species is sometimes planted as an ornamental tree.

SIMILAR SPECIES: Arizona Oak.

CONSERVATION STATUS: 🄴🄽
Engelmann Oak is rare and decreasing in California. The largest remaining stands of Engelmann Oaks in the wild are on the Santa Rosa Plateau and Black Mountain in San Diego County, with smaller, fragmented subpopulations in the surrounding counties. Suburban sprawl, increasing risk of human-induced wildfires, stress from pests, land use changes, and the effects of climate change combine to place compounded pressure on this species.

Diameter: 1–2½' (0.3–0.8 m)
Height: 20–60' (6–18 m)

■ DENSE POPULATION
■ LIGHT POPULATION
■ NATURALIZED

Southern Red Oak is found on well-drained lowland soils from southeastern Virginia to northwestern Florida and eastern Texas. This species is often called Spanish Oak, possibly because it commonly occurs in the areas of the early Spanish colonies. It is a member of the Red Oak group of related oak species, and the lumber of this commercially important tree is sometimes marketed as "Red Oak."

DESCRIPTION: This tree has a rounded, open crown of large spreading branches, and twigs with rust-colored hairs. Its leaves are 4–8" (10–20 cm) long, 2–6" (5–15 cm) wide, elliptical, deeply divided into a long, narrow end lobe and one to three shorter, mostly curved lobes on each side, with one to three bristle-tipped teeth; sometimes slightly triangular with bell-shaped base and three broad lobes. They are shiny green above, with rust-colored or gray soft hairs beneath, turning brown in fall. The bark is dark gray, becoming furrowed into broad ridges and plates.

FRUIT: Acorns: ½–⅝" (12–15 mm) long, elliptical or rounded, becoming brown, with ⅓ or more enclosed by a cup tapering to a broad stalklike base, maturing in the second year.

HABITAT: Dry, sandy loam and clay loam soils of uplands; in mixed forests.

RANGE: Long Island and New Jersey south to northern Florida, west to eastern Texas, and north to southern Missouri; to 2500' (762 m).

USES: This is an important timber species, particularly in the southeastern United States, often used for furniture, flooring, construction, and fuel.

SIMILAR SPECIES: Turkey Oak, Post Oak, Cherrybark Oak.

CONSERVATION STATUS: **LC**
Southern Red Oak has a widespread, stable population. Oak wilt is a threat, but it has not caused significant population decline.

Diameter: 1–2½' (0.3–0.8 m)
Height: 50–80' (15–24 m)

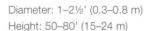

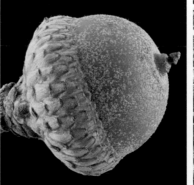

■ DENSE POPULATION
■ LIGHT POPULATION
■ NATURALIZED

Gambel's Oak is widespread in the foothills and lower mountain elevations of western North America. It is the common oak of the Rocky Mountains, and is notably abundant in Grand Canyon National Park. It is closely related to the White Oak of the eastern United States. This is an important food species for wildlife; deer browse the foliage, and many birds, small mammals, and livestock eat the sweetish acorns that are common in alternate years. The wood is used mainly for fence posts and fuel. This species is named for American naturalist and botanist William Gambel (1821–1849), who was among the first to collect botanical specimens from New Mexico to California.

DESCRIPTION: This tree has a rounded crown and is often in dense groves. It can also be a thicket-forming shrub. Its leaves are 2–6" (5–15 cm) long, 1¼–3¼" (3–8 cm) wide, elliptical or oblong, rounded at the tip, short-pointed at the base, deeply 7- to 11-lobed halfway or more to the middle, and with straight or wavy edges. They vary in size, lobing, and hairiness, and are shiny dark green and usually hairless above, and paler and with soft hairs below, turning yellow and reddish in autumn. The bark is gray, rough, thick, and deeply furrowed or scaly.

FRUIT: Acorns: ½–¾" (12–19 mm) long, egg-shaped, with about ⅓ enclosed by a deep, thick, scaly cup, with one or two on a short stalk (or nearly stalkless), maturing in the first year.

HABITAT: Slopes and valleys, in mountains, foothills, plateaus; scattered with Ponderosa Pine.

RANGE: Northern Utah east to extreme southern Wyoming, south to Trans-Pecos Texas, and west to Arizona; local in extreme northwestern Oklahoma and southern Nevada; it is also found in northern Mexico. It occurs at 5000–8000' (1524–2438 m).

USES: This species is an important food source for wildlife and livestock, and the wood is sometimes used for fence posts and firewood.

SIMILAR SPECIES: Similar to White Oak, although their natural ranges do not overlap.

CONSERVATION STATUS: LC
Gambel's Oak has a wide range and no significant threats.

Diameter: 1–2½' (0.3–0.8 m)
Height: 20–70' (6–21 m)

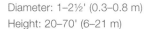

■ DENSE POPULATION
■ LIGHT POPULATION
■ NATURALIZED

Oregon White Oak is the only native oak tree species in Washington and British Columbia. This tree is of great commercial importance in the West; its wood is used for furniture, shipbuilding, construction, cabinetwork, interior finish, and fuel. The sweetish acorns, often common in alternate years, are relished by livestock and wildlife and were eaten by Native Americans. Planted for shade and ornament, it resembles the eastern White Oak (*Quercus alba*). Oregon White Oaks—called Garry Oaks in Canada—provide critical habitat for a variety of native wildlife, including the Propertius Duskywing butterfly (*Erynnis propertius*) and many bird species.

DESCRIPTION: This tree has a dense, rounded, spreading crown of stout branches; it is sometimes shrubby. The leaves are 3–6" (7.5–15 cm) long, 2–4" (5–10 cm) wide, elliptical, blunt or rounded at both ends, deeply lobed halfway or more to the midvein, with blunt or slightly toothed lobes, and slightly thickened. They are shiny dark green above, and light green and usually hairy beneath, sometimes turning reddish in autumn. The bark is light gray or whitish, thin, and scaly or furrowed into broad ridges.

FRUIT: Acorns: 1–1¼" (2.5–3 cm) long and elliptical, with ¼–⅓ enclosed by a shallow, thin, scaly cup; they are stalkless or short-stalked, sweetish, and edible.

HABITAT: In valleys and on mountain slopes; often in pure stands and with other oaks.

RANGE: Southwest British Columbia south to central California in the Coast Ranges and Sierra Nevada; to 3000' (914 m) in the north and at 1000–5000' (305–1524 m) in the south.

USES: This is an important timber species, used widely for furniture, cabinetwork, and many other purposes. It is also planted as a shade and ornamental tree.

SIMILAR SPECIES: White Oak (an eastern species). Oregon White Oak is the only native oak species in the Northwest.

CONSERVATION STATUS: LC

There are no serious threats to the Oregon White Oak population, but livestock overgrazing and various fungi and insects place pressure on this species. Filbertworm (*Cydia latiferreana*) damages seed crops, while shoestring root rot (*Armillaria mellea*) is one of the most damaging rots to afflict Oregon White Oak. The Western Oak Looper (*Lambdina fiscellaria*) is one of the most damaging insects. Oregon White Oak also hosts the Barred Fruit-tree Tortrix (*Pandemis cerasana*), a leafroller moth native to Europe that causes sporadic defoliation; the moth has become common in Oregon and is gradually spreading farther north.

Diameter: 1–2½' (0.3–0.8 m)
Height: 30–70' (9–21 m)

■ DENSE POPULATION
■ LIGHT POPULATION
■ NATURALIZED

This fairly small evergreen oak is closely related to Southern Live Oak, but is much less massive. This species is found in coastal woodlands, often in dry, sandy soils, in the southeastern United States. In Florida, where it is quite common, Sand Live Oak provides important habitat for the Florida Scrub-Jay (*Aphelocoma coerulescens*).

DESCRIPTION: This is a small evergreen tree or shrub. Its leaves are convex, with the edges strongly rolled under, and recessed veins above. The bark is dark brown or blackish with scaly plates. The twigs are tan or light gray, becoming smooth in the second year, with dark brown, ovoid buds.

FRUIT: Acorns are similar to those of Southern Live Oak, with a cup composed of grayish white, hairy scales.

HABITAT: Deep coastal plain sands, often found with pines. It occurs to an elevation of about 650' (200 m).

RANGE: North Carolina to Mississippi and south to southern Florida.

USES: This species is sometimes planted as an ornamental tree, particularly along roadsides.

SIMILAR SPECIES: Southern Live Oak.

CONSERVATION STATUS: LC

Sand Live Oak has an abundant and increasing population that is widely distributed, with no known threats.

Diameter: 1–2' (30–61 cm)
Height: to 15' (4.6 m)

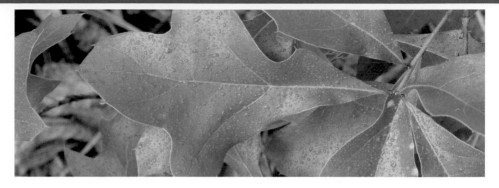

■ DENSE POPULATION
■ LIGHT POPULATION
■ NATURALIZED

This is a rare oak with a limited range in scattered areas of Alabama, Georgia, and South Carolina. It was first discovered in 1849 at Stone Mountain, Georgia, and then later found in the two adjacent states. It is still most easily seen at Stone Mountain.

DESCRIPTION: This is a shrub or sometimes a small tree, rare and very local in Piedmont of Georgia and adjacent states. Its leaves are mostly 2–4" (5–10 cm) long, 1–2" (2.5–5 cm) wide, elliptical, irregularly divided into three to five short-pointed, bristle-tipped lobes; larger lobes have one or two small teeth; the base is short-pointed and thick. They are shiny green above, and paler with tufts of hairs in vein angles beneath. The bark is gray or light brown, thin, and smooth, becoming scaly.

FRUIT: Acorns: ⅜–½" (10–12 mm) long, nearly round, with ¼–⅓ enclosed by a shallow cup, becoming brown, stalkless or short-stalked, and maturing in the second year.

Diameter: 1' (0.3 m)
Height: 6–10' (1.8–3 m),
sometimes to 30' (9 m)

HABITAT: Granite and sandstone outcrops of mountain slopes.

RANGE: Highly local in South Carolina, northern Georgia, and northern Alabama; at about 1700' (518 m).

USES: This species is occasionally planted as a landscaping shrub or tree.

SIMILAR SPECIES: Bear Oak.

CONSERVATION STATUS: **EN**

The Georgia Oak population is severely fragmented and decreasing, with most subpopulations too isolated to allow for seed dispersal. The species is confined to isolated "soil islands" on granite outcrops. Commercial logging and poor land management practices have caused population and habitat decline, and erosion, poor regeneration, and foot and vehicle traffic over its preferred soil places continued pressure on the species. Its limited dispersal ability, slow reproductive rates, specialized habitat requirements, restricted distribution, and rarity all suggest that this species would be highly vulnerable to the effects of climate change.

DENSE POPULATION
LIGHT POPULATION
NATURALIZED

This small species is found in dry areas of the northeastern United States, south to western North Carolina. A temporary scrub type after heavy cutting and repeated fires, Bear Oak is replaced by taller pines and oaks. The Latin species name means "holly-leaved," and refers to the foliage. It is commonly known as Bear Oak, supposedly because only bears like its bitter-tasting acorns. It is also called Scrub Oak.

DESCRIPTION: This is a much-branched shrub or small tree with a rounded crown. Its leaves are 2–4" (5–10 cm) long, 1½–3" (4–7.5 cm) wide, obovate, with three to seven (usually five) shallow short-pointed lobes ending in one to three bristle-tipped teeth, short-pointed at the base, and slightly thickened and firm. They are dull dark green above, and densely covered with light gray hairs beneath, turning dull red or yellow in autumn and often remaining attached in winter. The bark is dark gray, thin, and smooth, becoming fissured and scaly.

FRUIT: Acorns: ⅜–⅝" (10–15 mm) long, egg-shaped or rounded, brown-striped, with many clustered and mostly paired on twigs, with ½–¼ enclosed by a deep cup with a stalklike base and fringelike border of many overlapping brown scales, maturing in the second year.

HABITAT: Dry sandy barrens and rocky ridges in mountains, forming thickets or with pines and other oaks.

RANGE: Southern Maine southwest to western North Carolina; to 3000' (914 m).

USES: This species provides important habitat for a number of animal species and has been used in revegetation projects.

SIMILAR SPECIES: Georgia Oak.

CONSERVATION STATUS: LC

Bear Oak is locally abundant within its range, although it's threatened by the loss and degradation of pine barren habitat and changes in fire frequency. More than half of the species' range faces exposure to climate change pressure, but Bear Oak may be one of the tree species best suited to withstand the effects because of its dispersal ability and tolerance to disturbance.

Diameter: 5" (13 cm)
Height: 20' (6 m)

■ DENSE POPULATION
■ LIGHT POPULATION
■ NATURALIZED

This handsome tree is found in moist habitats in the Upper South region and throughout the Midwest of the United States. Its name refers to the use of this tree's wood for roof shingles by early American pioneers. Today this species is often cultivated as an ornamental and shade tree; it is also suitable for hedges, screens, and windbreaks.

DESCRIPTION: Shingle Oak has a symmetrical, conical to rounded crown. The leaves are 3–6" (7.5–15 cm) long, ¾–2" (2–5 cm) wide, oblong or lance-shaped, short-pointed or rounded at the ends, bristle-tipped, with edges straight or slightly wavy and turned under. They are shiny dark green above, and light gray-green with a yellow midvein and soft hairs beneath, turning yellow or reddish-brown in fall, often shedding late. The bark is brown to gray and smooth, becoming rough and furrowed into scaly ridges.

FRUIT: Acorns: ½–⅝" (12–15 mm) long; nearly round, ⅓–½ enclosed by a deep cup of blunt hairy scales, becoming brown, with one or two on stout stalks, maturing in the second year.

HABITAT: Moist soils along streams and in uplands; it is found scattered with Post and Black oaks.

RANGE: Pennsylvania south to North Carolina, west to Arkansas, and north to southern Iowa and southern Michigan; it is local in Louisiana and Alabama. It occurs to 2000' (610 m).

USES: The wood from this tree has long been used for making shingles. This species is commonly used in landscaping, often as an ornamental or shade tree.

SIMILAR SPECIES: Diamondleaf Oak.

CONSERVATION STATUS: LC

Shingle Oak has a wide geographic distribution throughout the central-eastern United States, and the population is stable. The species is susceptible to oak wilt but has not shown any population declines caused by the disease. Climate change poses a potential future risk. Shingle Oak is projected to lose up to 70% of its suitable habitat area to climate change by 2050, but its high dispersal ability and disturbance tolerance suggest the species is only moderately sensitive to the threat and may be able to withstand the changes to its habitat.

Diameter: 1–2' (0.3–0.6 m)
Height: 50–60' (15–18 m)

■ DENSE POPULATION
■ LIGHT POPULATION
■ NATURALIZED

This oak is easily recognized by its distinctive, bluish-gray deciduous leaves. The common name refers to the shiny blue-green foliage, while the Latin species name, meaning "hoary," describes the gray-green undersurface. Other common names for this species include Upland Willow Oak, Sandjack Oak, and Cinnamon Oak. It is found in dry, upland habitats in the southeastern United States.

DESCRIPTION: This is a thicket-forming shrub or small tree with an irregular crown of stout, crooked branches and distinctive, blue-green foliage. The leaves are 2–4" (5–10 cm) long, ½–1" (1.2–2.5 cm) wide, oblong, bristle-tipped, with straight edges (rarely, slightly lobed on young twigs), and slightly thickened and leathery. They are shiny blue-green above with a prominent network of veins, dull gray-green and finely hairy beneath, turning reddish before shedding in late fall or early winter. The bark is dark gray or blackish, rough, thick, and furrowed in nearly square plates.

FRUIT: Acorns: ½–⅝" (12–15 mm) long, rounded, with ¼–½ enclosed by a shallow or deep cup, becoming brown; they are nearly stalkless and mature in the second year.

HABITAT: Dry sandy uplands, with other oaks and pines.

RANGE: Southeast Virginia to central Florida, west to eastern and central Texas, and north to southeastern Oklahoma; to 500' (152 m).

USES: This species is a food source for a variety of native wildlife. Its hard wood is sometimes used for fence posts and firewood.

SIMILAR SPECIES: Laurel Oak.

CONSERVATION STATUS:

Bluejack Oak is widespread throughout its range and the population is stable, although extended fire suppression has caused minor decreases in population. Climate change poses a serious threat to the species; Bluejack Oak is projected to lose almost 60% of its habitat area by 2050 due to climate change.

Diameter: 6" (15 cm)
Height: 20' (6 m)

DENSE POPULATION
LIGHT POPULATION
NATURALIZED

California Black Oak is the common oak in the valleys and foothills of southwestern Oregon and in the Sierra Nevada. The large, deeply lobed leaves with bristle-tipped teeth differ from all other western oaks, but resemble those of Black Oak (*Q. velutina*) of the eastern United States. Woodpeckers drill holes in the bark and often cache acorns there for future use. Slow-growing and long-lived, California Black Oak is a popular and hardy shade tree often planted in dry soils. Deer and livestock browse the foliage.

DESCRIPTION: This tree has large branches and an irregular, broad, rounded crown of stout, spreading branches. The leaves are 3–8" (7.5–20 cm) long, 2–5" (5–13 cm) wide, elliptical, usually seven-lobed about halfway to the midvein, with each lobe having a few bristle-pointed teeth, and slightly thick. They are shiny dark green above, and light yellow-green and often hairy beneath, turning yellow or brown in autumn. The bark is dark brown and thick, becoming furrowed into irregular plates and ridges; on small trunks, it is smooth and light brown.

FRUIT: Acorns: 1–1½" (2.5–4 cm) long, elliptical, with ⅓–⅔ enclosed by a deep, thin, scaly cup, usually with 1 or a few on a short stalk, maturing in the second year.

HABITAT: Sandy, gravelly, and rocky soils of foothills and mountains; it is often found in nearly pure stands and in mixed coniferous forests.

RANGE: Southwestern Oregon south in the Coast Ranges and Sierra Nevada to southern California; it is found at 1000–8000' (305–2438 m).

USES: Many Native Americans used the acorns from this species as a food source. The wood is used for construction and making furniture. This species is also planted as an ornamental tree.

SIMILAR SPECIES: Black Oak (an eastern species). It is not commonly confused with other western oaks.

CONSERVATION STATUS: LC
The population is stable with no major threats.

Diameter: 1–3' (0.3–0.9 m)
Height: 30–80' (9–24 m)

TURKEY OAK *Quercus laevis*

- ◼ DENSE POPULATION
- ◼ LIGHT POPULATION
- ◼ NATURALIZED

Turkey Oak is native to the coastal plain of the southeastern United States. This species spreads by underground runners, especially after frequent fires. The plant's common name refers to the shape of its three-lobed leaves, which suggests a turkey's foot. The Latin species name, meaning "smooth," describes the nearly hairless leaves. It should not be confused with the Eurasian species *Quercus cerris*, which is also commonly called Turkey Oak; some sources refer to the North American species American Turkey Oak to avoid confusion.

DESCRIPTION: This tree has an irregular, open crown of crooked branches, and is often a shrub. The leaves are 4–8" (10–20 cm) long and 3–6' (7.5–15 cm) wide. They are nearly triangular, spreading from a pointed base into three to five (rarely seven) long, narrow lobes each with one to three long, pointed, bristle-tipped teeth, mostly turned on edge, and thick and stiff. They are shiny yellow-green above and light green beneath, with prominent veins and tufts of rust-colored hairs in vein angles, turning red before shedding in late fall or early winter. The bark is gray to blackish, becoming thick, rough, and deeply furrowed into irregular ridges; the inner bark is reddish.

FRUIT: Acorns: ¾–1" (2–2.5 cm) long, egg-shaped, about ½ enclosed by a top-shaped cup with hairy scales extending down the inner surface; they are short-stalked and becoming brown, maturing in the second year.

HABITAT: Dry sandy ridges and dunes, especially near the coast; it is often found in pure stands.

RANGE: Southeastern Virginia to central Florida and west to southeastern Louisiana; to 500' (152 m).

USES: Sometimes planted as an ornamental, this species also provides food for deer and other native wildlife.

SIMILAR SPECIES: Southern Red Oak, Cherrybark Oak.

CONSERVATION STATUS: (LC)

This species is common in its range and its population is considered stable. It is susceptible to oak wilt (*Ceratocystis fagacearum*) and other pests that also afflict Red Oaks. It may also be impacted by climate change, potentially losing more than 30% of its habitat by 2050 according to the prevailing climate model.

Diameter: 1' (0.3 m)
Height: 20–40' (6–12 m)

■ DENSE POPULATION
■ LIGHT POPULATION
■ NATURALIZED

This is a handsome shade tree, widely planted in the southeastern United States, where it is native. Its acorns provide food for a variety of wildlife, including deer, raccoons, squirrels, turkeys, and many other birds and small mammals. Its name is based on the resemblance of the tree's foliage to Grecian Laurel (*Laurus nobilis*) of the Mediterranean region.

DESCRIPTION: This is a large, nearly evergreen tree with a dense, broad, rounded crown. The leaves are 2–5½" (5–14 cm) long, ⅜–1½" (1–4 cm) wide, narrowly oblong, diamond- or lance-shaped, often broadest near the middle, bristle-tipped, with straight edges (rarely, with few lobes or teeth), thin or slightly thickened, and usually hairless. They are shiny green or dark green above and light green and slightly shiny beneath, shedding in early spring and nearly evergreen. The bark is brown to gray and smooth, becoming blackish, rough, and furrowed.

FRUIT: Acorns: ½" (12 mm) long, nearly round, with ¼ or less enclosed by a shallow cup of blunt hairy scales, short-stalked or nearly stalkless, becoming brown, and maturing in the second year.

HABITAT: Moist to wet well-drained sandy soil along rivers and swamps; it is sometimes found in pure stands.

RANGE: Southeast Virginia to southern Florida, west to southeast Texas, and north locally to southern Arkansas; to 500' (152 m).

USES: This species is used for pulpwood. It is also widely planted as a shade and ornamental tree.

SIMILAR SPECIES: Bluejack Oak.

CONSERVATION STATUS: Ⓛ Ⓒ

Swamp Laurel Oak is common throughout the coastal plain of the southeastern United States, but oak leaf blister (*Taphrina caerulescens*) is a threat, as are various other pests and pathogens common to Red Oaks.

Diameter: 1–2½' (0.3–0.8 m)
Height: 60–80' (18–24 m)

■ DENSE POPULATION
■ LIGHT POPULATION
■ NATURALIZED

Valley Oak is the largest of the western deciduous oaks and a handsome, graceful shade tree. This long-lived relative of the eastern White Oak (*Q. alba*) is endemic to California and common throughout the state's interior valleys. Many kinds of wildlife and domestic animals, especially hogs, consume the acorn crops that are often abundant. California's Native American groups roasted these large acorns and also ground the edible portion into meal, which they prepared as bread or mush.

DESCRIPTION: This is a large, handsome tree with a stout, short trunk and large, widely spreading branches drooping at the ends, forming a broad, open crown. The leaves are 2–4" (5–10 cm) long and 1¼–2½" (3–6 cm) wide. They are elliptical, rounded or blunt at both ends, deeply 7- to 11-lobed more than halfway to the midvein, with larger lobes broadest and notched at the end. They are dark green and nearly hairless above, and paler and finely hairy beneath. The bark is light gray or brown, thick, and deeply furrowed and broken horizontally into thick plates.

FRUIT: Acorns: 1¼–2¼" (3–6 cm) long, oblong, pointed, with ⅓ enclosed by a deep, half-round cup with light brown scales, the lowest ones thick and warty or knobby; they are sweetish and edible and mature in the first year.

HABITAT: Valleys and slopes on rich loam soils; forming groves in foothill woodland.

RANGE: California, also Santa Cruz and Santa Catalina islands; to 5000' (1524 m).

USES: Native Americans used this tree's sweet acorns as a food source. The wood is used for cabinetwork, flooring, and wine barrels.

SIMILAR SPECIES: White Oak, Oregon White Oak, Blue Oak.

CONSERVATION STATUS:

Once much more extensive in California, 150 years of widespread agricultural and residential development has caused Valley Oak population fragmentation and decline, especially in the Central Valley, where only a few groves remain intact. Climate change models project that Valley Oak habitat will likely constrict and shift northward.

Diameter: 3–4' (0.9–1.2 m), sometimes much greater
Height: 40–100' (12–30 m)

■ DENSE POPULATION
■ LIGHT POPULATION
■ NATURALIZED

This attractive oak is native to the lowlands of the southeastern United States. Its common name describes the acorns, which are almost enclosed in the cup. The Latin species name, meaning "lyre-shaped," refers to the leaves.

DESCRIPTION: This tree has a rounded crown of small, often drooping branches, with acorns almost covered by the cup, and narrow deeply lobed leaves. Those leaves are 5–8" (13–20 cm) long, 1½–4" (4–10 cm) wide, narrowly oblong, deeply divided into 7–11 rounded or short-pointed lobes, the longest near a short-pointed tip, and a pointed base. They are dark green and slightly shiny above, and gray-green and with soft hairs or nearly hairless beneath, turning yellow, brown, or red in fall. The bark is light gray and furrowed into scaly or slightly shaggy ridges or plates.

FRUIT: Acorns: ½–1" (1.2–2.5 cm) long, nearly round, almost enclosed by a large rounded cup of warty gray scales, the upper scales long-pointed; they are usually stalkless and mature in the first year.

HABITAT: Wet clay and silty clay soils, mostly on poorly drained flood plains and swamp borders; sometimes in pure stands.

RANGE: Delaware to northwest Florida, west to eastern Texas, and north to southern Illinois; to 500' (152 m), sometimes slightly higher.

USES: An important lumber tree, the wood from this species is useful for a variety of purposes. It is also cultivated as a landscaping ornamental.

SIMILAR SPECIES: Swamp White Oak, White Oak, Bluff Oak.

CONSERVATION STATUS: LC
Overcup Oak is abundant in its range and its population is stable. Climate change is a major threat to this species, which could lose more than 40% of its suitable habitat. Conservationists will closely monitor this species as the impacts of climate change are realized.

Diameter: 2–3' (0.6–0.9 m)
Height: 60–80' (18–24 m)

☐ DENSE POPULATION
☐ LIGHT POPULATION
☐ NATURALIZED

Bur Oak is primarily a northern species of oak. The majority of the tree's distribution occurs in the Great Lakes region and in eastern Great Plains, but it can occur as far east as New Brunswick and as far west as Alaska. In the West, it is a pioneer tree, bordering and sometimes invading prairie grasslands. The acorns of this species, distinguished by deeply fringed cups, are the largest among North America's native oaks. The common name (sometimes spelled "Burr") describes the cup of the acorn, which somewhat resembles the spiny bur of a chestnut.

DESCRIPTION: This tree has large acorns, a stout trunk, and a broad, rounded, open crown of stout, often crooked, spreading branches; it is sometimes a shrub. The leaves are 4–10" (10–25 cm) long, 2–5" (5–13 cm) wide, obovate, broadest beyond the middle, with the lower half deeply divided into two to three lobes on each side; the upper half usually has five to seven shallow rounded lobes on each side with a broad, rounded tip. They are dark green and slightly shiny above, and gray-green and with fine hairs beneath, turning yellow or brown in fall. The bark is light gray, thick, rough, and deeply furrowed into scaly ridges.

FRUIT: Acorns: ¾–2" (2–5 cm) long and wide, broadly elliptical, with ½–¾ enclosed by a large deep cup with hairy gray scales, (the upper scales are long-pointed) forming a fringelike border, maturing in the first year.

HABITAT: From dry uplands on limestone and gravelly ridges, sandy plains, and loamy slopes to moist floodplains of streams; it is often found in nearly pure stands.

RANGE: Extreme southeastern Saskatchewan east to southern New Brunswick, south to Tennessee, west to southeastern Texas, and north to North Dakota; it is local in Louisiana and Alabama. It is usually found at 300–2000' (91–610 m), and to 3000' (914 m) or above in the northwest.

USES: This species is often planted as an ornamental or shade tree.

SIMILAR SPECIES: Swamp White Oak, White Oak, Post Oak.

CONSERVATION STATUS: LC

This species has a widespread distribution throughout its range and it is known for having the largest distribution of all the North American oak species. There are no known threats.

Diameter: 2–4' (0.6–1.2 m)
Height: 50–80' (15–24 m)

■ DENSE POPULATION
■ LIGHT POPULATION
■ NATURALIZED

Blackjack Oak is a small tree able to grow in poor, thin, dry soils that few other oaks can tolerate. This species and Post Oak dominate the woodland and savanna portions of the Cross Timbers ecoregion in Texas and Oklahoma, forming the forest border of small trees and the transition zone to prairie grassland. The wood is dense and burns quite hot, making it useful for firewood and charcoal. This tree was first described in 1704 from a specimen in the colony of Maryland, referred to in the Latin species name.

DESCRIPTION: This tree has an open, irregular crown of crooked, spreading branches. The leaves are 2½–5" (6–13 cm) long, 2–4" (5–10 cm) wide, slightly triangular or broadly obovate, broadest near the tip; with three shallow, broad, bristle-tipped lobes; gradually narrowing to a rounded base, and slightly thickened. They are shiny yellow-green above, and light yellow-green with brownish hairs (especially along veins) beneath, turning brown or yellow in fall. The bark is blackish, rough, thick, and deeply furrowed into broad, nearly square plates.

FRUIT: Acorns: ⅝–¾" (15–19 mm) long, elliptical, and ending in a stout point, with ⅓–⅔ enclosed by a deep, thick, top-shaded cup of rusty-brown, hairy, loosely overlapping scales; they are short-stalked and mature in the second year.

HABITAT: Dry sandy and clay soils in upland ridges and slopes with other oaks and with pines.

RANGE: Long Island and New Jersey south to northwest Florida, west to central and southeast Texas, and north to southeast Iowa; local in southern Michigan; to 3000' (914 m).

USES: The dense wood of Blackjack Oak is often used as firewood and is excellent for barbecues and wood-burning stoves.

SIMILAR SPECIES: Arkansas Oak, Water Oak.

CONSERVATION STATUS: LC
Blackjack Oak is widely distributed with no reports of serious declines.

Diameter: 6–12" (15–30 cm)
Height: 20–50' (6–15 m)

■ DENSE POPULATION
■ LIGHT POPULATION
■ NATURALIZED

This tree is widely known as Swamp Chestnut Oak and can be found commonly in bottomlands and wetlands in the Southeast and Midwest in the United States. The species is commonly called Basket Oak because the wood is easily split into long, flexible strips that are ideal for basket weaving. Baskets woven from fibers and splints made from the wood of this tree make strong containers that were once used extensively in agriculture. This tree's sweetish acorns can be eaten raw, and cattle are known to readily consume them, resulting in the common name Cow Oak.

DESCRIPTION: This is a large tree with a compact, rounded crown and chestnut-like foliage. The leaves are 4–9" (10–23 cm) long, 2–5½" (5–14 cm) wide, obovate, broadest beyond the middle, with wavy edges with 10–14 rounded teeth on each side, abruptly pointed at the tip, gradually narrowing to the base. They are shiny dark green above and gray-green and with soft hairs beneath, turning brown or dark red in fall. The bark is light gray and fissured into scaly plates.

FRUIT: Acorns: 1–1¼" (2.5–3 cm) long, egg-shaped, with ⅓ or more enclosed by a deep, thick cup with a broad base, composed of many overlapping hairy brown scales; they are stalkless or short-stalked and mature in the first year.

HABITAT: Moist sites including well-drained, sandy loam and silty clay flood plains along streams; sometimes in pure stands.

RANGE: New Jersey south to northern Florida, west to eastern Texas, and north to southern Illinois; to 1000' (305 m).

USES: The wood of Swamp Chestnut Oak, along with that of many other white oaks, is used for a variety of purposes. In the case of this species, the wood is excellent for basket weaving. Basket Oak is sometimes cultivated as an ornamental tree.

SIMILAR SPECIES: White Oak, Bluff Oak.

CONSERVATION STATUS: 🄻🄲
Swamp Chestnut Oak is widely distributed and abundant throughout its range. Fungi and insects are minor threats but have not caused population decline.

Diameter: 2–3' (0.6–0.9 m)
Height: 60–80' (18–24 m)

■ DENSE POPULATION
■ LIGHT POPULATION
■ NATURALIZED

This species is characteristic of limestone uplands in the eastern and central United States. The sweet acorns are often eaten by people and provide a great source of food for wildlife, including squirrels, chipmunks, deer, and a variety of bird species. Its common name refers to the resemblance of the foliage to chinkapins (genus *Castanea*), and the Latin species name honors Pennsylvania botanist Gotthilf Heinrich Ernst Muhlenberg (1753–1815).

DESCRIPTION: This tree has a narrow, rounded crown. Its leaves are 4–6" (10–15 cm) long, 1½–3" (4–7.5 cm) wide, and narrowly elliptical to obovate; they are slightly thickened, pointed at the tip, and narrowed to the base, with many straight, parallel side veins, each ending in a curved tooth on wavy edges. They are shiny green above, and whitish-green and covered with tiny hairs beneath, turning brown or red in fall. The bark is light gray, thin, and fissured and scaly.

FRUIT: Acorns: ½–1" (1.2–2.5 cm) long, egg-shaped, with ⅓ or more enclosed by a deep, thin cup of many overlapping hairy long-pointed gray-brown scales; they are usually stalkless, and mature in the first year.

HABITAT: Mostly on limestone outcrops in alkaline soils, including dry bluffs and rocky river banks; it is often found with other oaks.

RANGE: Southern Ontario east to western Vermont, south to northwest Florida, west to central Texas, and north to Iowa; it is local in southeast New Mexico, Trans-Pecos Texas, and northeastern Mexico. It occurs at 400–3000' (122–914 m).

USES: The durable wood is used in construction and for a variety of other purposes.

SIMILAR SPECIES: Chestnut Oak, Swamp Chestnut Oak.

CONSERVATION STATUS: **LC**

Chinkapin Oak is usually found as a few scattered individuals throughout its wide range, and its population is presumed to be stable. Wildfires can kill young trees and create fire scars on older trees that serve as entry points for various pathogens and defoliating insects that can cause serious losses within a few years. Climate change poses a serious potential threat, as models project that this species could lose almost 60% of its suitable habitat by 2050.

Diameter: 2–3' (0.6–0.9 m)
Height: 50–80' (15–24 m)

■ DENSE POPULATION
■ LIGHT POPULATION
■ NATURALIZED

Myrtle Oak is a small tree of the southeastern United States. It is usually found in dry soils near the coast or on islands in Florida and several adjacent states. This plant's common and Latin species names refer to the resemblance of the leaves to those of Myrtle (*Myrtus communis*), an evergreen shrub from the Mediterranean region, introduced in Florida and California. Myrtle Oak is also commonly called Shrub Oak; the plant tends to grow as a shrub in drier sites.

DESCRIPTION: This is an evergreen, much-branched thicket-forming shrub or small tree with short, crooked branches and a rounded crown. Its leaves are evergreen, alternate, ¾–2" (2–5 cm) long, and ½–1" (1.2–2.5 cm) wide. They are usually elliptical to obovate but vary in shape; they can be rounded or sometimes pointed at the tip, gradually narrowing to a blunt or rounded base, with the edges turned under and sometimes wavy or toothed. They are thick and leathery, hairy, and with short leafstalks. They are shiny dark green and hairless with a prominent network of veins above, and dull light green beneath with tufts of hairs in vein angles. The bark is light gray and smooth, becoming furrowed.

FRUIT: Acorns: ⅜–½" (10–12 mm) long, nearly round, a quarter to a third enclosed by a shallow cup, becoming brown; they are stalkless or short-talked, and usually mature in the second year.

HABITAT: Dry sandy ridges and sand dunes, especially near the coast and on islands; it is usually found with other oaks and pines.

RANGE: Southern South Carolina to southern Florida and west to southern Mississippi; it occurs near sea level.

USES: This species does not have any significant commercial value, but it does provide an important food source for native wildlife.

SIMILAR SPECIES: Chapman Oak, Arkansas Oak.

CONSERVATION STATUS: ⓛⒸ

Myrtle Oak is common in its range. Development of coastal habitats and scrub oak communities that it occupies poses a local threat, and fire suppression allows succession of larger trees that can shade out this smaller species. Both of these threats have a fairly localized impact on the species, however, and there have been no reports of serious declines.

Diameter: 1' (0.3 m)
Height: 30' (9 m)

■ DENSE POPULATION
■ LIGHT POPULATION
■ NATURALIZED

This medium-sized, short-lived deciduous tree is found growing in moist soils in the southeastern United States. It is sometimes planted as a shade tree throughout its native range. Other common names for this species include Black Oak and Possum Oak.

DESCRIPTION: This tree has a conical or rounded crown of slender branches, and fine textured foliage of small leaves. Its leaves are 1½–5" (4–13 cm) long and ¾–2" (2–5 cm) wide. They are obovate or wedge-shaped, broadest near the rounded and slightly three-lobed tip; they are bristle-tipped and gradually narrowed to a long-pointed base, sometimes with small lobes on each side. They are dull blue-green above, and paler with tufts of hairs along vein angles beneath, turning yellow in late fall and shedding in winter. The bark is dark gray, smooth, becoming blackish, and furrowed into narrow scaly ridges.

Diameter: 1–2½' (0.3–0.8 m)
Height: 50–100' (15–30 m)

FRUIT: Acorns: ⅜–⅝" (10–15 mm) long and broad, nearly round, with a shallow, saucer-shaped cup, becoming brown, and maturing in the second year.

HABITAT: Moist or wet soils of lowlands, including flood plains or bottomlands of streams and borders or swamps; also moist uplands; it is often found with Sweetgum.

RANGE: Southern New Jersey south to central Florida, west to eastern Texas, and north to southeastern Missouri; to about 1000' (305 m).

USES: This species is used for timber and for fuel. It is also planted as an ornamental.

SIMILAR SPECIES: Durand Oak, Blackjack Oak, Arkansas Oak.

CONSERVATION STATUS: (LC)
The Water Oak population is currently stable, although the species is susceptible to various pests and pathogens.

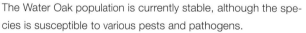

■ DENSE POPULATION
■ LIGHT POPULATION
■ NATURALIZED

This species can be found growing in wet lowlands in the southeastern United States. It was formerly considered to be a subspecies or variety of Southern Red Oak, but, unlike Southern Red, Cherrybark occurs in lowlands and lacks the U-shaped leaf base. It can be distinguished from Northern Red Oak by its densely hairy leaf underside. The leaf shape (upside down) resembles the outline of a pagoda, hence the Latin species name.

DESCRIPTION: This is a medium-sized to large tree with deciduous, simple leaves. It has a tall, largely branch-free trunk, and a high crown. Its leaves are similar to Northern Red Oak, but have five to eleven broad, shallow, usually single-toothed lobes, and they are whitish-hairy (not rusty) beneath. The bark is scaly and red-tinged as in Black Cherry.

FRUIT: Acorn: ½–1" (15–30 cm) long. The nut is nearly round, with about ⅓–½ enclosed within a deep cup of brown, hairy scales. They occur in fall and ripen in two years.

HABITAT: Floodplains and poorly drained bottomlands and lowlands.

RANGE: New Jersey south through Florida, west through Missouri and Texas. It occurs to 1000' (300 m).

USES: This species is a valuable timber tree; the heavy, hard wood is used for furniture, interior finishing, cabinetwork, and general construction. The acorns provide an important food source for a variety of native wildlife.

SIMILAR SPECIES: Southern Red Oak, Northern Red Oak, Overcup Oak, Black Cherry.

CONSERVATION STATUS: (LC)

Cherrybark Oak is common in its range and has a stable population. Like many species in the Red Oak group, Cherrybark Oak is susceptible to oak wilt (caused by *Ceratocystis fagacearum*), but its numbers are not seriously threatened by this fungus. Climate change may pose a much greater threat to the species; models project this species may lose more than 40% of its suitable habitat area by 2050.

Diameter: 2–4' (61–122 cm)
Height: 70–100' (21–31 m)

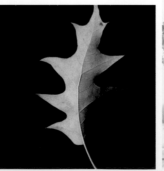

■ DENSE POPULATION
■ LIGHT POPULATION
■ NATURALIZED

Pin Oak is a fast-growing, medium-sized tree found mainly in the eastern and central United States. It is named for the many short side twigs or pin-like spurs. Pin Oaks are known for their attractive canopy shape, compact form, and fine-textured foliage. These trees are hardy and easily transplanted because the shallow, fibrous root system lacks tap roots. As such, Pin Oak has become a widely popular ornamental tree used in lawns, parks, and other residential environments.

DESCRIPTION: This is a straight-trunked tree with spreading to horizontal branches, slender, pin-like twigs, and a broadly conical crown. The leaves are 3–5" (7.5–13 cm) long, 2–4" (5–10 cm) wide, and elliptical, with five to seven deep lobes nearly to the midvein with a few bristle-tipped teeth and wide rounded sinuses; the base is short-pointed. They are shiny dark green above, and light green and slightly shiny with tufts of hairs in vein angles along the midvein beneath, turning red or brown in fall. The bark is dark gray, hard, and smooth, becoming fissured into short, broad, scaly ridges.

FRUIT: Acorns: ½" (12 mm) long and broad, nearly round, becoming brown, with ¼–⅓ enclosed by a thin saucer-shaped cup tapering to the base, maturing in the second year.

HABITAT: In nearly pure stands on poorly drained, wet sites, including clay soils on level uplands; it is less common on deep, well-drained bottomland soils.

RANGE: Extreme southern Ontario to Vermont, south to central North Carolina, west to northeastern Oklahoma, and north to southern Iowa; to 1000' (305 m).

USES: This species is commonly planted as an ornamental tree throughout its native range. The wood is sometimes used in construction and for firewood. Native Americans used the bark for medicinal purposes.

SIMILAR SPECIES: Black Oak, Northern Pin Oak, Scarlet Oak, Northern Red Oak.

CONSERVATION STATUS:

Pin Oak is believed to be stable. It is susceptible to the common pests and pathogens that affect other species in the Red Oak group, but its widespread population is not severely impacted by them. Climate change may threaten this species in the future; climate models suggest that this tree may lose more than 70% of its suitable habitat by 2050. Its wide area of distribution, dispersal ability, and disturbance tolerance may allow this species to persist, but careful monitoring of the species may be required.

Diameter: 1–2½' (0.3–0.8 m)
Height: 50–90' (15–27 m)

WILLOW OAK   *Quercus phellos*

■ DENSE POPULATION
■ LIGHT POPULATION
■ NATURALIZED

Willow Oak is a medium-sized tree found in the eastern and central United States. It is easily distinguishable from most other oaks by the narrow leaves without lobes or teeth. Although the foliage superficially resembles that of willows, it is recognized as an oak by the acorns and the tiny bristle-tip on the leaves. Wildlife, particularly squirrels, consume and spread the acorns. This species is a popular, widely planted street and shade tree with fine-textured foliage, but it becomes too large to be grown around houses. It is readily transplanted because of its shallow roots.

DESCRIPTION: This tree has a conical or rounded crown of many slender branches ending in very slender, pin-like twigs with willow-like foliage. The leaves are 2–4½" (5–11 cm) long, ⅜–¾" (10–19 mm) wide, narrowly oblong or lance-shaped, with a tiny bristle-tip and straight or slightly wavy edges. They are light green and slightly shiny above, and dull light green and sometimes with fine gray hairs beneath, turning pale yellow in fall. The bark is dark gray, smooth, and hard, becoming blackish, rough, and fissured into irregular narrow ridges and plates.

FRUIT: Acorns: ⅜–½" (10–12 mm) long and broad, nearly round, with a shallow, saucer-shaped cup; becoming brown and maturing in the second year.

HABITAT: Moist alluvial soils of lowlands, chiefly flood plains or bottomlands of streams; sometimes in pure stands.

RANGE: New Jersey south to northwest Florida, west to eastern Texas, and north to southern Illinois; to 1000' (305 m).

USES: This species is widely planted as an ornamental tree, particularly in urban landscaping. It is also used for paper production and lumber.

SIMILAR SPECIES: Shingle Oak, Laurel Oak.

CONSERVATION STATUS: LC

Willow Oak is common throughout its range. It is susceptible to various common pests and diseases, but there are no significant threats to the species and its population is stable.

Diameter: 1–2½' (0.3–0.8 m)
Height: 50–80' (15–24 m)

■ DENSE POPULATION
■ LIGHT POPULATION
■ NATURALIZED

This is a fairly large, long-lived tree native to the eastern United States, often in dry, upland soils. Because of its high tannin content, the bark of Chestnut Oak was once heavily used for tanning leather. The lumber is sometimes marketed as "White Oak," and because of its resistance to rot, the wood is often used for outdoor applications such as fence posts and railroad ties. As a shade tree, this species is adapted to dry, rocky soil; an alternate common name is Rock Oak.

ALTERNATE NAME: *Quercus montana*

DESCRIPTION: This large tree sports a broad, open, irregular crown of chestnut-like foliage. The leaves are 4–8" (10–20 cm) long, 2–4" (5–10 cm) wide, elliptical or obovate, broadest beyond the middle, short-pointed at the tip, with wavy edges with 10–16 rounded teeth on each side, gradually narrowing to the base. They are shiny green above, and dull gray-green and sparsely hairy beneath, turning yellow in fall. The bark is gray, becoming thick and deeply furrowed into broad or narrow ridges.

Diameter: 2–3' (0.6–0.9 m)
Height: 60–80' (18–24 m)

FRUIT: Acorns: ¾–1¼" (2–3 cm) long, egg-shaped, with ⅓ or more enclosed by a deep, thin cup narrowed at the base, composed of short, warty, hairy scales not overlapping, becoming brown, and short-stalked; they mature in the first year.

HABITAT: Sandy, gravelly, and rocky dry upland soils, but reaches greatest size on well-drained lowland sites; often in pure stands on dry rocky ridges.

RANGE: Extreme southern Ontario to southwest Maine, south to Georgia, west to northeastern Mississippi, and north to southeastern Michigan; at 1500–5000' (457–1524 m).

USES: Prior to the 20th century the bark of this species was heavily used in leather tanning. The wood is used for various purposes, including fencing, railroad ties, and firewood.

SIMILAR SPECIES: Swamp White Oak, Chinkapin Oak.

CONSERVATION STATUS: LC
Chestnut Oak is dominant in much of its range and its population is presumed to be large and stable. It is susceptible to various common diseases and insects, but these are not considered significant risks to the global population. Chestnut Oak is one of the preferred host species of the introduced Gypsy Moth (*Lymantria dispar*).

■ DENSE POPULATION
■ LIGHT POPULATION
■ NATURALIZED

English Oak, also called Common Oak, is widespread in Europe and widely planted elsewhere. This noble oak is one of the most characteristic native British trees, attaining an impressively large size with age. This species is both symbolically and historically important; it supplied the timbers for the wooden ships of the Royal Navy and oak paneling for many famous European buildings. Widely cultivated as an ornamental tree around the world, many horticultural varieties are distinguished by crown shape and leaf shape and color.

DESCRIPTION: This is an introduced tree with a short, stout trunk; wide-spreading branches; and a broad, rounded, open crown. The leaves are 2–5" (5–13 cm) long, 1¼–2½" (3–6 cm) wide, and oblong, with six to fourteen shallow rounded lobes, including two small ear-shaped lobes at a short-stalked base. They are dark green above and pale blue-green beneath. The bark is dark gray, and deeply and irregularly furrowed, becoming thick.

FRUIT: Acorns: ⅝–1" (1.5–2.5 cm) long, egg-shaped, with about ⅓ enclosed by a half-round cup, becoming brown, with one to five occurring on a long, slender stalk; they mature in the first year.

HABITAT: Spreading from cultivation in moist soil, along roadsides and forest edges.

RANGE: Native of Europe, northern Africa, and western Asia. It is naturalized locally in southeastern Canada and the northeastern United States and planted in the southeastern and Pacific states.

USES: This species is widely cultivated as an ornamental tree. The durable wood has been used for many purposes, including furniture making and shipbuilding, and the bark was once used as a source of tannin.

SIMILAR SPECIES: White Oak.

CONSERVATION STATUS: **LC**
English Oak is common to abundant throughout most of its wide native range. It is vulnerable to overharvesting from the timber trade, and to various pests and diseases. Climate change may threaten the species with a higher disease risk, loss of suitable habitat, and greater exposure to severe weather conditions.

Diameter: 2–3' (0.6–0.9 m)
Height: 80' (24 m), taller with age

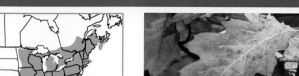

■ DENSE POPULATION
■ LIGHT POPULATION
■ NATURALIZED

The northernmost eastern oak, this is also the most important lumber species of red oak, used for flooring, furniture, millwork, railroad ties, mine timbers, fence posts, pilings, and pulpwood. One of the most rapid-growing oaks, Northern Red Oak transplants easily, is hardy in city conditions, and endures cold. It is a popular, handsome shade and street tree with good form and dense foliage.

DESCRIPTION: This is a large tree with a rounded crown of stout, spreading branches. The leaves are 4–9" (10–23 cm) long, 3–6" (7.5–15 cm) wide, elliptical, usually divided less than halfway to the midvein into 7–11 shallow wavy lobes with a few irregular bristle-tipped teeth. They are usually dull green above, and dull light green beneath with tufts of hairs in angles along the midvein, turning brown or dark red in fall. The bark is dark gray or blackish, rough, and furrowed into scaly ridges; the inner bark is reddish.

FRUIT: Acorns: ⅝–1⅛" (1.5–2.8 cm) long, egg-shaped, with less than ⅓ enclosed by a broad cup of reddish-brown, blunt, tightly overlapping scales; maturing in the second year.

CAUTION: Poisonous: Do not eat or ingest any part of this species. If accidentally ingested, contact your doctor or poison control center.

HABITAT: Moist, loamy, sandy, rocky, and clay soils; often forming pure stands.

RANGE: Western Ontario to Cape Breton Island, south to Georgia, west to eastern Oklahoma, and northern Minnesota; it occurs to 5500' (1676 m) in the south.

USES: Northern Red Oak is an important lumber species, and its wood is used for a variety of purposes. It is also commonly planted as an ornamental tree.

SIMILAR SPECIES: Southern Red Oak, Cherrybark Oak.

CONSERVATION STATUS: (LC)
Northern Red Oak is widespread and abundant throughout its range. It is often canopy-dominant in North American forests. This species is vulnerable to oak wilt and extreme weather conditions such as drought, but no significant range-wide declines have been reported.

Diameter: 1–2½' (0.3–0.8 m)
Height: 60–90' (18–27 m)

■ DENSE POPULATION
■ LIGHT POPULATION
■ NATURALIZED

This large oak occurs widely throughout central and southeastern North America, mainly in the Atlantic coastal plain and in the Mississippi River Valley. It is a handsome shade tree and valued both commercially and as a food source for various native birds and mammals. Texas Oak (var. *texana*), a variety in central Texas and southern Oklahoma, has small, usually five-lobed leaves, small acorns, and hairy red buds (instead of hairless brown buds). The species is named for Benjamin Franklin Shumard (1820–1869), a 19th century geologist and physician from Texas.

DESCRIPTION: This is a large tree with a straight axis and a broad, rounded, open crown. The leaves are 3–7" (7.5–18 cm) long, 2½–5" (6–13 cm) wide, and elliptical, usually deeply divided nearly to the midvein into 5–9 lobes becoming broadest toward the tip, with several spreading bristle-tipped teeth, and large rounded sinuses between the lobes, sometimes nearly closed. They are slightly shiny dark green and hairy above, and slightly shiny or dull green beneath with tufts of hairs in vein angles, turning red or brown in fall. The bark is gray and smooth, becoming dark gray and slightly furrowed into ridges.

FRUIT: Acorns: ⅝–1⅛" (1.5–2.8 cm) long and egg-shaped, with ¼–⅓ enclosed by a shallow cup of tightly overlapping blunt scales; they are green becoming brown and usually hairless, maturing in the second year.

HABITAT: Moist, well-drained soils including flood plains along streams; also on dry ridges and limestone hills.

RANGE: North Carolina to northern Florida, west to central Texas, and north to eastern Kansas; it is local north to southern Michigan and southern Pennsylvania. it occurs to 2500' (762 m).

USES: This species is a popular shade tree. The lumber, along with other red oak lumber, is used for furniture, paneling, flooring, and other purposes.

SIMILAR SPECIES: Pin Oak, Northern Red Oak.

CONSERVATION STATUS: LC

The Shumard Oak population is widespread and stable. It is susceptible to oak wilt and could be significantly impacted by climate change in the future, with models projecting this species may lose more than 60% of its suitable habitat by 2050.

Diameter: 1–2½' (0.3–0.8 m)
Height: 60–90' (18–27 m)

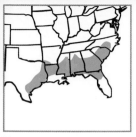

■ DENSE POPULATION
■ LIGHT POPULATION
■ NATURALIZED

This oak is native to the south-central and southeastern United States as well as northern Mexico. It is named for Elias Magloire Durand (1794–1873), Philadelphia pharmacist and botanist. It is also called Bastard Oak. There are two variations of the species. The typical var. *sinuata* tends to be a larger tree that prefers moist bottomlands and riparian habitats, whereas var. *breviloba* occurs farther west and generally prefers limestone soils; it can be a shrub or a small tree.

ALTERNATE NAMES: Bastard Oak, *Quercus durandii*

DESCRIPTION: Durand Oak can be a tree with a rounded crown, or a shrub. The leaves are 1½–4" (4–10 cm) long, and ⅝–1½" (1.5–4 cm) wide, sometimes larger. They are obovate to elliptical, and variable in shape. They are sometimes slightly three-lobed toward a rounded tip or wavy-lobed, and thin or slightly thickened. They are shiny dark green above, and gray-green and covered with tiny star-shaped hairs beneath. The bark is light gray, thin, and scaly.

FRUIT: Acorns: ½–⅝" (12–15 mm) long, egg-shaped, with less than ¼ enclosed by a saucer-shaped scaly cup, becoming brown, and maturing in the first year.

HABITAT: Various soils, including limestone hills; in hardwood forests.

RANGE: North Carolina to northern Florida, west to southern and central Texas, and north to southern Oklahoma; it is also found in Mexico. It occurs to 2000' (620 m).

USES: This species is sometimes planted as an ornamental.

SIMILAR SPECIES: White Oak, Bluff Oak, Boynton Oak, Water Oak.

CONSERVATION STATUS: LC
This species occurs regularly throughout its range, but some subpopulations are scattered. There are no known threats.

Diameter: 2' (0.6 m)
Height: 70' (21 m)

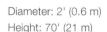

■ DENSE POPULATION
■ LIGHT POPULATION
■ NATURALIZED

This slow-growing oak is native to the southeastern United States, where it prefers dry, upland soils. Its rot-resistant wood is sometimes marketed as "White Oak" and is used for posts (hence its name) and a variety of other outdoor purposes. It grows to a large size in the lower Mississippi Valley, where it is known as "Delta Post Oak." Old-growth forests of Post Oak and Blackjack Oak form the Cross Timbers ecoregion in Texas and Oklahoma, the forest border of small trees and transition zone to the prairie grassland.

DESCRIPTION: This tree has a dense, rounded crown and distinctive leaves suggesting a Maltese cross; it is sometimes a shrub. The leaves are 3¼–6" (8–15 cm) long, 2–4" (5–10 cm) wide, and obovate; with five to seven deep, broad, rounded lobes, the two middle lobes being the largest, with a short-pointed base and a rounded tip; they are slightly thickened. The leaves are shiny dark green and slightly rough with scattered hairs above, and gray-green with tiny star-shaped hairs beneath, turning brown in fall. The bark is light gray and fissured into scaly ridges.

FRUIT: Acorns: ½–1" (1.2–2.5 cm) long, elliptical, with ⅓–½ enclosed by a deep cup; they are green becoming brown, and usually stalkless or short-stalked, maturing in the first year.

HABITAT: Sandy, gravelly, and rocky ridges, also moist loamy soils of flood plains along streams; sometimes found in pure stands.

RANGE: Southeastern Massachusetts south to central Florida, west to northwestern Texas, and north to southeastern Iowa; to 3000' (914 m).

USES: The wood from this species is used for posts, siding, railroad ties, construction timbers, and other purposes.

SIMILAR SPECIES: Sand Post Oak, Blackjack Oak, Bur Oak.

CONSERVATION STATUS: LC

This species is widespread and is especially abundant throughout the southeastern and south-central United States, where it is often dominant. Most insects and diseases that afflict other eastern oak species also attack Post Oak, including the chestnut blight fungus (*Cryphonectria parasitica*).

Diameter: 1–2' (0.3–0.6 m)
Height: 30–70' (9–21 m)

■ DENSE POPULATION
■ LIGHT POPULATION
■ NATURALIZED

One of the rarest of our native oaks, this species was discovered on Guadalupe Island off the coast of Mexico in 1875 and later found on five islands off southern California. Fossil records indicate it was once more widespread in mainland California, but today is found only in relict island populations. The Latin species name means "with fine, woolly hairs."

DESCRIPTION: This evergreen tree has a rounded crown, spreading branches, and fine white hairs on twigs, young leaves, and acorn cups. The leaves are evergreen, 1¼–3½" (3–9 cm) long, ¾–2" (2–5 cm) wide, oblong or lance-shaped, short-pointed at tip, blunt or rounded at the base, with edges mostly wavy-toothed and often turned under, and thick and leathery, with many parallel side veins. They are shiny dark green above, and paler and hairy with a prominent network of veins beneath. The bark is reddish-brown, thin, and scaly. The twigs are covered with white hairs, turning brown.

FRUIT: Acorns: 1–1½" (2.5–4 cm) long and egg-shaped, with a shallow, thin cup of densely hairy scales; they are stalkless or nearly so, and mature in the second year.

HABITAT: Canyons and ravines; in chaparral woodland.

RANGE: Santa Rosa, Santa Cruz, Anacapa, Santa Catalina, and San Clemente islands of southern California and Guadalupe Island of Baja California; it is absent from the mainland. It occurs near sea level.

USES: Due to its rarity and limited range, this species has not had any significant human use.

SIMILAR SPECIES: Canyon Live Oak.

CONSERVATION STATUS: **EN**
Island Live Oak is known to occur at just five sites in an area of occupancy totaling less than 100 square miles (250 km²). The population is small and declining. The introduction of agriculture and ranching on the Channel Islands and Guadalupe Island introduced several problematic non-native plants and grazing animals. Since the 1980s, systematic efforts have been made to remove feral livestock species such as sheep, goats, and pigs from the islands. The Channel Island subpopulations are more stable or perhaps increasing, but any gains are more than offset by continued Guadalupe Island population declines.

Diameter: 1–2' (0.3–0.6 m)
Height: 20–40' (6–12 m)

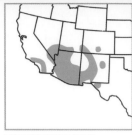

■ DENSE POPULATION
■ LIGHT POPULATION
■ NATURALIZED

Native to the American Southwest, Shrub Live Oak is the characteristic shrub in the chaparral vegetation of Arizona mountain slopes. Wildlife and occasionally livestock browse the foliage, and many small animals use these often-shrubby plants as cover. The species name *turbinella*, which means "like a little top," refers to this plant's acorns, which many native birds and mammals consume. Other common names for this species include Turbinella Oak and Gray Oak.

DESCRIPTION: This is an evergreen, much-branched, thicket-forming shrub or small tree with a spreading crown. The leaves are evergreen, ⅝–1½" (1.5–4 cm) long, and ⅜–¾" (10–19 mm) wide. They are elliptical or oblong, short-pointed at the tip, rounded or notched at the base, spiny-toothed, thick and stiff. They are blue-green with whitish bloom and nearly hairless above, and yellow-green and with fine hairs beneath. The bark is gray and fissured and scaly.

FRUIT: Acorns: ⅝–1" (1.5–2.5 cm) long and narrowly oblong, with ¼–⅓ enclosed by a shallow, scaly cup, with one or a few at the end of a stalk, maturing in the first year.

HABITAT: On mountain slopes, forming thickets; also with other oaks, pinyons, and junipers.

RANGE: Southwestern Colorado south to southern New Mexico, west to southern California, and south to Baja California; at 4000–8000' (1219–2438 m).

USES: This species is an important source of food and cover for wildlife. Many people, including some Native Americans, have used the acorns of this species as a food source.

SIMILAR SPECIES: Gambel's Oak, Arizona White Oak.

CONSERVATION STATUS: LC
Shrub Live Oak has a large geographical range and is not experiencing any threats.

Diameter: 4" (10 cm)
Height: 5–15' (1.5–4.6 m), sometimes larger

■ DENSE POPULATION
■ LIGHT POPULATION
■ NATURALIZED

This tree is found throughout the eastern United States, most often in dry upland soils. It remains a small tree in the northern part of its range but can grow much larger in the South. Once commonly called Yellow Oak, this species is easily distinguishable by the yellow or orange inner bark, formerly used as a source of tannin, medicine, and a yellow pigment that was sold commercially until the 1940s. This species readily hybridizes with other members of the Red Oak group.

DESCRIPTION: This is a medium-sized to large tree with an open, spreading crown. The leaves are 4–9" (10–23 cm) long and 3–6" (7.5–15 cm) wide. They are elliptical, usually with seven to nine lobes, either shallow or deep and narrow, ending in a few bristle-tipped teeth, and slightly thickened. They are shiny green above, and yellow-green and usually with brown hairs beneath, turning dull red or brown in fall. The bark is gray and smooth on small trunks, becoming blackish, thick and rough, and deeply furrowed into ridges; the inner bark is yellow or orange, and strongly bitter.

FRUIT: Acorns: ⅝–¾" (15–19 mm) long and elliptical, with ½ enclosed by a deep, thick, top-shaped cup narrowed at the base, with a fringed border of loose rust-brown hairy scales; maturing in the second year.

CAUTION: Poisonous: Do not eat or ingest any part of this species. If accidentally ingested, contact your doctor or poison control center.

HABITAT: Dry upland sandy and rocky ridges and slopes, also on clay hillsides; sometimes in pure stands.

RANGE: Extreme southern Ontario and southwestern Maine, south to northwest Florida, west to central Texas, and north to southeast Minnesota; to 5000' (1524 m).

USES: The wood from this species is used for fence posts and firewood.

SIMILAR SPECIES: Northern Red Oak, Scarlet Oak, Pin Oak.

CONSERVATION STATUS: LC
Black Oak has a wide range. Like its close relatives, it is susceptible to oak wilt, but the disease has not caused any major population decline.

Diameter: 1–2½' (0.3–0.8 m)
Height: 50–80' (15–24 m)

LIVE OAK *Quercus virginiana*

Live Oak, also known as Southern Live Oak, is a handsome shade tree, popular in the southeastern United States, where it attains an impressively large size. It is called Live Oak because of its evergreen foliage. The broad branches are usually draped with Spanish-moss (*Tillandsia usneoides*). This tree's dense timber was once important for building ships—the United States' first publicly owned timber lands were purchased as early as 1799 to preserve these trees for this purpose. Texas Live Oak (*Q. fusiformis*)—a related species found in central Texas and local in southwestern Oklahoma and northeastern Mexico—was formerly considered a variety of *Q. virginiana*; it has slightly smaller leaves that are broadest toward the base, and acorns with cups that are narrowed at the base.

DESCRIPTION: This medium-sized evergreen tree has a short, broad trunk buttressed at the base forking into a few nearly horizontal, long branches, and a very broad, spreading, dense crown. Its leaves are evergreen and 1½–4" (4–10 cm) long, ⅜–2" (1–5 cm) wide. They are elliptical or oblong, thick, with a rounded tip sometimes ending in a tiny tooth; the base is short-pointed and the edges are usually straight and slightly rolled under and, rarely, with a few spiny teeth. They are shiny dark green above, and gray-green and densely hairy beneath, shedding after new leaves appear in spring. The bark is dark brown, rough, and deeply furrowed into scaly ridges.

FRUIT: Acorns: ⅝–1" (1.5–2.5 cm) long, narrow and oblong, with ¼–½ enclosed by a deep cup; they are green becoming brown, long-stalked, maturing in the first year.

HABITAT: Sandy soils including coastal dunes and ridges near marshes; often in pure stands.

RANGE: Southeastern Virginia south to southern Florida and west to southern and central Texas; it is local in southwestern Oklahoma and northeastern Mexico. It occurs to 300' (91 m) and in Texas to 2000' (610 m).

USES: This tree's hard, heavy wood was once used for building ships but is not commonly used commercially today. The species provides an important food source for a wide variety of birds and mammals.

SIMILAR SPECIES: Texas Live Oak, Sand Live Oak.

CONSERVATION STATUS: LC

Live Oak is common throughout its range. Oak wilt is problematic for this species in Texas, where the disease is killing thousands of trees each year, but there are no reported range-wide threats.

Diameter: 2–4' (0.6–1.2 m)
Height: 40–50' (12–15 m)

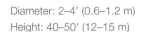

■ DENSE POPULATION
■ LIGHT POPULATION
■ NATURALIZED

Interior Live Oak is found in California and Baja California, mainly in the lower-elevation foothills of the Sierra Nevada and in the Coast Ranges. Deer browse the foliage, and the wood is used for fuel. Although slow-growing, Interior Live Oak is often planted as an ornamental tree. This species is named for its discoverer, Friedrich Adolph Wislizenus (1810–1889), a German-born physician of St. Louis, Missouri. The trees in the Coast Ranges are sometimes treated as a separate species called Coast Oak (*Q. parvula*).

DESCRIPTION: This evergreen tree has a short trunk and a broad, rounded crown of stout, spreading branches; it is sometimes a shrub. The leaves are evergreen, 1–2" (2.5–5 cm) long, and ½–1¼" (1.2–3 cm) wide. They are lance-shaped to elliptical, short-pointed at the tip, blunt or rounded at the base, often with short, spiny teeth, thick and leathery, and hairless. They are shiny dark green above, and light yellow-green with a prominent network of veins beneath. The bark is gray, becoming furrowed into narrow, scaly ridges.

Diameter: 1–3' (0.3–0.9 m)
Height: 30–70' (9–21 m)

FRUIT: Acorns: ¾–1½" (2–4 cm) long, egg-shaped, long-pointed, and often with long, dark lines, with about ½ enclosed by a deep, thin, scaly cup; there are typically one or two on short stalks (or stalkless), maturing in the second year.

HABITAT: Valleys and slopes in foothill woodlands; it is found with other oaks and Digger Pine.

RANGE: Northern to southern California, mostly in the foothills of the Sierra Nevada and inner Coast Ranges, and northern Baja California; it occurs at 1000–5000' (305–1524 m).

USES: This species is often planted as an ornamental tree. The wood is sometimes used for fuel.

SIMILAR SPECIES: Coast Live Oak.

CONSERVATION STATUS: LC

The population is abundant in California and is dominant or co-dominant in most of the habitats where it occurs. Changes in land use and development are the biggest threats to these trees, as native oak forests are cleared for agricultural or residential development projects.

PACIFIC WAX-MYRTLE *Morella californica*

■ DENSE POPULATION
■ LIGHT POPULATION
■ NATURALIZED

This small tree or shrub is native to the Pacific Coast of North America from Vancouver Island south. Pacific Wax-myrtle is sometimes planted as an ornamental shrub for the showy berries and dense, shiny evergreen foliage. The fruit is eaten in small quantities by warblers and many other birds. The wax from the fruits of related eastern bayberries or wax-myrtles was formerly extracted in boiling water and made into fragrant-burning candles; the waxy coating on Pacific Wax-myrtle fruit is much thinner and less useful for this purpose.

ALTERNATE NAME: *Myrica californica*

DESCRIPTION: This is an evergreen, much-branched shrub or small tree with a narrow, rounded crown and waxy brownish berries. The leaves are evergreen, 2–4½" (5–11 cm) long, ½–¾" (12–19 mm) wide, reverse lance-shaped, usually broadest near the short-pointed tip, saw-toothed except near the long-pointed base, aromatic, slightly thickened, and hairless. They are shiny dark green above, and yellow-green with tiny black gland-dots beneath. The bark is gray or brown, smooth, and thin. The twigs are green or brown, slender, and hairy when young.

FLOWERS: Early spring. Tiny yellowish male flowers occur in almost stalkless clusters ⅜–¾" (10–19 mm) long at the base of the lower leaves. Tiny reddish-green female flowers occur in clusters ⁵⁄₁₆–½" (8–12 mm) long at the base of the upper leaves of the same plant.

FRUIT: ¼–⁵⁄₁₆" (6–8 mm) in diameter, brownish-purple, warty with a whitish wax coat, and one-seeded, with several along the stalk at the leaf base; maturing in early fall.

HABITAT: Moist sand dunes, hillsides, and canyons; forming thickets with coastal scrub, Redwood, and Shore Pine.

RANGE: Southwest Washington south near the Pacific Coast to southern California; to 500' (152 m).

USES: This species is sometimes planted for windbreaks and shelter. Wax from the fruit of this and related species has been used for fragrant candles and in soaps.

SIMILAR SPECIES: Southern Bayberry.

CONSERVATION STATUS: LC

The Pacific Wax-myrtle population is stable with no major threats.

Diameter: 1' (0.3 m)
Height: 30' (9 m)

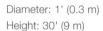

This is a small tree or a shrub found in a wide range from the southeastern United States into the Caribbean and Central America. This popular evergreen ornamental is used for screens, hedges, and landscaping, and as a source of honey. The fruits are a traditional source of the wax for fragrant-burning candles.

ALTERNATE NAME: *Myrica cerifera, Myrica pusilla*

DESCRIPTION: Southern Bayberry is an evergreen, aromatic, resinous shrub or small tree with a narrow, rounded crown. The leaves are 1½–3½" (4–9 cm) long and ¼–¾" (6–19 mm) wide; those toward the end of the twigs are often smaller. They are reverse lance-shaped, coarsely saw-toothed beyond the middle, slightly thickened and stiff, aromatic when crushed, and short-stalked. They are shiny yellow-green with tiny dark brown gland dots above, and paler with tiny orange gland-dots and often hairy beneath. The bark is light gray, smooth, and thin.

Diameter: 6" (15 cm)
Height: 30' (9 m)

FLOWERS: Spring. Tiny, yellow-green flowers in narrowly cylindrical clusters ¼–¾" (6–19 mm) long, at the base of the leaf. Male and female flowers are on separate trees.

FRUIT: One-seeded drupes, ⅛" (3 mm) in diameter, warty, light green, and covered with bluish-white wax; several are crowded in a cluster, maturing in fall and remaining attached in winter.

HABITAT: Moist, sandy soil, in fresh or slightly brackish banks, swamps, hammocks, flatwoods, pinelands, and upland hardwood forests.

RANGE: Southern New Jersey south to southern Florida, west to southern Texas, and north to extreme southeastern Oklahoma; to about 500' (152 m).

USES: This species is sometimes planted as an ornamental and for hedging. Wax from the fruit has been used to make candles.

SIMILAR SPECIES: Northern Bayberry, Pacific Wax-myrtle.

CONSERVATION STATUS: LC
The Southern Bayberry population is stable with no major threats.

■ DENSE POPULATION
■ LIGHT POPULATION
■ NATURALIZED

This generally small plant is native to the eastern coastal zone of North America, where it can grow in poor soils in dry, sandy areas as well as in boggy environments. The thick wax that coats the fruits of these plants is used to make fragrant candles. The closely related Southern Bayberry (*M. cerifera*) occurs from New Jersey to Florida and along the Gulf Coast to Texas; it is a much larger plant, reaching a height of up to 30' (10 m). The northern and southern species are similar in appearance, and distinguishing between them in regions where they both occur can be difficult, especially because they hybridize readily.

ALTERNATE NAME: *Myrica pensylvanica*

DESCRIPTION: This plant is usually a woody shrub with fragrant, dark green leaves and waxy berries. The leaves grow up to 4" (10 cm) long and are slightly toothed near the tip. They are sticky with a spicy scent when crushed.

FLOWERS: Flowers are tiny and not showy, occurring in early spring.

FRUIT: Round, grayish-white, wax-covered berries ⅛" (3–4 mm) in diameter, growing in clusters from a stem below the leaves.

HABITAT: Dry sandy areas, coastal and inland, including dunes, pine barrens, bogs, and watersides.

RANGE: Maritime Provinces south to northeastern North Carolina, inland to western New York and Pennsylvania.

USES: This species is often used in landscaping and for wildlife. The berries are readily eaten by a wide variety of birds, and the wax from the berries is used in candle making.

SIMILAR SPECIES: Southern Bayberry.

CONSERVATION STATUS: LC
Northern Bayberry has a wide distribution and is not currently experiencing any major threats.

Height: 1½–6' (0.45–1.8 m)

■ DENSE POPULATION
■ LIGHT POPULATION
■ NATURALIZED

This is the most abundant hickory, common in the eastern United States, where it prefers moist upland soils. The wood of this hickory and related species is prized for furniture, flooring, tool handles, baseball bats, skis, and veneer. Squirrels, chipmunks, birds, and many other forms of wildlife consume the small, edible hickory nuts. The former Latin species name, *tomentosa*, which means "densely covered with soft hairs," describes the undersurfaces of the leaflets, a diagnostic characteristic that helps in identifying this tree.

ALTERNATE NAME: *Carya tomentosa*

DESCRIPTION: Mockernut Hickory is a nut tree with a rounded crown and leaves that are very aromatic when crushed. Those leaves are pinnately compound, 8–20" (20–51 cm) long, with a hairy axis, 7 or 9 leaflets 2–8" (5–20 cm) long, elliptical or lance-shaped, finely saw-toothed, and nearly stalkless. They are shiny dark yellow-green above, and pale and densely hairy and glandular beneath, turning yellow in autumn. The bark is gray and irregularly furrowed into narrow forking ridges. The twigs are brown, stout, and hairy, ending in a large hairy bud.

FLOWERS: Tiny and greenish, appearing in early spring before the leaves. The male flower has four to five stamens, with many flowers in slender drooping catkins, with three hanging from a single stalk. Female flowers occur in bunches of two to five, occurring at the tip of the same twig.

FRUIT: 1½–2" (4–5 cm) long, elliptical or pear-shaped, becoming brown, with a thick husk splitting to the middle or nearly to the base. The hickory nut is rounded or elliptical, slightly four-angled, and thick-shelled, with an edible seed.

HABITAT: Moist uplands and less frequently on flood plains; usually with oaks, also pines.

RANGE: Extreme southern Ontario east to Massachusetts, south to northern Florida, west to eastern Texas, and north to southeast Iowa; to 3000' (914 m) in southern Appalachians.

USES: The wood of this species is used for furniture, flooring, and many other commercial products. Hickory wood has a high fuel value, both as firewood and as charcoal, and is preferred for smoking meats. The nuts are edible and eaten by some people.

SIMILAR SPECIES: Mockernut Hickory is not likely to be confused with other hickory species.

CONSERVATION STATUS: 🔵LC
Mockernut Hickory has a secure population with no major threats.

Diameter: 2' (0.6 m)
Height: 50–80' (15–24 m)

■ DENSE POPULATION
■ LIGHT POPULATION
■ NATURALIZED

Both the common and scientific names of this large native hickory describe this tree's ability to tolerate the wettest of soils. In the South it can be a dominant species on clay flats and backwater areas near rivers and other waterways, forming a major component of wetland forests. Ducks, squirrels, and other wildlife consume the tree's bitter-tasting nuts. Also known as Water Pecan or Bitter Pecan, Water Hickory is the tallest of all hickories, with exceptional specimens measuring 150' (45.7 m) tall.

DESCRIPTION: This is a large tree with a tall straight trunk, slender upright branches, a narrow crown, and bitter inedible nuts. The leaves are pinnately compound and 9–15" (23–38 cm) long, with a dark red, hairy axis. There are usually 9–13 leaflets 2–5" (5–13 cm) long; they are lance-shaped, long-pointed at the tip, slightly curved, finely saw-toothed, and mainly stalkless. They are dark green and hairless above, and often hairy beneath. The bark is light brown, thin, and fissured into long, platelike, red-tinged scales. The twigs are brown and slender, becoming hairless.

FLOWERS: Flowers appear in early spring, before the leaves. They are tiny and greenish. Male flowers have six or seven stamens, with many occurring in slender drooping catkins, and three hanging from a single stalk. Female flowers occur in groups of two to ten at the tip of the same twig.

FRUIT: 1–1½" (2.5–4 cm) long, broadly elliptical, much flattened, four-winged, becoming dark brown, with a thin husk splitting to the middle, with four or fewer in a cluster. The nut is flattened, four-angled, and thin-shelled, with a bitter seed.

HABITAT: Low wet flatlands, especially clay and flats, often submerged, in flood plains and swamps; bottomland hardwood forests.

RANGE: Southeastern Virginia south to central Florida, west to eastern Texas, and north to southern Illinois; to 400' (122 m).

USES: This species is occasionally planted as a shade tree and the wood is sometimes used for fence posts. Native wildlife uses the nuts as a food source.

SIMILAR SPECIES: Pecan, Bitternut Hickory.

CONSERVATION STATUS: (LC)
Water Hickory has a wide geographical range and is not experiencing any threats.

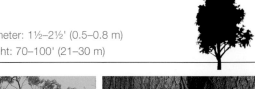

Diameter: 1½–2½' (0.5–0.8 m)
Height: 70–100' (21–30 m)

DENSE POPULATION
LIGHT POPULATION
NATURALIZED

One of the most widely distributed and most common hickories throughout eastern North America, Bitternut Hickory is also one of the easiest to identify because of its small, bright yellow buds. It thrives in moist mountain valleys and wet bottomlands along waterways and in swamps. It is closely related to the Pecan, but its bitter nuts are inedible. Although rabbits have been observed eating the seeds, they are unpalatable to most wildlife and to humans.

DESCRIPTION: This tree has a tall trunk, a broad and rounded crown, and bitter inedible nuts. Its leaves are pinnately compound and 6–10" (15–25 cm) long, with a slender hairy axis. There are seven to nine leaflets 2–6" (5–15 cm) long; they are stalkless, lance-shaped, and finely saw-toothed. They are yellow-green above, and light green and slightly hairy beneath, turning yellow in autumn. The bark is gray or light brown, and shallowly furrowed into narrow, forking, scaly ridges. The twigs are slender, ending in bright yellow, slightly flattened buds.

Diameter: 1–2' (0.3–0.6 m)
Height: 60–80' (18–24 m)

FLOWERS: In early spring before leaves. They are tiny and greenish. Male flowers have four or five stamens, with many arranged in slender, drooping catkins, and three hanging from a single stalk. One or two female flowers appear at the tip of the same twig.

FRUIT: ¾–1¼" (2–3 cm) long, nearly round or slightly flattened, short-pointed; the husk is thin, with tiny yellow scales, and splitting along four wings. The nut is nearly smooth and thin-shelled, with a bitter seed.

HABITAT: Moist soil of valleys and in the north also on dry upland soil; it is found in mixed hardwood forests.

RANGE: Southern Quebec and southwestern New Hampshire, south to northwestern Florida, west to eastern Texas, and north to Minnesota; to 2000' (610 m).

USES: The durable wood from this tree is used for furniture, paneling, tool handles, and other purposes. Some Native Americans used the wood to make bows and early European settlers extracted oil from the nuts for lamps.

SIMILAR SPECIES: Pecan, Water Hickory.

CONSERVATION STATUS: (LC)
Bitternut Hickory has a stable population and is not experiencing any threats.

PIGNUT HICKORY   *Carya glabra*

■ DENSE POPULATION
■ LIGHT POPULATION
■ NATURALIZED

Pignut Hickory is a widespread eastern species. It is one of the most common hickories in the southern Appalachians and an important commercial timber source. This species provides ample food for squirrels, Wild Turkey (*Meleagris gallopavo*), and many native songbirds. European settlers named this tree in colonial times based on the tendency of hogs to consume the tree's small nuts. Settlers also made brooms from narrow splits of the wood, and sometimes referred to this tree as Broom Hickory. Red Hickory (var. *odorata*), a variety with nearly the same range, has the fruit husk splitting to the base, usually seven leaflets, and often shaggy bark.

DESCRIPTION: This tree has an irregular, spreading crown and thick-shelled nuts. The leaves are pinnately compound, 6–10" (15–25 cm) long, with a slender, hairless axis. There are usually five leaflets 3–6" (7.5–15 cm) long; they are largest toward the tip, lance-shaped, nearly stalkless, finely saw-toothed, and hairless or hairy on veins beneath. They are light green, turning yellow in fall. The bark is light gray and smooth, or becoming furrowed with forking ridges. The twigs are brown, slender, and hairless.

FLOWERS: Early spring, before leaves. They are tiny and greenish. The male flower has four stamens, with many occurring in slender drooping catkins, and three hanging from a single stalk. The female flower occurs in groups of two to ten at the tip of the same twig.

FRUIT: 1–2" (2.5–5 cm) long, slightly pear-shaped or rounded; the husk is thin, becoming dark brown and opening late and splitting usually to the middle. The nut is usually not angled, and thick-shelled, with a small sweet or bitter seed.

HABITAT: Dry and moist uplands in hardwood forests with oaks and other hickories.

RANGE: Southern Ontario east to southern New England, south to central Florida, west to extreme eastern Texas, and north to Illinois; to 4800' (1463 m) in the southern Appalachians.

USES: This is an important timber species; the wood is used for tool handles, skis, and other items. It was formerly used for wagon wheels and textile loom picker sticks because it could sustain tremendous vibration.

SIMILAR SPECIES: Sand Hickory.

CONSERVATION STATUS: LC
The Pignut Hickory population is stable and the species is not experiencing any threats.

Diameter: 1–2' (0.3–0.6 m)
Height: 60–80' (18–24 m)

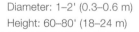

- ■ DENSE POPULATION
- ■ LIGHT POPULATION
- ■ NATURALIZED

Pecan is a native, medium- to large-sized deciduous tree, and is one of the most valuable cultivated plants originating in North America. Pecan is native from Iowa to Indiana to Alabama, Texas, and Mexico. This tree's range follows the river basins closely; Pecan trees commonly grow on rich, moist, well-drained soils. The Latin species name is from an old term, "Illinois nuts," and refers to the region where traders found wild trees and nuts. The word pecan is of Algonquin origin; Native Americans may have extended the important food tree's natural range by planting. This tree was unknown to British colonists on the Atlantic Coast, where this species is not native. Thomas Jefferson planted seeds at Monticello and gave some to George Washington; now these Pecans are the oldest trees in Mount Vernon.

ALTERNATE NAME: *Carya illinoensis*

DESCRIPTION: This large wild and planted tree sports a tall trunk, a broad, rounded crown of massive spreading branches, and familiar pecan nuts. Its leaves are pinnately compound and 12–20" (30–51 cm) long, with 11–17 slightly sickle-shaped leaflets 2–7" (5–18 cm) long; they are long-pointed at the the tip, finely saw-toothed, short-stalked, and hairless or slightly hairy. They are yellow-green above, and paler beneath, turning yellow in autumn. The bark is light brown or gray and deeply and irregularly furrowed into narrow forked scaly ridges.

FLOWERS: Early spring, before leaves. Tiny and greenish. The male flower has five or six stamens, with many clustered in slender drooping catkins, with three hanging from a single stalk. Females occur in groups of two to ten flowers at the tip of the same twig.

FRUIT: 1¼–2" (3–5 cm) long, oblong, short-pointed at the tip and rounded at the base, with a thin husk becoming dark brown, splitting to the base along four ridges, with three to as many as ten in a cluster. The pecan nut is light brown with darker markings and thin-shelled, with edible seed.

HABITAT: Moist, well-drained, loamy soils of river flood plains and valleys; found in mixed hardwood forests.

RANGE: Eastern Iowa east to Indiana, south to Louisiana, west to southern Texas; to 1600' (488 m); also occurs in the mountains of Mexico.

USES: Improved cultivated varieties with large, thin-shelled nuts are grown in plantations or orchards in the Southeast; pecans are also harvested locally from wild trees. The sweet nuts are widely eaten fresh and used in cooking, especially for sweet desserts such as pies and candies. The wood is used for furniture, flooring, veneer, and charcoal for smoking meats.

SIMILAR SPECIES: Arizona Walnut, Bitternut Hickory, Black Walnut.

CONSERVATION STATUS: **LC**
Pecan has a wide geographical range and is not experiencing any threats.

Diameter: 3' (0.9 m)
Height: 100' (30 m)

SHELLBARK HICKORY *Carya laciniosa*

■ DENSE POPULATION
■ LIGHT POPULATION
■ NATURALIZED

This uncommon, slow-growing and long-lived species is distinguished from other hickories by its large leaves, large nuts, and orange twigs. It differs from the closely related Shagbark Hickory by the larger number of leaflets (usually seven instead of five) and the thick-shelled nuts. Many wildlife species, including quail, turkeys, squirrels, foxes, and mice, consume the seeds found within the nuts. The Latin species name, which means "with flaps or folds," refers to the tree's shaggy bark.

DESCRIPTION: Shellbark Hickory is a large tree with a straight trunk, a narrow rounded crown, large leaves, and the largest hickory nuts. Its leaves are pinnately compound and 12–20" (30–51 cm) long. There are usually seven broadly lance-shaped leaflets 2–8" (5–20 cm) long; they are finely saw-toothed and nearly stalkless. They are shiny dark green above, and pale and covered with soft hairs beneath. The bark is light gray, becoming rough and shaggy, separating into long narrow strips loosely attached. The twigs are pale orange, stout, and hairy, ending in large brown hairy buds.

FLOWERS: Early spring. Tiny and greenish. The male flower has three to ten stamens, with many flowers in slender drooping catkins, with three hanging from a single stalk. Two to five female flowers occur at the tip of the same twig.

FRUIT: 1¾–2½" (4.5–6 cm) long, nearly round, and flattened, becoming light to dark brown, with a thick husk splitting to the base. The nut is nearly round and thick-shelled, with edible seeds.

HABITAT: Moist or wet soils of flood plains and valleys, with other hardwoods.

RANGE: Southeastern Iowa east to Ohio and southwest Pennsylvania, south to Tennessee, and west to northeastern Oklahoma; it is also local to extreme southern Ontario, New York, northern Georgia, and Mississippi. It occurs to 1000' (305 m).

USES: The wood from Shellbark Hickory is used for furniture, tool handles, drumsticks, firewood, and many other purposes.

SIMILAR SPECIES: Shagbark Hickory.

CONSERVATION STATUS: LC
Shellbark Hickory has a large, stable population and is not experiencing any threats.

Diameter: 2½' (0.8 m)
Height: 70–100' (21–20 m)

■ DENSE POPULATION
■ LIGHT POPULATION
■ NATURALIZED

The widespread eastern tree is distinguished by the 6–8" (15–20 cm) long plates of gray bark that cover its trunk. The name "hickory" is from the Native American word for the oily food removed from pounded kernels steeped in boiling water. This sweet hickory milk was used in cooking corn cakes and hominy. Pioneers made a yellow dye from the inner bark. Hickory nuts were an important food source to early settlers who would store them by the bushel, and are a delight to squirrels and chipmunks that collect the nuts from midsummer to late fall. It is virtually pest and disease free.

DESCRIPTION: This is a large tree with a tall trunk, a narrow irregular crown, and distinctive rough shaggy bark. Its leaves are pinnately compound, 8–14" (20–36 cm) long, with five (rarely seven) elliptical or ovate leaflets that are 3–7" (7.5–18 cm) long. They are stalkless and the edges are finely saw-toothed and hairy; they are yellow-green above, and paler (and hairy when young) beneath, turning golden-brown in autumn. The bark is light gray, separating into long narrow curved strips loosely attached at the middle. The twigs are brown and stout, ending in large brown hairy buds.

Diameter: 2½' (0.8 m)
Height: 70–100' (21–30 m)

FLOWERS: Tiny green flowers appear before the leaves in early spring. The male flower has four stamens, many in slender drooping catkins, three hanging from one stalk. Three to five female flowers form at the tip of the same twig.

FRUIT: 1¼–2½" (3–6 cm) long nearly round; flattened at tip; with husk thick, becoming dark brown or blackish and splitting to base. Hickory nut elliptical or rounded, slightly flattened and angled, light brown, with edible seed.

HABITAT: Moist soils of valleys and upland slopes in mixed hardwood forests.

RANGE: Extreme southern Quebec and southwestern Maine, south to Georgia, west to southeast Texas, and north to southeastern Minnesota; also northeastern Mexico; to 2000' (610 m) in the north and 3000' (914 m) in the southern Appalachians.

USES: Wild trees and improved cultivated varieties produce commercial hickory nuts. The exceptionally strong, hard hickory wood has been used for axe handles, baseball bats, and skis. It also makes excellent firewood.

SIMILAR SPECIES: Shellbark Hickory is a smaller tree with larger fruit, and grows in wetter soils.

CONSERVATION STATUS:
The population is stable and no major threats to the species are known.

■ DENSE POPULATION
■ LIGHT POPULATION
■ NATURALIZED

First found north of Los Angeles in 1850 by the survey of the Mexican Boundary Commission, this species is now established by planting beyond its local range. It is widely cultivated in California. Some authorities treat Northern California Walnut and Southern California Walnut as distinct species. Southern California Walnut (*J. californica*) is often small and shrublike; its native range is restricted to southwestern California, where it prefers moist soils. By contrast, Northern California Walnut (*J. hindsii*) is a rare and threatened plant endemic to northern California with only one confirmed native stand remaining; it becomes a tall tree with larger nuts to 2" (5 cm) in diameter.

DESCRIPTION: This is a small to medium-sized tree, usually forked near the base, with a rounded crown. The leaves are pinnately compound and 6–9" (15–23 cm) long, with 11–15 leaflets 1–2½" (2.5–6 cm) long; they are oblong lance-shaped, short- or long-pointed, finely saw-toothed, and stalkless. They are shiny green above, and paler, with tufts of hairs along veins beneath, turning yellow or brown in autumn. The bark is dark brown or blackish, and rough and furrowed into broad ridges. The twigs are brown and stout, with chambered pith.

FLOWERS: Early spring. Small and greenish. Male flowers have 30–40 stamens, with numerous flowers in catkins 2–3" (5–7.5 cm) long. A few female flowers with a two-lobed style occur at the tip of the same twig.

FRUIT: 1–1¼" (2.5–3 cm) in diameter; walnut with a thin, dark brown husk; a thick, grooved shell; and an edible seed; maturing in early fall.

HABITAT: Moist soils along streams in canyons and foothills; with cottonwoods, willows, and sycamores.

RANGE: Coastal southern California only; to 2500' (762 m).

USES: This species is cultivated as a rootstock to support commercial English Walnut orchards, and is planted as an ornamental. Some people eat the nuts locally.

SIMILAR SPECIES: Other walnut species.

CONSERVATION STATUS: NT

California Walnut occurs in a limited range but can be dominant in suitable woodlands. Its population size isn't fully known but is believed to be declining due to habitat loss, overgrazing, introduced invasive species, and an increasing frequency of fires. Human population growth may place additional pressure on this species and its habitat; urbanization has led to the clearing of large stands of this tree.

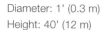

Diameter: 1' (0.3 m)
Height: 40' (12 m)

■ DENSE POPULATION
■ LIGHT POPULATION
■ NATURALIZED

This tree, also known as the White Walnut, is native to the eastern United States and southeastern Canada. It is a slow-growing tree that thrives on stream banks and on well-drained soils. The edible butternuts become rancid soon after maturing, and so must be harvested quickly. A variety of native wildlife readily consumes the namesake oily nuts. The husks of the nuts, which contain a brown stain that colors the fingers, yield a yellow or orange dye.

DESCRIPTION: This tree has a short straight trunk, stout branches, a broad open crown, and butternut fruit with a sticky husk. The leaves are pinnately compound, 15–24" (38–61 cm) long, with a hairy axis, and are sticky when young. There are 11–17 leaflets 2–4½" (5–11 cm) long; they are broadly lance-shaped, pointed at the tip, unequal and rounded at the base, finely saw-toothed, and stalkless. They are yellow-green and slightly hairy above, and paler and covered with soft hairs beneath, turning yellow or brown in autumn. The bark is light gray and smooth, becoming rough and furrowed. The twigs are brown and stout, with sticky hairs and a hairy fringe above leaf-scars, and with chambered pith.

Diameter: 1–2' (0.3–0.6 m)
Height: 40–70' (12–21 m)

FLOWERS: Early spring. The flowers are small and greenish. The male has eight to twelve stamens, with many in catkins. The female, with a two-lobed ovary, and six to eight flowers at the tip of the same twig.

FRUIT: 1½–2½" (4–6 cm) long with three to five in drooping clusters; they are narrowly egg-shaped, long-pointed, with two ridges, rust-colored sticky hairs, and a thick husk. The shell of the nut is thick, light brown, rough with eight ridges, and contains an oily, edible seed (the butternut).

HABITAT: Moist soils of valleys and slopes; also dry rocky soils; in hardwood forests.

RANGE: Southern Quebec east to southwestern New Brunswick, south to extreme northern Georgia, west to Missouri and Arkansas, and north to eastern Minnesota; to 4800' (1463 m).

USES: The lightweight wood of Butternut is used to make furniture and cabinets. Oil from the butternuts was used by Native Americans for various purposes.

SIMILAR SPECIES: Black Walnut.

CONSERVATION STATUS: 🔘 **EN**
The Butternut population has been rapidly declining due to the invasion of the lethal fungus Butternut canker (*Ophiognomonia clavigignenti-juglandacearum*), first reported in North America in 1967. Although large numbers of this tree persist across its range, its population has declined by as much as 80% from its historic numbers due to the spread of this pathogen, which has infected more than 90% of the population is some states. The high rate of infection and high mortality rate suggest further declines are likely.

ARIZONA WALNUT *Juglans major*

■ DENSE POPULATION
■ LIGHT POPULATION
■ NATURALIZED

This tree has a somewhat limited range in western Mexico and scattered areas in the southwestern United States. It produces small walnuts, known in Spanish as *nogales*, which are gathered locally. It often has several slender trunks in dry conditions but may have a single trunk in moister conditions. The restricted distribution of this tree has limited its commercial uses, but the wood—especially the enlarged burls and bases of the trunks— is beautifully patterned and valuable.

DESCRIPTION: This tree often has a forked trunk and a rounded crown of widely spreading branches, with a distinct walnut odor. The leaves are pinnately compound, 7–14" (18–36 cm) long, usually with 9–13 leaflets 2–4" (5–10 cm) long, broadly lance-shaped, often slightly curved, coarsely saw-toothed, and covered with scurfy hairs when young, becoming hairless or nearly so. They are yellow-green in color. The bark is gray-brown, smoothish, and becoming thick and deeply furrowed into ridges. The twigs are brown and stout, with chambered pith.

FLOWERS: Early spring. Small and greenish. The male has 30–40 stamens, with numerous flowers in drooping catkins. Female flowers have a two-lobed style, with only a few occurring at the tip of the same twig.

FRUIT: 1–1½" (2.5–4 cm) in diameter walnut with a thin, densely hairy, brown husk, and a thick, grooved, hard shell. It has a small, edible seed, and matures in early fall.

HABITAT: Moist soils along streams and canyons, mostly in the mountains, in desert, desert grassland, and oak woodland zones.

RANGE: Central Texas west to Arizona; it is also found in Mexico. It occurs at 2000–7000' (610–2134 m).

USES: The walnuts from this species are gathered locally, and the beautifully patterned wood is used for furniture, gunstocks, and other purposes.

SIMILAR SPECIES: Little Walnut, Western Soapberry.

CONSERVATION STATUS:
Arizona Walnut has a large range in the southern U.S. and Mexico, and can be dominant in some habitats; its global population is considered large and secure. Its status in Mexico is less certain, with some subpopulations there possibly extinct, and little population information available.

Diameter: 1–2' (0.3–0.6 m)
Height: 30–50' (9–15 m)

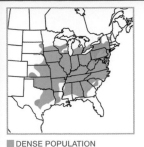

■ DENSE POPULATION
■ LIGHT POPULATION
■ NATURALIZED

One of the most coveted and valuable native hardwoods, Black Walnut heartwood is attractively straight-grained and dark, heavy, durable, and strong while also being easy to split and work. Individual trees fetch attractive prices and a few prized trees have even been stolen. Since antiquity, Black Walnut has provided humans with edible nuts and a blackish dye made from the husks. The delicious nuts must be gathered early before squirrels and other wildlife can consume them. Black Walnuts excrete chemicals into the soil that discourage competition; many other plants—notably tomatoes, apples, pines, and birches—do not survive near mature trees.

DESCRIPTION: This large walnut tree has an open, rounded crown of dark green, aromatic foliage. The leaves are pinnately compound and 12–24" (30–61 cm) long, with 9–21 leaflets 2½–5" (6–13 cm) long; they are broadly lance-shaped, finely saw-toothed, long-pointed, stalkless, nearly hairless above, and covered with soft hairs beneath. They are green or dark green, turning yellow in autumn. The bark is dark brown, deeply furrowed into scaly ridges. The twigs are brown, stout, with brown chambered pith.

FLOWERS: Early spring. Small and greenish; the male has 20–30 stamens, with many flowers in catkins. The female flowers have a two-lobed style, and occur in groups of two to five at the tip of the same twig.

FRUIT: Single or paired, 1½–2½" (4–6 cm) in diameter, thick green or brown husk, irregularly ridged, with a thick-shelled inner layer covering a sweet, edible seed.

HABITAT: Moist well-drained soils, especially along streams, scattered in mixed forests.

RANGE: Eastern half of the United States except the northern border; New York south to northwestern Florida, west to central Texas, north to southeastern South Dakota; it is local in southern New England and southern Ontario. It occurs to 4000' (1219 m).

USES: Black Walnut wood is used for furniture, gunstocks, veneer, and other purposes. Many people consume this tree's edible nuts.

SIMILAR SPECIES: Butternut.

CONSERVATION STATUS: LC
Black Walnut has a large, stable population and wide range. There are no significant threats to the species.

Diameter: 2–4' (0.6–1.2 m)
Height: 70–90' (21–27 m)

ENGLISH WALNUT *Juglans regia*

■ DENSE POPULATION
■ LIGHT POPULATION
■ NATURALIZED

This introduced planted tree yields the familiar walnuts that are most commonly sold commercially. It is a large, deciduous tree native to Eurasia that is quite similar to the Black Walnut. In North America, English Walnut is most commonly grown in Oregon and California. The species goes by several common names including Persian Walnut, Carpathian Walnut, or simply Common Walnut.

DESCRIPTION: This wide-spreading tree is similar to Black Walnut, but has shorter leaves and only 7–9 leaflets. The leaves are strongly aromatic when crushed; they are late to emerge in spring, and turn yellow and drop early in fall. The bark is gray and smooth.

FLOWERS: Male flowers are in spikes and have 6–40 stamens. Female flowers occur in short terminal spikes, giving way to edible nuts.

FRUIT: The large nuts are thin-shelled with green husks, maturing in fall. This tree is the main source of the familiar walnuts that are sold in stores.

HABITAT: Fertile soils in orchards and residential areas.

RANGE: Native to Eurasia from the Balkans to southwestern China. It is widely planted in North America, especially in Oregon and California, and naturalized in parts of the continent's interior.

USES: This species is widely cultivated across Europe and also in North America for its walnuts. The durable, hard wood is used for furniture, flooring, guitars, gunstocks, and many other purposes.

SIMILAR SPECIES: Black Walnut, Butternut.

CONSERVATION STATUS: LC

English Walnut has a large, stable population and an extensive global range, although it is difficult to distinguish between native and cultivated populations in Eurasia. It is introduced and naturalized in parts of North America.

Height: to 70' (21.3 m)

 DENSE POPULATION
LIGHT POPULATION
NATURALIZED

Introduced from Australia, the Australian-pine is a serious pest in Florida, crowding out native plants, depleting soil of moisture, and destroying breeding habitat for threatened species such as the Loggerhead Sea Turtle (*Caretta caretta*) and Gopher Tortoise (*Gopherus polyphemus*). It has also been planted in other areas of the United States, and is classified as an invasive weed. Conservation and government organizations have plans under way to eradicate all *Casuarina* species from North America.

DESCRIPTION: This is an introduced species. It is pine-like, with a shaggy open crown. The leaves are tiny, scale-like, and whorled on jointed, quill-like twigs. The bark is dark brown to light gray, and furrowed into thin strips.

FLOWERS: Male and female flowers are on the same tree, with male flowers in spikes and female flowers in rounded clusters on short stalks.

FRUIT: ¾" (2 cm) yellowish-brown, woody balls.

HABITAT: Beaches, pinelands, disturbed areas.

RANGE: Australian-pine has a wide geographic range, located in much of Southeast Asia and Oceania, including the Pacific Islands. It is abundant as a non-native tree in coastal peninsular Florida.

USES: This species is used for bonsai, and the wood is used for shingles, fencing, and other purposes. It has been widely planted as an ornamental.

SIMILAR SPECIES: This tree is not likely to be confused with any native species in North America. It resembles other exotic *Casuarina* species native to Australia, including River She-oak (*C. cunninghamiana*) and Gray She-oak (*C. glauca*).

CONSERVATION STATUS: (LC)
This species is stable in its native range. In North America, this species can be highly invasive, and is linked to severe ecological damage of native wildlife communities and individual species. It is listed as a priority for removal in Florida.

Height: 100' (30 m)

■ DENSE POPULATION
■ LIGHT POPULATION
■ NATURALIZED

European Alder was introduced in colonial times and planted for shade and ornament. It is considered invasive in some areas in North America. Other common names for this species are Common Alder and Black Alder. The Latin species name, which means "gummy" or "gluey," describes the young twigs and leaves of this tree.

DESCRIPTION: This is a large tree with a rounded or oblong crown and gummy twigs and foliage. The leaves are arranged in three rows; they are 1½–4" (4–10 cm) long, 1–2½" (2.5–6 cm) wide, elliptical to nearly round, doubly saw-toothed (also lobed in cultivated varieties), with five to seven parallel veins on each side, and gummy when young. They are shiny dark green above, and light green with tufts of rusty hairs beneath, remaining green and shedding late. The bark is brown, and smooth becoming furrowed into broad plates. The twigs are mostly hairless, and gummy when young, with three-angled pith.

FLOWERS: Early spring before leaves. Tiny. Male flowers are in catkins 1–1½" (2.5–4 cm) long, upright and later drooping. Female flowers are in cones ¼" (6 mm) long.

FRUIT: Cones: ⅝–⅞" (15–22 mm) long, in clusters of three to five, complete or egg-shaped, black, hard, gummy, long-stalked, and remaining attached. The nutlets are rounded, nearly ⅛" (3 mm) long, and flattened.

HABITAT: Wet soils in humid, cool temperate regions.

RANGE: A native of Europe, Northern Africa, and Asia. It is naturalized locally in southeastern Canada and the northeastern United States.

USES: This species has long been used for a variety of purposes. It has been planted to stabilize river banks, as a windbreak, and as an ornamental tree. The soft wood is used to make furniture, toys, blocks, pencils, and other items. The bark is used in tanning and dyeing.

SIMILAR SPECIES: Hazel Alder, Speckled Alder.

CONSERVATION STATUS: LC

In its native range, European Alder occurs in small, isolated subpopulations that are vulnerable to changes in local conditions or climate, but globally its population is considered stable. In North America, this species is an invasive species of fairly high concern, in part because it is a nitrogen fixer that can form monospecific stands. It is known to escape into natural areas and has already become widely naturalized in the temperate Northeast, where it inhabits early successional forest, edge habitats, floodplain forests, forested wetlands, roadsides, and shrub wetlands.

Diameter: 1–2' (0.3–0.6 m)
Height: 50–70' (15–21 m)

■ DENSE POPULATION
■ LIGHT POPULATION
■ NATURALIZED

Also widely known as Thinleaf Alder, this is the common alder throughout the Rockies. It can be a small to medium-sized tree or a shrub and is found across the Northern Hemisphere. There are several recognized subspecies, some of which are considered full species by some authorities.

ALTERNATE NAMES: Thinleaf Alder, *Alnus incana* ssp. *tenuifolia*

DESCRIPTION: Mountain Alder is a shrub with spreading, slender branches or sometimes a small tree with several trunks and a rounded crown, often forming thickets. The leaves are arranged in three rows, 1½–4" (4–10 cm) long, 1–2½" (2.5–6 cm) wide, ovate or elliptical, wavy-lobed and doubly saw-toothed, rounded at the base, with six to nine nearly straight parallel veins on each side. They are dull dark green above, and light yellow-green and sometimes finely hairy beneath. The bark is gray, thin, and smooth, becoming reddish-gray and scaly. The twigs are slender, reddish, and hairy when young, becoming gray, with three-angled pith.

FLOWERS: Early spring before leaves. Tiny; male flowers are yellowish and in catkins 1–2¾" (2.5–7 cm) long. Female flowers are brownish and in narrow cones ¼" (6 mm) long.

FRUIT: Cones: ⅜–⅝" (10–15 mm) long, 3–9 clustered on short stalks, elliptical, with many hard black scales, maturing in late summer and remaining attached, with tiny, elliptical, flat nutlets.

HABITAT: Banks of streams, swamps, and mountain canyons in moist soils.

RANGE: Central Alaska, Yukon, and Mackenzie southeast mostly in the mountains to New Mexico and central California; it occurs near sea level in the north; to 9000' (2743 m) in the south.

USES: The bark of Mountain Alder has been used for medicinal purposes. This plant is used for afforestation and is also cultivated as an ornamental.

SIMILAR SPECIES: Speckled Alder, Sitka Alder.

CONSERVATION STATUS: **LC**
Mountain Alder has a wide distribution and there are no significant threats noted for this species. It grows abundantly in various soil types and can be weedy in wet, disturbed habitats.

Diameter: 6" (15 cm)
Height: 30' (9 m)

SPECKLED ALDER *Alnus rugosa*

■ DENSE POPULATION
■ LIGHT POPULATION
■ NATURALIZED

This widespread species is found growing in wet areas across Canada and into the northeastern United States. It is often planted as an ornamental at water edges. Alder thickets provide important habitat for a variety of mammals and birds, including the Alder Flycatcher (*Empidonax alnorum*). The Latin species name, which means "wrinkled," refers to the network of sunken veins prominent on the lower leaf surfaces.

ALTERNATE NAME: *Alnus incana* ssp. *rugosa*

DESCRIPTION: This is a low and clump-forming shrub; it is sometimes a small tree. The leaves are arranged in three rows; they are 2–4" (5–10 cm) long, 1¼–3" (3–7.5 cm) wide, elliptical or ovate, broadest near or below the middle; doubly and irregularly saw-toothed and wavy-lobed, with 9–12 nearly straight parallel veins on each side, on short, hairy stalks. They are dull dark green with a network of sunken veins above, and whitish-green and often with soft hairs, and with prominent veins and veinlets arranged in rows like a ladder beneath. The bark is gray and smooth. The twigs are gray-brown, slender, and slightly hairy when young, with three-angled pith.

FLOWERS: Early spring, before leaves. Males are in drooping catkins 1½–3" (4–7.5 cm) long. Females are in cones ¼" (6 mm) long.

FRUIT: Cones: ½–⅝" (12–15 mm) long, elliptical, blackish, hard, and short-stalked, maturing in autumn, with tiny rounded flat nutlets.

HABITAT: Wet soil along streams and lakes, and in swamps.

RANGE: Widespread across Canada from Yukon and British Columbia to Newfoundland, south to West Virginia, west to northeastern Iowa, and north to northeastern North Dakota; it occurs almost to the northern limit of the trees, and in the south to 2600' (792 m).

USES: This species is often planted as an ornamental, especially along water in parks and gardens.

SIMILAR SPECIES: Hazel Alder, European Alder.

CONSERVATION STATUS:
The Speckled Alder population is widespread and dense throughout its range.

Diameter: 4" (10 cm)
Height: 20' (6 m)

- DENSE POPULATION
- LIGHT POPULATION
- NATURALIZED

This small tree or shrub is found naturally in isolated areas near the Atlantic Coast in Maryland and Delaware, in northern Georgia, and in southern Oklahoma. Seaside Alder is unique in that it is the only fall-flowering native alder in North America—all other species bloom in spring. Its name is a reference to its larger population near the coast.

DESCRIPTION: This is a small, autumn flowering tree with a straight trunk and a narrow, rounded crown; it is sometimes a much-branched shrub. The leaves are arranged in three rows, 2½–4" (6–10 cm) long, 1–2" (2.5–5 cm) wide, elliptical, and finely saw-toothed, with five to eight curved veins on each side. They are shiny dark green above, and dull light green, sometimes slightly hairy and with tiny gland-dots beneath. The bark is light brown or gray, smooth, and thin. The twigs are slender, slightly zigzag, and hairy when young, with three-angled pith.

FLOWERS: Tiny; opening in fall. Male flowers are yellowish and in narrowly cylindrical catkins 1½–2½" (4–6 cm) long, drooping from near the ends of the twigs. Female flowers are in cones ¼" (6 mm) long.

FRUIT: Cones: ⅝–¾" (15–19 mm) long, single or in clusters of two or three, elliptical, dark brown, hard, short-stalked, maturing in the second autumn and remaining attached, with tiny flat egg-shaped nutlets having a membranous border.

HABITAT: Wet soil bordering streams and ponds.

RANGE: Found on the Atlantic Coast of Maryland, and in southern Delaware near sea level; an isolated population occurs in southern Oklahoma at 700' (213 m).

USES: This fast growing species is cultivated as a landscaping ornamental.

SIMILAR SPECIES: Hazel Alder, Speckled Alder, European Alder.

CONSERVATION STATUS: EN
The population of this species is small but stable. It occurs in a scattered range at few known sites. Climate change, grazing and agricultural run-off, and low genetic diversity are the biggest threats to this species. Rising sea levels would create saltwater intrusions into the freshwater tidal systems in which Seaside Alder occurs, putting these trees at risk.

Diameter: 4" (10 cm)
Height: 30' (9 m)

■ DENSE POPULATION
■ LIGHT POPULATION
■ NATURALIZED

Named for its pale green foliage, White Alder is a medium-sized to large deciduous tree found in far western North America. It is the only native alder found in southern California and common throughout its range. Its distribution is generally limited to permanent streams, making the presence of White Alder trees a good indicator of nearby water. It is sometimes planted as an ornamental in wet sites and can tolerate infertile soils.

DESCRIPTION: This medium-sized to large tree has a tall, straight trunk and an open, rounded crown, showy in winter with long, golden-colored male catkins hanging from slender, leafless twigs. The leaves are arranged in three rows; they are 2–3½" (5–9 cm) long, 1½–2" (4–5 cm) wide, ovate or elliptical, finely saw-toothed but not lobed, and slightly thickened, with 9–12 nearly straight, parallel veins on each side. They are dull dark green, hairless or nearly so, and often with tiny gland-dots above, and light yellow-green and slightly hairy beneath. The bark is light or dark brown, and fissured into flat, scaly ridges. The twigs are slender, light green, and finely hairy when young, with three-angled pith.

FLOWERS: Winter–early spring, before leaves. Tiny. Male flowers are yellowish in drooping, narrowly cylindrical catkins 1½–5" (4–13 cm) long. Female flowers are reddish in narrow cones ⅜" (10 mm) long.

FRUIT: Cones: ⅜–¾" (10–19 mm) long, elliptical; with three to seven clustered on short stalks; with many hard, black scales, remaining closed until early spring. They produce tiny, elliptical, flat nutlets, maturing in late summer.

HABITAT: With chaparral and Ponderosa Pine in foothill woodlands.

RANGE: Western Idaho and Washington south in the mountains to western Nevada and southern California. It occurs at 100–8000' (30–2438 m), generally below 5000' (1524 m).

USES: This species is sometimes planted as an ornamental tree.

SIMILAR SPECIES: Red Alder.

CONSERVATION STATUS: (LC)
White Alder is common in the western United States, with no known or anticipated threats to the species.

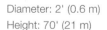

Diameter: 2' (0.6 m)
Height: 70' (21 m)

■ DENSE POPULATION
■ LIGHT POPULATION
■ NATURALIZED

The largest North American alder species, Red Alder is an ecologically important species native to the Pacific Northwest. Red Alder thickets are short-lived and serve as a cover for the seedlings of future coniferous forests. Alder roots, like those of legumes, often have swellings or root nodules containing nitrogen-fixing bacteria, which enrich poor soil by converting nitrogen from the air into fertilizer-like chemicals. Red Alder is also planted as an ornamental in wet soils and is a pioneer species on moist disturbed sites, such as those formed by landslides, road cuts, and after logging or fire. The common name of this species describes the reddish-brown inner bark and heartwood.

DESCRIPTION: Red Alder is a graceful tree with a straight trunk, a pointed or rounded crown, and mottled, light gray to whitish, smooth bark. The leaves are arranged in three rows; they are 3–5" (7.5–13 cm) long, 1¾–3" (4.5–7.5 cm) wide, ovate to elliptical, short-pointed at both ends, slightly thickened, wavy-lobed and doubly saw-toothed, with the edges slightly turned under, and with 10–15 nearly straight parallel veins on each side. They are dark green and usually hairless above, and gray-green with rust-colored hairs beneath. The bark is mottled light gray to whitish, smooth or becoming slightly scaly, and thin; the inner bark is reddish-brown. The twigs are slender, light green, and covered with gray hairs when young, with three-angled pith.

FLOWERS: Tiny, occurring in spring before the leaves. Male flowers are yellowish, appearing in drooping, narrowly cylindrical catkins 4–6" (10–15 cm) long and ¼" (6 mm) wide. Female flowers are reddish, occurring in narrow cones ⅜–½" (10–12 mm) long.

FRUIT: Cones: ½–1" (1.2–2.5 cm) long; four to eight occurring on short stalks. They are elliptical, with many hard black scales, remaining attached; with tiny, rounded, flat nutlets with two narrow wings, maturing in late summer.

HABITAT: Moist soils including loam, gravel, sand, and clay, along streams and lower slopes; it is often found in nearly pure stands.

RANGE: Southeastern Alaska southeast to central California; it is also local in northern Idaho. It occurs to 2500' (762 m).

USES: This species is used for pulpwood and to make furniture, cabinets, and tool handles. It is often planted as an ornamental tree.

SIMILAR SPECIES: White Alder.

CONSERVATION STATUS: LC
Red Alder is common in North America. There are no known threats to the species.

Diameter: 2½' (0.8 m), sometimes larger
Height: 40–100' (12–30 m)

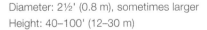

- ■ DENSE POPULATION
- ■ LIGHT POPULATION
- ■ NATURALIZED

Smooth Alder is widespread in eastern North America, and is the only alder native to the southeastern United States, where it is common and widespread, often forming thickets as a shrub. It prefers wet soils near streams, rivers, and lakes. Another common name for this species is Hazel Alder.

DESCRIPTION: This is a large, spreading shrub with several trunks, or sometimes a small tree, commonly found at the edge of water. The leaves are arranged in three rows and gummy and aromatic when immature; they are 2–4½" (5–11 cm) long, 1¼–2¾" (3–7 cm) wide, obovate to elliptical, broadest usually beyond the middle, finely saw-toothed with regular sharp teeth, sometimes also slightly wavy, with 9–12 nearly straight parallel veins on each side. They are dull green above, and light green and often hairy on veins beneath, turning red-brown in autumn. The bark is dark gray or brown, and smooth. The twigs are covered with rust-colored hairs when young, with three-angled pith.

FLOWERS: Early spring. Tiny. Male flowers occur in drooping catkins. Female flowers occur in cones ¼" (6 mm) long.

FRUIT: Cones: ⅜–⅝" (10–15 mm) long, in clusters of four to ten, elliptical, dark brown, hard, and short-stalked; maturing in late summer or fall and remaining attached, with tiny egg-shaped flat brown nutlets.

HABITAT: Wet soil bordering streams and lakes, and in swamps.

RANGE: Southwest Nova Scotia and southern New Brunswick south to northern Florida, west to eastern Texas and north to southeastern Kansas; to 3000' (914 m).

USES: This species is often used to stabilize riversides or streambanks. It has also been used to treat various ailments, usually as a tea made from the bark.

SIMILAR SPECIES: Seaside Alder, Speckled Alder, European Alder.

CONSERVATION STATUS: LC

Smooth Alder has a wide geographic distribution in eastern North America. There are no known threats.

Diameter: 4" (10 cm)
Height: 20' (6 m)

■ DENSE POPULATION
■ LIGHT POPULATION
■ NATURALIZED

In Alaska, Green Alder is a pioneer in disturbed areas, following landslides, logging, and glacial retreat. Adapted to soils too barren for other trees, this species improves soil conditions by adding organic matter and nitrogen from bacteria in its root nodules. It acts as a short-lived nurse tree for Sitka Spruce, later dying when shaded by the larger conifer. Sitka Alder (*Alnus viridis* ssp. *sinuata*) is a subspecies of Green Alder that is sometimes treated as a separate species; it is the form most likely to occur as a tree.

DESCRIPTION: This is a thicket-forming shrub or small tree, often with several trunks, and with shiny yellow-green leaves, gummy when young. The leaves are 2½–5" (6–13 cm) long, 1½–3" (4–7.5 cm) wide, ovate, shallowly wavy-lobed and doubly saw-toothed with long-pointed teeth and six to ten nearly straight parallel veins on each side; they are gummy or sticky when young, and shiny, speckled yellow-green on both surfaces, and paler and often slightly hairy beneath. The bark is gray to light gray, smooth, and thin; the inner bark is red. The twigs are gummy, finely hairy, and orange-brown when young, becoming light gray, slender, and slightly zigzag.

FLOWERS: Tiny; in spring with or after leaves. Male flowers are yellowish, drooping, and narrowly cylindrical, in catkins 3–5" (7.5–13 cm) long, ⅜" (10 mm) wide. Female flowers are reddish, in narrow cones ⅜" (10 mm) long.

FRUIT: Cones: ½–¾" (12–19 mm) long, with three to six clustered on slender, spreading, long stalks; they are elliptical, with many hard, black scales, remaining attached, with tiny, elliptical, flat nutlets with two broad wings; they mature in summer.

HABITAT: Along streams and lakes and in valleys.

RANGE: Southwestern and central Alaska and Yukon southeast to northwest California and central Montana; in Alaska to the alpine zone above timberline; in northwest California to 7000' (2134 m).

USES: This species is sometimes used for afforestation on infertile soils.

SIMILAR SPECIES: Sitka Alder.

CONSERVATION STATUS: (LC)
Green Alder is secure throughout its range with no known threats.

Diameter: 8" (20 cm)
Height: 30' (9 m)

■ DENSE POPULATION
■ LIGHT POPULATION
■ NATURALIZED

Arizona Alder is a large native tree of the southwestern United States. This is one of the largest native alders and a handsome tree of rocky canyon bottoms. It is common in scattered sites on sandy or rocky stream banks and moist slopes. It is recognized by its smooth, dark gray bark.

DESCRIPTION: This is a tall, straight-trunked tree with an open, rounded crown. Its leaves occur in three rows, 1½–3¼" (4–8 cm) long, 1–1½ (2.5–4 cm) wide, ovate or elliptical, usually double saw-toothed but not lobed, with seven to ten nearly straight parallel veins on each side. They are dark green and almost hairless above, pale green and often finely hairy with tufts of rust-colored hairs in the vein angles below. The bark on mature trees is dark gray, smooth, thin, becoming fissured and scaly. The twigs are slender, reddish-brown, finely hairy when young, with three-angled pith.

FLOWERS: Tiny, in early spring before the leaves. Male flowers are yellowish, and occur in three or four drooping, narrowly cylindrical catkins 2–3" (5–7.5 cm) long. Female flowers are reddish, in narrow cones ¼" (6 mm) long.

FRUIT: Cones: ½–¾" (12–19 mm) long, with three to eight clustered on short stalks; they are elliptical, with many hard, black scales, remaining attached, with tiny, elliptical, flat nutlets; they mature in late summer.

HABITAT: Wet canyon soils in mountains and along streams.

RANGE: Southwestern New Mexico and Arizona; it is local in northern New Mexico and also in northern Mexico. It occurs at 4500–7500' (1371–2286 m).

USES: This tree's soft wood is sometimes used as firewood. It generally is not cultivated.

SIMILAR SPECIES: Mountain Alder, Speckled Alder.

CONSERVATION STATUS:

The Arizona Alder population is small and scattered but appears to be stable. There are no known threats, but such isolated populations are vulnerable to unforeseen local events such as wildfires or habitat destruction.

Diameter: 2' (61 cm)
Height: 80' (24 m)

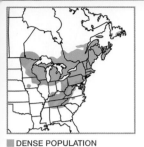

■ DENSE POPULATION
■ LIGHT POPULATION
■ NATURALIZED

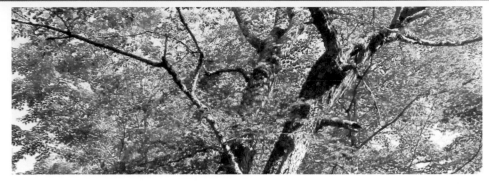

This is one of the most valuable birches and one of the largest hardwoods in northeastern North America. Yellow Birch is easily recognized by its distinctive yellow-bronze bark once it becomes fairly mature. Younger trees may be mistaken for Sweet Birch, but are most readily identified by their hairy twigs and buds and most persistently hairy leaves with mostly unbranched side veins. Yellow Birch is the provincial tree of Quebec.

DESCRIPTION: This is a large, aromatic tree with a broad, rounded crown of drooping branches and a slight odor of wintergreen in crushed twigs and foliage. The leaves are 3–5" (7.5–13 cm) long, 1½–2" (4–5 cm) wide, elliptical, short-pointed or rounded at the base, sharply and doubly saw-toothed, mostly with 9–11 veins on each side, and hairy when young. They are dark dull green above, and light yellow-green beneath, turning bright yellow in autumn. The bark is shiny yellowish or silvery-gray, separating into papery curly strips, becoming reddish-brown and fissured into scaly plates. The twigs are greenish-brown, slender, and hairy.

FLOWERS: Early spring. Tiny. The male flower is yellowish, with two stamens, with many occurring in long drooping catkins near the tip of twigs. The female flower is greenish, in short upright catkins at the tip of the same twig.

FRUIT: Cones: ¾–1¼" (2–3 cm) long, oblong, hairy, brownish, upright, and nearly stalkless, with many hairy scales and two-winged nutlets, maturing in fall.

HABITAT: Cool moist uplands including mountain ravines; it is found with hardwoods and conifers.

RANGE: Extreme southeastern Manitoba east to southern Newfoundland, south to extreme northeastern Georgia, and west to northeastern Iowa. It occurs to 2500' (762 m) in the north and 3000–6000' (914–1829 m) or higher in the south.

USES: This is an important timber species, particularly in Canada. The strong, heavy wood is widely used for flooring, furniture, cabinets, gun stocks, and other purposes. The sap is used to make syrup.

SIMILAR SPECIES: Young trees especially may be mistaken for Sweet Birch.

CONSERVATION STATUS: LC
Yellow Birch has a widespread distribution and is common in eastern North America and Canada. There are no known threats.

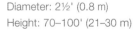

Diameter: 2½' (0.8 m)
Height: 70–100' (21–30 m)

■ DENSE POPULATION
■ LIGHT POPULATION
■ NATURALIZED

Also called Black Birch, this species is native to eastern North America. The twigs possess a strong scent of wintergreen. The trees can be tapped like Sugar Maples in early spring and the fermented sap made into birch beer. Sweet Birch was nearly endangered in the mid-20th century by the overharvesting of trees for birch oil, also called "oil of wintergreen," used to flavor medicines and candy. The oil was obtained from the bark and wood of young Sweet Birch trees through an unsustainable, wasteful process that is no longer used now that a synthetic method of manufacturing the same oil has been developed.

DESCRIPTION: This aromatic tree has a rounded crown of spreading branches and an odor of wintergreen in crushed twigs and foliage. The leaves are 2½–5" (6–13 cm) long, 1½–3" (4–7.5 cm) wide, elliptical, long-pointed, often notched at the base, sharply and doubly saw-toothed, mostly with 9–11 veins on each side, and becoming nearly hairless. They are dull dark green above, and light yellow-green beneath, turning bright yellow in autumn. The bark is shiny dark brown or blackish, and smooth but not papery; on large trunks, the bark is fissured into scaly plates like Black Cherry. The twigs are dark brown, slender, and hairless.

FLOWERS: Tiny flowers in early spring. The male flowers are yellowish, with two stamens, with many occurring in long, drooping catkins near the tips of twigs. Female flowers are greenish, in short upright catkins at the tip of the same twig.

FRUIT: Cones: ¾–1½" (2–4 cm) long; oblong, brownish, upright, nearly stalkless; with hairless scales and many two-winged nutlets, maturing in fall.

HABITAT: Cool, moist uplands; with hardwoods and conifers.

RANGE: Southern Maine southwest to northern Alabama and north to Ohio; it is local in extreme southern Quebec and southeastern Ontario. It occurs nearly to sea level in the north and at 2000–6000' (610–1829 m) in the southern Appalachians.

USES: The sap from this tree is used to make syrup and birch beer. Oil of wintergreen was formerly obtained from the bark and wood. The wood is used for furniture, millwork, and cabinetry.

SIMILAR SPECIES: Yellow Birch.

CONSERVATION STATUS:
Sweet Birch is a widespread tree, commonly found in the eastern United States and Canada. The population is stable and there are no known threats. Climate change may affect precipitation patterns and cause the species' suitable habitat range to shift northward.

Diameter: 1–2½' (0.3–0.8 m)
Height: 50–80' (15–24 m)

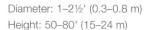

- ■ DENSE POPULATION
- ■ LIGHT POPULATION
- ■ NATURALIZED

This is an unusually heat-tolerant New World birch and the only birch that occurs at low altitudes in the southeastern United States. Its ability to thrive on moist sites makes it useful for erosion control. River Birch provides important habitat for a wide variety of birds, including waterfowl, grouse, and turkeys. Native Americans used the sap to sweeten food and sometimes ate the inner bark.

DESCRIPTION: This is often a slightly leaning and forked tree with an irregular, spreading crown. The leaves are 1½–3" (4–7.5 cm) long and 1–2¼" (2.5–6 cm) wide. They are ovate or nearly four-sided, coarsely doubly saw-toothed or slightly lobed, and usually with seven to nine veins on each side. They are shiny dark green above, and whitish and usually hairy beneath, turning dull yellow in fall. The bark is shiny pinkish-brown or silvery-gray, separating into papery scales, and becoming thick, fissured, and shaggy. The twigs are reddish-brown, slender, and hairy.

FLOWERS: Early spring. Tiny. The male is yellowish, with two stamens, with many in long drooping catkins near the tip of twigs. The female is greenish, in short upright catkins at the tip of the same twig.

FRUIT: Cones: 1–1½" (2.5–4 cm) long, cylindrical, brownish, upright, and short-stalked, with many hairy scales and hairy two-winged nutlets, maturing in late spring or early summer.

HABITAT: Wet soil of stream banks, lakes, swamps, and floodplains; it is often found with other hardwoods.

RANGE: Southwest Connecticut south to northern Florida, west to eastern Texas, and north to southeastern Minnesota; it is local in Massachusetts and southern New Hampshire. It generally occurs to 1000' (305 m), and to 2500' (762 m) in the southern Appalachians.

USES: This species is often planted as an ornamental tree. The wood is occasionally used locally for furniture and firewood.

SIMILAR SPECIES: American Hornbeam, Eastern Hophornbeam.

CONSERVATION STATUS: LC

River Birch is commonly seen and has a wide distribution across the United States; it is most abundant in the Southeast. There are no range-wide threats, but climate change may put pressure on the species at its westernmost distribution.

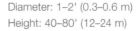

Diameter: 1–2' (0.3–0.6 m)
Height: 40–80' (12–24 m)

■ DENSE POPULATION
■ LIGHT POPULATION
■ NATURALIZED

This uncommon but widespread western species is the only native birch in the southwestern United States and the southern Rocky Mountains, and it occupies a wide stretch of western Canada. Another common name for this species is Red Birch, because of the reddish-brown bark. It is typically found along streams in mountainous areas, where sheep, goats, and other large mammals browse the foliage.

DESCRIPTION: This shrub or small tree has a rounded crown of spreading and drooping branches, usually forming clumps and often in thickets. The leaves are ¾–2" (2–5 cm) long, ¾–1" (2–2.5 cm) wide, ovate, sharply and often doubly saw-toothed, usually with four or five veins on each side. They are dark green above, and pale yellow-green with tiny gland-dots beneath, turning dull yellow in autumn. The bark is shiny, dark reddish-brown, and smooth, with horizontal lines, not peeling. The twigs are greenish and slender, with gland-dots.

FLOWERS: Early spring. Tiny. Male flowers are yellowish, with two stamens, with many occurring in long drooping catkins near the tip of the twigs. Female flowers are greenish, in short upright catkins at the back of the tip of the same twig.

FRUIT: Cones: 1–1¼" (2.5–3 cm) long, cylindrical, brownish, and upright or spreading on slender stalk, with many two-winged nutlets, maturing in late summer.

HABITAT: Moist soil along streams in mountain canyons, usually in coniferous forests and with cottonwoods and willows.

RANGE: Northeastern British Columbia, east to southern Manitoba, and south to northern New Mexico and California; at 2000–8000' (610–2438 m).

USES: This species is planted to stabilize streambanks. The hard, heavy wood is sometimes used for posts and firewood. Beavers often use the wood to construct their dams.

SIMILAR SPECIES: Paper Birch.

CONSERVATION STATUS: LC
Water Birch is a common species throughout a large geographic range. There are no known threats to this species.

Diameter: 6–12" (15–30 cm)
Height: 25' (7.6 m)

■ DENSE POPULATION
■ LIGHT POPULATION
■ NATURALIZED

Paper Birch is named for its characteristic thin, white bark, which often peels off the trunk in papery layers. Souvenirs of birch bark should always be taken from a fallen log, because stripping bark from living trees leaves permanent ugly black scars. It is a pioneer species, often becoming one of the first trees to take hold in disturbed areas. It grows best in well-drained acid, sandy, or silty loam; in cold soil with ample moisture; and in full sunlight. It does not tolerate shade, harsh conditions, high heat, or pollution.

DESCRIPTION: One of our most beautiful native trees, Paper Birch has a narrow, open crown of slightly drooping to nearly horizontal branches. It is sometimes a shrub. Its leaves are 2–4" (5–10 cm) long, 1½–2" (4–5 cm) wide, ovate, long-pointed, coarsely and doubly saw-toothed, usually with five to nine veins on each side. They are dull dark green above, and light yellow-green and nearly hairless beneath, turning light yellow in fall. The bark is chalky to creamy white; smooth, thin, with long horizontal lines; separating into papery strips to reveal orange inner bark; becoming brown, furrowed, and scaly at base; and bronze to purplish in some varieties. The twigs are reddish-brown, slender, and mostly hairless.

FLOWERS: Early spring. Tiny. Male is yellowish, with 2 stamens, many in long drooping catkins near the tip of twigs. Female is greenish, in short upright catkins back of the tip of the same twig.

FRUIT: Cones: 1½–2" (4–5 cm) long, narrowly cylindrical, brownish, hanging on slender stalk, with many two-winged nutlets, maturing in fall.

HABITAT: Occurs in moist upland soils and cutover lands, often in nearly pure stands.

RANGE: Transcontinental across North America near the northern limit of trees from northwest Alaska east to Labrador, south to New York, and west to Oregon; it is local south to northern Colorado and western North Carolina; to 4000' (1219 m), higher in the southern mountains.

USES: Paper Birch trees can be tapped in the spring to obtain sap from which beer, syrup, wine, or vinegar is made. The inner bark can be dried and ground into a meal and used as a thickener in soups or added to flour used in making bread. A tea is made from the root bark and young leaves of white birch. The Shuswap made soap and shampoo from the leaves. Paper Birch wood is used for specialty products such as ice cream sticks, toothpicks, bobbins, clothespins, spools, broom handles, and toys, as well as pulpwood. Native Americans made their lightweight birch-bark canoes by stretching the stripped bark over frames of Northern White-cedar, sewing it with thread from Tamarack roots, and caulking the seams with pine or Balsam Fir resin.

SIMILAR SPECIES: American Hazelnut, American Hornbeam, Beaked Hazel.

CONSERVATION STATUS: 🔵 LC

Paper Birch is widespread and common throughout Canada and in the east and east-central United States. Increasing temperatures caused by climate change may adversely affect Paper Birch by shrinking the southernmost part of its range, causing southern subpopulations to shift northward and compete with existing subpopulations in more northerly areas.

Diameter: 1–2' (0.3–0.6 m)
Height: 50–70' (15–21 m)

EUROPEAN WHITE BIRCH *Betula pendula*

European White Birch is a graceful short-lived ornamental, grown for its white papery bark and drooping branches. It is named for the smooth white bark on the trunk that peels off in papery strips. Commonly cultivated varieties of this tree have long branches and finely divided or lobed leaves. This non-native species has naturalized in the northeastern United States and several other areas in North America, and efforts are under way in some localities to eradicate it from native habitats.

DESCRIPTION: This introduced ornamental tree has an open, pyramid-shaped or spreading crown of long, drooping branches. The leaves are 1¼–2¾" (3–7 cm) long, 1–1½" (2.5–4 cm) wide, ovate or nearly triangular, long-pointed at the tip, blunt or almost straight at the base, doubly saw-toothed with six to nine veins on each side, sticky when young, and long-stalked. They are dull green above, and paler beneath, turning yellow in fall. The bark is white, smooth, and flaky, peeling in papery strips. The twigs are slender and drooping, with tiny resin gland-dots.

FLOWERS: Early spring. Tiny. The male flower is yellowish, with two stamens, with many occurring in long, drooping catkins near the tip of the twigs. Female flowers are greenish, appearing in short upright catkins at the tip of the same twig.

FRUIT: Cones: ¾–1¼" (2–3 cm) long, cylindrical, hanging on slender stalks, composed of many small two-winged nutlets and three-lobed bracts, maturing in fall.

HABITAT: Moist soils on lawns, and in parks and cemeteries. Sometimes an escapee in thickets and open forest areas.

RANGE: Native of Europe and Asia Minor. It is planted across the United States.

USES: European White Birch is commonly planted as an ornamental tree. The wood is used for furniture, plywood, skis, and many other purposes. The sap is used to make syrup and sometimes fermented into wine.

SIMILAR SPECIES: Gray Birch, Paper Birch.

CONSERVATION STATUS: LC
European White Birch has a widespread, stable population. In North America, it has been widely planted and is considered invasive in some states and provinces.

Diameter: 1' (0.3 m)
Height: 50' (15 m)

■ DENSE POPULATION
■ LIGHT POPULATION
■ NATURALIZED

This birch is native to the northeastern United States and southeastern Canada, where it is often a pioneer tree on clearings, abandoned farms, and burned areas. Gray Birch grows rapidly but is short-lived, and serves as a nurse tree, shading and protecting seedlings of the larger, long-lived forest trees that succeed it. Gray Birch trunks are highly flexible; when weighted with snow, the upper branches may bend to the ground without breaking. The attractive, long-stalked leaves seem to dance in the slightest breeze.

DESCRIPTION: This is a small, bushy tree with an open, conical crown of short, slender branches reaching nearly to the ground. The leaves are 2–3" (5–7.5 cm) long, 1½–2½" (4–6 cm) wide. They are triangular, tapering from near the base to a long-pointed tip, sharply and doubly saw-toothed, usually with four to eight veins on each side, with slender leafstalks, and with black gland-dots. They are shiny dark green above, and paler with tufts of hairs along midvein beneath, turning pale yellow in autumn. The bark is chalky or grayish-white; smooth, thin, and not papery; and becoming darker and fissured at the base. The twigs are reddish-brown and slender, with warty gland-dots.

FLOWERS: Early spring. Tiny. The males are yellowish, with two stamens, with many occurring in long drooping catkins near the tip of the twigs. The females are greenish, in short upright catkins at the back of the tip of the same twig.

FRUIT: Cones: ¾–1¼" (2–3 cm) long, cylindrical, brownish, spreading, and short-stalked, with many hairy scales and hairy, two-winged nutlets, maturing in fall.

HABITAT: Dry barren uplands, also on moist soils, in mixed woodlands.

RANGE: Southern Ontario east to Cape Breton Island, south to Pennsylvania and New Jersey; local to western North Carolina and northwestern Indiana; to 2000' (610 m).

USES: Gray Birch wood is sometimes used for spools and similar items, and for firewood.

SIMILAR SPECIES: European White Birch, Paper Birch.

CONSERVATION STATUS: (LC)

Gray Birch is common in the northeastern portion of its geographic range. Disjunct subpopulations of this species are seen in Indiana, West Virginia, and Virginia.

Diameter: 1' (0.3 m)
Height: 30' (9 m)

■ DENSE POPULATION
■ LIGHT POPULATION
■ NATURALIZED

This small, rare tree has an extremely limited range, found naturally only in Smyth County, Virginia. It was first discovered in the early 1900s, and then was not seen for several decades and presumed to be extinct. It was rediscovered in 1975. Virginia Roundleaf Birch is federally listed as an endangered species in the United States.

DESCRIPTION: This is a small tree with a narrow trunk and many slender spreading branches and twigs. The leaves are 1½–2½" (4–6 cm) long, 1¼–2" (3–5 cm) wide, nearly round, ovate, or broadly elliptical, blunt or rounded at the tip, slightly notched at the base, irregularly saw-toothed, with three to six veins on each side, and short, slender, hairy stalks. They are dark green and hairless except along the veins, and paler beneath. The bark is gray, thin, smooth with horizontal lines, and aromatic, with a strong odor of wintergreen. The twigs are brown, slender, and hairless.

FLOWERS: Tiny. Male flowers are yellowish, with two stamens, with many in long drooping catkins near the tip of twigs. Female flowers are greenish, in short upright catkins at the tip of the same twig.

FRUIT: Cones: ½" (12 mm) long, elliptical, upright, and nearly stalkless, with many dark brown nutlets.

HABITAT: This tree is rare; it occurs along a stream in the understory of mixed hardwood forest.

RANGE: Only in southwestern Virginia in Smyth County; at about 2750' (838 m).

USES: This species is sometimes planted outside its natural range, but, due to its rarity, it is not widely available commercially.

SIMILAR SPECIES: The rounded leaves are distinctive and help to separate this species from other birches.

CONSERVATION STATUS: **CR**
The Virginia Roundleaf Birch is on the U.S. Endangered Species List. This extremely rare species is known from a single natural, native population that declined to just eight individual trees in 2003. An aggressive recovery plan to plant seedlings grown in a greenhouse has been in place since the 1980s, and today there are almost 1000 planted trees still alive. This species is seriously threatened by a lack of natural reproduction, and it remains unclear whether the greenhouse-grown trees are able to compete and reproduce successfully.

Diameter: 4" (10 cm)
Height: 15–33' (4.6–10 m)

This species, native to Europe and southwestern Asia, has been introduced and widely cultivated as an ornamental tree in North America. It is a small to medium-sized deciduous tree, larger than American Hornbeam but otherwise quite similar. The word hornbeam is from the words "horn" (for toughness) and "beam" (for tree) and refers to the hard, tough wood.

DESCRIPTION: This nonnative tree is much larger than American Hornbeam. It is pyramidal to rounded in shape and often planted in tall hedges. The leaves are similar to American Hornbeam, although the leaflike bracts attached to the fruits are narrower. The long leaf buds hug the twig (they do this less so in American Hornbeam). The bark is steel gray.

FLOWERS: Early spring before leaves. Tiny. The male is greenish and in drooping catkins 1¼–1½" (3–4 cm) long. The female is reddish-green, and paired in narrow catkins ½–¾" (12–19 mm) long.

FRUIT: ¼" (6 mm) long, paired, egg-shaped, hairy greenish nutlets, with leaflike three-pointed, toothed, greenish scale in clusters 2–4" (5–10 cm) long, hanging on slender stalks, and maturing in late summer.

HABITAT: Introduced in North America; it is often planted in hedgerows.

RANGE: Native from Europe to southwestern Asia. It occurs as an introduced tree locally in North America.

USES: This species is widely cultivated as an ornamental tree. The wood is used for tool handles and other items.

SIMILAR SPECIES: American Hornbeam.

CONSERVATION STATUS: LC

European Hornbeam is common throughout its range, where there are no major threats to the species.

Height: 40–60' (12–18 m)

AMERICAN HORNBEAM *Carpinus caroliniana*

■ DENSE POPULATION
■ LIGHT POPULATION
■ NATURALIZED

American Hornbeam is a small hardwood tree native to eastern North America. Deer browse the twigs and foliage, and grouse, pheasants, and quail eat the nutlets. The name "beech" or "Blue-beech" has sometimes been misapplied to this member of the Birch family, because this tree's smooth, blue-gray bark resembles that of beeches. It is also sometimes called Musclewood for its exceptionally hard, heavy wood.

DESCRIPTION: This is a small, shrubby tree with one or more short trunks angled or fluted; long, slender, spreading branches; and a broad, rounded crown. The leaves are 2–4½" (5–11 cm) long, 1–2½" (2.5–6 cm) wide, elliptical, long-pointed at the tip, sharply doubly saw-toothed, with many nearly straight parallel side veins; they are dull dark blue-green above, and paler with hairs on veins and vein angles below, turning orange to red in fall. The bark is blue-gray, thin, and smooth. The twigs are brown, slender, and slightly zigzag.

FLOWERS: Early spring before leaves. Tiny. The male is greenish and in drooping catkins 1¼–1½" (3–4 cm) long. The female is reddish-green, and paired in narrow catkins ½–¾" (12–19 mm) long.

FRUIT: ¼" (6 mm) long, paired, egg-shaped, hairy greenish nutlets, with leaflike three-pointed, toothed, greenish scale in clusters 2–4" (5–10 cm) long, hanging on slender stalks, and maturing in late summer.

HABITAT: Moist rich soils, mainly along streams and in ravines; in understory of hardwood forests.

RANGE: Ontario east to Quebec and central Maine, south to central Florida, west to Texas, and north to Minnesota; to 3000' (914 m). It is also found in Mexico.

USES: The hard, heavy wood is often used for tool handles, walking canes, golf clubs, and other small items.

SIMILAR SPECIES: European Hornbeam.

CONSERVATION STATUS: LC
American Hornbeam is common throughout its range and the population is globally stable, although the subpopulations in Quebec are listed as vulnerable. There are no range-wide threats to this species.

Diameter: 1' (0.3 m)
Height: 30' (9 m)

AMERICAN HAZELNUT
Corylus americana

■ DENSE POPULATION
■ LIGHT POPULATION
■ NATURALIZED

American Hazelnut is native to eastern North America. Usually a medium to large shrub, it spreads mainly by creeping rhizomes and can form thickets, and under some conditions can take the form of a small tree. Its leaves and catkins resemble those of other birches. It is a fast-growing species, often planted as an ornamental shrub and to attract wildlife. Ground birds such as grouse and Northern Bobwhite (*Colinus virginianus*) eat the nuts.

DESCRIPTION: This is a spreading deciduous shrub that bears long male catkins and clusters of small, edible nuts. The plant grows on straight main stems or sometimes a trunk, with spreading, ascending branches. Its leaves are 3–5" (7.5–12.5 cm) long, broad, elliptical, sharp-pointed, and toothed.

FLOWERS: Male catkins to 8" (20 cm) long are spaced along the branches. The female flowers are tiny and hidden by bracts.

FRUIT: Small nuts enclosed in large, leafy bracts.

HABITAT: Open woods, thickets, roadsides, fields, and disturbed areas.

RANGE: Eastern North America, from Maine to southern Georgia and west to eastern Oklahoma, western North Dakota, and extreme southeastern Saskatchewan.

USES: Many people eat the edible nuts from this species, and Native Americans used it for medicinal purposes. It is often cultivated as an ornamental plant and to attract wildlife.

SIMILAR SPECIES: Beaked Hazelnut.

CONSERVATION STATUS: LC

The American Hazelnut population is stable and there are no current threats. Climate change may locally impact this species' habitat in the future.

Height: 3–10' (1–3 m)

BEAKED HAZELNUT *Corylus cornuta*

■ DENSE POPULATION
■ LIGHT POPULATION
■ NATURALIZED

Beaked Hazelnut has a wide, scattered distribution in woodlands and edge habitats across North America but is especially common in southern Canada. The nuts—similar to commercially sold hazelnuts—are a favorite of squirrels, which may hang precariously from the plant's slender branches to reach them. Deer and livestock often browse the foliage. The western variety (var. *californica*), sometimes called California Hazelnut, is a common understory species in cool, moist forests of the Pacific states. The eastern variety (var. *cornuta*) tends to be smaller and is found in drier sites in the East, where it can become weedy.

DESCRIPTION: This is an open, spreading shrub or a small tree with numerous ascending stems; thin, downy, double-toothed leaves; and nuts wrapped in an "envelope" of leafy bracts. Its leaves are alternate, 1½–3" (4–7.5 cm) long, rounded to obovate, more or less unequal at the base, doubly toothed and sometimes slightly 3-lobed. They are glandular-hairy until early summer, becoming hairless or nearly so with age, and paler below. The bark is smooth but with more or less down-covered branchlets.

FLOWERS: Spring, before leaves. Male and female are on the same plant. Male flowers are tiny, numerous, in elongated, drooping, stalkless catkins at the ends of the previous years' branchlets. Female flowers are fewer, but borne severally in small, rounded, scaly buds.

FRUIT: The nut is ovoid to nearly globular, ⅜–⅝" (6–15 mm) long, hard-shelled, enclosed by two or three stiff-hairy bracts united to form an "envelope" extending to form a fringed tube longer than the nut.

HABITAT: On the Pacific coast, moist, well-drained sites in woodlands and forests at lower elevations. Elsewhere, moist to dry roadsides, woodland edges, open woodlands, and thickets.

RANGE: Occurs from Newfoundland south through New England to Georgia and Alabama; west across southern Canada to central Alberta and into east-central British Columbia; south into Idaho, the Black Hills, and the Great Lakes states.

USES: This species is an important food source for jays, squirrels, and a variety of other wildlife.

SIMILAR SPECIES: American Hazelnut.

CONSERVATION STATUS: 🔘 LC

Beaked Hazelnut has a wide distribution and there are no known threats to this species.

Height: 3–20' (0.9–6 m)

■ DENSE POPULATION
■ LIGHT POPULATION
■ NATURALIZED

Eastern Hophornbeam is a small understory tree native to eastern North America. Wildlife including Northern Bobwhites (*Colinus virginianus*), pheasants, grouse, deer, and rabbits eat the nutlets and buds. This tree's common name refers to the resemblance of the fruit clusters to hops (*Humulus lupulus*). It is also called Hardhack and Ironwood for its extremely hard, tough wood, which is used to make tool handles, small wooden items, and fence posts. Although slow-growing, it is often planted as an ornamental tree.

DESCRIPTION: This is a small deciduous tree. Its leaves are 2–5" (5–13 cm) long, 1–2" (2.5–5 cm) wide, ovate or elliptical, sharply doubly saw-toothed, with many nearly straight parallel side veins, and short, hairy leafstalks. They are dull yellow-green and nearly hairless above, and paler and hairy chiefly on veins beneath, turning yellow in fall. The bark is light brown, thin, and finely fissured into long narrow scaly ridges.

Diameter: 1' (0.3 m)
Height: 20–50' (6–15 m)

FLOWERS: Early spring. Tiny. Male flowers are greenish and in one to three drooping, narrowly cylindrical clusters 1½–2½" (4–6 cm) long. Females are reddish-green and in narrowly cylindrical clusters ½–¾" (12–19 mm) long.

FRUIT: 1½–2" (4–5 cm) long, ¾–1" (2–2.5 cm) wide, cone-like hanging clusters maturing in late summer and composed of many flattened, small, egg-shaped brown nutlets, each within a swollen egg-shaped flattened light brown cover that is papery and sac-like.

HABITAT: Moist soil in understory of upland hardwood forests.

RANGE: Manitoba east to Cape Breton Island, south to northern Florida, and west to eastern Texas; to 4500' (1372 m).

USES: This species is cultivated as an ornamental tree and the wood is used for tool handles, fence posts, wooden longbows, and other purposes.

SIMILAR SPECIES: American Hornbeam.

CONSERVATION STATUS:  LC
Eastern Hophornbeam has a widespread and large population. The tree is sensitive to pollution but relatively free of insect pests or diseases, and the population in North America is secure. Deforestation is causing declines among the smaller subpopulations in Mexico and Central America.

■ DENSE POPULATION
■ LIGHT POPULATION
■ NATURALIZED

This shrub or (rarely) small tree is found growing in moist soils in the eastern United States, especially in the Midwest. Native Americans and early settlers used the powdered bark from this tree as a purgative, and the plant is still used for medicinal purposes today. "Wahoo" was the native term for the plant. The Latin species name, meaning "dark purple," refers to the color of the fruit.

ALTERNATE NAMES: Wahoo, Eastern Wahoo

DESCRIPTION: This is a shrub or small tree with a spreading, irregular crown and red or purple capsules suggesting a burning bush. The leaves are opposite, 2–4½" (5–11 cm) long, 1–2" (2.5–5 cm) wide, elliptical, abruptly long-pointed at the tip, and finely saw-toothed. They are green above, and paler and often with fine hairs beneath, turning light yellow in autumn. The bark is gray and smooth, becoming slightly fissured. The twigs are dark purplish-brown, slender, and sometimes four-angled or slightly winged.

FLOWERS: Late spring–early summer. ⅜" (10 mm) wide with four dark red or purple petals, with seven to fifteen flowers clustered on slender, widely forking stalks.

FRUIT: ⅝" (15 mm) wide, red or purple capsules deeply four-lobed and four-celled, each lobe splitting open; they are smooth, with several hanging on a slender stalk, maturing in fall, and remaining attached into winter; in each cell there are one or two rounded, light brown seeds with a red covering.

CAUTION: Poisonous: Do not eat or ingest any part of this species. The fruit of this plant is poisonous to humans. If accidentally ingested, contact your doctor or poison control center.

HABITAT: Moist soils, especially in thickets, valleys, and forest edges.

RANGE: Extreme southern Ontario to central New York, south to northern Georgia, west to central Texas, and north to southeastern North Dakota; to 2000' (610 m).

USES: This species is used medicinally in both the United States and Canada.

SIMILAR SPECIES: Strawberry Bush (*Euonymus americana*), a related native shrub.

CONSERVATION STATUS:
Eastern Burningbush has a wide distribution and large population. There are no major threats known or anticipated, although changes in land use, habitat fragmentation, and forest management practices have a minor impact on this species.

Diameter: 4" (10 cm)
Height: 20' (6 m)

■ DENSE POPULATION
■ LIGHT POPULATION
■ NATURALIZED

Red Mangroves are unusual tropical and subtropical plants found in Florida's estuarine ecosystems in North America as well as in Central and South America and West Africa. Like other mangroves, they do not propagate using ordinary means of dispersal. Rather, young plants begin to grow while still attached to the parent tree, developing into propagules that drop into the water and float along until an appropriate substrate is encountered. The germinating fruits of the Red Mangrove form impressive, pendulous, torpedo-like seedlings often seen dangling from the branches. Once they fall, these propagules take root mainly in sheltered areas or on shorelines with relatively calm water.

DESCRIPTION: This species can be a shrub or a small tree supported by interlacing "prop roots" from the trunk or large branches. The leaves are 6" (15 cm) long, opposite, elliptical, leathery, and evergreen. The bark is gray, furrowed, and scaly-ridged.

FLOWERS: Year-round. ½" (1.3 cm), funnel- to star-shaped, white or yellowish, and hairy.

FRUIT: 1" (2.5 cm), egg-shaped, rusty brown, and germinates on the parent tree, with a long, green, torpedo-shaped root.

HABITAT: Tidal swamps.

RANGE: Coastal peninsular Florida.

USES: The wood from Red Mangrove is used for construction materials, fencing, and firewood. This species has also been used for medicinal purposes.

SIMILAR SPECIES: White Mangrove, as well as Black Mangrove (*Avicennia germinans*), a shrub that shares similar habitat.

CONSERVATION STATUS:
Red Mangrove is common throughout much of its range and is often the dominant species. It is expanding its range in northern Florida but declining globally, faced with habitat destruction due to development and pollution. Because it occurs along coasts, sea-level rise and an increase in destructive weather events caused by climate change constitute serious threats to this species.

Height: 50' (15 m)

■ DENSE POPULATION
■ LIGHT POPULATION
■ NATURALIZED

This plant is endemic to the Caribbean islands and the southern tip of Florida. Other names for this species include Clam Cherry, Gooseberry, and Long Key Locustberry. Like so many south Florida plants, it has suffered from the loss of habitat to development, and it is listed as a threatened species in Florida under the Preservation of Native Flora of Florida Act.

DESCRIPTION: This is a multi-trunked shrub or small tree with an irregular crown and erect branches. Its leaves are 2" (5 cm) long, opposite, obovate, leathery, and evergreen. The bark is pale brown and smooth.

FLOWERS: March–June. ½" (1.3 cm), white to pinkish to crimson, with five stalked petals around a cup-like center.

FRUIT: ½" (1.3 cm), berrylike, and red.

HABITAT: Coastal pinelands.

RANGE: The southeastern tip of Florida and the Florida Keys.

USES: Locust Berry is sometimes planted as an ornamental plant and to attract butterflies.

SIMILAR SPECIES: This plant is distinct and rarely confused with other species.

CONSERVATION STATUS: NT

Locust Berry is locally common but scattered in Florida and the Caribbean islands. It has lost some of its habitat due to development. This species is listed as state-threatened in Florida.

Height: 20' (6 m)

■ DENSE POPULATION
■ LIGHT POPULATION
■ NATURALIZED

This small, tropical plant is native to Florida, the Caribbean, and Central and South America. The fruits of this tree are often consumed by wildlife and also by people; they are commonly made into preserves. It is sometimes planted as an ornamental shrub in subtropical regions; in some areas outside of North America this species has become a problematic invasive.

DESCRIPTION: This is a spreading shrub to a small tree with a dense, rounded crown. The leaves are 3" (7.6 cm) long, obovate to round, leathery, erect, arranged in two rows, and evergreen. Leaf colors range from green to pinkish. The bark is smooth and brownish with white specks.

FLOWERS: 1/3" (2.5 cm), bell-shaped, whitish, hairy, and in short clusters; they bloom year-round.

FRUIT: 1½" (3.8 cm), berrylike, and dark purple; they are ripe year-round.

HABITAT: Hammocks, canal edges, and beaches.

RANGE: Central Florida and south into the tropics. It is found mainly along the coast.

USES: The edible fruit is often used in preserves, and this plant has also been used for medicinal purposes. It is sometimes planted as a hedge in Florida.

SIMILAR SPECIES: This plant is distinct and rarely confused with other species.

CONSERVATION STATUS: LC
Coco-plum has a widespread distribution and a large population. There are no major threats to this species.

Height: 16' (4.9 m)

WHITE POPLAR *Populus alba*

■ DENSE POPULATION
■ LIGHT POPULATION
■ NATURALIZED

This tree has long been cultivated and was introduced to North America in the 18th century. White Poplar, also called Silver Poplar, is hardy in cities and in dry areas. Handsome varieties, some with silvery leaves and one with a columnar form, are planted for shade and ornament. It grows rapidly and spreads readily by cottony, wind-dispersed seeds and via root sprouts that can create extensive clonal colonies. It sometimes becomes an undesirable weed, and many agencies and organizations in North America now classify this tree as invasive.

DESCRIPTION: This is a large, much-branched tree with leaves that toss in the slightest breeze to reveal silvery-white lower surfaces. The leaves are 2½–4" (6–10 cm) long and nearly as wide, ovate, three- or five-lobed and maplelike, and blunt-tipped with scattered small teeth. They are dark green above, and densely white hairy and felt-like beneath, turning reddish in fall. The long leafstalks are covered with white hairs. The bark is whitish-gray and smooth, becoming rough and furrowed at the base. The twigs are densely covered with white hairs.

FLOWERS: Early spring, before leaves. Catkins are 1½–3" (4–7.5 cm) long and densely covered with white hairs; male and female are on separate trees.

FRUIT: ³⁄₁₆" (5 mm) long, egg-shaped capsules with many tiny, cottony seeds.

HABITAT: Moist soils, especially along roadsides and borders of fields.

RANGE: Native of Europe and Asia. Planted and widely naturalized in southern Canada and across the United States.

USES: This species is sometimes planted as an ornamental tree. A yellow dye is sometimes produced from the bark.

SIMILAR SPECIES: Quaking Aspen.

CONSERVATION STATUS: 🔵 LC

White Poplar has a wide geographic range in Europe, North Africa and Asia. The population is large but locally threatened in southern Europe, where the riparian forest has been altered by hydraulic engineering, agriculture and urbanization of floodplain areas. The tree's wind-dispersed seeds have allowed this species to escape cultivation into natural habitats in North America, where it can become dense and crowd out native species.

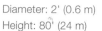

Diameter: 2' (0.6 m)
Height: 80' (24 m)

■ DENSE POPULATION
■ LIGHT POPULATION
■ NATURALIZED

This willow native to western North America is the characteristic cottonwood of the northern Rocky Mountains. Lewis and Clark discovered the species in 1805 on their expedition to explore the Northwest. It is easily distinguishable from related species by the narrow, short-stalked, willow-like leaves. Its is a slim tree and can grow in tightly packed clusters. It occurs near streams and other water sources, and its root system makes it suitable for erosion control.

DESCRIPTION: This tree has a narrow, conical crown of slender, upright branches and with resinous, balsam-scented buds. The leaves are 2–5" (5–13 cm) long, ½–1" (1.2–2.5 cm) wide, lance-shaped, long-pointed at the tip, rounded at the base, finely saw-toothed, hairless or nearly so, and short-stalked. They are shiny green above, and paler beneath, turning dull yellow in autumn. The bark is yellow-green and smooth, becoming gray-brown and furrowed into flat ridges at the base. The twigs are yellow-green, slender, and hairless.

FLOWERS: Early spring before leaves. Catkins are 1½–3" (4–7.5 cm) long and reddish; male and female flowers are on separate trees.

FRUIT: About ¼" (6 mm) long, broadly egg-shaped capsules that are light brown and hairless; they mature in spring and split into two parts, with many cottony seeds.

HABITAT: Moist soils along streams in mountains; with willows and alders in coniferous forests.

RANGE: Mountains from southern Alberta and extreme southwestern Saskatchewan south to Trans-Pecos Texas and California; also northern Mexico; at 3000–8000' (914–2438 m).

USES: The sticky, gummy buds were used as chewing gum by some Native Americans.

SIMILAR SPECIES: Easily distinguished from similar species by the narrow, short-stalked, willow-like leaves.

CONSERVATION STATUS: 🅛🅒

Narrowleaf Cottonwood has a widespread distribution and large population. There are no known threats to this species.

Diameter: 1½' (0.5 m)
Height: 50' (15 m)

■ DENSE POPULATION
■ LIGHT POPULATION
■ NATURALIZED

The northernmost New World hardwood, Balsam Poplar extends in scattered groves to Alaska's Arctic Slope. It grows best in floodplains but can be found in a variety of habitats. These trees are fragrant thanks to the resin produced by buds. Balm-of-Gilead Poplar, an ornamental tree with a broad, open crown and larger, heart-shaped leaves, is a hybrid between Balsam Poplar and Eastern Cottonwood.

DESCRIPTION: This is a large tree with a narrow, open crown of upright branches and fragrant, resinous buds with a strong balsam odor. The leaves are 3–5" (7.5–13 cm) long, 1½–3" (4–7.5 cm) wide, ovate, pointed at the tip, rounded or slightly notched at the base, finely wavy-toothed, slightly thickened, and hairless or nearly so. They are shiny dark green above, and whitish, often with rusty veins beneath. The leafstalks are slender, round, and hairy. The bark is light brown and smooth, becoming gray, and furrowed into flat, scaly ridges. The twigs are brownish and stout, with large, gummy or sticky buds producing fragrant yellowish resin.

FLOWERS: Early spring. Catkins are 2–3½" (5–9 cm) long and brownish; male and female flowers are on separate trees.

FRUIT: ⁵⁄₁₆" (8 mm) long, egg-shaped capsules; they are pointed, light brown, and hairless, maturing in spring and splitting into two parts, with many tiny, cottony seeds.

HABITAT: Moist soils of valleys, mainly stream banks, sandbars, and floodplains, also lower slopes; it is often found in pure stands.

RANGE: Across northern North America along the northern limit of trees from northwest Alaska east to Newfoundland, south to California in the West, and south to Pennsylvania and west to Iowa in the East; it is local south to Colorado and in the eastern mountains to West Virginia.

USES: The light, soft wood from this tree is used as pulpwood and in construction. Balm of Gilead, a fragrant resin derived from the buds, has been used in home remedies.

SIMILAR SPECIES: Swamp Cottonwood.

CONSERVATION STATUS: LC

The Balsam Poplar population is abundant and far-reaching. There are no known threats to this species.

Diameter: 1–3' (0.3–0.9 m)
Height: 60–80' (18–24 m)

■ DENSE POPULATION
■ LIGHT POPULATION
■ NATURALIZED

Black Cottonwood is the tallest native western hardwood species, with champion specimens in Oregon measuring up to 141' (43 m) in height, 31.6' (9.6 m) in trunk circumference, and 96' (29.3 m) in crown spread. The wood is used for boxes and crates, pulpwood, and excelsior (wood wool). Black Cottonwood is sometimes considered a subspecies of its northern relative, Balsam Poplar, and the two species intergrade where both meet in southern Alaska and elsewhere.

ALTERNATE NAME: *Populus balsamifera* ssp. *trichocarpa*

DESCRIPTION: The tallest native cottonwood, Black Cottonwood has an open crown of erect branches and sticky, resinous buds with a balsam odor. The leaves are 3–6" (7.5–15 cm) long and 2–4" (5–10 cm) wide, larger on young twigs. They are broadly ovate, short- or long-pointed at the tip, rounded or slightly notched at the base, finely wavy-toothed, slightly thickened, and hairless or nearly so. They are shiny dark green above, and whitish and often with rusty veins beneath, turning yellow in fall. The leafstalks are slender, round, and hairy. The bark is gray and smooth, becoming thick and deeply furrowed into flat, scaly ridges. The twigs are brownish, stout, and often hairy when young.

FLOWERS: Early spring. Catkins are 1½–3¼" (4–8 cm) long and reddish-purple; male and female flowers are on separate trees.

FRUIT: Light brown, round, and hairy capsules ¼" (6 mm) in diameter, maturing in spring and splitting into 3 parts, with many cottony seeds.

HABITAT: Moist to wet soils of valleys, mainly on stream banks and floodplains, and also on upland slopes. It is often found in pure stands and with willows and Red Alder.

RANGE: Southern Alaska south to southern California and east in the mountains to extreme southwestern Alberta and Montana; it is also local in southwest North Dakota and Baja California. It occurs to 2000' (610 m) in the north, and to 9000' (2743 m) in the south.

USES: This fast-growing tree is widely used for lumber, the lightweight and relatively strong wood most often used for pulpwood, plywood, and boxes and crates. Extracts from the fragrant buds are used commercially as perfumes in cosmetic products.

SIMILAR SPECIES: Balsam Poplar, Swamp Cottonwood.

CONSERVATION STATUS: **LC**
The Black Cottonwood population is stable and often dense in moist areas.

Diameter: 1–3' (0.3–0.9 m), sometimes much larger
Height: 60–120' (18–37 m)

EASTERN COTTONWOOD *Populus deltoides*

■ DENSE POPULATION
■ LIGHT POPULATION
■ NATURALIZED

This species is fairly widespread across North America and one of our largest eastern hardwoods. Although short-lived, it is one of the fastest-growing native trees; on favorable sites in the Mississippi Valley, trees average 5' (1.5 m) in height growth annually with as much as 13' (4 m) the first year. The common name refers to the abundant, cottony seeds. It is also sometimes called Necklace Poplar because its long, narrow lines of seed capsules resemble strings of beads. Classification of this tree is uncertain, and different sources divide it into three subspecies or as many as five varieties.

DESCRIPTION: This is a large tree with a massive trunk often forked into stout branches, and a broad, open crown of spreading and slightly drooping branches. The leaves are 3–7" (7.5–18 cm) long, 3–5" (7.5–13 cm) wide, triangular, long-pointed, usually straight at the base, curved, with coarse teeth, and slightly thickened; they are shiny green, turning yellow in fall. The leafstalks are long, slender, and flattened. The bark is yellowish-green and smooth, becoming light gray, thick, rough, and deeply furrowed. The twigs are brownish and stout, with large resinous or sticky buds.

FLOWERS: Early spring. Catkins are 2–3½" (5–9 cm) long and brownish, with male and female on separate trees.

FRUIT: ⅜" (10 mm) long, light brown, elliptical capsules, maturing in spring and splitting into three or four parts, with many on slender stalks in catkin to 8" (20 cm) long, and many tiny cottony seeds.

HABITAT: Bordering streams and in wet soils in valleys; it is found in pure stands or often with willows. It often pioneers on new sandbars and bare flood plains.

RANGE: Widespread southern Alberta east to extreme southern Quebec and New Hampshire, south to northwest Florida, west to Texas, and north to central Montana. it occurs to 1000' (305 m) in the east, and to 5000' (1524 m) in the west.

USES: This large tree is often planted as a shade tree and for shelterbelts. The wood is used for furniture, boxes and crates, plywood, woodenware, matches and pulpwood.

SIMILAR SPECIES: Swamp Cottonwood.

CONSERVATION STATUS: (LC)
Eastern Cottonwood is fast-growing and dominant or codominant in woodlands across its far-stretching range. There are no major threats to this species.

Diameter: 3–4' (0.9–1.2 m), often larger
Height: 100' (30 m)

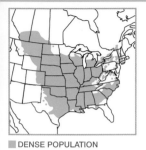

■ DENSE POPULATION
■ LIGHT POPULATION
■ NATURALIZED

The distinctive western subspecies of Eastern Cottonwood, Plains Cottonwood has slightly smaller leaves that are often broader than long and more coarsely toothed and is adapted to tolerate annual drought. It thrives in moist bottomlands and can tolerate a wide range of soils. The root system of these trees penetrates deep underground in search of moisture. Plains Cottonwood and the nominate Eastern Cottonwood (ssp. *deltoides*) subspecies intergrade where their ranges meet.

ALTERNATE NAMES: *Populus sargentii, Populus deltoides* ssp. *occidentalis*

DESCRIPTION: Plains Cottonwood is a riverside tree with a spreading crown; rough, dark gray bark; and yellowish twigs. The leaves are, on average, 3" (7.5 cm) long, flat-based, heart-shaped, long-pointed, and coarse-toothed, turning yellow.

FLOWERS: Early spring. Catkins are 2–3½" (5–9 cm) long and brownish, with male and female on separate trees.

FRUIT: 2½" (6.5 cm) catkins growing to 5" (12.5 cm) long when in fruit, with tiny cottony seeds that emerge when the capsules open.

HABITAT: Along streams and rivers, below 7000' (2130 m).

RANGE: Alberta, Saskatchewan, and Manitoba south through the Great Plains to eastern Colorado, northern Texas and northeastern New Mexico, and west into Montana.

USES: This tree is sometimes planted for shade, and the wood is used for a variety of purposes.

SIMILAR SPECIES: Eastern Cottonwood, Swamp Cottonwood.

CONSERVATION STATUS: LC
Plains Cottonwood is widespread and the population is stable.

Diameter: 36" (90 cm)
Height: to 75' (22.5 m)

■ DENSE POPULATION
■ LIGHT POPULATION
■ NATURALIZED

This fast-growing, short-lived species is the common cottonwood found at low altitudes along the Rio Grande and Colorado River and in the rest of the Southwest and California. Fremont Cottonwood grows only on wet soil and is an indicator of permanent water and shade. It is extensively planted in its range for erosion control along irrigation ditches. It is an ecologically important species, providing food and shelter for various wildlife. The species is named General John Charles Fremont (1813–90), an American explorer of the West who collected many yet-undescribed plants during his expeditions in the 1840s and 1850s.

DESCRIPTION: This tree has a broad, flattened, open crown of large, widely spreading branches. The leaves are 2–3" (5–7.5 cm) long and wide, broadly triangular, often broader than long, sharp-pointed, and nearly straight at the base, with coarse, irregular, curved teeth; they are thick and hairless, and the leafstalks are long and flattened. They are shiny yellow-green, turning bright yellow in fall. The bark is gray, thick, rough, and deeply furrowed. The twigs are light green, stout, and hairless.

FLOWERS: Early spring. Catkins 2–3½" (5–9 cm) long and reddish; male and female flowers are on separate trees.

FRUIT: Light brown, hairless, egg-shaped capsules about ½" (12 mm) long, maturing in spring and splitting into three parts, with many cottony seeds.

HABITAT: Wet soils along streams, often with sycamores, willows, and alders; it is found in deserts, grasslands, and woodlands.

RANGE: Southern and western Colorado west to northern California and southeast to Trans-Pecos Texas; it is also found in northern Mexico. It occurs to 6500' (1981 m).

USES: This species is widely planted as an ornamental tree and is also used for habitat restoration, erosion control, and for wildlife. Early settlers and ranchers used the wood for fence posts and fuel. To this day, the Hopi of the Southwest carve cottonwood roots into kachina dolls, the representations of supernatural beings, that have become valuable collectors' items.

SIMILAR SPECIES: Plains Cottonwood.

CONSERVATION STATUS: LC
The Fremont Cottonwood population is thought to be secure throughout its natural range.

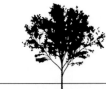

Diameter: 2–4' (0.6–1.2 m)
Height: 40–80' (12–24 m)

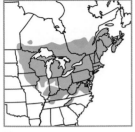

- ■ DENSE POPULATION
- ■ LIGHT POPULATION
- ■ NATURALIZED

Bigtooth Aspen is native to eastern North America, and is especially common in the northeastern United States and southeastern Canada. It is easily distinguished from the related Quaking Aspen by the large, curved teeth of its leaf edges, for which it is named. Like Quaking Aspen, Bigtooth Aspen is tolerant of a wide range of soils and is often a pioneer tree on disturbed sites, but it is short-lived and soon replaced by conifers. A variety of wildlife readily consumes the foliage, twig buds, and bark.

DESCRIPTION: This is a medium-sized tree with a narrow, rounded crown. The leaves are 2½–4" (6–10 cm) long, 1¾–3½" (4.5–9 cm) wide, broadly ovate, short-pointed at the tip, rounded at the base, with coarse, curved teeth, and white hairs when young. They are dull green above, and paler beneath, turning pale yellow in autumn. The leafstalks are long, slender, and flattened. The bark is greenish, smooth, and thin, becoming dark brown and furrowed into flat, scaly ridges. The twigs are brown, slender, and hairy when young.

FLOWERS: Early spring. Catkins are 1½–2½" (4–6 cm) long and brownish; with male and female on separate trees.

FRUIT: ¼" (6 mm) long, narrowly conical capsules; they are light green, slightly curved, and finely hairy, maturing in spring and splitting into two parts, with many tiny, cottony seeds.

HABITAT: It is found on sandy upland soils and also in the floodplains of streams, often in even-aged mixed stands with Quaking Aspen.

RANGE: Manitoba east to Cape Breton Island, south to Virginia, and west to northeastern Missouri; it is local south to western North Carolina. It occurs to 2000' (610 m), or to 3000' (914 m) in the south.

USES: This species is valuable to wildlife, and the wood is used for pulpwood, and sometimes for pallets, boxes, chopsticks, ladders, and other various purposes.

SIMILAR SPECIES: Quaking Aspen.

CONSERVATION STATUS: LC
Bigtooth Aspen is abundant in central and eastern regions of the United States and Canada. There are no major threats to the population.

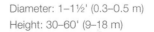
Diameter: 1–1½' (0.3–0.5 m)
Height: 30–60' (9–18 m)

■ DENSE POPULATION
■ LIGHT POPULATION
■ NATURALIZED

This large, deciduous tree is found in scattered areas in the easterrn United States, where it prefers wet soils. Because of its rapid growth, Swamp Cottonwood is sometimes planted as a shade tree. Other names for this species include Downy Poplar and Swamp Poplar.

DESCRIPTION: This tree has a narrow, rounded crown and stout branches. Its leaves are 4–7" (10–18 cm) long, 3–6" (7.5–15 cm) wide, broadly ovate, blunt or rounded at the tip, and heart-shaped or rounded at the base, with fine, curved teeth. They are densely covered with white hairs when unfolding, becoming hairless and dark green above, remaining woolly and pale beneath. The leafstalks are slender, rounded, and hairless. The bark is brown, and furrowed into scaly ridges. The twigs are stout, and covered with white hairs when young.

FLOWERS: Early spring. Catkins are 1–2½" (2.5–6 cm) long and brownish; male and female flowers are on separate trees.

FRUIT: ½" (12 mm) long, egg-shaped capsules; they are brown and long-stalked, maturing in spring and splitting into two or three parts, with tiny cottony seeds.

HABITAT: Wet sites, often submerged in flood plains and edges of swamps, with willows, Bald Cypress, and Water Tupelo.

RANGE: Connecticut south to eastern Georgia and from northwestern Florida west to eastern Louisiana, north to southern Michigan; to 800' (244 m).

USES: This species is sometimes planted as an ornamental or shade tree.

SIMILAR SPECIES: Eastern Cottonwood, Balsam Poplar.

CONSERVATION STATUS: LC

Swamp Cottonwood has a widespread distribution but is generally uncommon in its range. The population is declining in some parts of its range, but there are no range-wide threats identified. The limited populations in Canada (found only in Ontario) are critically imperiled.

Diameter: 2' (0.6 m)
Height: 80' (24 m)

■ DENSE POPULATION
■ LIGHT POPULATION
■ NATURALIZED

The Quaking Aspen may be the most widely distributed tree in North America. The name refers to the leaves, which in the slightest breeze tremble on their flattened leafstalks. It produces the most colorful foliage of any tree in the Western states and provinces. The soft, smooth bark is sometimes marked by bear claws. Fast growing, with a short lifespan, it is often a pioneer tree after fires and logging and on abandoned fields, later replaced by conifers. In the far north it is the only hardwood among the conifers. The twigs and foliage are browsed by deer, elk, and moose, also by sheep and goats. Beavers, rabbits, and other mammals eat the bark, foliage, and buds, and grouse and quail feed on the winter buds. The Quaking Aspen propagates principally through its roots, and large colonies are often a single plant—clones sharing common roots.

DESCRIPTION: Quaking Aspen is a tall, fast-growing tree with a narrow, rounded crown of thin foliage. Its leaves are 1¼–3" (3–7.5 cm) long, nearly round, abruptly short-pointed, rounded at the base, finely saw-toothed, and thin. They are shiny green above, and dull green beneath, turning golden-yellow in autumn before shedding. The leafstalks are slender and flattened. The bark is whitish, smooth, and thin, becoming dark gray, furrowed, and thick on large trunks. The twigs are shiny brown, slender, and hairless.

FLOWERS: Early spring before leaves. Catkins are 1–2½" (2.5–6 cm) long and brownish, with male and female on separate trees.

FRUIT: ¼" (6 mm) long, narrowly conical light green capsules in drooping catkins to 4" (10 cm) long, maturing in late spring and splitting in two parts, with many tiny cottony seeds; propagation is often by root sprouts rather than seeds.

HABITAT: Many soil types, especially sandy and gravelly slopes; it is often found in pure stands and in western mountains in an altitudinal zone below the spruce-fir forest.

RANGE: Across northern North America from Alaska to Newfoundland, south to Virginia, and in the Rocky Mountains south to southern Arizona and northern Mexico; it occurs from near sea level in the north to 6500–10,000' (1981–3048 m) farther south.

USES: Principal uses of the wood include pulpwood, boxes, furniture parts, matches, excelsior, and particle-board.

SIMILAR SPECIES: Bigtooth Aspen.

CONSERVATION STATUS: 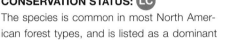 LC

The species is common in most North American forest types, and is listed as a dominant species in more than 100 habitats. Conifers outcompete this species in some places. Because aspens sprout from roots that survive fires while competing conifers do not, a reduction in the number of trees appears to be the result of fire suppression programs. Browsing of young shoots by deer, elk, moose, and cattle also impacts the aspen population. A reduction in the numbers of wolves contributes to this outcome. This tree's thin bark makes it susceptible to some pests and diseases, but its population is large and stable.

Diameter: 1–1½' (0.3–0.5 m)
Height: 40–70' (12–21 m)

WHITE WILLOW *Salix alba*

☐ DENSE POPULATION
☐ LIGHT POPULATION
☐ NATURALIZED

White Willow is native to Eurasia, and was introduced to North America in the 18th century. This attractive willow has been widely planted as a shade and ornamental tree and for shelterbelts, and grown for timber and firewood. Some cultivated varieties have golden-yellow or reddish twigs that have been used in basketmaking. Like many exotic species in North America, however, it has shown invasive qualities. In some areas efforts are under way to eradicate it from the landscape. These trees tend to be fast-growing and short-lived.

DESCRIPTION: White Willow is a naturalized tree with one to four trunks and an open crown of spreading branches. Its leaves are 2–4½" (5–11 cm) long, ⅜–1¼" (1–3 cm) wide, lance-shaped to elliptical, finely saw-toothed, and firm. They are shiny dark green above, and whitish and silky beneath, turning yellow in autumn. The bark is gray, rough, and furrowed into narrow ridges. The twigs are yellow to brown, flexible and often slightly drooping, and silky when young.

FLOWERS: Early spring. Catkins are 1¼–2¼" (3–6 cm) long with yellow, hairy scales, occurring at the end of short, leafy twigs.

FRUIT: ³⁄₁₆" (5 mm) long, light brown and hairless capsules, maturing in late spring or early summer.

HABITAT: Wet soils of stream banks and valleys near cities.

RANGE: Native from Europe and northern Africa to central Asia. Naturalized widely in southeastern Canada and the eastern United States, and more recently in the West.

USES: This tree has been widely planted as an ornamental and shade tree, and the wood is used for fence posts and firewood.

SIMILAR SPECIES: Black Willow, Sandbar Willow.

CONSERVATION STATUS:

White Willow has a widespread, stable population and does not face any major threats in its native range.

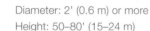

Diameter: 2' (0.6 m) or more
Height: 50–80' (15–24 m)

■ DENSE POPULATION
■ LIGHT POPULATION
■ NATURALIZED

This widespread, small- to medium-sized tree is the common willow across the northern plains, where it is important for protecting river banks from erosion. It can be found on the northern prairies, often near streams, and is the largest tree native to that region. It is named for the shape of its leaves, which suggest those of Peach trees.

DESCRIPTION: This tree has one or sometimes several straight trunks, upright branches, and a spreading crown. The leaves are 2–4½" (5–11 cm) long, ½–1¼" (1.2–3 cm) wide, lance-shaped, often slightly curved to one side, and tapering to a long, narrow point; they are finely saw-toothed, with long, slender leafstalks, becoming hairless; they are shiny green above, and whitish beneath. The bark is dark brown, rough, and furrowed into flat, scaly ridges. The twigs are shiny orange or brown and hairless.

FLOWERS: Spring. Catkins are 1¼–3" (3–7.5 cm) long, with yellow, hairy scales, occurring on short, leafy twigs.

FRUIT: ¼" (6 mm) long, long-stalked, reddish-yellow, hairless capsules, maturing in late spring or early summer.

HABITAT: Wet soils of valleys, often bordering stream banks with cottonwoods.

RANGE: Southeast British Columbia east to extreme southern Quebec and New York, south to northwest Pennsylvania and west to western Texas; it is also found in northern Mexico. It occurs at 500–7000' (152–2134 m).

USES: This fast-growing yet short-lived species is important to riparian ecosystems in its range. Tea made from the leaves, bark, twigs, or roots of this and other *Salix* species has been used for thousands of years as a remedy for inflammation, headache, fever, sore throat, and other minor ailments; aspirin is a derivative of the active ingredient, salicylic acid.

SIMILAR SPECIES: Coastal Plain Willow.

CONSERVATION STATUS: LC

Peachleaf Willow is widespread and its population is considered secure. Communities are vulnerable to overgrazing and trampling by livestock and wild ungulates.

Diameter: 2' (0.6 m)
Height: 60' (18 m)

■ DENSE POPULATION
■ LIGHT POPULATION
■ NATURALIZED

This species is widespread across Canada and the northern and western United States. Bebb Willow is the most important of the "diamond willows," a term applied to several species whose wood is sometimes deformed into diamond-shaped patterns as a result of fungal attack. The contrasting whitish and brownish stems are carved into canes, lamps, furniture, candleholders, and other decorative pieces. These plants form willow thickets as a weed on uplands after forest fires. The species name honors Michael Schuck Bebb (1833–95), an American botanist who researched willows extensively.

DESCRIPTION: This is a much-branched shrub or small tree with a broad, rounded crown. The leaves are 1–3½" (2.5–9 cm) long, ⅜–1" (1–2.5 cm) wide, elliptical, often broadest beyond the middle, sharp-pointed at the ends, slightly saw-toothed or wavy, firm, and slightly hairy. They are dull green above, and gray or whitish and net-veined beneath. The bark is gray and smooth, becoming rough and furrowed. The twigs are reddish-purple, slender, and widely forking, with pressed hairs when young.

FLOWERS: Spring, before or with leaves. Catkins are ¾–1½" (2–4 cm) long, with yellow or brown scales, occurring on short, leafy stalks.

FRUIT: ⅜" (10 mm) long, hairy, light brown, and slender capsules ending in a long point; they are long-stalked, and mature in early summer.

HABITAT: Moist open uplands and borders of streams, lakes, and swamps.

RANGE: Central and southwestern Alaska south to British Columbia and east to Newfoundland, south to Maryland, west to Iowa, and south in the Rocky Mountains to southern New Mexico; it occurs to 11,000' (3353 m) in the south of its range.

USES: This species is used for canes, lamps, posts, and many other items. Some Native Americans used the branches for basket weaving and to make arrows.

SIMILAR SPECIES: Balsam Willow, Goat Willow.

CONSERVATION STATUS: LC
The overall Bebb Willow population is stable, although there have been localized declines due to overgrazing, prolonged fire suppression, declining water availability, and fungal pathogens including the rust fungus *Melampsora epitea* and the polypore *Phellinus punctatus*. The species is listed as state-threatened in several U.S. states.

Diameter: 6" (15 cm)
Height: 10–25' (3–7.6 m)

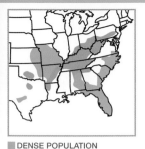

■ DENSE POPULATION
■ LIGHT POPULATION
■ NATURALIZED

Also known as Carolina Willow, this is the common small willow found at low altitudes in the southeastern United States. It can be a shrub or a small tree, preferring wet soils. It was first described in 1803 by André Michaux (1746–1802), a French botanist and explorer.

DESCRIPTION: This is a shrub or small tree with spreading or slightly drooping branches. The leaves are 2–4" (5–10 cm) long, ½–¾" (12–19 mm) wide, lance-shaped, finely saw-toothed, and densely hairy when young. They are green above and whitish and nearly hairless beneath. The leafstalks are hairy. The bark is gray to blackish, fairly smooth, and furrowed into broad scaly ridges. The twigs are brown, slender, limber, and hairy when young.

FLOWERS: Spring. Catkins are 3–4" (7.5–10 cm) long and greenish or yellowish, occurring at the ends of leafy twigs.

FRUIT: ¼" (6 mm) long, long-pointed and light reddish-brown capsules, maturing in late spring or early summer.

HABITAT: Wet soils of stream banks and swamps.

RANGE: Southern Pennsylvania south through Florida, west to central Texas, and north to southeastern Nebraska; to 2000' (610 m).

USES: This species is an important larval host for several moth species, and the male flowers produce spring pollen for bees.

SIMILAR SPECIES: Florida Willow, Black Willow, Peachleaf Willow.

CONSERVATION STATUS:
The Coastal Plain Willow population is abundant and stable. There are no known threats to this species.

Diameter: 1' (0.3 m)
Height: 30' (9 m)

■ DENSE POPULATION
■ LIGHT POPULATION
■ NATURALIZED

Pussy Willow is native to the northern forests of North America, found extensively in the eastern and central United States and across most of Canada. This plant is named for its large flower buds, which burst and expose their soft, furry catkins early in the year. It is often grown for cut flowers. In winter, cut Pussy Willow twigs can be put in water and the flowers can be forced at warm temperatures—some twigs will produce beautiful golden stamens, while others will bear slender greenish pistils. The Latin species name refers to the contrasting colors of the leaf surfaces, which can be an identification field mark.

DESCRIPTION: This is a many-stemmed shrub or small tree with an open rounded crown and silky, furry catkins that appear in late winter and early spring. The leaves are 1½–4¼" (4–11 cm) long, ⅜–1¼" (1–3 cm) wide, lance-shaped or narrowly elliptical, irregularly wavy-toothed, stiff, hairy when young, and slender-stalked. They are shiny green above, and whitish beneath. The bark is gray, fissured, and scaly. The twigs are brown, stout, and hairy when young.

FLOWERS: Late winter and early spring long before leaves. Catkins are 1–2½" (2.5–6 cm) long, cylindrical, thick with blackish scales, and covered with silky whitish hairs.

FRUIT: ⁵⁄₁₆–½" (8–12 mm) long, narrow capsules that are light brown and finely hairy, occurring in early spring before leaves.

HABITAT: Wet meadows and borders of streams and lakes; usually in coniferous forests.

RANGE: Northern British Columbia to Labrador, south to Delaware, west to northeastern Missouri, and north to northern Wyoming and North Dakota; to 4000' (1219 m).

USES: Some Native Americans used this plant as a painkiller. It is cultivated for its flowers.

SIMILAR SPECIES: Rusty Willow, Large Gray Willow.

CONSERVATION STATUS: LC
Pussy Willow has a wide geographical range and the population is stable. There are no known threats.

Diameter: 8" (20 cm)
Height: 20' (6 m)

- ■ DENSE POPULATION
- ■ LIGHT POPULATION
- ■ NATURALIZED

This hardy species has an extremely wide range and is adapted to many wet habitats. It is found from the Yukon River in central Alaska to the Mississippi River in southern Louisiana. Throughout the interior of North America, it is typically a common and characteristic shrub found along streams, especially in the Great Plains and Southwest. It is drought-resistant and suitable for planting on stream bottoms to prevent surface erosion. It spreads by basal shoots to form dense clonal colonies. It is cultivated as an ornamental tree.

ALTERNATE NAME: *Salix exigua* ssp. *interior*

DESCRIPTION: This is a thicket-forming shrub with clustered stems or rarely a tree, with narrow leaves. Those leaves are 1½–4" (4–10 cm) long, ¼" (6 mm) wide, linear, very long-pointed at the ends, with a few tiny, scattered teeth or none, varying from hairless to densely hairy with pressed, silky hairs, and almost stalkless. They are yellow-green to gray-green on both surfaces. The bark is gray, and smooth or becoming fissured. The twigs are reddish- or yellowish-brown, slender, upright, and hairless or with gray hairs.

FLOWERS: Spring. Catkins are 1–2½" (2.5–6 cm) long with hairy yellow scales, at the end of leafy twigs.

FRUIT: ¼" (6 mm) long, light brown and usually hairy capsules, maturing in early summer.

HABITAT: Wet soils, especially riverbanks, sandbars, and silt flats.

RANGE: Central Alaska east to Ontario and New York, southwest to Mississippi, and west to southern California; it is also local east to Quebec and Virginia and in northern Mexico. It occurs to 8000' (2438 m).

USES: Sandbar Willow is planted as an ornamental and to help to prevent erosion. Livestock browse the foliage, and some Native Americans made baskets from the twigs and bark.

SIMILAR SPECIES: Elaeagnus Willow.

CONSERVATION STATUS: LC
Sandbar Willow is widespread and has a stable population, with no significant threats to the species.

Diameter: 5" (13 cm)
Height: 3–10' (1–3 m),
sometimes to 20' (6 m)

■ DENSE POPULATION
■ LIGHT POPULATION
■ NATURALIZED

This tree was introduced to the colonies from Europe and western Asia to provide charcoal for gunpowder and to serve as a shade tree. It is called Crack Willow, and sometimes Brittle Willow, because the twigs easily break off at the base, especially in spring; the Latin species name also refers to the fragile twigs. If they are partly covered by soil, detached twigs or cuttings will establish roots and grow into new plants. This species readily hybridizes with native Black Willows in eastern North America and with introduced White Willows from Eurasia.

DESCRIPTION: This is a large, naturalized tree with widely forking branches and brittle twigs. The leaves are 4–6" (10–15 cm) long, 1–1½" (2.5–4 cm) wide, lance-shaped, ending in a long point turned to one side, coarsely saw-toothed, with gland-tipped teeth, and hairless. They are shiny green above and whitish beneath. The bark is gray, and rough, thick, and deeply furrowed into narrow ridges. The twigs are shiny brownish, erect or spreading, and easily broken at the base, with gummy buds.

FLOWERS: Spring. Catkins are 1–2¼" (2.5–6 cm) long with yellow or greenish hairy scales, occurring at the ends of leafy twigs.

FRUIT: ³⁄₁₆" (5 mm) long, light brown capsules, maturing in late spring and early summer.

HABITAT: It is often found escaping in moist soil along roadsides and streams and in clearings.

RANGE: Native to Europe and western Asia. It is naturalized from Newfoundland to Virginia, west to Kansas, and north to South Dakota.

USES: This introduced, fast-growing species is planted as an ornamental and shade tree.

SIMILAR SPECIES: White Willow.

CONSERVATION STATUS:
The conservation status of this species has not yet been assessed.

Diameter: 2½' (0.8 m)
Height: 80' (24 m)

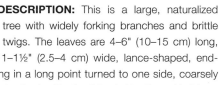

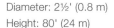

■ DENSE POPULATION
■ LIGHT POPULATION
■ NATURALIZED

Many sources treat Hinds Willow as a western subspecies of Sandbar Willow. This willow native to southern Oregon and California is useful for erosion control; it holds soil banks well but is considered a weed when it clogs irrigation ditches. Hinds Willow is named for Richard Brinsley Hinds (1812–1847), a British botanist who collected plant specimens along the West Coast on a surveying expedition in 1836–42.

ALTERNATE NAME: *Salix exigua* ssp. *hindsiana*

DESCRIPTION: This is a thicket-forming shrub or small tree with many trunks and very narrow leaves covered with silky hairs. The leaves are 1½–3¼" (4–8 cm) long, usually ⅛–¼" (3–6 mm) wide, very narrow or linear, tapering at both ends, usually without teeth, covered with gray, silky hairs on both surfaces, and almost stalkless. The bark is gray and furrowed. The twigs are gray or silvery, and covered with woolly hairs when young.

FLOWERS: Catkins are ¾–1½" (2–4 cm) long and covered with yellow, densely hairy scales, occurring on leafy twigs after the leaves appear in spring.

FRUIT: ¼" (6 mm) long, light brown and densely hairy capsules; they are almost stalkless, and mature in late spring or early summer.

HABITAT: Moist soils of ditches, sandbars, and stream banks.

RANGE: Southwest Oregon to southern California and Baja California; to 3000' (914 m).

USES: Hinds Willow is often planted to help to prevent erosion.

SIMILAR SPECIES: Sandbar Willow.

CONSERVATION STATUS:
The global population of Hinds Willow has not yet been assessed. It is endangered in Canada, where it occurs only in British Columbia.

Diameter: 10" (25 cm)
Height: 7–23' (2–7 m)

ARROYO WILLOW *Salix lasiolepis*

- ■ DENSE POPULATION
- ■ LIGHT POPULATION
- ■ NATURALIZED

This common species is native to the western United States and northwestern Mexico, where it prefers wet soils. The alternate name "White Willow" may come from the light-colored bark and leaves with whitish lower surfaces, and should not be confused with the introduced White Willow (*S. alba*). The scientific name, which means "shaggy scale," refers to the white hairs on the scales of the flowers.

DESCRIPTION: This is usually a thicket-forming shrub with clustered stems; it is sometimes a small tree with slender, erect branches forming a narrow, irregular crown. The leaves are 2½–4" (6–10 cm) long, ⅜–¾" (1–2 cm) wide, narrow, reverse lance-shaped, short- or blunt-pointed at the ends, broadest beyond the middle, without teeth or slightly wavy with a few small teeth, and thick and leathery. They are dark green and hairless above, and whitish and usually hairy beneath. The bark is pale gray-brown with whitish areas, and smooth, becoming darker, rough, and furrowed into broad ridges. The twigs are yellow to brown and finely hairy.

FLOWERS: Early spring before or with leaves. Catkins are 1–2" (2.5–5 cm) long, with black or brown scales and dense, long, white hairs; they are almost stalkless.

FRUIT: ¼" (6 mm) long, light reddish-brown, hairless capsules; they are crowded and mature in late spring.

HABITAT: Wet soils along streams and arroyos, or gullies in valleys, foothills, and mountains.

RANGE: Washington and Idaho south to southern California and New Mexico; it is also found in northern Mexico. It occurs to 7500' (2286 m).

USES: Native Americans used this species for medicinal purposes and to make rope and baskets.

SIMILAR SPECIES: Pacific Willow, Sitka Willow.

CONSERVATION STATUS: ⬤ LC

Arroyo Willow has a stable population and there are no known threats to the species.

Diameter: 6" (15 cm)
Height: 30' (9 m)

■ DENSE POPULATION
■ LIGHT POPULATION
■ NATURALIZED

Pacific Willow is a tall, slender, plant that grows as a large shrub or a small tree. As the common name suggests, Pacific Willow is typically found along riverbanks and in valleys throughout the Pacific states. Pacific Willow is sometimes treated as a distinct species but is often considered conspecific with *Salix lucida*, Shining Willow or Greenleaf Willow, found in northern North America, which is usually a shrub. It is closely related to Bay Willow (*Salix pentandra*), a species native to Europe and Asia.

ALTERNATE NAME: *Salix lasiandra*

DESCRIPTION: This tree has an open, irregular crown; it is sometimes a thicket-forming shrub. The leaves are 2–5" (5–13 cm) long, ½–1" (1.2–2.5 cm) wide, narrowly lance-shaped, very long-pointed, mostly rounded at the base, finely saw-toothed, becoming almost hairless; the leafstalks are slender, with glands at the upper end. They are shiny green above, and whitish beneath. The bark is gray or dark brown, becoming rough and deeply furrowed into flat, scaly ridges. The twigs are shiny reddish to brownish or yellow, and hairless.

FLOWERS: Spring. Catkins are 1½–4" (4–10 cm) long, with hairy, yellow or brown scales, occurring at the ends of leafy twigs.

FRUIT: ¼" (6 mm) long, light reddish-brown and hairless capsules, maturing in early summer.

HABITAT: Wet soils along streams, lakes, and roadsides; it is found in valleys and on mountains.

RANGE: Central and southeastern Alaska east to Saskatchewan and south mostly in the mountains to southern New Mexico and southern California; to 8000' (2438 m).

USES: This species has been used for a variety of medicinal purposes, including to relieve pain and to treat diarrhea.

SIMILAR SPECIES: Arroyo Willow, Sitka Willow.

CONSERVATION STATUS: 🄻🄲
The Pacific Willow population is widespread and stable across its range. There are no known threats.

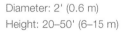

Diameter: 2' (0.6 m)
Height: 20–50' (6–15 m)

■ DENSE POPULATION
■ LIGHT POPULATION
■ NATURALIZED

Black Willow is the largest New World willow, and one of the most ecologically important, with a wide range across the United States. In the lower Mississippi Valley it attains commercial timber size, reaching 100–140' (30–42 m) in height and 4' (1.2 m) in diameter. This tree thrives in swampy conditions and is sometimes colloquially called "Swamp Willow," not to be confused with the European species also called Swamp Willow (*S. myrtilloides*). Large trees are valuable in binding soil banks and for marshland stabilization, preventing soil erosion and flood damage; mats and poles made from Black Willow trunks and branches provide further protection of riverbanks and levees.

DESCRIPTION: This is a large tree with one or more straight and usually leaning trunks, upright branches, and a narrow or irregular crown. The leaves are 3–5" (7.5–13 cm) long, ⅜–¾" (10–19 mm) wide, narrowly lance-shaped, often slightly curved to one side, long-pointed, finely saw-toothed, and hairless or nearly so; they are shiny green above and paler beneath. The bark is dark brown or blackish, and deeply furrowed into scaly, forking ridges. The twigs are brownish, slender, and easily detached at the base.

FLOWERS: Spring. Catkins are 1–3" (2.5–7.5 cm) long with yellow hairy scales, occurring at the end of leafy twigs.

FRUIT: ³⁄₁₆" (5 mm) long, reddish-brown, and hairless capsules, maturing in late spring.

HABITAT: Wet soils of banks of streams and lakes, especially floodplains; it is often found in pure stands and with cottonwoods.

RANGE: Southern New Brunswick and Maine south to northwest Florida, west to southern Texas, and north to southeast Minnesota; also from western Texas west to northern California; local in northern Mexico; to 5000' (1524 m).

USES: This is an important timber species, used for a wide variety of purposes, including furniture, doors, cabinets, toys, and pulpwood. It is also used to help to prevent soil erosion.

SIMILAR SPECIES: Crack Willow, Coastal Plain Willow.

CONSERVATION STATUS: (LC)
Black Willow is widespread geographically and the population is stable. There are no known threats.

Diameter: 1½–2½' (0.5–0.8 m)
Height: 60–100' (18–30 m)

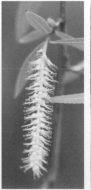

■ DENSE POPULATION
■ LIGHT POPULATION
■ NATURALIZED

This species is native to Western North America, where it is sometimes called Fire Willow because it rapidly and abundantly occupies burned areas, forming blue-green thickets. Scouler Willow is a common upland willow through most of its range. It is one of the earliest flowering species, and is an important browse plant for moose in Alaska and for livestock. It is one of several species sometimes called "diamond willow," whose stems develop diamond-shaped patterns caused by fungi. The species is named for Scottish naturalist John Scouler (1804–71), who collected a large number of specimens from western North America.

DESCRIPTION: This is usually a shrub or small tree with an erect trunk and a compact, rounded crown; it is sometimes medium-sized. Freshly stripped bark of the twigs often has a skunk-like odor. The leaves are spreading and fanlike; they are 2–5" (5–13 cm) long, ½–1½" (1.2–4 cm) wide, and variable in shape; they are mostly obovate to elliptical and broadest beyond the middle, short-pointed at the tip and tapering to the base, without teeth to sparsely wavy-toothed. They are dark green and nearly hairless above, and whitish with gray hairs or with few reddish hairs beneath. The bark is gray, smooth, and thin, becoming dark brown and fissured into broad, flat ridges. The twigs are yellow to reddish-brown, stout, and densely hairy when young, with reddish buds.

FLOWERS: Abundant in early spring before leaves. Catkins are 1–2" (2.5–5 cm) long, stout, and stalkless or nearly so, with black, long-haired scales.

FRUIT: ⅜" (10 mm) long, narrow, and stalkless light brown capsules with gray, woolly hairs, maturing in early summer.

HABITAT: Upland coniferous forests under larger trees, in cutover areas, and in clearings, including dry sites.

RANGE: Central Alaska east to Manitoba and southwest to Idaho and California; it is also found in the Black Hills of South Dakota and in the mountains to southern New Mexico. It occurs to 10,000' (3048 m) in the mountains.

USES: The wood from this "diamond willow" species is used for canes, furniture posts, and other carved items. It has also been used in traditional medicine.

SIMILAR SPECIES: Bebb Willow, Sitka Willow.

CONSERVATION STATUS:
Scouler Willow is widespread and stable throughout its range. There are no known threats to this species.

Diameter: 1½' (0.5 m)
Height: 15–50' (4.6–15 m)

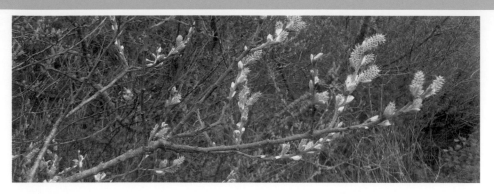

■ DENSE POPULATION
■ LIGHT POPULATION
■ NATURALIZED

Sitka Willow is common to abundant in far northwestern North America, where it is native to a variety of coastal and inland habitats from southern Alaska to central California. This variable species may be a large, bushy shrub or an erect tree, but it is easily recognized by the satiny sheen on the lower leaf surfaces. This species is named for the small city of Sitka in southeastern Alaska, where it was first collected.

DESCRIPTION: This is usually a large shrub or small tree, much-branched with a rounded crown; it can be a low shrub in exposed places. The leaves are 2–4" (5–10 cm) long, ¾–1½" (2–4 cm) wide, reverse lance-shaped or elliptical, blunt-pointed, broadest beyond the middle, tapering to a narrow base, and mostly without teeth or wavy-edged. They are shiny dark green and with sparse, short hairs when young above, and paler and with short, silvery, silky hairs beneath. The bark is gray and smooth, becoming slightly furrowed and scaly. The twigs are reddish-brown, slender, and brittle, and hairy when young.

FLOWERS: Spring. Catkins are 1½–4" (4–10 cm) long and slender, with densely hairy, black or brown scales, occurring on short, leafy stalks.

FRUIT: ¼" (6 mm) long, light brown capsules covered with silvery hairs, maturing in late spring.

HABITAT: Moist soils, along beaches and streams and in open areas through coastal coniferous forests.

RANGE: Pacific Coast from southwest Alaska southeast to central California and east in the mountains to western Montana and central Alberta. It occurs mostly near sea level, but inland locally to 5000' (1524 m).

USES: Some Native Americans used the twigs from this species for basket making, for stretching skins, and other purposes. Because the smoke from Sitka Willow fires does not have a bad odor, the wood is often used for drying fish.

SIMILAR SPECIES: Bebb Willow, Scouler Willow.

CONSERVATION STATUS: LC
Sitka Willow spreads a wide range geographically and the population is stable. There are no known threats to this species.

Diameter: 4–12' (10–30 cm)
Height: 10–30' (3–9 m)

■ DENSE POPULATION
■ LIGHT POPULATION
■ NATURALIZED

The *Salix* Sepulcralis Group is a cultivar group containing all cultivars of hybrids between the introduced White Willow (*Salix alba*) and the Asian species *Salix babylonica*, which was traded along the Silk Road that connected East Asia to southern Europe as early as 200 BCE. Many of these willows are well known for their distinctive weeping foliage. Weeping Willows are among the first willows to bear leaves in spring and among the last to shed them in fall.

ALTERNATE NAME: *Salix babylonica x Salix alba*

DESCRIPTION: Weeping Willow is a handsome, naturalized tree in North America with a short trunk and a broad, open, irregular crown of drooping branches. The leaves are 2½–5" (6–13 cm) long, ¼–½" (6–12 mm) wide, narrowly lance-shaped, and finely saw-toothed, with long-pointed tips. They are dark green above, and whitish or gray beneath, hanging from short leafstalks. The bark is gray, rough, thick, and deeply furrowed in long, branching ridges. The twigs are yellowish-green to brownish, slender, and unbranched, drooping vertically.

FLOWERS: Early spring. Catkins are ⅜–1" (1–2.5 cm) long and greenish, occurring at the end of short leafy twigs; the plants are mostly female.

FRUIT: Light brown capsules ¹⁄₁₆" (1.5 mm) long, maturing in late spring or early summer.

HABITAT: Parks, gardens, and cemeteries, especially near water.

RANGE: This tree is naturalized locally from extreme southern Quebec and southern Ontario south to Georgia and west to Missouri. It is also planted in several western states.

USES: This tree has been widely cultivated as an ornamental.

SIMILAR SPECIES: White Willow.

CONSERVATION STATUS:

This hybrid is known mainly from cultivation and has naturalized locally throughout North America. Its status in the wild has not been assessed.

Diameter: 2' (0.6 m), sometimes much larger
Height: 30–40' (9–12 m)

■ DENSE POPULATION
■ LIGHT POPULATION
■ NATURALIZED

This exotic plant escapes from cultivation and becomes a weedy pest, invading open disturbed habitats in North America. Because of its invasive qualities and its extreme toxicity, Castor Bean is considered a noxious or invasive weed. Most members of the family are poisonous, and their milky sap will irritate the membranes of the eyes and mouth. The seeds of this plant contain ricin, a potent toxin, and the plant is highly allergenic and a strong trigger for asthma in sensitive individuals.

DESCRIPTION: This is an introduced tree-like shrub. The leaves are 16" (41 cm) long, green to reddish, and deeply-palmately five- to eleven-lobed.

FLOWERS: This plant blooms all year in warmer regions. Its flowers are ½" (1.27 cm) pink, spiny balls.

FRUIT: ½" (1.27 cm) green to brown, spiny balls, clustered at the branch tips.

CAUTION: Poisonous: Do not eat or ingest any part of this species. If accidentally ingested, contact your doctor or poison control center.

HABITAT: Coastal sage scrub, thickets, lowlands, waste areas, building sites, and other disturbed areas.

RANGE: Probably native to Africa. It is naturalized in California, Arizona, and Utah in the West, and across much of the East from Michigan and New England south to Florida and west to Texas and Kansas.

USES: This species is cultivated for a wide variety of purposes, including to produce castor oil, which has been used medicinally, as lubricant in airplane and automobile engines, as an ingredient in cosmetics, and in other ways.

SIMILAR SPECIES: This plant is not likely to be confused with any native North American species.

CONSERVATION STATUS:

The conservation status of Castor Bean has not yet been fully assessed. It is considered a weedy or invasive pest where it occurs outside of cultivation in North America.

Height: 10–16' (3–4.8 m)

■ DENSE POPULATION
■ LIGHT POPULATION
■ NATURALIZED

Also known as Button Mangrove, this tropical shrub is found throughout the world, including the Caribbean, Central and South America, and western Africa. Its natural range in the United States is limited to coastal Florida.

DESCRIPTION: This is a spreading, thicket-forming shrub to low-branching tree with a narrow, rounded crown. The leaves are 4" (10 cm) long, elliptical to obovate, and evergreen. The bark is gray to dark brown, rough, and ridged.

FLOWERS: Year-round. Minute and greenish, occurring in tiny, dense, egg-shaped heads.

FRUIT: Tiny and seed-like, occurring in reddish-green, cone-like 1¼" (3 cm) clusters.

HABITAT: Swamps, hammocks, towns.

RANGE: Coastal peninsular Florida.

USES: This species is popular as an ornamental plant and is used in bonsai. The wood is sometimes used to make cabinets and for firewood. The bark is sometimes harvested for its tannin.

SIMILAR SPECIES: Black Mangrove.

CONSERVATION STATUS: LC

Buttonwood has a wide distribution in coastal Central and South America and West Africa. It can exist in a number of habitats, but it is threatened by extraction and urban development within its range.

Height: 50' (15.2 m)

WHITE MANGROVE *Laguncularia racemosa*

■ DENSE POPULATION
■ LIGHT POPULATION
■ NATURALIZED

This tropical species is found on the shores of central and southern Florida, including the Florida Keys, and occurs in tropical and subtropical regions around the world. It is commonly a small evergreen tree closely associated with marshes. This plant's oval-shaped leaves that are rounded at both apices differentiate it from other mangrove species. White mangroves also have unique glands on either side of the stem at the leaf base that excrete sugars which may attract ants, which in turn protect the plant from harmful insect pests. The plant is more tolerant of saltwater than many other mangrove species, taking in saltwater through its roots and excreting out salt through pores on the surface of its leaves. Although these mangroves can survive in fresh water as well, they generally cannot compete with hardier plants that grow there.

DESCRIPTION: This shrub or tree has a narrow rounded crown. Its trunk is short, crooked, often with "knees." It is thicket-forming. The leaves are 3" (7.6 cm), opposite, oblong to ovate, leathery, and evergreen. The bark is reddish brown and scaly-ridged.

FLOWERS: April–June. Tiny, greenish, and in hairy spikes.

FRUIT: ½" (1.3 cm), leathery, flattish, ribbed, and reddish.

HABITAT: Tidal swamps.

RANGE: Coastal peninsular Florida.

USES: This species provides important habitat for a variety of fish, birds, and other wildlife. Historically, the bark and leaves were used for medicinal purposes.

SIMILAR SPECIES: Black Mangrove (a shrub) and Red Mangrove.

CONSERVATION STATUS: **LC**
The White Mangrove population is declining. Its coastal habitat is coveted for residential and commercial development. Rising sea levels caused by climate change may make this species vulnerable to coastal flooding and salt intrusion.

Height: 50' (15 m)

■ DENSE POPULATION
■ LIGHT POPULATION
■ NATURALIZED

This large tropical tree is found growing in parts of Asia, Africa, and Australia. It is also found in Florida in the United States. The many common names for this tree include Sea Almond, Tropical Almond, and False Kamani.

DESCRIPTION: This is a large tree with whorled spreading branches. The leaves are 12" (30 cm) long, obovate, leathery, and crowded at the twig ends, turning deep red in fall. The bark is scaly and dark gray.

FLOWERS: June–August. Flowers are ⅓" (8.5 mm), greenish to whitish, in narrow spikes.

FRUIT: 2½" (6.4 cm), almond-shaped, woody, yellowish green; the kernel is edible, ripening in August–September.

HABITAT: Mangrove swamps, disturbed hammocks, and beaches.

RANGE: Coastal southern Florida and the Keys.

USES: This species is widely cultivated as an ornamental tree. The fruit is edible and eaten by some people. The wood is sometimes used to make canoes.

SIMILAR SPECIES: Wild Almond.

CONSERVATION STATUS: LC

Indian Almond is widespread in tropical and subtropical coastal habitats worldwide. Most populations of this tree occur near sea level and are at risk from rising sea levels and storm surge.

Height: 50' (15 m)

■ DENSE POPULATION
■ LIGHT POPULATION
■ NATURALIZED

Native to India, Crapemyrtle is a popular ornamental plant grown for its profuse, showy, late-summer blossoms. It is a popular low-maintenance plant in North America, used as a street tree in states with mild winters. Well-established plants are extremely drought tolerant. Many varieties with different flower colors are grown from cuttings as well as from seed. This plant's common name refers to the wrinkled petals. It is not related to Myrtle (*Myrtus communis*) of the Mediterranean region.

DESCRIPTION: This is a cultivated ornamental shrub or a small tree often branching near the base, with slightly angled and curved or crooked trunks and an open, spreading, rounded crown. The leaves are deciduous or evergreen in tropical climates, mostly opposite or upper leaves alternate, often appearing in two rows; they are 1–2" (2.5–5 cm) long, ½–⅞" (12–22 mm) wide, elliptical, without teeth, and nearly stalkless. They are dull green above, and paler and sometimes hairy on the midvein beneath. The bark is mottled gray and brown, and smooth and flaking off in patches. The twigs are light green, turning light brown, long and slender, slightly four-angled, and hairless or nearly so.

FLOWERS: Mid–late summer. 1¼–1½" (3–4 cm) wide with six spreading, rounded, crapelike, fringed, stalked petals, commonly pink but varying from white to red, purple, and bluish; they are odorless, and abundant in showy masses in upright, branched clusters 2½–6" (6–15 cm) long.

FRUIT: ⅜–½" (10–12 mm) in diameter, rounded, brown capsule splitting into six parts, with many small, winged seeds, maturing in fall and remaining attached.

HABITAT: Around houses and long persisting at old home sites, sometimes escaped but not naturalized; in humid, warm temperate to tropical regions.

RANGE: Maryland to Florida and Texas and on the Pacific Coast.

USES: This flowering shrub is commonly planted as an ornamental.

SIMILAR SPECIES: This species is distinct and unlikely to be confused with other plants.

CONSERVATION STATUS: LC

Crapemyrtle has a wide native distribution and no major threats have been identified. It is known to escape cultivation and has naturalized locally in the U.S., mainly in the Southeast, but it is not associated with any significant ecological effects.

Diameter: 4" (10 cm)
Height: 20' (6 m)

- ■ DENSE POPULATION
- ■ LIGHT POPULATION
- ■ NATURALIZED

Eucalyptus trees originated roughly 40 million years ago. The bluegum is one of many species of eucalyptus native to Australia. It is one of the most extensively cultivated species of eucalyptus in subtropical regions of the world and the most common eucalyptus in California among the numerous introduced species. It grows rapidly and sprouts from stumps. It was brought to the West in 1770 by Sir Joseph Banks, who served as botanist on Captain Cook's expedition. Eucalyptus trees contain volatile, combustible oils and are consequently highly inflammable. Burning trees can occasionally explode. Burning Bluegum Eucalyptus has resulted in catastrophic fires, including the Oakland firestorm of 1990.

DESCRIPTION: This exceptionally tall, straight-trunked, introduced tree has a narrow, irregular crown of drooping, evergreen foliage, with an odor of camphor. Its leaves are evergreen, 4–12" (10–30 cm) long, 1–2" (2.5–5 cm) wide, narrowly lance-shaped, long-pointed, usually curved, without teeth, thick and leathery, and hairless; they are dull green on both surfaces. Leaves on young plants are opposite, ovate, stalkless, with bluish or whitish bloom beneath. The bark is mottled gray, brown, and greenish; smooth, peeling in strips; and becoming gray, thick, rough, furrowed, and shaggy. The twigs are yellow-green, slender, angled, hairless, and drooping.

FLOWERS: Winter and spring. 2" (5 cm) wide with many white stamens and without petals with an odor of camphor; they are scattered, single, and almost stalkless at leaf base.

FRUIT: ¾–1" (2–2.5 cm) wide, broad, top-shaped, angled capsules, warty and bluish-white, with three to five narrow openings at the top and many tiny seeds, ripening in spring.

HABITAT: Moist soils in subtropical regions.

RANGE: Native of Australia. Widely planted along streets and roads and in parks in southern and coastal California, with some becoming naturalized in that area.

USES: Eucalyptus trees are used as street trees, for windbreaks and screens, and in forest plantations for fuel and pulpwood. The tree consumes a large volume of water and can be used to dry out mosquito breeding swamps, bogs, and other wet areas. As a fast-growing tree with a straight trunk, it is used as construction timber in South America and elsewhere. It is also grown for the oil distilled from the leaves, is used as a disinfectant in cleaning, for its fragrance, as a decongestant, and as an insecticide.

SIMILAR SPECIES: There are more than 700 eucalypt species, a number of which have been propagated in California. The Bluegum is by far the most numerous.

CONSERVATION STATUS: **LC**
The population is stable, but the danger of catastrophic fire is a continuing concern, especially in Bluegum Eucalyptus groves.

Diameter: 3' (0.9 m)
Height: 120' (37 m)

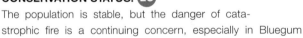

■ DENSE POPULATION
■ LIGHT POPULATION
■ NATURALIZED

This tree, also known as Broad-leaved Paperbark and Punk Tree, was introduced to Florida from Australia in the early 1900s in a misguided effort to help drain wetlands. The species proved exceedingly effective at wetland depletion, as it quickly forms dense, impenetrable hammocks that replace natural freshwater marsh habitat. In addition to damaging native flora, Melaleuca stands are unproductive as wildlife habitat. Ecologists are now battling its spread in the Everglades; the species is classified as a noxious weed by the U.S. government and by many states.

DESCRIPTION: This is a highly invasive, hammock-forming tree with layers of thick, papery, peeling white bark; it may have multiple trunks. The leaves are 1–5" (2.5–12.5 cm) long, narrow, evergreen, and alternate.

FLOWERS: The creamy white flowers are shaped like bottle brushes and occur in elongated clusters at the branch tips.

FRUIT: Small, cylindrical woody capsules densely borne along the twigs.

HABITAT: Aquatic to upland areas of marshes and wet prairies.

RANGE: Native of Australia and Pacific islands. Introduced and naturalized in Florida, mainly in the Everglades; it is also planted in Louisiana, California, and Texas.

USES: This species has been used for medicinal purposes in Australia, and the wood is sometimes used for fencing. It is widely planted globally as an ornamental or street tree. It is a problematic invasive in the United States.

SIMILAR SPECIES: This distinct tree is not likely to be confused with other species.

CONSERVATION STATUS: (LC)
Melaleuca is common throughout its range despite local deforestation and minor threats from various pests and diseases. In North America, the spread of this exotic species can have severe ecological impacts. The U.S. Department of Agriculture classifies this species as a noxious weed, and its expansion in the Everglades in southern Florida is one of the most serious threats to that important native ecosystem.

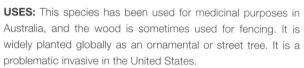

Height: usually 50–70' (15–21 m);
maximum 90' (27 m)

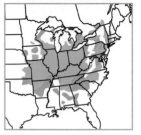

■ DENSE POPULATION
■ LIGHT POPULATION
■ NATURALIZED

This species, usually a medium-sized shrub, is found in moist soils in many parts of the eastern United States and extreme southeastern Canada. This medium- to fast-growing species is sometimes used as an ornamental plant. The genus name *Staphylea* is from Greek and means "cluster of grapes," referring to the flowers. The Latin species name *trifolia*, which means "three-leaf," refers to the leaflets.

DESCRIPTION: This is a shrub or small tree with paired leaves of three leaflets, striped twigs, and drooping, bladderlike seed capsules. The leaves are opposite, palmately compound, 6–9" (15–23 cm) long, with a long slender stalk and three leaflets, with two paired and nearly stalkless and one long-stalked. Leaflets are 1½–3¼" (4–8 cm) long, elliptical, long-pointed at the tip, finely saw-toothed, green and hairy on veins above, and paler and hairy beneath. The bark is gray and smooth, becoming slightly fissured. The twigs are green to brown or gray and often striped, with rings at nodes.

FLOWERS: Spring. ½" (12 mm) long, bell-shaped with five white petals, and long-stalked, in terminal drooping clusters to 4" (10 cm) long.

FRUIT: 1¼–2" (3–5 cm) long, drooping elliptical capsule, swollen, slightly three-lobed and three-celled, green to brown, opening at long-pointed tip, few rounded shiny yellow-brown seeds, maturing in late summer.

HABITAT: Moist soils in the understory of hardwood forests.

RANGE: Ontario to extreme southern Quebec and New Hampshire, south to Florida, west to eastern Oklahoma, and north to Minnesota; to 2000' (610 m).

USES: This species is sometimes planted as an ornamental shrub or tree, and some people consume the nut-like fruit.

SIMILAR SPECIES: Common Hoptree.

CONSERVATION STATUS: 🄛🄒
This species has a wide distribution and large population. No range-wide threats have been identified.

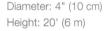
Diameter: 4" (10 cm)
Height: 20' (6 m)

■ DENSE POPULATION
■ LIGHT POPULATION
■ NATURALIZED

Elephant Tree is found mainly in the desert regions of California and Arizona in the United States as well as in northwestern Mexico. The thickened trunk and stout branches of this small, desert-adapted tree recall the legs and trunk of an elephant. This species is the northernmost representative of a small tropical family, and as such it is highly susceptible to frost; cold weather kills back young plants. The innermost bark exudes a reddish sap used as dye and tannin.

DESCRIPTION: This aromatic shrub or tree has a short, thick, sharply tapered trunk with stout, crooked, tapering branches and a widely spreading but sparse, open crown. The leaves are pinnately compound, 1–1¼" (2.5–3 cm) long with a winged axis, and aromatic. There are 15–30 leaflets about ¼" (6 mm) long; they are narrowly oblong, short-pointed at the base, stalkless, and not toothed; they are dull light green on both surfaces. The bark is papery, peeling in thin flakes, and white on the outside; the next thin layers are green, and the inner layers are red and corky. The twigs are reddish-brown.

Diameter: 1' (0.3 m)
Height: 16' (5 m)

FLOWERS: Early summer. Less than ¼" (6 mm) wide with five whitish petals, and short-stalked, with one to three at the leaf base; male and female are on the same tree.

FRUIT: ¼" (6 mm) long, elliptical, red, aromatic, and three-angled, splitting into three parts, drooping on a slender, curved stalk, with one nutlet, and maturing in fall.

HABITAT: Dry rocky slopes of desert mountains.

RANGE: Southwestern Arizona and extreme southern California; also northwestern Mexico; to 2500' (762 m).

USES: This is a popular landscaping plant in the southwestern United States. Some Native Americans used the sap for medicinal purposes.

SIMILAR SPECIES: Unlikely to be confused with other species in its North American range.

CONSERVATION STATUS:

Although Elephant Tree is relatively rare in its U.S. range, its population appears to be stable and does not face an imminent threat of habitat loss because its native range in the desert is fairly remote. Harvesting of branches to collect resin for medicinal and aromatic uses are currently sustainable because demand is low and resin collection is non-destructive.

 DENSE POPULATION
 LIGHT POPULATION
■ NATURALIZED

Gumbo Limbo is one of only three native tree species in the torchwood or bursera family (family Burseraceae) of trees and shrubs in North America. It has several common names including West Indian Birch, Copperwood, and Turpentine Tree (a common name assigned to several unrelated species elsewhere in the world). This tropical tree is ecologically important, adapted to a variety of habitats. It produces fruit year-round and is an important food source for many migratory bird species. Its resistance to wind damage makes it well adapted to surviving tropical storms and hurricanes.

DESCRIPTION: This tree has a spreading, rounded crown; large, crooked branches, and a stout trunk. The leaves are 8" (20 cm) long and pinnately compound, with five to nine ovate, 3" (7.5 cm) leaflets. The bark is reddish brown to green, peeling in papery sheets, and resinous, with a turpentine odor.

FLOWERS: February–April. Tiny, creamy to greenish white, in 4" (10 cm) clusters.

FRUIT: A ⅗" (1.5 cm) long, elliptical, one-seeded drupe; ripe and unripe fruits are present on the tree year-round, but the most fruit is produced from March to April in the North American part of its range.

HABITAT: Hammocks and shell mounds.

RANGE: Central Florida and southward into the tropics; it is mainly coastal.

USES: This species has been used for medicinal purposes, treating a variety of ailments. It is sometimes planted as a wind barrier and the wood may be used for light construction and firewood. The resin is used as glue and varnish.

SIMILAR SPECIES: Unlikely to be confused with other species in its Floridian range.

CONSERVATION STATUS:
The Gumbo Limbo population has a wide distribution and is considered stable. Habitat fragmentation and changing land uses can negatively affect these trees, but are not believed to threaten the species' survival.

Height: 80' (24 m)

■ DENSE POPULATION
■ LIGHT POPULATION
■ NATURALIZED

This is a rare species found in scattered areas of the south-central and southeastern United States, in limited parts of Oklahoma, Texas, Arkansas, Missouri, Tennessee, and Alabama. It is named for the masses of smoke-like fruit clusters with hairy stalks found on sterile flowers. American Smoketree is sometimes grown for its attractive fall foliage and unusual fruit, but the related European Smoketree (*Cotinus coggygria*)—a native of Eurasia—is more commonly planted; it has elliptical, hairless leaves abruptly narrowed at the base.

DESCRIPTION: This is a shrub or small tree with a short trunk, an open crown of spreading branches, resinous sap with a strong odor, and deep orange-yellow heartwood. Its leaves are 2–6" (5–15 cm) long, 1½–3" (4–7.5 cm) wide, obovate or elliptical, short-stalked, rounded or blunt at the tip with straight or slightly wavy edges, dull green above, and paler (and covered with silky hairs when young) beneath, turning orange and scarlet in autumn. The bark is gray to blackish, thin, and scaly. The twigs are slender with a whitish bloom when young, turning brown.

FLOWERS: Spring. ⅛" (3 mm) wide with five greenish-white petals in branched clusters 6" (15 cm) long, with male and female on separate plants. The stalks are covered with short purplish or brownish hairs. The flowers are mostly sterile.

FRUIT: The fruit is ³⁄₁₆" (5 mm) long, oblong, flat, pale brown, dry, one-seeded, and scattered in large branched clusters to 8' (20 cm) long. It matures in summer.

HABITAT: Limestone rocky uplands and ravines; it is found in hardwood forests.

RANGE: Scattered in southeastern Tennessee and northern Alabama, the Ozark Plateau of southwestern Missouri, Arkansas, and eastern Oklahoma, and the Edwards Plateau of central Texas; at 700–2000' (213–610 m).

USES: This species is sometimes planted as an ornamental. The wood was once used for making a yellow dye.

SIMILAR SPECIES: European Smoketree.

CONSERVATION STATUS: **NT**
American Smoketree is relatively rare in its narrow native range. The small population is believed to be stable, and no specific threats to this species have been identified.

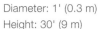

Diameter: 1' (0.3 m)
Height: 30' (9 m)

■ DENSE POPULATION
■ LIGHT POPULATION
■ NATURALIZED

Native to a limited range in southern California and Baja California, Laurel Sumac produces fragrant new leaves and stems even during dry periods. It grows rapidly and is drought-resistant but is damaged by cold winters. Often planted as an ornamental in California, many bird species eat the small, whitish fruits. It is named for the resemblance of its foliage to that of Laurel (*Laurus nobilis*), an unrelated small tree of the Mediterranean region.

ALTERNATE NAME: *Rhus laurina*

DESCRIPTION: This is an aromatic evergreen shrub or small tree with a dense, rounded crown of glossy foliage and an odor of bitter almonds. The leaves are evergreen, 2–4" (5–10 cm) long, ¾–2" (2–5 cm) wide, lance-shaped or ovate, short and sharp-pointed at the tip, rounded at the base, not toothed, and thick and leathery with the edges folded upward. They are shiny green with reddish veins above, and pale green and whitish beneath. The bark is brown or reddish, and smooth. The twigs are reddish, turning brown, and slender and hairless.

Diameter: 6" (15 cm)
Height: 16' (5 m)

FLOWERS: Spring–summer. Less than ¹⁄₁₆" (1.5 mm) wide with five tiny white petals. They are numerous, crowded in many-branched clusters 2–6" (5–15 cm) long, at the end of twigs.

FRUIT: Less than ⅛" (3 mm) in diameter, rounded, whitish, hairless, and waxy, with one flattish seed, maturing in late summer.

HABITAT: Dry slopes, in chaparral and coastal sage scrub.

RANGE: Southern California including Santa Catalina and San Clemente islands, south to Baja California Sur; to 3000' (914 m).

USES: Some Native Americans used this species as a food source and for medicinal purposes. It is commonly planted as an ornamental.

SIMILAR SPECIES: Sugar Sumac, Lemonade Berry.

CONSERVATION STATUS: (NT)

This species occupies a limited range but its numbers are apparently stable. Laurel Sumac is one of the first plants to return to an area following fire. Because of this, the species is commonly impacted where it shares chaparral habitat with the parasitic plant California dodder (*Cuscuta californica*). Sensitive to cold, the Laurel Sumac cannot tolerate prolonged frosts; unusually cold winters in 1949 and 1978 resulted in high mortality rates of this species.

■ DENSE POPULATION
■ LIGHT POPULATION
■ NATURALIZED

This domesticated tree species that originated on the Indian subcontinent is the source of the familiar mango fruit. Introduced to North America as a fruit tree, Mango has escaped from cultivation in some parts of Florida. This species has been grown for thousands of years, first in Southeast Asia and later in Africa and South America, and hundreds of cultivated varieties exist today. Its quality wood is valuable for general carpentry.

DESCRIPTION: This introduced fruit tree has a dense, spreading crown. Its leaves are up to 8" (20 cm) long, crowded (appear whorled), lanceolate to narrowly elliptical, leathery, and evergreen. The bark is gray or dark brown, and rough.

FLOWERS: January–March. ⅓" (8 mm), cup-shaped, yellowish-green, greenish-white, or reddish flowers, occurring in large upright clusters.

FRUIT: May–June. 6" (15 cm), green, one-seeded fruits hang from long stems and turn red or pink, tinged with yellow or green; they are edible.

CAUTION: Contact irritant: This species produces chemicals that may irritate the skin. Although the fruit of the Mango is delicious and commonly eaten, the skin of the fruit contains low levels of urushiols, organic compounds similar to those of Poison Ivy. Wash your hands after enjoying a mango.

HABITAT: Hammocks, disturbed areas.

RANGE: Native to the Indian subcontinent; it is established in Florida locally along the coast from central Florida southward.

USES: This species has long been cultivated for its fruit. It has also been used in traditional medicine and for its wood, which is sometimes used for plywood, furniture, and even ukuleles.

SIMILAR SPECIES: Loquat.

CONSERVATION STATUS:
This domesticated species has not been formally evaluated for conservation threats.

Height: to 50' (15 m)

■ **DENSE POPULATION**
■ **LIGHT POPULATION**
■ **NATURALIZED**

Native to southern Florida and southward, Poisonwood is a member of the cashew family (family Anacardiaceae), which includes trees, shrubs, and a few woody vines with resinous sap, often found in bark and other parts. This species is related to Poison Sumac and Poison Ivy and, like these familiar plants, produces the contact irritant urushiol that can cause a skin rash. This species can grow abundantly as an understory shrub in southern Florida and the Keys; it is also commonly known as Florida Poisontree.

DESCRIPTION: This is a shrub to a small tree with a low, broad crown. The leaves are up to 8" (20 cm) long, black-spotted, leathery, evergreen, and pinnately compound, with three to seven ovate, 3" (7.5 cm) leaflets. The bark is reddish-brown and scaly.

FLOWERS: April–June. Tiny, greenish yellow flowers arranged in 8" (20 cm) clusters.

FRUIT: ½" (12 mm), berrylike, and orange-yellow.

CAUTION: Contact irritant: This species produces chemicals that irritate the skin. Do not touch. If you accidentally touch this species, wash the affected area thoroughly as soon as possible.

HABITAT: Pinelands, hammocks, and dunes.

RANGE: Coastal southern Florida and throughout the Caribbean.

USES: This species is sometimes planted as an ornamental but generally is not recommended for landscaping due to its toxicity. It is useful for natural habitat restorations.

SIMILAR SPECIES: Black Poisonwood.

CONSERVATION STATUS: NT
This species occurs in a limited range in North America; more research is needed to understand possible threats to this species.

Height: 40' (12 m)

■ DENSE POPULATION
■ LIGHT POPULATION
■ NATURALIZED

This widespread eastern species is easily distinguishable from other sumacs by the winged leaf axes and watery sap. It often forms thickets. Also known as Shining Sumac, this species is sometimes planted as an ornamental for its shiny leaves and showy fruit. The fruit is edible but sour, and can be made into a drink that resembles lemonade. It is a good plant for wildlife, providing dense cover for birds and small mammals, as well as food; many forms of native wildlife eat the fruit, and deer sometimes browse the twigs.

ALTERNATE NAME: *Rhus copallina*

DESCRIPTION: This is a shrub or small tree with a short trunk and an open crown of stout, spreading branches. The leaves are pinnately compound and up to 12" (30 cm) long, with a flat, broad-winged axis. There are 7–17 leaflets (27 in the southeastern variety) 1–3¼" (2.5–8 cm) long; they are lance-shaped, usually without teeth, and slightly thickened. They are shiny dark green and nearly hairless above, and paler and covered with fine hairs beneath, turning dark reddish-purple in autumn, and stalkless. The bark is light brown or gray and scaly. The twigs are brown, stout, slightly zigzag, and covered with fine hairs, with watery sap.

FLOWERS: Late summer. ⅛" (3 mm) wide with five greenish-white petals, crowded in spreading clusters to 3" (13 cm) wide, with hairy branches; male and female flowers are usually on separate plants.

FRUIT: More than ⅛" (3 mm) in diameter, one-seeded, crowded in clusters, rounded and slightly flattened, dark red, covered with short sticky red hairs, and maturing in the fall and remaining attached in the winter.

HABITAT: Open uplands, valleys, edges of forests, grasslands, clearings, roadsides, and waste places.

RANGE: Southern Ontario east to southwestern Maine, south to Florida, west to central Texas, and north to Wisconsin; to 4500' (1372 m) in the Southeast.

USES: Winged Sumac is sometimes planted as an ornamental.

SIMILAR SPECIES: Prairie Sumac, Brazilian Pepper.

CONSERVATION STATUS: 🅛🅒
The Winged Sumac has a large population that spans a wide range. The species is considered stable and does not face major threats.

Diameter: 6" (15 cm)
Height: 25' (7.6 m)

■ DENSE POPULATION
■ LIGHT POPULATION
■ NATURALIZED

Smooth Sumac has the distinction of being the only shrub or tree species native to all 48 contiguous U.S. states. It also occurs in the southern tier of Canada and the northern tier of Mexico. The sour fruit is mostly a large seed, but it can be chewed or prepared as a drink similar to lemonade. Birds of many kinds as well as small mammals consume the fruits, mainly in winter. Deer browse the twigs and fruit throughout the year.

DESCRIPTION: This is the most common sumac, a large shrub or sometimes a small tree with an open, flattened crown of a few stout, spreading branches and with whitish sap. Its leaves are pinnately compound 12" (30 cm) long, with a slender axis. There are 11–31 leaflets 2–4" (5–10 cm) long, lance-shaped, saw-toothed, hairless, and almost stalkless. They are shiny green above and whitish beneath, turning reddish in fall. The bark is brown and smooth or becoming scaly. The twigs are gray with a whitish bloom, stout, and hairless.

FLOWERS: Early summer. Less than ⅛" (3 mm) wide with five whitish petals, crowded in large upright clusters to 8" (20 cm) long, with hairless branches; male and female flowers are usually on separate plants.

FRUIT: The one-seeded fruits are more than ⅛" (3 mm) in diameter, rounded, and numerous, crowded in upright clusters; they are dark red and covered with short, sticky, red hairs. Fruits mature in late summer and remain attached in winter.

HABITAT: Open uplands including edges of forests, grasslands, clearings, roadsides, and waste places, especially in sandy soils.

RANGE: Eastern Saskatchewan east to southern Ontario and Maine, south to northwest Florida, and west to central Texas; it is also found in the mountains from southern British Columbia south to southeast Arizona and in northern Mexico. It occurs to 4500' (1372 m) in the East.

USES: This species is an important food source for a variety of wildlife. Native Americans ate the raw young sprouts as salad. It is also planted as an ornamental, particularly popular for its red fall foliage.

SIMILAR SPECIES: Staghorn Sumac, Winged Sumac.

CONSERVATION STATUS: 🅛🅒
Smooth Sumac's large population spans a wide native range and is considered stable. Although native to North America, this species can encroach on wetlands and prairies, especially when facilitated by agricultural runoff and grazing in these habitats. It is sometimes subject to control measures.

Diameter: 4" (10 cm)
Height: 20' (6 m)

■ DENSE POPULATION
■ LIGHT POPULATION
■ NATURALIZED

This plant's native range is limited to southern California and Baja California, where it usually occurs in canyons and on dry slopes. Another common name for this species is Lemonade Sumac. As its name implies, a sour, refreshing drink can be made from the acidic fruit. Greater Roadrunners (*Geococcyx californianus*) and other birds often consume the fruits, and the plant is an important nectar source for hummingbirds and butterflies.

DESCRIPTION: This is an evergreen, aromatic, rounded, thicket-forming shrub, or, rarely, a small tree with a short, stout trunk and many branches. The leaves are evergreen, 1–2½" (2.5–6 cm) long, ¾–1½" (2–4 cm) wide, elliptical, mostly without teeth or with short, sharp teeth, sometimes with one or two lobes at the base or, rarely, with three leaflets; they are thick and stiff and nearly hairless; they are shiny dark green above and paler with raised veins beneath. The bark is reddish-brown and shedding in large plates or scales. The twigs are reddish, stout, and finely hairy.

FLOWERS: Late winter–early spring. ¼" (6 mm) wide with five pink to white petals, with many crowded in much-branched clusters 1–3" (2.5–7.5 cm) long, occurring upright at the end of the twig. Male and female are usually on different plants.

FRUIT: ½" (12 mm) long, elliptical, berrylike, flattened, sour, dark red, densely hairy, and covered with whitish secretion; they are resinous and sticky and one-seeded, maturing in summer.

HABITAT: Dry, sandy and rocky soils of beaches, ocean bluffs, and canyons; it is found in coastal sage scrub and chaparral communities.

RANGE: Coastal southern California including islands; it is also found in Baja California. It occurs to 2500' (762 m).

USES: The berries of this species have been used to make lemonade flavored drinks. This plant is sometimes used as an ornamental shrub.

SIMILAR SPECIES: Laurel Sumac, Sugarbush.

CONSERVATION STATUS: NT
This species occurs in a limited range with specific habitat needs, but its population is apparently secure. Although sensitive to cold, these plants typically regenerate well after frost damage. Additional research is needed to identify other potential threats to this species.

Diameter: 8" (20 cm)
Height: 20' (6 m)

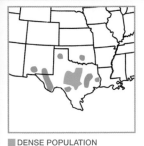

■ DENSE POPULATION
■ LIGHT POPULATION
■ NATURALIZED

This plant, also called Flameleaf Sumac, has a somewhat limited native range in Texas, Oklahoma, Arizona, New Mexico, and northeastern Mexico. Birds—especially bobwhites, grouse, and pheasants—consume large quantities of the fruit in winter; deer browse the foliage. The leaves contain tannin and were once used as a replacement for oak bark in tanning leather.

DESCRIPTION: This is a large shrub or small tree with a short trunk and an open, rounded crown of foliage turning flame-colored in autumn. Its leaves are pinnately compound and 9" (23 cm) long, with a flat, narrowly winged axis. There are usually 13–19 leaflets 1–2½" (2.5–6 cm) long, less than ½" (12 mm) wide, paired (except at the end), narrowly lance-shaped, slightly curved, long-pointed at the tip, blunt and unequal at the base, and usually without teeth. They are shiny dark green above, and paler, covered with fine hairs, and with prominent veins beneath, turning reddish-purple in fall. The bark is gray or brown and smooth or becoming scaly. The twigs are green or reddish and hairy when young, becoming gray and hairless, and stout, ending in a whitish hairy bud.

FLOWERS: Summer. ³⁄₁₆" (5 mm) wide with 5 greenish-white petals, crowded in upright clusters to 6" (15 cm) long, composed of many flowers, and occurring at the ends of the twigs; male and female flowers are usually on separate plants.

FRUIT: About ³⁄₁₆" (5 mm) in diameter, slightly flattened, dark red, covered with short sticky red hairs, and one-seeded; they are numerous and crowded in clusters, maturing in early fall and falling in early winter.

HABITAT: Dry rocky slopes and hills, especially limestone; often forming thickets.

RANGE: Texas and southern New Mexico, local in southern Oklahoma and in northeastern Mexico; to 2500' (762 m); locally to 4000' (1219 m).

USES: The tannin from the leaves of this plant has been used to tan leather.

SIMILAR SPECIES: Winged Sumac.

CONSERVATION STATUS: NT
This species occupies a limited native range. Although the population is believed to be secure, it occurs in scattered, disjunct populations that could be vulnerable. Additional research is needed to understand its status in the wild and to identify potential threats.

Diameter: 6" (15 cm)
Height: 25' (7.6 m)

■ DENSE POPULATION
■ LIGHT POPULATION
■ NATURALIZED

This evergreen species commonly grows in chaparral in dry canyons and on low-elevation slopes in Southern California, Arizona, and Baja California. The edible fruit contains a large, inedible seed, but the thin pulp is sweet. This is an important plant for wildlife, providing food and cover for numerous native species of birds and butterflies. It is sometimes planted for erosion control and landscaping in mountainous areas, and makes for an attractive ornamental.

DESCRIPTION: This is an evergreen shrub or small tree with a rounded crown. Its leaves are evergreen, 1½–3¼" (4–8 cm) long, 1–2" (2.5–5 cm) wide, ovate, short-pointed at the tip, rounded at the base, without teeth, and thick and leathery; they are curved or folded up at the midvein, and shiny light green on both surfaces. The bark is gray-brown, rough, shaggy, and extremely scaly. The twigs are reddish, stout, and hairless.

FLOWERS: Early spring. ¼" (6 mm) wide with five rounded, whitish petals, from pink or reddish buds, with many crowded in clusters 2" (5 cm) long, occurring at the end of the twig.

FRUIT: More than ¼" (6 mm) in diameter, slightly flattened, one-seeded, red, and covered with short, sticky, red hairs, with many occurring in clusters, and maturing in summer.

HABITAT: Dry slopes in chaparral zones.

RANGE: The mountains of central Arizona and southern California including Santa Cruz and Santa Catalina islands south to northern Baja California. It occurs from near sea level to 2500' (762 m), and in Arizona to 5000' (1524 m).

USES: Some Native Americans used this plant for medicinal purposes and as a sweetener. It is sometimes planted as an ornamental. The edible fruit is used to make lemonade-like drinks.

SIMILAR SPECIES: Lemonade Berry.

CONSERVATION STATUS: NT

The Sugarbush is highly drought-tolerant and recovers well after fire. Although additional information is needed on the nature and level of risks, the Sugarbush is likely to be threatened by habitat loss related to climate change. The species is also susceptible to some fungal diseases.

Diameter: 5" (13 cm)
Height: 15' (4.6 m)

■ DENSE POPULATION
■ LIGHT POPULATION
■ NATURALIZED

Native to eastern North America, Staghorn Sumac reaches tree size more often than many other sumacs, and commonly forms thickets. In winter, its leafless, forking, stout, twigs covered in long hairs resemble the antlers of a male deer "in velvet." Staghorn Sumac is widely grown as an ornamental for the bright fall foliage and showy fruit; a popular cultivated variety 'Dissecta,' sometimes called Cutleaf Staghorn Sumac, has dissected leaves.

ALTERNATE NAME: *Rhus hirta*

DESCRIPTION: This is a tall shrub or small tree with an irregular, open, flat crown of a few stout, spreading branches, and whitish, sticky sap that turns black on exposure. The leaves are pinnately compound and 12–24" (30–61 cm) long, with a stout, soft, and hairy reddish-tinged axis. There are 11–31 leaflets 2–4" (5–10 cm) long; they are lance-shaped, often slightly curved, saw-toothed, and nearly stalkless. They are dark green above and whitish (with reddish hairs when young) beneath, turning bright red with purple and orange in fall. The bark is dark brown, thin, and smooth or becoming scaly. The twigs are few, rather stout, brittle, and densely covered with long, velvety, brown hairs.

Diameter: 8" (20 cm), sometimes larger
Height: 30' (9 m)

FLOWERS: Early summer. ⅛–³⁄₁₆" (3–5 mm) wide with greenish petals, crowded in upright clusters to 8" (20 cm) long; male and female flowers are usually on separate plants.

FRUIT: ³⁄₁₆" (5 mm) in diameter, rounded, one-seeded, dark red, and covered with long dark red hairs; they are numerous and crowded in upright clusters, maturing in late summer and fall, and remaining attached in winter.

HABITAT: Open uplands, edges of forests, roadsides, and old fields.

RANGE: Southern Ontario east to Nova Scotia, south to northwestern South Carolina, west to Tennessee, and north to Minnesota; to 5000' (1524 m) in the southeastern United States.

USES: This species is an important food source for a variety of wildlife. Native Americans made a lemonade-like drink from the crushed fruit of this and related species. Sumac bark and foliage are rich in tannin and were used to tan leather. It is also planted as an ornamental, particularly popular for its red fall foliage.

SIMILAR SPECIES: Smooth Sumac.

CONSERVATION STATUS: LC
The Staghorn Sumac has a stable, widespread population. The species faces no major threats.

■ DENSE POPULATION
■ LIGHT POPULATION
■ NATURALIZED

This exotic ornamental from Peru and nearby countries in north-ern South America is sometimes called California Peppertree because of its popularity for planting along streets and highways in that state. The abundant red fruit has a peppery taste and is showy in fall. Although heat- and drought-resistant, Peruvian Peppertree has shallow roots that grow to crack pavement and damage sewers, and the species has a habit of escaping from cultivation and becoming invasive. The bark, gum from the trunk, leaves, and fruit have a long history of use in traditional medicine in Latin America.

DESCRIPTION: This cultivated and natural-ized evergreen tree has a short, often gnarled trunk and a widely spreading, rounded crown of drooping branches and fine foliage. It is aro-matic and resinous, with milky sap. The leaves are pinnately compound, 6–12" (15–30 cm) long, and droop-ing, with milky sap. They have 19–41 leaflets that are generally paired and 1–2" (2.5–5 cm) long, narrowly lance-shaped, slightly curved at the tip, sometimes slightly toothed, hairless or nearly so, stalkless, and yellow-green on both surfaces. The bark is light brown and scaly. The twigs are brownish, slender, and hair-less or finely hairy.

FLOWERS: ⅛" (3 mm) wide with five yellowish-white petals, with many in branched clusters 5–6" (13–15 cm) long; male and female are on separate trees.

FRUIT: ³⁄₁₆–¼" (5–6 mm) in diameter, berrylike, reddish or pink-ish, shiny, resembling beads, resinous, juicy, and one-seeded, with many hanging down on short stalks. They remain attached in winter.

HABITAT: Hardy in a wide range of soils including alkaline; it is found in subtropical regions.

RANGE: A native of South America. It has been planted in southern Texas, Arizona, and California, and has become natu-ralized in extreme southern Texas and California.

USES: The berries are sometimes used as peppercorns, blended with commercial pepper. This species has also been used in traditional medicine, and its wood used to fashion sad-dles and for fuel.

SIMILAR SPECIES: This distinct species is not likely to be con-fused with other trees in North America.

CONSERVATION STATUS: LC

The fast-growing Peruvian Peppertree is a hardy pioneer spe-cies that can become seriously invasive outside of its native range. Due to its rate of growth, the Peruvian Peppertree can pose a risk to native vegetation in savannas, grasslands, and disturbed habitats. The species has no significant conservation threats in the wild.

Diameter: 1' (0.3 m)
Height: 40' (12 m)

■ DENSE POPULATION
■ LIGHT POPULATION
■ NATURALIZED

This small tree native to tropical and subtropical South America is among the most widespread and damaging exotic plants in Florida. It forms dense monocultures that crowd out native flora, creating habitats that don't support native wildlife. Its leaves exude toxins that prohibit the growth of other plants. Although it was once planted as an ornamental, it is now considered a noxious weed and its use is prohibited by law in some states.

DESCRIPTION: This introduced, highly invasive understory tree or shrub has an uneven crown of arched interlaced branches; it forms densely vegetated thickets. The leaves are 6" (15 cm) long, and pinnately compound, with three to eleven elliptical to lanceolate, toothed, 2" (5 cm) leaflets; they are evergreen and exude an odor resembling turpentine when crushed. The bark is grayish brown and smooth.

FLOWERS: September–October. Yellowish clusters.

FRUIT: Tiny, berrylike, and red, in abundant clusters on female plants, persisting through winter.

Height: to 30' (9 m)

CAUTION: Contact irritant: This species produces chemicals that irritate the skin. Do not touch. If you accidentally touch this species, wash the affected area thoroughly as soon as possible.

HABITAT: Hammocks, mangrove swamps, pinelands, and disturbed areas.

RANGE: Native of Brazil, Argentina, and Paraguay; introduced in the United States and naturalized in Florida, less so in Texas and California.

USES: This South American native has been widely planted as an ornamental tree and is now considered an invasive pest in many areas.

SIMILAR SPECIES: Winged Sumac.

CONSERVATION STATUS: (NT)

Although a problematic invasive in North America, the Brazilian Pepper is rare and sparse in dry savannas in its native range. Because of its hardiness, the Brazilian Pepper has few threats. Female trees are capable of abundant seed production and can form dense monospecific stands. In North America, the invasion of Brazilian Pepper into rare ecological communities in Florida and Hawaii threatens multiple native species, including several that are considered endangered.

■ DENSE POPULATION
■ LIGHT POPULATION
■ NATURALIZED

The notorious Poison Sumac may be the most toxic plant species in North America; all parts of the plant contain a high concentration of the contact irritant urushiol, which causes skin and mucous membrane irritation in humans. Smoke produced by burning this plant can be extremely painful and even deadly if inhaled. The fruit does not seem to be toxic to birds or other animals, however, and many kinds of wildlife—including bobwhites, pheasants, grouse, and rabbits—eat the fruit, especially in winter when other food is scarce. This plant is also commonly known as Thunderwood.

ALTERNATE NAME: *Rhus vernix*

DESCRIPTION: This is a poisonous yet attractive narrow-crowned shrub or small tree with waxy whitish berries and dramatic fall foliage. The leaves are pinnately compound, 7–12" (18–30 cm) long with a reddish axis, with five to thirteen leaflets 2½–3½" (6–9 cm) long and paired except at the end; they are ovate or elliptical, without teeth, and short-stalked. They are shiny dark green above, and paler and slightly hairy beneath, turning scarlet or orange in early fall. The bark is gray or blackish, thin, and smooth or slightly fissured. The twigs are hairless, and reddish when young, turning gray with many orange dots.

FLOWERS: Early summer. ⅛" (3 mm) long with five greenish petals, with many arranged, in long, open, branching clusters to 8" (20 cm) long; male and female may be on the same or separate plants.

FRUIT: ¼" (6 mm) in diameter, rounded and slightly flat, whitish, one-seeded, shiny, and hairless, with many arranged in drooping branched clusters, and maturing in early fall and often remaining attached until spring.

CAUTION: Contact irritant: This species produces chemicals that irritate the skin. Do not touch. If you accidentally touch this species, wash the affected area thoroughly as soon as possible.

HABITAT: Wet soil of swamps, bogs, seepage slopes, and frequently flooded areas; in shady hardwood forests.

RANGE: Extreme southern Quebec and Maine south to central Florida, west to eastern Texas, and north to southeastern Minnesota; mostly confined to Atlantic and Gulf coastal plains and Great Lakes region; to 1000' (305 m).

USES: This species is an important food source for a variety of wildlife. A black varnish can be made from the sap.

SIMILAR SPECIES: May be confused with other sumac species, but note the whitish (rather than red) fruit of Poison Sumac.

CONSERVATION STATUS: LC
Poison Sumac has an extensive range with no threats known to the species. Although some areas of its population are thought to be in decline, the population on a global scale is stable.

Diameter: 6" (15 cm)
Height: 25' (7.6 m)

Commonly known as Field Maple in its native range, this small tree originated in Europe and southwestern Asia. It has been widely planted as a landscaping ornamental tree in the United States and Australia, and has naturalized locally in some areas.

DESCRIPTION: This is a small tree with one or more short, often twisted trunks and a dense, low, rounded crown. The leaves are 3–4" (7.6–10 cm) long with three or five deeply divided, rounded lobes. They are leathery, dark green above, and pale and hairy beneath. Their autumn color is gold, then red or purple. The leafstalk is red and it exudes milky sap when broken. The bark is corky and pale brown, cracking into squares and wide, orange fissures; the branchlets develop corky ridges.

FLOWERS: Spring. Yellow-green.

FRUIT: 1" (2.5 cm) long, paired samaras, spread at almost 180-degree angle, and red-tinged, maturing in late summer.

Diameter: 2–3' (61–91 cm)
Height: 35–50' (11–15 m)

HABITAT: Introduced in North America; it is planted along streets and in landscapes in the East and Northwest.

RANGE: Native to Europe and Asia. It is introduced in North America, where it has been widely planted especially in the East and Northwest.

USES: This introduced maple is widely cultivated as an ornamental tree. The wood has been used for furniture, flooring, and musical instruments.

SIMILAR SPECIES: The rounded lobes of the leaves help to distinguish this tree from native maples.

CONSERVATION STATUS: LC
Although its population is stable in its native range, the species is highly susceptible to the Asian Long-horned Beetle (*Anoplophora glabripennis*), which may threaten the species in the future if infestations are not adequately managed. Hedge Maple is also vulnerable to certain fungal diseases and other pests, including aphids and gall mites. In North America, it rarely escapes from cultivation and tends to be less invasive than other exotic maple species.

■ DENSE POPULATION
■ LIGHT POPULATION
■ NATURALIZED

This native maple of the Pacific Northwest is dramatically colored in most seasons with bright green foliage turning orange and red in fall, purple and white flowers in spring, and young red fruit in summer. A wide variety of birds and mammals consume the seeds of this and related maples. It is often planted as an ornamental. The scientific name of the species, which means "rounded" or "circular," refers to the shape of the leaves.

DESCRIPTION: This is a shrub or small tree with a short trunk or several branches turning and twisting from the base; it is often vine-like and leaning or sprawling. The leaves are opposite, 2½–4½" (6–11 cm) long and wide, rounded, with seven to eleven long-pointed lobes, sharply doubly toothed, with seven to eleven main veins from the notched base, and with long leafstalks. They are bright green above and paler with tufts of hairs in vein angles beneath, turning orange and red in fall. The bark is gray or brown and smooth or finely fissured. The twigs are green to reddish-brown with a whitish bloom, and slender.

FLOWERS: Early spring. ½" (12 mm) wide, spreading purple sepals and whitish petals in broad, branched clusters at the end of short twigs. Male and female are usually on the same plant.

FRUIT: 1½" (4 cm) long, paired, long-winged keys spreading almost horizontally; it is reddish when young, one-seeded, and matures in fall.

HABITAT: Moist soils, especially along shaded stream banks; in understory of coniferous forests.

RANGE: Pacific Coast from southwest British Columbia south to northern California; to 5000' (1524 m).

USES: This species is an important food source for a variety of native wildlife species. It is sometimes cultivated as an ornamental tree.

SIMILAR SPECIES: Rocky Mountain Maple, Fullmoon Maple, Korean Maple. The latter two are native to East Asia and may be confused with Vine Maple in cultivation.

CONSERVATION STATUS:
The widespread Vine Maple is considered stable. There are currently no known threats to the species.

Diameter: 8" (20 cm)
Height: 25' (7.6 m)

■ DENSE POPULATION
■ LIGHT POPULATION
■ NATURALIZED

This shade-tolerant plant is native to northeastern Asia. With its seeds carried by the wind and randomly deposited, this exotic maple has the potential to spread into native habitats. It is considered an invasive plant in many parts of North America, where it has infiltrated some native prairie and woodland habitats. It is sometimes grown as an ornamental plant, but North America's many native maples make excellent additions to the garden while also being better for native habitats and wildlife.

DESCRIPTION: This is a multi-stemmed invasive shrub or small tree with smooth gray-brown bark, becoming rough and furrowed with age. Its leaves are 1½–3" (4–7.5 cm) long and palmately lobed, with a very long-pointed center lobe, coarsely-toothed leaf edges, and dark green, turning red in fall.

FLOWERS: Spring. Tiny, yellow, sweet-smelling flowers in small clusters.

FRUIT: ¾–1" (2–2.5 cm) long, paired, parallel-winged keys; they are pink to red in summer and remain on the tree into winter.

HABITAT: Disturbed habitats, prairies, woodlands.

RANGE: Native to China and Japan; it has escaped from cultivation in the northeastern United States, reported from Maine to Kentucky, Iowa, and North Dakota.

USES: This species has been cultivated as an ornamental but is now considered an invasive pest. It is also used for bonsai.

SIMILAR SPECIES: Tatarian Maple (Amur Maple is sometimes considered a subspecies of Tatarian Maple).

CONSERVATION STATUS:
The species has not yet been formally evaluated in the wild. In North America, Amur Maple has escaped from cultivation in New England, the Mid-Atlantic, and the Midwest. Its shade tolerance and abundant seed production has allowed it to invade mature forests and colonize aggressively.

Height: 15–20' (4.5–6 m)

■ DENSE POPULATION
■ LIGHT POPULATION
■ NATURALIZED

The Rocky Mountain Maple's range is scattered throughout western North America and extends through southeastern Alaska, making it the northernmost native maple on the continent. Deer, elk, and livestock often browse the foliage. There are four to six geographic varieties that some authorities treat as subspecies. Other common names for this plant and its varieties include Douglas Maple and Torrey Maple. The Latin species name means "hairless," in reference to the plant's leaves.

DESCRIPTION: This is a shrub or small tree with a short trunk and slender, upright branches, hairless throughout. The leaves are opposite, 1½–4½" (4–11 cm) long and wide, with three short-pointed lobes (sometimes five) or divided into three lance-shaped leaflets; they are doubly saw-toothed, with three or five main veins from the base, with long, reddish leafstalks. They are shiny dark green above and paler or whitish beneath, turning red or yellow in fall. The bark is gray or brown, smooth, and thin. The twigs are reddish-brown and slender.

FLOWERS: Spring. ¼" (6 mm) wide and greenish-yellow, with four narrow sepals and four petals on drooping stalks, occurring in branched clusters, usually with male and female on separate plants.

FRUIT: ¾–1" (2–2.5 cm) long, paired, forking, long-winged keys; they are reddish, turning light brown, and one-seeded, maturing in late summer or fall.

HABITAT: Moist soils, especially along canyons and mountain slopes in coniferous forests.

RANGE: Southeast Alaska, British Columbia, and southwestern Alberta, south mostly in the mountains to southern New Mexico and southern California; to 5000–9000' (1524–2743 m) in the south.

USES: Some Native Americans used this plant for medicinal purposes.

SIMILAR SPECIES: Vine Maple.

CONSERVATION STATUS: 🔵 LC

The species has a large population with a widespread native range that is considered stable. Although currently unthreatened, the Rocky Mountain Maple is a fire-dependent species that could see a population decrease in the future due to fire suppression and exclusion techniques. The species also faces potential future threats from climate-change related habitat loss.

Diameter: 1' (0.3 m)
Height: 30' (9 m)

■ DENSE POPULATION
■ LIGHT POPULATION
■ NATURALIZED

The western relative of Sugar Maple, Canyon Maple has sweetish sap used locally to prepare maple sugar. It occurs in scattered populations throughout the interior western United States. The showy fall foliage makes it an attractive ornamental. Canyon Maple grows well in cultivation and has few pests, and is tolerant to both drought and cold temperatures. The scientific name means "large-toothed," a reference to the leaves. It is sometimes called Bigtooth Maple.

DESCRIPTION: This is a small to medium-sized tree with a short trunk and a spreading, rounded, dense crown; it is often a shrub. The leaves are opposite, 2–3¼" (5–8 cm) long and wide, with three broad, blunt lobes and two small basal lobes, a few blunt teeth, and three or five main veins from a notched or straight base; they are slightly thickened with long leafstalks, shiny dark green above and pale and finely hairy beneath, and turning red or yellow in autumn. The bark is gray or dark brown, thin, and smooth or scaly. The twigs are reddish, slender, and hairless.

FLOWERS: With new leaves in early spring. ³⁄₁₆" (5 mm) long with bell-shaped, five-lobed, yellow calyx, occurring in drooping clusters on long, slender, hairy stalks; male and female may be in the same or different clusters.

FRUIT: 1–1¼" (2.5–3 cm) long, paired, forking, long-winged keys; they are reddish or green, mostly hairless, and one-seeded, maturing in fall.

HABITAT: Moist soils of canyons in mountains and plateaus; it is found in woodlands.

RANGE: Southeastern Idaho south to Arizona and east to southern New Mexico and Trans-Pecos Texas; it is local in Edwards Plateau of south-central Texas, southwest Oklahoma, and northern Mexico. It occurs at 4000–7000' (1219–2134 m), and locally to 1500' (457 m).

USES: This maple is sometimes planted as an ornamental tree. The sap is used locally to make maple sugar. The wood provides good fuel.

SIMILAR SPECIES: Sugar Maple, Rocky Mountain Maple.

CONSERVATION STATUS: LC

Canyon Maple faces habitat loss in parts of its range, but the population is still considered widespread and stable. Current habitat loss is primarily the result of human activities; however, the species potentially faces an increased risk of habitat loss due to the effects of climate change.

Diameter: 8" (20 cm)
Height: 40' (12 m)

BIGLEAF MAPLE *Acer macrophyllum*

■ DENSE POPULATION
■ LIGHT POPULATION
■ NATURALIZED

As its name implies, Bigleaf Maple has the largest leaves of any of our maples. This species can form pure stands in moist soil near water, and is common in various types of mixed woodlands. Grosbeaks and other wildlife species often eat the winged fruits. It is a handsome shade tree and makes an excellent native planting on the Pacific Coast; it is particularly showy in the fall when its leaves turn bright gold and yellow.

DESCRIPTION: This is a small to large tree with a broad, rounded crown of spreading or drooping branches and the largest leaves of all maples. Those leaves are opposite, 6–10" (15–25 cm) long and wide, and rounded, with five deep, long-pointed lobes (sometimes three); the edges have a few small, blunt lobes and teeth; they have five main veins and are slightly thickened; they are shiny dark green above and paler and hairy beneath, turning orange or yellow in autumn. The leafstalks are up to 10" (25 cm) and stout, with milky sap when broken. The bark is brown and furrowed into small four-sided plates. The twigs are green, stout, and hairless.

FLOWERS: Spring. ¼" (6 mm) long, with many occurring on slender stalks; they are yellow and fragrant, with the male and female together in narrow, drooping clusters to 6" (15 cm) long at the end of a leafy twig.

FRUIT: 1–1½" (2.5–4 cm) long, paired, long-winged keys; they are brown with stiff yellowish hairs, and one-seeded, maturing in fall.

HABITAT: Stream banks and in moist canyon soils; sometimes in pure stands.

RANGE: British Columbia to southern California; it occurs to 1000' (305 m) in the north and at 3000–5500' (914–1676 m) in the south.

USES: This is the only western maple species with wood of significant commercial importance, commonly used for veneer, furniture, handles, woodenware, and novelties. The sap is used to make maple syrup.

SIMILAR SPECIES: Vine Maple.

CONSERVATION STATUS: LC
The Bigleaf Maple does not face any major threats and has a widespread population. Several pests and diseases can affect these trees, but none appear to threaten their numbers. Habitat loss due to climate change poses a possible future threat to the species.

Diameter: 1–2½' (0.3–0.8 m)
Height: 30–70' (9–21 m)

■ DENSE POPULATION
■ LIGHT POPULATION
■ NATURALIZED

Box Elder is among the most widespread tree species in North America. It is classed with maples, having similar, paired key fruits, but is easily distinguishable by the pinnately compound leaves.This common and widely distributed native species is spreading in eastern North America as a weed tree, and can become invasive in some situations. Hardy and fast-growing, it has been planted for shade and shelterbelts; despite being short-lived and easily broken in storms, it remains popular in cultivation. The common name indicates the resemblance of the foliage to that of elders (*Sambucus*) and the whitish wood to that of Boxwood (*Buxus sempervirens*).

DESCRIPTION: This is a small to medium-sized tree with a short trunk and a broad, rounded crown of light green foliage. The leaves are opposite, pinnately compound, 6" (15 cm) long, with a slender axis and three to seven leaflets that are sometimes slightly lobed and 2–4" (5–10 cm) long and 1–1½" (2.5–4 cm) wide; they are paired and short-stalked (except at the end), ovate or elliptical, long-pointed at the tip, short-pointed at the base, coarsely saw-toothed, and sometimes lobed. They are light green and mostly hairless above and paler and varying in hairiness beneath, turning yellow (or sometimes red) in autumn. The bark is light gray-brown with many narrow ridges and fissures, becoming deeply furrowed. The twigs are green, often whitish or purplish, slender, ringed at the nodes, and mostly hairless.

FLOWERS: Before leaves in early spring. ³⁄₁₆" (5 mm) long with very small yellow-green calyx of five lobes or sepals, with several clustered on slender drooping stalks; male and female are on separate trees.

FRUIT: 1–1½" (2.5–4 cm) long, paired, slightly forking keys with a flat, narrow body and a long, curved wing; they are pale yellow and one-seeded, maturing in summer and remaining attached in winter.

HABITAT: Wet or moist soils along stream banks and in valleys, with various hardwoods; also naturalized in waste places and roadsides.

RANGE: Southern Alberta east to extreme southern Ontario and New York, south to central Florida, and west to southern Texas; also scattered from New Mexico to California and naturalized in New England; to 8000' (2438 m) in the Southwest.

USES: The wood from this tree is soft and relatively worthless, but is sometimes used commercially for fiberboard. Some Native Americans used the wood to make bowls, drums, and other items, and made sugar from the sap. This tree is sometimes planted for shade.

SIMILAR SPECIES: Poison Ivy has similar leaves.

CONSERVATION STATUS: 🔵 **LC**

The Box Elder population is large and considered stable. This tree can be weedy or invasive, especially in areas where it has been introduced. Box Elder is highly susceptible to the invasive Asian Long-horned Beetle (*Anoplophora glabripennis*). Habitat changes for the Box Elder's native range have been predicted to decline based on climate changes which also creates a potential threat.

Diameter: 2½' (0.8 m)
Height: 30–60' (9–18 m)

- ■ DENSE POPULATION
- ■ LIGHT POPULATION
- ■ NATURALIZED

Black Maple is closely related to Sugar Maple and is sometimes treated as a variety of that species. The ranges are similar, except that Black Maple occurs farther west in Iowa, and Sugar Maple's range extends farther into Canada in the Northeast. The two species are known to hybridize, which can sometimes make identification difficult in the field. Leaves provide the most reliable identification clues, with Black Maple usually having three-lobed leaves and sugar maple typically showing five-lobed leaves, although this feature can be variable. Like Sugar Maple, Black Maple's sweet sap is tapped for maple syrup.

DESCRIPTION: This large tree has a rounded, dense crown and paired, palmately three-lobed leaves. Those leaves are opposite, 4–5½ (10–14 cm) long and wide, with three (sometimes five) broad long-pointed lobes, with wavy edges or a few blunt teeth, three or five main veins from the base; often drooping sides, and long, hairy leafstalks. They are dull green above and yellow-green with soft hairs beneath, turning yellow in fall. The bark is dark gray or blackish, becoming deeply furrowed. The twigs are brown, slender, and hairy when young.

FLOWERS: Early spring. ³⁄₁₆" (5 mm) long with a bell-shaped, five-lobed yellow calyx, occurring in drooping clusters on long, slender hairy stalks; male and female may be on the same or separate trees.

FRUIT: 1–1¼" (2.5–3.2 cm) long including a long wing, with paired forking keys; they are brown, and one-seeded, maturing in fall.

HABITAT: Moist soils of valleys and uplands; in mixed hardwood forests.

RANGE: Southern Ontario east to southern Quebec and Vermont, southwest to Tennessee and Missouri and north to southeast Minnesota; local in adjacent states; to 2500' (762 m), slightly higher in the south.

USES: Black Maple is often planted as an ornamental, and the sap is used to make maple syrup.

SIMILAR SPECIES: Sugar Maple.

CONSERVATION STATUS:

The Black Maple population is large and does not currently face major threats. However, like others in the genus, the Black Maple is among the preferred hosts of the invasive Asian Long-horned Beetle (*Anoplophora glabripennis*) in North America. If pest numbers grow, the Black Maple could be among the species that face decline. The species is also potentially threatened by habitat loss and by extensive hybridization with Sugar Maple.

Diameter: 2–3' (0.6–0.9 m)
Height: 80' (24 m)

Japanese Maple is a native of far-eastern Asia, and in North America it is a widely planted ornamental in gardens and parks. There are more than 1000 cultivars available that possess a wide range of leaf shapes, colors, and forms, including weeping, dwarf, and bonsai-like plants. Many commercially sold "Japanese Maple" cultivars are actually of Fullmoon Maple (*A. japonicum*), a native of Japan distinguished from Japanese Maple by its downy leafstalks and more circular leaves. Japanese Maple has leaves with deeply divided, long-pointed lobes.

DESCRIPTION: This is a small tree or large shrub with a short, stout trunk and a spreading crown; some cultivated varieties have a low, domed crown of contorted branches. The leaves are 2–4" (5–10 cm) long and wide, with seven to eleven shallowly to deeply divided, long-pointed lobes (in some cases, leaflets), and double-toothed margins. They are pale or bright green, green-gold, or maroon, with intense autumn colors ranging from scarlet, ruby red, and purple to pink, bright gold, and orange.

Diameter: 8–12" (20–30 cm)
Height: 15–20' (4.5–6 m)

FLOWERS: Spring. Purple-red flowers arranged in upright or drooping clusters on long stalks.

FRUIT: 1" (2.5 cm) long, paired samaras that make an obtuse angle; the wings are often reddish. It ripens in autumn.

HABITAT: Introduced; this plant is widely cultivated in gardens and parks.

RANGE: Native to Japan, China, and Korea. It is introduced in North America and occasionally escapes cultivation. It has naturalized locally in parts of Ontario, Ohio, Pennsylvania, Delaware, New York, and Washington D.C.

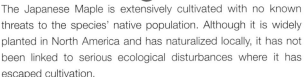

USES: This species is widely cultivated as an ornamental throughout eastern North America.

SIMILAR SPECIES: Sweetgum, Fullmoon Maple.

CONSERVATION STATUS:  LC

The Japanese Maple is extensively cultivated with no known threats to the species' native population. Although it is widely planted in North America and has naturalized locally, it has not been linked to serious ecological disturbances where it has escaped cultivation.

■ DENSE POPULATION
■ LIGHT POPULATION
■ NATURALIZED

This small maple is native to eastern North America. It is easy to recognize by its green-and-white striped twigs and bark, which make it a popular ornamental. This shade-tolerant understory species can remain as a shrub for many years until an opening in the canopy allows it to reach its full height as a tree. Wildlife such as rabbits, beavers, deer, and moose may eat the bark, particularly during winter. Other common names for this tree include Moosewood, Moose Maple, and Goosefoot Maple.

DESCRIPTION: This is a small tree with a short trunk and an open crown of striped, upright branches and coarse foliage; it is often a shrub. The leaves are opposite, 5–7" (13–18 cm) long and nearly as wide, with three short, broad long-pointed lobes at the tip; they are doubly saw-toothed, with three main veins from the base, with rust-colored hairs when young, and a stout leafstalk. They are light green above and paler beneath, turning yellow in fall. The bark is bright green with white stripes, becoming reddish-brown with long pale vertical lines, thin, and smooth or warty. The twigs are green, becoming striped with whitish lines.

FLOWERS: Late spring. ⅜" (10 mm) wide and bell-shaped with five bright yellow petals; they are slender-stalked and the male and female are usually in separate clusters to 6" (15 cm) long, drooping at the end of leafy twigs.

FRUIT: 1¼" (3 cm) long, many paired, widely forking keys; they are long-winged, light brown, and one-seeded, maturing in fall.

HABITAT: Moist upland soils in understory of hardwood forests.

RANGE: Southern Ontario east to Nova Scotia, south to northern Georgia, and west to southern Minnesota; to 5500' (1676 m).

USES: Striped Maple is frequently planted as an ornamental tree.

SIMILAR SPECIES: Mountain Maple has similar leaves but lacks the striped bark that distinguishes Striped Maple.

CONSERVATION STATUS: LC

Striped Maple is a common species whose population is presumed to be large and stable. There are no serious threats to the species, although it is susceptible to a number of diseases and pests. The species is believed to be less abundant in the southern part of its range, and the loss of habitat due to climate change could result in serious population decline.

Diameter: 8" (20 cm)
Height: 30' (9 m)

■ DENSE POPULATION
■ LIGHT POPULATION
■ NATURALIZED

This fast-growing maple was introduced from Europe as a popular street tree capable of tolerating urban pollution. Unfortunately the species has been escaping from cultivation and increasing its presence in natural areas. It has begun invading forests, where it is considered one of the worst invasive trees, reproducing vigorously by seed and resprouting after cutting, making it extremely difficult to eradicate. Norway Maple outcompetes native trees such as Sugar Maples, shades out all understory growth, hosts several invasive pest species, and changes the forest makeup. By the time many states began banning Norway Maple's use as a street tree, introduced populations had already become widespread.

DESCRIPTION: This introduced, invasive shade tree has a rounded crown of dense foliage and milky sap in its leafstalks. The leaves are opposite, 4–7" (10–18 cm) long and wide, and palmately five-lobed, with five or seven main veins from notched base. They are dull green with sunken veins above and paler and hairless (except in vein angles) beneath, turning bright yellow in fall. The leafstalk is long and slender, with milky sap at the end when broken off. The bark is gray or brown, becoming rough and furrowed into narrow ridges. The twigs are brown and hairless.

FLOWERS: Early spring before leaves. 5⁄16" (8 mm) wide with five greenish-yellow petals, occurring in upright or spreading clusters, usually with male and female on separate trees.

FRUIT: 1½–2" (4–5 cm) long, paired keys with a long wing and flattened body, spreading widely; they are light brown and hang on a long stalk, maturing in summer.

HABITAT: A street tree, escaping along roadsides and into second-growth forest; in humid temperate regions.

RANGE: Native across Europe from Norway to Caucasus and northern Turkey. It is widely planted in North America and has naturalized in the Northeast south to North Carolina and west to Minnesota, and in the Northwest from central Montana to coastal Oregon and British Columbia.

USES: This species was once widely used as an ornamental shade tree and street tree, but is now considered an invasive pest in many areas.

SIMILAR SPECIES: Sycamore Maple, Sugar Maple.

CONSERVATION STATUS: **LC**
The Norway Maple population is large and considered secure in its native range. In North America, it is an invasive tree of high ecological significance, with a tendency to turn native habitats into unproductive monoculture stands if left unchecked.

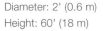

Diameter: 2' (0.6 m)
Height: 60' (18 m)

■ DENSE POPULATION
■ LIGHT POPULATION
■ NATURALIZED

This maple is an important timber and shade tree in Europe, where it is called "Sycamore"; in North America it is sometimes called Sycamore Maple. This fast-growing species is often planted in urban areas, but it has escaped cultivation and invaded road-sides and woodlands. It is considered an invasive species in New England and other areas of the Northeast. The species name, which means "false Platanus," refers to the resemblance of the foliage to that of the sycamore or planetree genus.

DESCRIPTION: This tree is introduced and sometimes invasive. It is a large shade tree with a widely spreading, rounded crown and large, paired, palmately five-lobed leaves. The leaves are opposite; 3½–6" (9–15 cm) long and wide. The five shallow lobes are short-pointed and wavy saw-toothed, with five main veins from notched base, and a long slender leafstalk. They are dull dark green with sunken veins above, and pale with raised, sometimes hairy veins beneath, turning brown in fall. The bark is gray and smooth or with broad flaky scales. The twigs are gray and hairless.

FLOWERS: Early spring. ³⁄₁₆" (5 mm) wide with five greenish-yellow petals; male and bisexual; occurring in narrow branched drooping clusters 5" (13 cm) long and appearing in early spring.

FRUIT: 1¼–2" (3–5 cm) long, paired keys with an elliptical body and a long wing; they are light brown, maturing in summer.

HABITAT: Hardy in exposed places and adapted to seashore gardens, tolerant of salt spray; sometimes escaping along roadsides.

RANGE: Native of Europe and western Asia; planted across the United States, and becoming established in the East.

USES: This species has been widely used as a windbreak and ornamental tree, but is now considered an invasive pest in many areas.

SIMILAR SPECIES: Norway Maple.

CONSERVATION STATUS: 🅛🅒

Planetree Maple has a large population and is thought to be stable in the wild. While the species has not yet faced population decline, it is highly susceptible to the Asian Long-horned Beetle (*Anoplophora glabripennis*). In North America, this species has escaped cultivation and invaded natural areas, especially disturbed sites. This species tends to be a more destructive invasive in other parts of the world.

Diameter: 2' (0.6 m)
Height: 60' (21 m)

■ DENSE POPULATION
■ LIGHT POPULATION
■ NATURALIZED

This large tree may be known as Swamp Maple, Soft Maple, or Red Maple. It is among the most common and widespread trees of eastern and central North America, with a wide north-south distribution along the East Coast. It is a handsome shade tree and popular landscaping tree because of its rapid growth, attractive colors and form, and value for native wildlife.

DESCRIPTION: This is a large tree with a narrow or rounded, compact crown and red flowers, fruit, leafstalks, and autumn foliage. The leaves are opposite and 2½–4" (6–10 cm) long and nearly as wide. They are broadly ovate, with three shallow short-pointed lobes (sometimes with two smaller lobes near the base), irregularly and wavy saw-toothed, with five main veins from base, and a long red or green leafstalk. They are dull green above and whitish and hairy beneath, turning red, orange, and yellow in fall. The bark is gray; thin, and smooth, becoming fissured into long thin scaly ridges. The twigs are reddish, slender, and hairless.

FLOWERS: Late winter or early spring before leaves. ⅛" (3 mm) long and reddish, crowded in nearly stalkless clusters along the twigs, with male and female occurring in separate clusters.

FRUIT: ¾–1" (2–2.5 cm) long including long wing, paired forking keys; they are red turning reddish-brown and one-seeded, maturing in spring.

HABITAT: Wet or moist soils of stream banks, valleys, swamps, and uplands and sometimes on dry ridges; in mixed hardwood forests.

RANGE: Extreme southeast Manitoba east to eastern Newfoundland, south to southern Florida, west to eastern Texas; to 6000' (1829 m).

USES: This species is widely planted as an ornamental tree, both throughout its native eastern range and in other parts of North America and in Europe. The wood is sometimes used for furniture and musical instruments. The sap is used to make syrup, though not as widely as that of Sugar Maple and Black Maple. Pioneers made ink as well as brown and black dyes from an extract of the bark.

SIMILAR SPECIES: Sugar Maple, Silver Maple.

CONSERVATION STATUS: 🔵LC
In eastern North America, the Red Maple is currently one of the most abundant tree species. It is highly vulnerable to various fungi and to the Asian Long-horned Beetle (*Anoplophora glabripennis*) and other pests, but its numbers appear to be secure. Future habitat loss caused by climate change may stress the species further.

Diameter: 2½' (0.8 m)
Height: 60–90' (18–27 m)

■ DENSE POPULATION
■ LIGHT POPULATION
■ NATURALIZED

This widespread eastern tree species is an important food source for a variety of wildlife, particularly squirrels. Its rapid growth has made Silver Maple a popular shade tree, but it has fallen out of favor in some localities because it can be a messy tree—its branches are brittle and easily broken by wind, and it produces abundant fruit that litters the ground and sometimes produces many volunteer seedlings. Sugar can be obtained from the sweetish sap, but the yield is low and less sweet than other maples. Other common names for this tree include Water Maple, Creek Maple, Large Maple, and Soft Maple.

DESCRIPTION: This is a large tree with a short, stout trunk; a few large forks; a spreading, open, irregular crown of long, curving branches; and graceful cut-leaves. The leaves are opposite and 4–6" (10–15 cm) long and nearly as wide. They are broadly ovate, deeply five-lobed and long-pointed (the middle lobe is often three-lobed), doubly saw-toothed, with five main veins from the base, and becoming hairless, with a slender, drooping reddish leafstalk. They are dull green above and silvery-white beneath, turning pale yellow in autumn. The bark is gray, becoming furrowed into long, scaly, shaggy ridges. The twigs are light green to brown; long, spreading and often slightly drooping; and hairless, with a slightly unpleasant odor when crushed.

FLOWERS: Late winter–early spring. ¼" (6 mm) long, reddish buds turning greenish-yellow and crowded in nearly stalkless clusters; male and female are in separate clusters.

FRUIT: 1½–2½" (4–6 cm) long including a long, broad wing and with paired, widely forking keys; they are light brown and one-seeded, maturing in spring.

HABITAT: Wet soils of stream banks, flood plains, and swamps; with other hardwoods.

RANGE: Southern Ontario east to New Brunswick, south to northwest Florida, west to eastern Oklahoma, north to northern Minnesota; to 2000' (610 m), higher in mountains.

USES: Some Native Americans used the Silver Maple's sap to make sugar and used the wood to make furniture. The wood is now used for pulpwood, furniture, flooring, musical instruments, and other purposes. This species is planted as an ornamental or street tree.

SIMILAR SPECIES: Red Maple.

CONSERVATION STATUS: LC

The Silver Maple is vulnerable to several fungal diseases that can kill both seedlings and mature trees. Boring pests, including the Asian Long-horned Beetle (*Anoplophora glabripennis*), also pose a minor threat but do not typically increase mortality rates of the Silver Maple.

Diameter: 3' (0.9 m)
Height: 50–80' (15–24 m)

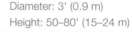

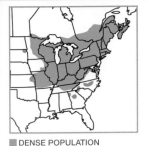

■ DENSE POPULATION
■ LIGHT POPULATION
■ NATURALIZED

Sugar maples were planted in most New England villages, where they can live for more than 400 years despite local traffic and disturbances such as sidewalks and underground pipes. The species is especially noted for its full spread of foliage, providing deep shade, and for its autumn display of many shades of yellow, orange, scarlet, and crimson. The wood is particularly hard, strong, and durable, known to have outlasted marble on some century-old floors.

DESCRIPTION: This is a large tree with a rounded, dense crown and striking, multicolored foliage in autumn. The leaves are opposite, 3½–5½" (9–14 cm) long and wide, and palmately lobed with five deep long-pointed lobes; they have a few narrow, long-pointed teeth and five main veins from the base; the leafstalks are long and often hairy. They are dull dark green above and paler and often hairy on the veins beneath, turning deep red, orange, and yellow in fall. The bark is light gray, becoming rough and deeply furrowed into narrow, scaly ridges. The twigs are greenish to brown or gray, and slender.

FLOWERS: Early spring. ³⁄₁₆" (5 mm) long with a bell-shaped, five-lobed, yellowish-green calyx; male and female occur in drooping clusters on long, slender, hairy stalks.

FRUIT: 1–1¼" (2.5–3 cm) long including a long wing, paired forking keys, brown, and one-seeded, maturing in fall.

HABITAT: Moist soils of uplands and valleys, sometimes in pure stands.

RANGE: Extreme southeastern Manitoba east to Nova Scotia, south to North Carolina, and west to eastern Kansas; it is local in northwestern South Carolina and northern Georgia. It occurs to 2500' (762 m) in the north and 3000–5500' (914–1676 m) in the southern Appalachians.

USES: Maples, particularly Sugar Maple, are among the leading furniture woods. This species is also used for flooring, furniture, and veneer, violin backs and gun stocks. Some trees develop special grain patterns, including bird's-eye maple with dots suggesting the eyes of birds, and curly and fiddleback maple, with wavy annual rings. Such variations in grain are in great demand. The boiled concentrated sap is the commercial source of maple sugar and syrup, a use colonists learned from the Native Americans. With the sap running in early spring, each tree yields between 5 and 60 gallons of sap per year; about 32 gallons of sap make 1 gallon of syrup or 4½ pounds of sugar.

SIMILAR SPECIES: Norway Maple, a non-native species often planted in cities and suburbs in North America.

CONSERVATION STATUS: 🇱🇨
An important tree in the forests of the northeastern United States and southeastern Canada. It is particularly susceptible to air pollution and acid rain. The species is already being affected by climate change as conditions in some parts of its range become wetter than it can tolerate; climate change can also make these trees increasingly susceptible to freeze-thaw injury or to scorching.

Diameter: 2–3' (0.6–0.9 m)
Height: 70–100' (21–30 m)

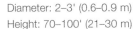

MOUNTAIN MAPLE *Acer spicatum*

■ DENSE POPULATION
■ LIGHT POPULATION
■ NATURALIZED

This small maple is also commonly known as Dwarf Maple. Mountain Maple is hardy and adapted to high latitudes and elevations, and to partial shade. The Latin species name, meaning "spiked," refers to the long, spike-like flower clusters. An important wildlife tree, many native herbivores such as rabbits, beavers, deer, and moose browse the bark, and Ruffed Grouse (*Bonasa umbellus*) eat the buds.

DESCRIPTION: This is a shrub or small tree with a short trunk and slender, upright branches. The leaves are opposite and 2½–4½" (6–11 cm) long and wide, with three (sometimes five) short broad lobes; they are short-pointed and coarsely saw-toothed, with three or five main veins from base and long leafstalks that often turn red. They are light green and becoming hairless above, and hairy beneath, turning bright red and orange in fall. The bark is brown, thin, and scaly or slightly furrowed. The twigs are light gray and slender, and hairy when young.

FLOWERS: Early summer. ¼" (6 mm) wide, greenish-yellow, and short-stalked, occurring in narrow upright hairy clusters to 5" (13 cm) at the end of a leafy twig; male and female are in separate clusters.

FRUIT: ¾–1" (2–2.5 cm) long, paired forking long-winged keys clustered along a slender stalk; they are one-seeded, and red or yellow when immature, turning brown in fall.

HABITAT: Moist rocky uplands, especially mountains; in understory of hardwood forests.

RANGE: Saskatchewan east to Newfoundland, south to Pennsylvania, and west to northeastern Iowa; it is also found in the southern Appalachians to northern Georgia. It occurs to 6000' (1829 m).

USES: The sap from Mountain Maple is sometimes used to make syrup and the tannin-containing leaves have been used to tan leather. Some Native Americans used this species for medicinal purposes.

SIMILAR SPECIES: Striped Maple.

CONSERVATION STATUS:
The widespread Mountain Maple faces no major threats and has a large, stable population. Like other members of the genus, Mountain Maple is a favorite host of the invasive pest the Asian Long-horned Beetle (*Anoplophora glabripennis*). Habitat loss caused by climate change could place further stress on this species.

Diameter: 6" (15 cm)
Height: 25' (7.6 m)

■ DENSE POPULATION
■ LIGHT POPULATION
■ NATURALIZED

California Buckeye is widely distributed in California from the coast to the lower elevations in the mountains. This species, the only native buckeye in the West, is sometimes grown as an ornamental for its attractive foliage and flowers. Its large seeds are poisonous to many animals, but chipmunks and squirrels appear to consume the seeds with no ill effect. Bees may be poisoned by the nectar and pollen.

DESCRIPTION: This is a thicket-forming shrub or small tree with a short trunk often enlarged at the base; a broad, rounded crown of crooked branches; and many showy flowers. Its leaves are opposite, palmately compound, and long-stalked, generally with five leaflets (sometimes four to seven) 3–6" (7.5–15 cm) long and 1–2" (2.5–5 cm) wide; they are narrowly elliptical, finely saw-toothed, and short-stalked. They are dark green above and paler with whitish hairs on veins beneath, turning dull brown and shedding in late summer. The bark is light gray, smooth, and thin. The twigs are reddish-brown and stout, ending in a resinous bud.

FLOWERS: Late spring–early summer. 1–1¼" (2.5–3 cm) long with four or five nearly equal, white or sometimes pale pink petals and five to seven much longer stamens; they are fragrant and occur in upright, narrow clusters 4–8" (10–20 cm) long.

FRUIT: 2–3" (5–7.5 cm) long, pear-shaped capsules that are pale brown, smooth, and splitting usually on three lines, usually with one large, rounded, shiny brown, poisonous seed, maturing in late summer.

CAUTION: Poisonous: Do not eat or ingest any part of this species. If accidentally ingested, contact your doctor or poison control center.

HABITAT: Moist soils of canyons and on hillsides in chaparral and oak woodland.

RANGE: California in Coast Ranges and Sierra Nevada foothills; to 4000' (1219 m).

USES: This species is sometimes used as an ornamental; it is also planted to prevent erosion. Native Californians made flour from the poisonous seeds after leaching out the toxic element with boiling water; they also ground up untreated seeds to throw into pools of water to stupefy fish, which then rose to the surface and were easily caught.

SIMILAR SPECIES: California Buckeye is the only native western buckeye, but it may be confused with the introduced Horse-chestnut.

CONSERVATION STATUS: The California Buckeye faces no major threats. Like other trees in this genus, the California Buckeye is susceptible to a number of pests, including borer beetles, the Japanese beetle (*Popillia japonica*), and bagworms (moths in the family Psychidae), as well as fungal diseases including anthracnose, rust, and powdery mildew.

Diameter: 1' (0.3 m)
Height: 25' (7.6 m)

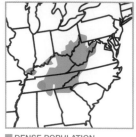

■ DENSE POPULATION
■ LIGHT POPULATION
■ NATURALIZED

The largest of North America's buckeyes, Yellow Buckeye is native to the Ohio Valley and Appalachian Mountains, where it is typically found in rich soils on river bottoms, stream banks, and mountain slopes. It is particularly abundant in Great Smoky Mountains National Park, which straddles the border between Tennessee and North Carolina. Other common names for this tree include Sweet Buckeye and Big Buckeye.

ALTERNATE NAME: *Aesculus octandra*

DESCRIPTION: This tree has a rounded crown and upright clusters of showy yellow flowers. The leaves are opposite and palmately compound, with slender leafstalks 3½–7" (9–18 cm) long and five to seven leaflets 4–8" (10–20 cm) long and 1½–3" (4–7.5 cm) wide; they are elliptical to obovate, evenly saw-toothed, and short-stalked. They are dark green and usually hairless above, and yellow-green and often hairy beneath. The bark is brown to gray, thin, and fissured into large scaly plates. The twigs are light brown, stout, and often hairy.

FLOWERS: Spring. 1¼" (3 cm) long with 4 very unequal yellow petals and seven or eight shorter stamens, occurring in upright branched terminal clusters 4–6" (10–15 cm) long.

FRUIT: A pale brown, smooth or slightly pitted capsule 2–3" (5–7.5 cm) in diameter, splitting on two or three lines, with one to three large shiny brown poisonous seeds, maturing in early autumn.

CAUTION: Poisonous: Do not eat or ingest any part of this species. If accidentally ingested, contact your doctor or poison control center.

HABITAT: Rich, moist, deep soils from river bottoms to deep mountain valleys or slopes; in mixed forests.

RANGE: Southwestern Pennsylvania south to northern Alabama and northern Georgia and north to extreme southern Illinois; at 500–6300' (152–1920 m).

USES: Some Native Americans used the seeds as a food source after removing the toxic element by roasting and soaking them. Saponins in the seeds have been used as a soap substitute.

SIMILAR SPECIES: Painted Buckeye.

CONSERVATION STATUS:
The Yellow Buckeye has a stable, widespread population. It faces no serious current or anticipated threats.

Diameter: 2–3' (0.6–0.9 m)
Height: 70–90' (21–27 m)

■ DENSE POPULATION
■ LIGHT POPULATION
■ NATURALIZED

This is the state tree of Ohio, which is nicknamed the Buckeye State. It is a small tree that usually stays under 50' (15 m) tall with a dense oval to round crown, and branching rather low. Pioneers carried a buckeye seed in their pockets to ward off rheumatism, and are considered by some to be good-luck charms. The seeds and young foliage are poisonous if ingested, however, and the toxic bark was formerly used medicinally.

DESCRIPTION: This tree has upright flowers at the ends of the twigs and an irregular, rounded crown; it is sometimes shrubby. The twigs and leaves often have a slightly unpleasant odor when crushed. The leaves are opposite and palmately compound with slender leafstalks 2–6" (5–15 cm) long. There are five to seven leaflets 2½–6" (6–15 cm) long and ¾–2¼" (2–6 cm) wide; they are elliptical, unevenly saw-toothed, and nearly stalkless. They are yellow-green above and paler and often hairy beneath, turning orange or yellow in autumn. The bark is ashy-gray, scaly, and becoming rough and furrowed into thick scaly plates, with an unpleasant odor. The twigs are reddish-brown and stout, becoming hairless.

FLOWERS: Spring. ¾–1" (2–2.5 cm) long and narrowly bell-shaped, with four nearly equal pale yellow or greenish-yellow petals and seven longer stamens, occurring in upright branched clusters 4–6" (10–15 cm) long, and giving off an unpleasant odor.

FRUIT: A pale brown spiny capsule 1–2" (2.5–5 cm) in diameter and splitting on two or three lines, with one to three large, dark brown, poisonous seeds and maturing in summer and fall.

CAUTION: Poisonous—do not eat or ingest any part of this species. If accidentally ingested, contact your doctor or poison control center.

HABITAT: Rich, moist soils of valleys and mountain slopes, also found on drier sites. It is sometimes a thicket-forming shrub on stream banks. It is found in mixed hardwood forests.

RANGE: Western Pennsylvania south to central Alabama, west to southeastern Oklahoma, and north to central Iowa; at 500–2000' (152–610 m).

USES: Sometimes planted as an ornamental for the showy autumn foliage. Native Americans extracted tannic acid from the nuts to make leather, and ground Buckeye to use as a powder on ponds to stun fish. The soft, lightweight wood is used for furniture, boxes, flooring, and musical instruments.

SIMILAR SPECIES: California Buckeye, Horse-chestnut, Painted Buckeye.

CONSERVATION STATUS:

Quite common in its native range, the Ohio Buckeye is considered stable with no major threats. Like other trees in the genus, it is susceptible to various fungal diseases, including leaf blotch and powdery mildew, as well as pests including the Japanese beetle (*Popillia japonica*) and bagworms.

Diameter: 1–2' (0.3–0.6 m)
Height: 30–70' (9–21 m)

- ■ DENSE POPULATION
- ■ LIGHT POPULATION
- ■ NATURALIZED

Texas Buckeye, sometimes called White Buckeye, was formerly considered a full species (*A. arguta*), but is now classified as a western shrubby variety of Ohio Buckeye. It replaces Ohio Buckeye in the Great Plains in southeastern Nebraska, eastern Kansas, western Missouri, Oklahoma, and Texas. Texas Buckeye is distinguishable by the large number of narrow leaflets and its usually shrubby size. It grows best on well-drained, slightly acid soil, and is able to tolerate drought to a degree. All parts of this plant are considered poisonous.

ALTERNATE NAME: *Aesculus arguta*

DESCRIPTION: This is a shrub or small tree with a rounded to oblong crown of stout branches. The leaves are opposite and palmately compound, with slender leafstalks 2–5" (5–13 cm) long and seven to eleven leaflets 2½–5" (6–13 cm) wide; they are lance-shaped, long-pointed at both ends, unevenly saw-toothed, and nearly stalkless. They are shiny yellow-green above and paler and hairy beneath. The bark is light gray and smooth, becoming scaly or furrowed. The twigs are reddish-brown and stout, becoming hairless.

FLOWERS: Spring. ½–¾" (12–19 mm) long and narrowly bell-shaped, with four unequal pale yellow petals and 7–8 longer stamens; occurring in upright branched clusters 4–6" (10–15 cm) long.

FRUIT: A light brown spiny capsule ¾–1¾" (2–4 cm) in diameter and splitting on two or three lines, with one or two large shiny brown poisonous seeds, maturing in summer.

CAUTION: Poisonous: Do not eat or ingest any part of this species. If accidentally ingested, contact your doctor or poison control center.

HABITAT: Sandy soils of slopes, hills, and stream bluffs; forming thickets and in forest understory.

RANGE: Extreme southeastern Nebraska south to western Missouri, south and southwest Oklahoma, and central Texas; at 500–2000' (152–610 m).

USES: This tree is sometimes planted as an ornamental.

SIMILAR SPECIES: Ohio Buckeye, Painted Buckeye.

CONSERVATION STATUS: LC

The Texas Buckeye population is stable, and there are no known serious threats to the species.

Diameter: 6" (15 cm), sometimes larger
Height: 20' (6 m)

■ DENSE POPULATION
■ LIGHT POPULATION
■ NATURALIZED

Native to a small area in southeastern Europe, the Horse-chestnut has been widely planted in North America. It is easily propagated from seed and tolerant of city conditions, but its branches may break in high winds. The tree is showy when it bears masses of whitish flowers for a few weeks in spring. The species is so named because the seeds were once used to concoct a remedy for horses suffering from respiratory issues.

DESCRIPTION: This is an introduced shade and ornamental tree with a spreading, elliptical to rounded crown of stout branches and coarse foliage. The leaves are opposite and palmately compound, with leafstalks 3–7" (7.5–18 cm) long and seven leaflets (sometimes five) spreading fingerlike, 4–10" (10–25 cm) long and 1–3½" (2.5–9 cm) wide; they are obovate or elliptical, broadest toward an abrupt point, and tapering to a stalkless base; they are saw-toothed, and dull dark green above and paler beneath. The bark is gray or brown, thin, and smooth, becoming fissured and scaly. Twigs are light brown, stout, and hairless, ending in a large blackish sticky bud.

FLOWERS: Late spring. 1" (2.5 cm) long and narrowly bell-shaped with four or five spreading narrow white petals that are red- and yellow-spotted at the base, with many flowers in upright branched clusters 10" (25 cm) long.

FRUIT: A brown spiny or warty capsule 2–2½" (5–6 cm) in diameter, splitting into two or three parts, with one or two large, rounded, shiny brown, poisonous seeds, maturing in late summer.

CAUTION: Poisonous: Do not eat or ingest any part of this species. If accidentally ingested, contact your doctor or poison control center.

HABITAT: A shade and street tree in rich moist soils.

RANGE: Native of southeastern Europe; widely planted across the United States and escaped in the Northeast.

USES: This species is widely planted as an ornamental and shade tree.

SIMILAR SPECIES: California Buckeye, Ohio Buckeye.

CONSERVATION STATUS:
The Horse-chestnut faces a number of threats in its limited native range in southeastern Europe; deforestation, soil erosion, forest fires, and other human factors that damage or reduce habitat have all contributed to the population's decreased stability. The species has been significantly impacted by the leafminer moth (*Cameraria ohridella*) which has led to population decline. In North America, this introduced tree has escaped from cultivation, mainly in the Northeast and Pacific Northwest, and is considered invasive in some locales where it competes with native species for sunlight, water, and nutrients.

Diameter: 2' (0.6 m)
Height: 70' (21 m)

■ DENSE POPULATION
■ LIGHT POPULATION
■ NATURALIZED

This small tree is native to the southeastern United States but is hardy farther north of its natural range. Red Buckeye is frequently planted as a handsome ornamental for its showy red flowers, which have given rise to a number of common names for the species, including Scarlet Buckeye and Firecracker Plant. Sporting bright red flowers in late spring, it is an attractive nectar plant for hummingbirds and bees. A yellow-flowered variety (var. *flavescens*) can be found in Texas along with several intermediate hybrids. The scientific name *pavia* is an old word for "buckeye."

DESCRIPTION: This is a shrub or small tree with bright red flowers and an irregular crown of short, crooked branches. The leaves are opposite and palmately compound, with slender leafstalks 3–6" (7.5–15 cm) long and 5 (sometimes 7) leaflets 2½–6" (6–15 cm) long and 1¼–2½" (3–6 cm) wide; they are narrowly elliptical, irregularly saw-toothed, and short-stalked. They are dark green with sunken veins and nearly hairless above, and dull green and sometimes densely covered with whitish hairs beneath. The bark is brown-gray to light gray and smooth. The twigs are reddish-brown and stout.

FLOWERS: Late spring. 1¼" (3 cm) long with four unequal bright red (sometimes yellow-red or yellow) petals and six to eight stamens about as long as the petals, with many arranged in narrow upright branched clusters 4–8" (10–20 cm) long.

FRUIT: A smooth light brown capsule 1½–2" (4–5 cm) in diameter and splitting on two or three lines, with one to three large, rounded, shiny brown, poisonous seeds, maturing in early fall.

CAUTION: Poisonous: Do not eat or ingest any part of this species. If accidentally ingested, contact your doctor or poison control center.

HABITAT: Moist soils, especially along river bluffs, borders of streams, swamps, and in flood plains; in understory of mixed forests.

RANGE: Southeastern North Carolina southwest to northern Florida, west to central Texas, and northern Illinois; to 1500' (457 m).

USES: This flowering plant is a popular ornamental. Native Americans threw powdered seeds and crushed branches of this and other buckeyes into pools of water to stupefy fish, which then rose to the surface and were easily caught. Pioneers used the gummy roots as a soap substitute and made home remedies from the bitter bark.

SIMILAR SPECIES: The red flowers help to distinguish this species from other buckeyes.

CONSERVATION STATUS: LC
The Red Buckeye is considered stable. The species faces no significant threats from diseases or pests. It is susceptible to leaf blotch, a fungal disease that is not fatal to these trees.

Diameter: 8" (20 cm)
Height: 25' (7.6 m)

■ DENSE POPULATION
■ LIGHT POPULATION
■ NATURALIZED

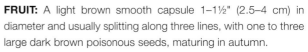

Painted Buckeye is commonly found in forests and along stream banks in a somewhat limited range in the southeastern United States, where it usually grows as a shrub but can be a small tree. It has become a popular ornamental, grown for the large, showy flowers for which the species is named. The flowers are variable in color, ranging from yellows to reds, suggesting an artist's paintbrush.

DESCRIPTION: This is a rounded shrub or sometimes a small tree, with a spreading crown and showy yellow or reddish flowers. The leaves are opposite and palmately compound, with a slender leafstalk 4–6" (10–15 cm) long and five (sometimes seven) leaflets 4–6" (10–15 cm) long and 1½–2½" (4–6 cm) wide; they are narrowly elliptical or lance-shaped, finely saw-toothed, and short-stalked. They are yellow-green above, and green and often hairy beneath. The bark is gray or brown, thin, and scaly. The twigs are light reddish-brown, stout, and hairless.

FLOWERS: Spring. 1–1¼" (2.5–3 cm) long with four unequal bright yellow or sometimes reddish petals and six or seven shorter stamens, occurring in upright branched clusters 4–6" (10–15 cm) long.

FRUIT: A light brown smooth capsule 1–1½" (2.5–4 cm) in diameter and usually splitting along three lines, with one to three large dark brown poisonous seeds, maturing in autumn.

CAUTION: Poisonous: Do not eat or ingest any part of this species. If accidentally ingested, contact your doctor or poison control center.

HABITAT: Moist well-drained slopes in understory of mixed forests.

RANGE: Southeastern Virginia southeast to northern Alabama; at 100–1200' (30–366 m).

USES: Painted Buckeye is a popular ornamental.

SIMILAR SPECIES: Yellow Buckeye, Ohio Buckeye.

CONSERVATION STATUS: 🔘 LC

The Painted Buckeye has a stable population with no major conservation concerns. Mild threats include common pests such as Japanese beetles, borers, and bagworms, and diseases including anthracnose, rust, and powdery mildew.

Diameter: 8" (20 cm)
Height: 25' (7.6 m)

■ DENSE POPULATION
■ LIGHT POPULATION
■ NATURALIZED

This evergreen shrub or tree is native to southernmost Florida and the Florida Keys, where it prefers rocky areas and coastal hammocks. The bark and berries have been used to make ink, hence the common name. Wildlife, particularly birds, eat the berries and spread the seeds, and its flowers attract various pollinators. The species is also called Butterbough.

DESCRIPTION: This is a small evergreen tree with pinnately compound leaves, a tall trunk, and a dense crown of upright branches. The leaf has a few, even-numbered leaflets, and the leaflet tip is blunt or notched. The leaves are 3–8" (7.6–20 cm) long with one to three pairs of leaflets that are 2–5" (5–13 cm) long; they are elliptic to lance-shaped and blunt or notched at the tip, with untoothed margins. They are shiny, dark green. The bark is mottled reddish brown, separating into large scales.

FLOWERS: Spring. Tiny, fragrant, white flowers with an orange center are arranged in branched clusters.

FRUIT: ½" (1.3 cm) wide berry that is orange in summer and ripening to dark purple in fall.

HABITAT: Coastal hammocks and shell mounds of Florida. It occurs at elevations near sea level.

RANGE: Florida south through the American tropics.

USES: This native species is popular as a landscaping ornamental, especially along the coast. The bark and berries have been used to make ink.

SIMILAR SPECIES: Wingleaf Soapberry, Spanish Lime.

CONSERVATION STATUS:

Globally this species is widespread through the American tropics and faces no range-wide threats. In North America, however, Inkwood is found only in Florida, where it is listed as vulnerable.

Diameter: 6–15" (15–38 cm)
Height: 25–45' (7.6–13.7 m

Goldenrain Tree is native to eastern Asia and is a popular ornamental tree grown in temperate regions around the world. It is one of the few trees to flower in midsummer in northern climates. The flowers are showy, yellow, foot-long clusters appearing after most other trees have bloomed and lasting about two weeks. The ornamental fruits resemble paper lanterns. The common name is sometimes applied to a related species (*K. bipinnata*) that has more upright branches, bipinnately compound leaves, and pink fruits.

DESCRIPTION: This is a small tree with deciduous, pinnately (sometimes bipinnately) compound leaves. It may have a single trunk or be multi-stemmed, and has a rounded to vase-shaped, dense crown of upright, spreading branches. The leaves are deciduous and 6–15" (15–38 cm) long with seven to fifteen leaflets that are 1–4" (2.5–10 cm) long and ovate to elliptic, with a pointed tip and coarsely toothed margins. They are dark green above and paler beneath. Their fall color is yellow. The bark is silvery gray and ridged.

FLOWERS: Summer. ½" (1.3 cm) wide yellow flowers with a red center, occurring in 10–15" (25–38 cm) long, many-branched clusters. They are slightly fragrant.

FRUIT: 1–2" (2.5–5 cm) long, triangular, three-part capsule; it is yellow to brown, inflated, and papery. Ripens in autumn and persists until spring.

HABITAT: Introduced in North America; it is planted along streets and as an ornamental.

RANGE: Native to China and Korea; it is introduced in North America.

USES: This species is a popular ornamental tree. In some parts of the eastern United States it is considered an invasive species.

SIMILAR SPECIES: Chinese Flame Tree and Chinese Rain Tree are similar species sometimes cultivated in North America.

CONSERVATION STATUS: (LC)
The large native range of the Goldenrain Tree makes it a stable species. No current or anticipated threats have been identified. It is widely planted in North America and occasionally escapes cultivation into adjacent early-successional or open disturbed habitats, but it has not shown significant impact on native plant communities.

Diameter: 20" (51 cm)
Height: 20–40' (6–12 m)

■ DENSE POPULATION
■ LIGHT POPULATION
■ NATURALIZED

This small to medium-sized tree (sometimes a large shrub) is native to Florida and south into tropical America. It is common in Central and South America and has been spread farther by cultivation. The poisonous fruit, containing the alkaloid saponin, has been used as a soap substitute for washing clothes.

DESCRIPTION: This is a large shrub or tree with poisonous fruit. Its leaves are pinnately compound and 5–12" (13–30 cm) long, with a long, slender axis and 6–19 leaflets 1½–7" (4–17.5 cm) long; they are paired (except at the end), lance-shaped, curved and slightly one-sided, long-pointed at the tip, blunt and unequal at the base, without teeth, slightly thickened, and short-stalked. They are dull grayish to yellow-green above, and slightly hairy with prominent veins beneath.

FLOWERS: Late spring–fall. ⅛" (3 mm) wide, usually with five rounded yellowish-white petals, and almost stalkless in upright clusters 6–9" (15–23 cm) long.

FRUIT: ⅜–¾" (10–18 mm) in diameter, berrylike, sometimes paired, yellow or orange to brownish, nearly transparent, one-seeded, and leathery, maturing in autumn, turning black and remaining attached in winter. Fruit is poisonous and has a single round, dark brown seed.

CAUTION: Poisonous: Do not eat or ingest any part of this species. If accidentally ingested, contact your doctor or poison control center.

HABITAT: Moist soils along streams and on limestone uplands, in and bordering hardwood forests; westward in plains and mountains, grassland, upper desert, and oak woodland zones.

RANGE: Native to far southern Florida; it also occurs in an isolated area of coastal southeastern Georgia.

USES: The fruit has been used to wash clothing. The seeds are made into necklaces and buttons, and the wood is easy to split and is used to make baskets.

SIMILAR SPECIES: Florida Soapberry, Western Soapberry.

CONSERVATION STATUS: **LC**
The Wingleaf Soapberry has a large wild population. The species is considered stable and faces no major threats.

Diameter: to 1' (0.3 m)
Height: 20–50' (6–15 m)

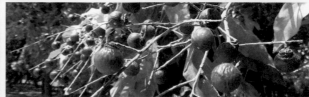

■ DENSE POPULATION
■ LIGHT POPULATION
■ NATURALIZED

This species is closely related and quite similar to Wingleaf Soapberry, and is sometimes considered a variety of that species. Western Soapberry is found in the south-central and southwestern United States and in northern Mexico. Like Wingleaf Soapberry, it has poisonous fruit that contains the alkaloid saponin, from which a soapy lather can be produced.

ALTERNATE NAME: *Sapindus saponaria* var. *drummondii*

DESCRIPTION: This poisonous tree has a rounded crown of upright branches; it is sometimes a large, spreading shrub. The leaves are pinnately compound and 5–8" (13–20 cm) long, with a long slender axis and 11–19 leaflets 1½–3" (4–7.5 cm) long and ⅜–¾" (10–19 mm) wide; they are paired (except at the end), lance-shaped, curved and slightly one-sided, long-pointed at the tip, blunt and unequal at the base, without teeth, slightly thickened, and short-stalked. They are dull yellow-green above, and slightly hairy with prominent veins beneath. The bark is light gray, becoming rough and furrowed. The twigs are yellow-green and covered with fine hairs.

FLOWERS: Late spring–summer. ⅛" (3 mm) wide, usually with five rounded, yellowish-white petals, and almost stalkless in upright clusters 6–9" (15–23 cm) long.

FRUIT: ⅜–½" (10–12 mm) in diameter, berrylike, sometimes paired, yellow or orange, nearly transparent, one-seeded, and leathery, maturing in fall, turning black and remaining attached in winter. The fruit is poisonous and has a single round, dark brown seed.

CAUTION: Poisonous: Do not eat or ingest any part of this species. If accidentally ingested, contact your doctor or poison control center.

HABITAT: Moist soils along streams and on limestone uplands, in and bordering hardwood forests; westward in plains and mountains, grassland, upper desert, and oak woodland zones.

RANGE: Southwest Missouri south to Louisiana, west to southern Arizona, and northeast to southeast Colorado; also northern Mexico; to 6000' (1829 m).

USES: The fruit has been used to wash clothing. The seeds are made into necklaces and buttons, and the wood is used to make baskets.

SIMILAR SPECIES: Wingleaf Soapberry.

CONSERVATION STATUS: 🔵LC
The Western Soapberry faces minor threats from overbrowsing by cattle. Low genetic variability, floods, and fires also pose a potential risk to the species. Overall, however, the Western Soapberry is believed to be stable.

Diameter: to 1' (0.3 m)
Height: 20–40' (6–12 m)

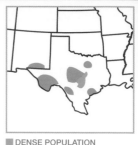

■ DENSE POPULATION
■ LIGHT POPULATION
■ NATURALIZED

This plant, native to Texas and southern New Mexico as well as northern Mexico, is common in rocky areas in canyons and on slopes and ridges in its range. It is showy in full flower and makes a good ornamental, grown for its attractive foliage and flowers and dense branching. Livestock seldom browse the toxic foliage, but bees produce fragrant honey from the flowers. This species is named for the large capsules and seeds, which resemble those of true buckeyes (genus *Aesculus*). The seeds of this species are sweetish but mildly poisonous.

DESCRIPTION: This is a shrub or small tree with an irregular crown of upright branches and showy pink flowers. Its leaves are pinnately compound and 5–12" (13–30 cm) long, with a slender axis and five to nine leaflets 3–5" (7.5–13 cm) long and 1–1½" (2.5–4 cm) wide; they are paired (except at the end), ovate, long-pointed at the tip, rounded and often unequal at the base, wavy saw-toothed, hairy when young, leathery, and stalkless or nearly so. They are shiny dark green above and light green and nearly hairless beneath. The bark is light gray, thin, and fissured. The twigs are light brown, slender, slightly zigzag, and covered with fine hairs.

FLOWERS: Before or with leaves in spring. The fragrant flowers are 1" (2.5 cm) wide with four or five unequal pink or purplish-pink petals; they are male, female, and bisexual, and slender-stalked, occurring in clusters crowded along the twigs.

FRUIT: A long-stalked drooping capsule 1½–2" (4–5 cm) wide, broadly pear-shaped, three-lobed, reddish-brown, rough and leathery, and three-celled, with three round, poisonous seeds, maturing and splitting open in fall.

CAUTION: Poisonous: Do not eat or ingest any part of this species. If accidentally ingested, contact your doctor or poison control center.

HABITAT: Moist soils of rocky canyons, slopes, and ridges.

RANGE: Central Texas west to Trans-Pecos Texas and southern New Mexico; it is also found in northeast Mexico. It occurs to 5000' (1524 m).

USES: Mexican-buckeye is sometimes planted as an ornamental and to attract butterflies and other wildlife.

SIMILAR SPECIES: This species is not likely to be confused with other plants; it does not closely resemble true buckeyes.

CONSERVATION STATUS:
Although the Mexican-buckeye is common, its range is limited. The species faces potential risks from browsing by herbivores.

Diameter: 8" (20 cm)
Height: 25' (7.6 m)

■ DENSE POPULATION
■ LIGHT POPULATION
■ NATURALIZED

Sour Orange is one of several *Citrus* species cultivated in the United States. Sour Orange is preferred as an ornamental not only for its hardiness but because the inedible, showy fruit remains on the tree indefinitely. It is native to southeast Asia but has been cultivated and spread by humans for centuries. The naturalized population in Florida and Georgia likely originated from plants introduced by Spanish colonists.

DESCRIPTION: This small ornamental evergreen tree has a rounded crown of glossy foliage. It is aromatic (and spicy in taste), with fragrant, white blossoms and sour, orange-colored fruit. The leaves are evergreen, 2½–5½" (6–14 cm) long, 1½–4" (4–10 cm) wide, ovate, and long-pointed, with many tiny, rounded teeth; they are slightly leathery, with tiny gland-dots and broadly winged leafstalks; they are shiny green above and pale light green beneath. The bark is brown and smooth. The twigs are green, and angled when young, with sharp spines to 1" (2.5 cm) long, occurring singly at the leaf bases.

FLOWERS: Spring. 1" (2.5 cm) wide with five narrow, white, curved petals and many stamens; they are very fragrant, with few occurring in short-stalked clusters at the base of the leaf.

FRUIT: More or less a rough orange 2½–4½" (6–11 cm) in diameter, with a thick peel and partly hollow, pulpy core; it is sour, bitter, and inedible, with many whitish seeds, ripening in early fall.

HABITAT: Moist soils in subtropical regions.

RANGE: Native of southeastern Asia. Naturalized in Florida and Georgia; also planted as an ornamental in southern Texas, southern Arizona, California, Hawaii, and Puerto Rico.

USES: This species is commonly planted as an ornamental. Orange marmalade is made from peels and the juice has been used in home remedies as an antiseptic and to check bleeding. This species has been employed as the stock for budding the other species, particularly cultivated Sweet Oranges (*C. x sinensis*), that produce the familiar, commercially sold fruits.

SIMILAR SPECIES: Bergamot Orange.

CONSERVATION STATUS:

Although often cultivated, the natural Sour Orange population has not been formally evaluated in the wild.

Diameter: 1' (0.3 m)
Height: 30' (9 m)

■ DENSE POPULATION
■ LIGHT POPULATION
■ NATURALIZED

Key Lime is a *Citrus* hybrid native to Southeast Asia and spread around the world by cultivation. Once introduced in North America, these trees quickly naturalized on the Florida Keys; the limes from this tree are the source of the distinctive flavoring in the local specialty Key Lime pie. The scientific name *aurantiifolia* refers to the resemblance of this tree's leaves to that of the related Sour Orange, *C. aurantium*.

DESCRIPTION: This is a tall shrub to a small tree with spiny twigs. The leaves are 3" (7.6 cm) long, elliptical to ovate, and evergreen. The bark is greenish brown, and smooth to rough.

FLOWERS: March–April. 1¼" (3 cm), star-shaped, and white.

FRUIT: Limes are 2½" (6.4 cm) in diameter, greenish-yellow, and edible (sour). They ripen August–October.

HABITAT: Hammocks, shell mounds, disturbed areas.

RANGE: Coastal southern Florida.

USES: This plant is grown for its citrus fruits.

SIMILAR SPECIES: Key Lime may be confused with Persian Lime and other *Citrus* species or hybrids.

CONSERVATION STATUS:
Although commonly cultivated, the natural Key Lime population has not been formally evaluated in the wild.

Height: Up to 25' (7.6 m)

■ DENSE POPULATION
■ LIGHT POPULATION
■ NATURALIZED

This small tree is native to eastern Asia. It has become an invasive pest in many areas in North America, particularly in Massachusetts, where it was first introduced in 1906. The gray bark is thick, deeply, furrowed, and corky, and the twigs are covered with corky spots. The leaves, twigs, and fruit have a pungent, turpentine-like odor when crushed.

DESCRIPTION: This is a small tree with deciduous, pinnately compound leaves, a short trunk, and an open crown of spreading branches. The leaves are 10–15" (25.4–38 cm) long, with five to thirteen leaflets 2–4½" (5–11.4 cm) long; they are ovate to elliptic, with a pointed tip, and untoothed margins; they are shiny, dark green, and their fall color is yellow to copper. The gray bark is thick, deeply furrowed, and corky, and the twigs are covered with corky spots.

FLOWERS: Late spring. Tiny, greenish-yellow to maroon flowers in upright, 2–3½" (5–8.9 cm) clusters.

FRUIT: ⅓–½" (8.5–12.7 mm) long, ovoid drupe, greenish yellow, turning blue-black in clusters; ripens in autumn and persists into winter, long after leaves have fallen.

HABITAT: Introduced in North America; this tree escapes cultivation, especially in woodlands, and is considered an invasive species.

RANGE: Native to eastern Asia. Introduced populations can be found growing wild in parts of the midwestern and northeastern United States and eastern Canada.

USES: This tree has long been used in Chinese traditional medicine.

SIMILAR SPECIES: Chinese Corktree.

CONSERVATION STATUS:
The conservation status of this species in its native range has not been fully assessed. In North America, this tree can be invasive, spreading vigorously in disturbed areas. Although it rarely penetrates healthy forests, it can outcompete native tree and understory species in disturbed forests and natural areas if not controlled.

Diameter: 2' (61 cm)
Height: 30–50' (9–15 m)

■ DENSE POPULATION
■ LIGHT POPULATION
■ NATURALIZED

Common Hoptree is a widespread species native to North America that includes perhaps four subspecies and many varieties with leaflets of differing sizes and shapes. The common name refers to the use of this plant's bitter fruits as a substitute for hops in brewing beer in the 19th century. Other common names include Water Ash and Hop Tree.

DESCRIPTION: This is an aromatic shrub or small tree with a rounded crown. The bark, crushed foliage, and twigs have a slightly lemonlike, unpleasant odor. The leaves are palmately compound and 4–7" (10–18 cm) long, with three leaflets at the end of a long leafstalk. The leaflets are 2–4" (5–10 cm) long, ¾–2" (2–5 cm) wide, ovate or elliptical, long-pointed at the tip, finely wavy-toothed or not toothed, with tiny gland-dots, and hairy when young. They are shiny dark green above, and paler and sometimes hairy beneath, turning yellow in fall. The bark is brownish-gray, thin, smooth or slightly scaly, and bitter. The twigs are brown, slender, covered with fine hairs, and slightly warty.

FLOWERS: Spring. ⅜" (10 mm) wide, four or five narrow greenish-white petals in terminal branched clusters including male, female, and bisexual flowers.

FRUIT: ⅞" (22 mm) in diameter; they are yellow-brown, numerous, disk-shaped, and wafer-like with a rounded wing, occurring in drooping clusters; they mature in summer and remain closed and attached in winter; they have two or three reddish-brown, long-pointed seeds.

HABITAT: Dry rocky uplands; it is als o found in valleys and canyons.

RANGE: Extreme southern Ontario east to New York and New Jersey, south to Florida, west to Texas, and north to southern Wisconsin; it is also local west to Arizona and southern Utah and in Mexico. It occurs to 8500' (2591 m) in the West.

USES: These plants are frequently planted as ornamentals. Early German immigrants used the seeds in place of hops to brew beer. The bitter bark of the root, like other aromatic barks, has been used for herbal remedies.

SIMILAR SPECIES: California Hoptree.

CONSERVATION STATUS: LC
Although the Common Hoptree is not long-lived, mature tree populations have increased, strengthening the stability of the species. The species is listed as one of Special Concern in Canada, however, where the native population is limited to southwestern Ontario along the sandy shorelines of Lake Erie, where it is threatened by the degradation of habitat quality resulting from sand mining and shoreline hardening.

Diameter: 6" (15 cm)
Height: 20' (6 m)

■ DENSE POPULATION
■ LIGHT POPULATION
■ NATURALIZED

Common Prickly-ash is native to the central and eastern United States and southeastern Canada, and is more common in the northern part of its range. The plant has aromatic, somewhat bitter-scented leaves and is an important larval food source for various butterflies. It is the northernmost representative of a tropical genus named from Greek words meaning "yellow" and "wood."

ALTERNATE NAME: *Zanthoxylum americanum*

DESCRIPTION: This is a much-branched shrub, often forming thickets, and rarely a round-crowned tree; it is aromatic and spiny, with tiny gland-dots on foliage, flowers, and fruit. The leaves are pinnately compound and 5–10" (13–25 cm) long, with five to eleven paired leaflets, 1–2" (2.5–5 cm) long; they are elliptical or ovate, blunt-pointed at the ends, with straight or slightly wavy edges, hairy when young, and stalkless. They are dull green with sunken veins above and paler and hairy on veins beneath. The bark is gray to brown and smooth. The twigs are brown or gray, and hairy when young, often with paired short, stout spines less than ⅜" (10 mm) long.

FLOWERS: In spring before leaves. Less than ³⁄₁₆" (5 mm) wide, with five spreading, fringed, yellow-green petals, arranged in short-stalked clusters and with male and female on separate plants.

FRUIT: ³⁄₁₆" (5 mm) long and podlike; elliptical, brown, and slightly fleshy; maturing in late summer and splitting open.

HABITAT: Moist soils in valleys and rocky uplands.

RANGE: Southern Ontario east to southern Quebec, south to Pennsylvania, west to central Oklahoma, and north to North Dakota; it is also local to Georgia. It occurs to 2000' (610 m).

USES: An oil extracted from the dried, bitter, aromatic bark has been used for medicinal purposes, the extract being used to treat a variety of common ailments. The fresh bark is sometimes chewed as an herbal remedy for relieving toothaches.

SIMILAR SPECIES: Hercules'-club.

CONSERVATION STATUS: LC

The Common Prickly-ash has a large population that is widely distributed. The species does not currently face any significant threats.

Diameter: 6" (15 cm)
Height: 20' (6 m) or more

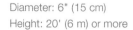

■ DENSE POPULATION
■ LIGHT POPULATION
■ NATURALIZED

Hercules'-club is a spiny tree, or sometimes a shrub, native to to the eastern coastal plain of the United States. It is an important food source for wildlife including deer, birds, and butterflies. The bark has characteristic corky, triangular knobs, and it is named for its spiny branches. It is also called "Toothache Tree" or "Tingle Tongue" because chewing the bitter, aromatic bark is used as a home remedy for numbing toothache pain.

DESCRIPTION: This is an aromatic, spiny, round-crowned tree with spreading branches and tiny gland-dots on foliage, flowers, and fruit. The leaves are pinnately compound and 5–10" (13–25 cm) long, usually with 7–17 leaflets 1½–3" (4–7.5 cm) long; they are narrowly ovate, pointed at the tip, often with small paired spines at the base, wavy edges, and almost stalk-less; they remain attached into winter and are shiny green above, and paler and sometimes hairy beneath. The bark is light gray and thin, with conspicuous corky-based conical knobs terminating in a stout spine that eventually drops off. The twigs are brown, and hairy when young, with scattered stout straight spines to ½" (12 mm) long.

FLOWERS: Spring. ³⁄₁₆" (5 mm) wide with five yellow-green petals, occurring in large much-branched clusters at the ends of leafy twigs; male and female are usually on separate plants.

FRUIT: ¼" (6 mm) long, brown elliptical pods, with three or fewer from a flower, maturing in early summer, with one shiny black seed hanging from a split pod.

HABITAT: Moist sandy soils near coast and along streams; often along fences, apparently spread by birds.

RANGE: Central Virginia through Florida, west to eastern Texas, and north to southeastern Oklahoma; to 500' (152 m).

USES: This tree has been used for medicinal purposes, particularly to treat toothaches.

SIMILAR SPECIES: Common Prickly-ash.

CONSERVATION STATUS: NT

Although its population appears to have stabilized, this species is threatened by land conversion to pine plantations in its native range. There is limited commercial demand for this medicinal plant, which is collected from the wild.

Diameter: 1' (0.3 m)
Height: 40' (12 m)

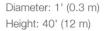

■ DENSE POPULATION
■ LIGHT POPULATION
■ NATURALIZED

Lime Prickly-ash can be found in southern Texas and Florida as well as throughout the American tropics, where it provides vital food and cover for native wildlife such as birds and butterflies. Its powdered bark and leaves have a sharp taste and have been used as a spice. The crushed foliage smells like limes. It is also sometimes called Wild Lime. The fruit is edible and numbs the mouth, with a citrusy taste reminiscent of the related Szechuan pepper.

DESCRIPTION: This is an evergreen shrub or small tree with a spreading rounded crown, often with a leaning trunk; it is aromatic and spiny, with tiny gland-dots on foliage, flowers, and fruit. The leaves are evergreen, pinnately compound, and 3–4" (7.5–10 cm) long, with a flat or winged axis. There are five to thirteen leaflets (usually 7–9) ⅜–1" (1–2.5 cm) long; they are elliptical, wavy-toothed beyond the middle, narrowed to the base, thick and leathery, and stalkless. They are shiny green above and paler beneath. The bark is gray, thin, and smooth or warty, becoming scaly. The twigs are dark gray, slender, zigzag, and hairless or nearly so, with paired hooked sharp spines less than ¼" (6 mm) long.

FLOWERS: Early spring. ⅛" (3 mm) wide with four yellow-green petals in small clusters to ½" (12 mm) wide, occurring on old twigs; male and female are on separate plants.

FRUIT: 3/16" (5 mm) long, podlike, rounded, brown, one-seeded, warty, and clustered along the twig, with one or two from a flower; maturing in fall and splitting open.

HABITAT: Moist soil mostly near the coast and on the plains.

RANGE: Central and southern Florida, Florida Keys, and southern Texas; it is also found in northern Mexico and throughout the American tropics. It occurs to 500' (152 m).

USES: This plant's powdered bark and leaves, which smell like limes, have been used as a spice.

SIMILAR SPECIES: Common Prickly-ash, Hercules'-club.

CONSERVATION STATUS: LC

The Lime Prickly-ash does not face any major threats. The species has a large population that is widespread and stable.

Diameter: 8" (20 cm)
Height: 25' (7.6 m)

■ DENSE POPULATION
■ LIGHT POPULATION
■ NATURALIZED

Ailanthus, also known as Tree-of-heaven, is native to China and grows in temperate regions, where it has been widely planted as an ornamental for shade and in shelterbelts for its rapid growth and coarse foliage that are reminiscent of tropical trees. However, it is now widely considered an invasive plant and is no longer recommended for planting. Today it occurs in nearly every U.S. state and several Canadaian provinces. It is an opportunistic and fast-growing plant that can quickly overtake disturbed habitats and sometimes higher quality areas. It is a prolific seed producer, its roots spread aggressively, and the plant produces toxins that prevent the establishment of other species.

DESCRIPTION: This is a hardy, introduced tree with a spreading, rounded, open crown of stout branches and coarse foliage. The male flowers and crushed leaves have a disagreeable odor. The leaves are pinnately compound and 12–24" (30–61 cm) long, with 13–25 leaflets (sometimes more). They are 3–5" (7.5–13 cm) long, 1–2" (2.5–5 cm) wide, paired (except at the end), and broadly lance-shaped, with two to five teeth near the broad one-sided base and a gland-dot beneath each tooth, and covered with fine hairs when young; they are green above and paler beneath. The bark is light brown and smooth, becoming rough and fissured. The twigs are light brown, stout, and covered with fine hairs when young, with brown pith.

FLOWERS: Late spring and early summer. ¼" (6 mm) long, five yellowish-green petals in terminal branched clusters 6–10" (15–25 cm) long, usually with male and female on separate trees.

FRUIT: 1½" (4 cm) long, showy, reddish-green or reddish-brown key; it is narrow, flat, winged, and one-seeded, with one to six from a flower, maturing in late summer and fall.

CAUTION: Poisonous parts: While some parts of this species are edible, other parts are toxic to humans. Eat only if you are absolutely sure which parts are edible.

HABITAT: Widespread in waste places, roadsides, old fields, and other disturbed habitats, spreading rapidly by suckers.

RANGE: Native of China but widely naturalized across temperate North America; from near sea level to high mountains.

USES: This tree has been widely planted as an ornamental. The wood is sometimes used for cabinetry and other purposes. This species has also been used in Chinese traditional medicine.

SIMILAR SPECIES: This species is rather distinct, although it may be confused with sumacs, ashes, or walnuts.

CONSERVATION STATUS:

Although this species has not been formally evaluated in its native range, it is recognized as an important exotic weed in North America. The plant is drought-resistant, has high rates of germination, and resprouts quickly and heavily after cutting. The species is difficult to eradicate because it can regrow from root fragments.

Diameter: 1–2' (0.3–0.6 m)
Height: 50–80' (15–24 m)

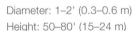

- ■ DENSE POPULATION
- ■ LIGHT POPULATION
- ■ NATURALIZED

A native of southern Asia and Australia, Chinaberry grows rapidly but is short-lived, and is often grown for timber around the world. A cultivated variety known as Umbrella-tree—which has a dense, compact, flattened crown like an umbrella—is a popular ornamental. In North America, this tree has become invasive, forming dense thickets that overtake native habitats. Its fruit is highly toxic to humans and other mammals, but not to birds, which eat the fruits and disperse the seeds of this plant into new areas.

DESCRIPTION: This is a naturalized shade tree, deciduous or nearly evergreen in the far South, with a dense, spreading crown and clusters of round, yellow, poisonous fruit; the foliage has a bitter taste and a strong odor when crushed. Its leaves are bipinnately compound and 8–18" (20–46 cm) long, with a slender green axis and a few paired forks. It has numerous leaflets 1–2" (2.5–5 cm) long, ⅜–¾" (10–19 mm) wide, paired (except at the end), lance-shaped or ovate, saw-toothed, wavy or lobed, and hairless or nearly so. They are dark green above and paler beneath. The bark is dark brown or reddish-brown and furrowed, with broad ridges. The twigs are green, stout, and hairless or nearly so.

FLOWERS: Spring. ¾" (19 mm) wide with 5 pale purple petals and a narrow violet tube; they are fragrant and occur on slender stalks in showy branched clusters 4–10" (10–25 cm) long.

FRUIT: A yellow poisonous berry ⅝" (15 mm) in diameter, becoming slightly wrinkled, with thin juicy pulp, maturing in autumn and remaining attached in winter, with a hard stone containing 3–5 seeds.

CAUTION: Poisonous: Do not eat or ingest any part of this species. If accidentally ingested, contact your doctor or poison control center.

HABITAT: Dry soils near dwellings, in open areas and clearings; sometimes within forests.

RANGE: Native of Asia. Naturalized from Virginia to Florida, west to Texas, and north to Oklahoma; also in California; usually below 1000' (305 m).

USES: This species has been used for medicinal purposes, to treat a variety of skin conditions. It is also planted as an ornamental tree and for timber. The stones inside the tree's toxic fruits can be made into beads.

SIMILAR SPECIES: Chinaberry is distinct and not likely to be confused with other species in its North American range.

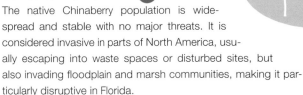

CONSERVATION STATUS: 🄻🄲
The native Chinaberry population is widespread and stable with no major threats. It is considered invasive in parts of North America, usually escaping into waste spaces or disturbed sites, but also invading floodplain and marsh communities, making it particularly disruptive in Florida.

Diameter: 1' (0.3 m)
Height: 40' (12 m)

■ DENSE POPULATION
■ LIGHT POPULATION
■ NATURALIZED

Also called American Mahogany, this small tree is a native of Florida, the Caribbean, and South America. Once prized for its rich, hard wood, this tree's numbers have declined due to a long history of exploitation and overharvesting. Additionally the species has lost considerable habitat to development. When forced to grow in poor habitat conditions, it often doesn't reach tree-size, with much of the remaining wild population growing as shrubs instead. Today, most commercially sold "mahogany" wood is from related species.

DESCRIPTION: This small tree has a large, dense, and rounded crown; the trunk is buttressed. The leaves are 7" (17.8 cm), pinnately compound, and with 8–20 ovate, long-pointed, 2½" (6.4 cm) leaflets.

FLOWERS: June–August. ⅓" (25 mm), star-shaped, and pale yellow to orange-yellow.

FRUIT: 4" (10 cm), pear-shaped, and brown.

HABITAT: Towns and hammocks (it is now uncommon in the wild).

RANGE: Extreme southern Florida (mainly coastal).

USES: This tree's hard wood has long been used for a variety of purposes, including ship building, furniture, and musical instruments. It is also grown as an ornamental tree.

SIMILAR SPECIES: Gumbo Limbo.

CONSERVATION STATUS:

Mahogany wood was used extensively for commercial and industrial purposes starting in the 16th century, and demand continued to rise through the 20th century, leading to overharvesting. Today the natural stands of Mahogany are extensively exhausted, with eroded genetic diversity in the remaining, depleted forest stands.

Height: 40' (12.2 m)

■ DENSE POPULATION
■ LIGHT POPULATION
■ NATURALIZED

American Basswood is native to eastern North America, and is found as far west as Nebraska. The northernmost basswood species, it is the dominant tree in the sugar maple–basswood forests common in western Wisconsin and central Minnesota. American Basswood is a handsome shade and street tree. When flowering, the trees are full of bees, earning it the nickname "Bee-tree." Bees seem to favor this species over others, and they produce a strongly flavored honey from its nectar.

DESCRIPTION: This is a large tree with a long trunk and a dense crown of many small, often drooping branches and large leaves, frequently with two or more trunks, and sprouts in a circle from a stump. The leaves are arranged in two rows; they are 3–6" (7.5–15 cm) long and almost as wide, broadly ovate or rounded, long-pointed at the tip, notched at the base, coarsely saw-toothed, and palmately veined, with long slender leafstalks; they are shiny dark green above, light green and nearly hairless with tufts of hairs in vein angles beneath, turning pale yellow or brown in fall. The bark is dark gray and smooth, becoming furrowed into narrow scaly ridges. The twigs are reddish or green, slender, slightly zigzag, and hairless.

FLOWERS: Early summer. ½–⅝" (12–15 mm) wide with five yellowish-white petals, fragrant, in long-stalked clusters hanging from the middle of leafy greenish bract.

FRUIT: ⅜" (10 mm) in diameter, nutlike, elliptical or rounded, gray, covered with fine hairs, and hard, with one or two seeds. It matures in late summer and fall, often persisting into winter.

HABITAT: Moist soils of valleys and uplands, in hardwood forests.

RANGE: Southeastern Manitoba east to southwestern New Brunswick and Maine, south to western North Carolina, and west to Oklahoma; to 3200' (975 m).

USES: American Basswood is a popular shade and ornamental tree. The soft, light wood is used for furniture, pulpwood, and a variety of other purposes. Native Americans made ropes and woven mats from the tough, fibrous inner bark. This species has also been used for medicinal purposes.

SIMILAR SPECIES: Red Mulberry, White Mulberry, Littleleaf Linden.

CONSERVATION STATUS: LC
The American Basswood has a large population with a wide native range. It does not currently face any major threats, although it can be affected by various minor pests and diseases. Climate change may cause habitat loss that would place new pressure on this species.

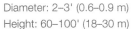

Diameter: 2–3' (0.6–0.9 m)
Height: 60–100' (18–30 m)

■ DENSE POPULATION
■ LIGHT POPULATION
■ NATURALIZED

This small to large deciduous tree, native to Europe and western Asia, has been widely cultivated in North America as an ornamental tree, often as a substitute for American Basswood. It grows best in moist, well-drained soil, but can tolerate sandy, infertile soils as well as flooding, but it is not tolerant of drought.

DESCRIPTION: This is a small to large tree with deciduous, simple leaves; a straight, short, thick trunk; and a dense, broad, rounded or sometimes columnar crown of stout branches. Older trees have a buttressed trunk base with many sprouts around it and a dome-shaped crown. The leaves can be more rounded than in other lindens. This tree often has many suckers at the base of the trunk; the branch tips touch the ground. The leaves are 1½–3" (4–8 cm) long, circular to asymmetrically heart-shaped, with an abruptly pointed tip, and finely toothed margins. They are shiny, dark green above, and pale, sometimes whitish, with whitish or rusty hairs in vein angles beneath. The fall color is yellow. The leafstalk is long, slender, and hairless. The bark is brown to dark gray, with narrow, flat fibrous ridges and long, thin, shallow fissures.

FLOWERS: Early summer. ½" (1.3 cm) wide, pale yellow to cream flowers with five petals, occurring in a cluster of five to twelve, attached to leaflike bract; they are fragrant.

FRUIT: ¼" (6.4 mm) wide, spherical nutlet; it is hard, dry, and four-ribbed, with fine gray-brown hairs, and leaflike bract. It ripens in fall and often persists into winter.

HABITAT: Introduced in North America; planted in parks and along streets.

RANGE: Native to Europe and Siberia, it is introduced into Maine, New York, Massachusetts, Connecticut, and Maryland, as well as in New Brunswick and Ontario.

USES: This species is widely cultivated as an ornamental tree or street tree. It is also used for bonsai.

SIMILAR SPECIES: American Basswood, Largeleaf Linden.

CONSERVATION STATUS: LC

Littleleaf Linden is still a widespread species across much of Europe, but its numbers are declining. Because the species is sensitive to both drought and frost, climate change is believed to have impacted forest stand health and may pose additional future risks. In North America it has naturalized locally, particularly in the Northeast.

Diameter: 1–3' (30–91 cm)
Height: 30–100' (9–30 m)

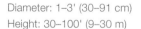

- DENSE POPULATION
- LIGHT POPULATION
- NATURALIZED

Largeleaf Linden is native to much of Europe. Species of the genus *Tilia* are called "limes" in Europe; in North America they are called "lindens" or "basswoods." It is the first linden to blossom in the summer and is planted as a street tree. Cultivated varieties include trees with red twigs, yellow branches, and twisted, sharply lobed leaves. Unlike European Linden, Largeleaf is not susceptible to aphid attack.

DESCRIPTION: This is a medium-sized tree with deciduous, simple leaves. It has a tall, straight, slender trunk, and a high, dense, rounded to broadly columnar crown of stout branches ascending at narrow angles. Largeleaf Linden is the earliest linden to flower. The leaves are the largest among European lindens, but in North America, the leaves of American Linden are larger. The leaves are 3½–6" (9–15 cm) long, and nearly as wide. They are asymmetrically heart-shaped to circular, with an abruptly pointed tip and coarsely toothed margins. They are shiny, dark green, and slightly hairy above, and paler and densely hairy in vein angles beneath. The autumn color is yellow. The leafstalk is long, slender, and white-hairy. The bark is light to dark gray, and in a ribbed pattern or in flat-topped ridges with long, narrow, shallow fissures.

FLOWERS: Summer. ¾" (1.9 cm) wide, pale yellow flowers with 5 petals, occurring in hanging, branched clusters of usually 3 flowers; the clusters are at the end of a long stalk that is attached to the middle of a 4–5" (10–13 cm) long, narrow, leaflike bract. They are very fragrant.

FRUIT: ½" (1.3 cm) wide, ovoid or spherical nutlet; it is gray-green, hard, dry, and five-ribbed, containing one or two seeds. The fruit is on a wiry stalk attached to the middle of a leaflike bract. It ripens in fall and often persists into winter.

HABITAT: Introduced in North America, where it is often planted along streets.

RANGE: Native to Europe and western Turkey.

USES: This tree has been used in traditional medicine. The wood is used for carving, firewood, and other purposes.

SIMILAR SPECIES: Silver Linden, Littleleaf Linden.

CONSERVATION STATUS: LC

Largeleaf Linden is widespread and is most common in the mountainous regions of central and eastern Europe. It is widely cultivated in residential and urban areas in North America.

Diameter: 2–3' (61–91 cm)
Height: 50–75' (15–23 m)

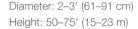

EUROPEAN LINDEN *Tilia x europaea*

Sometimes called Common Linden (or Common Lime in Europe), this hybrid of Largeleaf and Littleleaf lindens occurs naturally where the trees' European ranges overlap. It is a popular shade and lawn tree. Unlike Largeleaf Linden, European Linden is susceptible to aphid attacks, which cause the tree to exude its sticky sap, eventually causing mold. The fragrant flowers are favorites of bees.

DESCRIPTION: This is a hybrid cross between Largeleaf and Littleleaf lindens. The fruit is faintly five-ribbed. The tree often has a ring of suckers around the trunk base; burls form on the trunk. The leaves are 4" (10 cm) long (smaller than Largeleaf Linden and larger than Littleleaf Linden).

FLOWERS: Showy and fragrant clusters borne in early summer.

FRUIT: ¼" (6 mm) in diameter, round, and faintly five-ribbed, attached to a long stem.

HABITAT: Introduced in North America; it is planted along roadsides.

RANGE: Native to Europe; it is introduced in North America.

USES: These trees are often planted along roadsides or elsewhere as ornamentals or shade trees.

SIMILAR SPECIES: Littleleaf Linden, Largeleaf Linden.

CONSERVATION STATUS:
The European Linden population in the wild has not been formally evaluated. The species is widely cultivated.

Diameter: 2' (1.3 cm)
Height: 70' (21.3 m)

■ DENSE POPULATION
■ LIGHT POPULATION
■ NATURALIZED

This small, deciduous tree is native to subtropical Asia and has been cultivated in North America as an ornamental for its large leaves as far north as Washington, D.C. Opening its fruit releases a brownish-black liquid. This fast-growing tree has been widely planted, but it can be an aggressive, invasive weed in the warmer regions of North America. This tree is self-fertile and produces seeds prolifically from a single plant.

DESCRIPTION: This ornamental and naturalized tree has a crown rounded like an umbrella, with large, lobed leaves that are 6–12" (15–30 cm) long and wide. They are heart-shaped or rounded, with three or five long-pointed lobes, five main veins from the base; they are without teeth and are very long-stalked. They are dull green above and often hairy beneath. The bark is gray-green and smooth. The twigs are gray-green and stout.

FLOWERS: Spring–summer. ½" (12 mm) long and numerous, with five yellow sepals, turning red (petals absent); male and female occur in separate upright clusters 8–16" (20–41 cm) long, with scurfy, hairy branched stems.

FRUIT: 2–4" (5–10 cm) long, five produced from one flower; it is podlike, long-pointed, greenish turning light brown, and stalked, splitting open like a leaf, exposing on the edges several pealike seeds; it matures in late summer.

HABITAT: In North America, this species is increasingly escaping along roadsides and other disturbed habitats and in mixed hardwood forests.

RANGE: Native to East Asia, it is introduced in the southern United States from North Carolina south to northern Florida and west to California.

USES: Chinese Parasoltree is a popular ornamental, particularly in the southeastern United States. Its wood is used for musical instruments in China.

SIMILAR SPECIES: This tree is not likely to be confused with other species in North America.

CONSERVATION STATUS: 🅛🅒
The Chinese Parasoltree has a large population that is widely distributed in East Asia, where it faces no major threats. In North America it is introduced and naturalized in many southern U.S. states, where it can be an aggressive, invasive weed.

Diameter: 6" (15 cm)
Height: 30' (9 mm)

CALIFORNIA FLANNELBUSH *Fremontodendron californicum*

■ DENSE POPULATION
■ LIGHT POPULATION
■ NATURALIZED

California Flannelbush is widespread in California, and present but less prevalent in Arizona. This is a species of chaparral, pinyon scrublands, and sagebrush scrublands that occurs on nutrient-poor, dry, rocky soils of foothills and slopes. When flowering, the showy plants are conspicuous from a distance. The beautiful masses of large, yellow flowers and odd, small leaves covered with scurfy hairs make this plant an attractive ornamental. The name "Flannelbush" refers to the plant's densely hairy foliage.

DESCRIPTION: This is an evergreen, many-branched, thicket-forming shrub or small tree with a short trunk, an open crown, and large, showy, bright yellow flowers. The leaves are evergreen, ½–1½" (1.2–4 cm) long and wide, rounded or broadly ovate, usually with three blunt lobes and three main veins, thick, long-stalked, and mostly occurring on short side-twigs. They are dull dark green and sparsely hairy above, and covered with rust-colored and scurfy hairs and showing raised veins beneath. The bark is brownish-gray, fissured and scaly; the inner bark is mucilaginous. The twigs are stout and stiff, and densely covered with rust-colored hairs when young, becoming reddish-brown.

FLOWERS: Spring–early summer. 1¼–2" (3–5 cm) wide with five broad, bright yellow calyx lobes and without petals; they are borne singly on short side-twigs or opposite leaves.

FRUIT: 1–1¼" (2.5–3 cm) long, pointed, densely hairy, egg-shaped capsule; it is four- or five-celled, ripening in late summer and splitting open on four or five lines, with many elliptical, dark reddish-brown seeds.

HABITAT: Dry, rocky mountain slopes and canyons; with chaparral, pinyons, junipers, and Ponderosa Pine.

RANGE: California, central Arizona, and northern Baja California; at 3000–6500' (914–1981 m).

USES: This species is grown as an ornamental plant and is sometimes used in habitat restoration projects. Some Native Americans used the inner bark's sap for medicinal purposes.

SIMILAR SPECIES: Southern Flannelbush.

CONSERVATION STATUS: NT

The natural population of California Flannelbush is not large—known from less than 300 occurrences, most of which are in California—but appears stable. Because of the species' fast growth rate and plentiful seed production, the California Flannelbush is more resistant to common threats, including fires. The Pine Hill Flannelbush, a shrub sometimes considered a subspecies of the California Flannelbush, is on the U.S. Endangered Species List and is classified as endangered in California.

Diameter: 8" (20 cm)
Height: 20' (6 m)

■ **DENSE POPULATION**
■ **LIGHT POPULATION**
■ **NATURALIZED**

This long domesticated species likely originated in the eastern valleys of the Andes. This tropical fruit has become widely cultivated and naturalized in southern Florida, Bermuda, and throughout the West Indies, Mexico, Central America, and northern South America. In Florida, this rapidly growing tree often grows as a weed.

DESCRIPTION: This fast-growing, short-lived tree has a narrow crown and a tall, soft trunk. Its leaves are 24" (61 cm) long, evergreen, and deeply palmately seven-lobed, with main lobes pinnately lobed. The bark is green to grayish-brown and smooth.

FLOWERS: Year-round. 2" (5 cm), trumpet-shaped, yellow flowers.

Height: 10' (3 m)

FRUIT: Papayas: to 4" (10 cm), round to pear-shaped, green to orange, and clustered near the treetop; they are edible and ripe year-round.

HABITAT: Hammocks, shell mounds, and thickets.

RANGE: Found in central Florida and southward. It is local and mainly coastal.

USES: This tree is widely cultivated for its fruit, which is eaten raw, or cooked, in many parts of the world.

SIMILAR SPECIES: Papaya is not likely to be confused with other trees in North America.

CONSERVATION STATUS:
Papaya does not exist in a natural state in the wild, and is the product of centuries of domestication, introduction, and naturalization across much of the American tropics.

■ DENSE POPULATION
■ LIGHT POPULATION
■ NATURALIZED

This attractive plant is native to Florida and is also found in the Caribbean. Its beautiful, fragrant flowers make Jamaica Caper Tree a popular ornamental, and are also known to attract hummingbirds and butterflies.

DESCRIPTION: This is a sprawling shrub to a small tree with arched branches. The leaves are 4" (10 cm), elliptical to obovate, and leathery; they are yellow-green above and silvery brown below. The bark is dark red-brown, thin, and cracked.

FLOWERS: April–July. ½" (1.3 cm), white to bright purple.

FRUIT: 12" (30 cm) pods are scaly and reddish.

HABITAT: Coastal hammocks, and shell mounds.

RANGE: Coastal from central Florida south throughout the Caribbean and Central America.

USES: This species is planted as an ornamental and to attract wildlife. It is sometimes used locally for medicinal purposes.

SIMILAR SPECIES: Bay-leaved Caper.

CONSERVATION STATUS: LC

With a large and widely distributed population in the American tropics, the Jamaica Caper Tree is considered stable. The species faces no major threats.

Height: 20' (6 m)

■ DENSE POPULATION
■ LIGHT POPULATION
■ NATURALIZED

Native to China and Korea, Five-stamen Tamarisk was introduced to North America in the early 1900s as an ornamental and for erosion control. Thickets of this plant can provide cover and nesting sites for birds and nectar for pollinators. It has become an undesirable weed in many places, especially in the Southwest, known for having deep roots and consuming large amounts of water. Eradication of the plant is difficult because it spreads by both seeds and cuttings and grows rapidly.

DESCRIPTION: This is a naturalized shrub or small tree with slender, upright or spreading branches and a narrow or rounded crown; it resembles a juniper, but it is not evergreen. The leaves are about 1/16" (1.5 mm) long, scalelike, crowded, and narrow and pointed; they are dull blue-green. The bark is reddish-brown and smooth, becoming furrowed and ridged. The twigs are green, becoming purplish, and long, slender, and hairless, usually shedding with the leaves.

FLOWERS: Spring and summer. Less than 1/8" (3 mm) long and wide with five pink petals; they are numerous and crowded together in narrow clusters 3/4–2" (2–5 cm) long at the ends of the twigs.

FRUIT: 1/8" (3 mm) long, narrow, pointed, reddish-brown capsules, splitting into three to five parts, with many tiny, hairy seeds; they mature in summer.

HABITAT: Wet, open areas along streams, irrigation ditches, and reservoirs, including sand banks and alkali and salty soils.

RANGE: Extensively naturalized from southwestern Nebraska west to Nevada and south to southern California and southern Texas; local beyond; also in northern Mexico; to 5000' (1524 m).

USES: This species was once a popular ornamental plant but is now considered an invasive pest in many places.

SIMILAR SPECIES: Smallflower Tamarisk.

CONSERVATION STATUS:
The conservation status of Five-stamen Tamarisk has not been fully assessed in its native range. This species spreads aggressively in parts of North America and is sometimes considered a noxious weed.

Diameter: 4" (10 cm)
Height: 16' (5 m)

■ DENSE POPULATION
■ LIGHT POPULATION
■ NATURALIZED

Pigeon-plum is native to southern Florida, the Caribbean, and Central America. It is able to withstand winds and saline soils characteristic of the coasts. The large leathery leaves of this tree drop off all at once in early spring in Florida, and are soon followed by new leaves, which are red when they first appear. It can make an attractive shade tree, but the abundant fruit it drops can pose a litter problem on streets and sidewalks. The small, fleshy fruits attract birds.

DESCRIPTION: This is a broadleaf semi-evergreen tree with a rounded crown and dense, spreading branches. It may have more than one trunk. The leaves are 4" (10 cm) long, ovate to lance-olate, and evergreen. The bark is peeling and gray mottled with brown.

FLOWERS: June. Greenish-white flowers in 2–3" (5–7.5 cm) clusters.

FRUIT: ½" (12 mm), berrylike, and dark red.

HABITAT: Coastal hammocks.

RANGE: East-central Florida and south to the Keys; it is also found in the Caribbean and Central America.

USES: This species has been used for medicinal purposes, and the edible fruits are used to make jelly and wine.

SIMILAR SPECIES: Sea Grape.

CONSERVATION STATUS: LC
The population is stable and has a widespread distribution in its native range. The species faces no significant threats.

Height: usually 15–25' (4.5–7.5 m);
crown spreads to 35' (10.5) wide

■ DENSE POPULATION
■ LIGHT POPULATION
■ NATURALIZED

Commonly called Sea Grape or Bay Grape, this native flowering plant is found in Florida and throughout the Caribbean. It is planted as an ornamental and to control beach erosion and as a windbreak. The fruits are edible raw and can be used to make jam or wine, and are a valuable food for wildlife.

DESCRIPTION: This is a sprawling shrub to a compact tree, with a short, gnarled trunk (or trunks). Its leaves are up to 10" (25 cm) in diameter, roundish, pink-veined, leathery, and evergreen. The bark is light brown with pale blotches, and smooth.

FLOWERS: Tiny, white, fragrant flowers in slender spikes to 10" (25 cm) long.

FRUIT: ¾" (2 cm) diameter fruits, green ripening to red, in grape-like clusters; they are ripe August–September.

HABITAT: Beaches, hammocks, scrub, dunes, towns.

RANGE: Coastal, from north-central Florida southward.

USES: This plant is a popular ornamental, particularly in southern Florida landscaping. The wood is sometimes used for furniture or firewood.

SIMILAR SPECIES: Pigeon-plum.

CONSERVATION STATUS:
The Sea Grape is common along its native range. There are currently no known significant threats to the species.

Height: to 50' (15 m);
usually 12–36' (3.6–10.8 m)

■ DENSE POPULATION
■ LIGHT POPULATION
■ NATURALIZED

This species is native to eastern North America, where it is more common farther north, as far as Newfoundland. Alternate-leaf Dogwood is so named because, unlike all other native dogwoods, this species has alternate rather than opposite leaves. Another of its common names, Pagoda Dogwood, alludes to the flat-topped crown with horizontal layers of branches. A variety of wildlife consumes the bitter, berrylike fruits of this and other dogwoods in fall and winter.

DESCRIPTION: This is a shrub or small tree with a short trunk and a flat-topped, spreading crown of long, horizontal branches. Its leaves are alternate and clustered at the ends of twigs; they are 2½–4½" (6–11 cm) long, 1–2" (2.5–5 cm) wide, elliptical, appearing not toothed but with tiny teeth on the edges, with five or six long, curved veins on each side of the midvein; they are slender-stalked, green, and nearly hairless above, and paler or whitish with pressed hairs beneath, turning yellow or red in fall. The bark is gray or brown, and smooth or fissured into narrow ridges. The twigs are greenish and slender, branching singly.

FLOWERS: May. ¼" (6 mm) wide with four spreading white petals in upright branched and flat clusters about 2" (5 cm) wide at the end of leafy twigs.

FRUIT: About ¼" (6 mm) in diameter, berrylike, blue-black, and numerous, occurring on red stalks, with thin bitter pulp and a stone containing one or two seeds; it matures in late summer.

HABITAT: Moist soils in understory of hardwood and coniferous forests.

RANGE: Southern Manitoba east to Newfoundland, south to northwestern Florida and west to northern Arkansas; to 6500' (1981 m) in the southern Appalachians.

USES: This dogwood is often planted as an ornamental. It has also been used in traditional medicine.

SIMILAR SPECIES: Flowering Dogwood.

CONSERVATION STATUS: LC

The Alternate-leaf Dogwood is common and covers an expansive native range, although it is rarer and state-endangered in parts of the South. The species is stable and faces no known current or future threats.

Diameter: 6" (15 cm)
Height: 25' (7.6 m)

■ DENSE POPULATION
■ LIGHT POPULATION
■ NATURALIZED

This dogwood, native to the midwestern and south-central United States, is easily recognized by the rough, upper leaf surfaces, and white fruit. Its fruit is an important food source for dozens of species of native songbirds. This plant spreads from root sprouts and can form dense thickets that provide cover for wildlife; various small birds such as Bell's Vireo (*Vireo bellii*) nest in thickets.

DESCRIPTION: This is a thicket-forming shrub or sometimes a small tree with a short trunk and an open, spreading crown. The leaves are opposite, 1½–3½" (4–9 cm) long, 1¼–2" (3–5 cm) wide, elliptical; they are without teeth, have three to five long curved veins on each side of the midvein, and are densely covered with white hairs when young. They are green and rough with tiny stiff hairs above, and pale and covered with soft hairs beneath. The bark is gray-brown or reddish brown, thin, and finely fissured. The twigs are light green and covered with fine hairs when young, turning gray or reddish-brown; they are slender with a brownish pith.

FLOWERS: Late spring. ¼" (6 mm) wide with four spreading white petals, occurring in upright, branched, somewhat flat clusters 2–3" (5–7.5 cm) wide, at the ends of leafy twigs.

FRUIT: ¼" (6 mm) in diameter, berrylike, and white, with thin bitter pulp and arranged in loose clusters on red stalks; it has a stone containing one or two seeds and matures in late summer and fall.

HABITAT: Along streams and in dry uplands, forming thickets at forest borders in prairie grasslands and in understory of hardwood forests.

RANGE: Extreme southern Ontario and southern Michigan south to Louisiana, west to central Texas, and north to southeast South Dakota; to 2000' (610 m).

USES: This species is often used in landscaping as an ornamental. It is also an important food source and nesting site for birds and other wildlife.

SIMILAR SPECIES: Toughleaf Dogwood.

CONSERVATION STATUS: 🔵LC
The Roughleaf Dogwood is widespread and considered common. The species has no known conservation threats.

Diameter: 6" (15 cm)
Height: 20' (6 m)

■ DENSE POPULATION
■ LIGHT POPULATION
■ NATURALIZED

Flowering Dogwood is one of the most beautiful eastern North American trees with showy early spring flowers, red fruit, and scarlet fall foliage. It is found in a variety of habitats and is an important understory species in the eastern deciduous and southern coniferous forests. It is often planted as an ornamental in residential areas and around office buildings for its attractive flowers, foliage, and bark. Although poisonous to humans, its fruit is a favorite fall and winter food for squirrels and many native birds, and deer eagerly consume the leaves and twigs.

DESCRIPTION: Flowering Dogwood is a lovely, small, flowering tree with a short trunk and a crown of spreading or nearly horizontal branches. The leaves are opposite, 2½–5" (6–13 cm) long, 1½–2½" (4–6 cm) wide, and elliptical; they have slightly wavy edges, appearing not toothed but with tiny teeth visible under a lens; they have six or seven long, curved veins on each side of the midvein, and are short-stalked. They are green and nearly hairless above, and paler and covered with fine hairs beneath, turning bright red above in fall. The bark is dark reddish-brown and rough, broken into small square plates. The twigs are green or reddish and slender, becoming hairless.

FLOWERS: Early spring before leaves. The flowers are ³⁄₁₆" (5 mm) wide with four yellowish-green petals, tightly crowded in a head ¾" (19 mm) wide, bordered by four large broadly elliptical white petal-like bracts (pink in some cultivated varieties) 1½–2" (4–5 cm) long.

FRUIT: ⅜–⅝" (10–15 mm) long, berrylike, elliptical, and shiny red, with several occurring at the end of a long stalk; it has thin mealy bitter pulp and a stone containing one or two seeds; they mature in fall.

CAUTION: Poisonous: The fruits of this species are poisonous to humans. If accidentally ingested, contact your doctor or poison control center.

HABITAT: Both moist and dry soils of valleys and uplands in understory of hardwood forests; also in old fields and along roadsides.

RANGE: Southern Ontario east to southwest Maine, south to northern Florida, west to central Texas, and north to central Michigan; it occurs to 4000' (1219 m), and to almost 5000' (1524 m) in the southern Appalachians.

USES: This species is a popular ornamental tree. Its hard wood is strong and shock-resistant, making it useful for a variety of specialty items such as golf club heads, spools, pulleys, mallet heads, butcher blocks, and jeweler's blocks. Native Americans used the aromatic bark and roots as a remedy for malaria and extracted a red dye from the roots.

SIMILAR SPECIES: Rusty Blackhaw, Alternate-leaf Dogwood.

CONSERVATION STATUS:
Common throughout its vast native range, the Flowering Dogwood has a large population that is considered stable. Although the species faces minor threats from fungal diseases and pests, they have not led to decreases in population size.

Diameter: 8" (20 cm)
Height: 30' (9 m)

Kousa Dogwood is a small tree native to eastern Asia. It is widely planted as an ornamental in North America, where its abundant, pinwheel-like white flowers typically blossom two to three weeks later than the native Flowering Dogwoods. Other common names for this tree include Chinese Dogwood, Korean Dogwood, and Japanese Dogwood.

DESCRIPTION: This is a small tree with deciduous, simple leaves; a short trunk, usually multi-stemmed; and a broad crown of low, slender, spreading branches. In spring, the trees blossom so profusely that the crowns look like large, white mounds. Unlike North American and European dogwoods, Kousa Dogwood has hanging, raspberry-like fruits; the bark and tree shape are also distinctive. The leaves are 2½–4" (6.4–10 cm) long and broadly elliptic, with a pointed tip, slightly wavy margins, and six or seven parallel veins on each side, curved toward the leaf tip. They are dark green above and paler beneath. The fall colors are scarlet, purple, and bronze. The bark is smooth, and mottled tan and brown. The twigs are purple to green, turning brown.

FLOWERS: Late spring. Tiny, yellowish green flowers in a tightly packed, ¾" (1.9 cm) wide cluster, surrounded by four large, pale yellow to creamy white (sometimes pink-tinged), pointed, petal-like bracts that form what appears to be a showy flower, 2–3½" (5–8.9 cm) wide.

FRUIT: ⅝–¾" (1.5–1.9 cm) wide, raspberry-like, rounded cluster of fused drupes; they are pink to rose-red. The thin, rough rind contains thick, juicy, orange pulp. Each cluster hangs at the end of a long stalk. They ripen in late summer and persist into fall.

HABITAT: Introduced in North America and widely planted for ornament; the tree is native to mountain forests.

RANGE: Native to the mountain forests of Japan, Korea, and China's Sichuan Province. It is introduced in North America and has naturalized locally in southern New York and northern New Jersey.

USES: This small tree is a popular ornamental. The edible berries are sometimes made into jam.

SIMILAR SPECIES: Flowering Dogwood.

CONSERVATION STATUS: LC
The Kousa Dogwood has a large population with wide distribution. It does not face any current or future threats.

Diameter: 6–8" (15–20 cm)
Height: 15–20' (4.5–6 m)

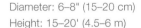

Native to southern Europe and southwestern Asia, this dogwood is widely planted as an ornamental in North America. Some cultivated varieties show leaves with pink, yellow, or white margins. It is hardy north to southern Canada. The flowers appear before the leaves and the fruit is an oblong, dark red, fleshy, edible drupe.

DESCRIPTION: This is a small, widely planted ornamental dogwood with attractive, early-blooming flower clusters. The leaves are 2½–5" (6.4–13 cm) long with an abruptly pointed tip. The fall color is reddish-purple.

FLOWERS: Early spring. Yellow flowers in 1" (2.5 cm) wide clusters, lacking bracts.

FRUIT: An oblong, red drupe, ripening in mid- to late summer. It is fleshy and edible, reminiscent of a coffee berry.

HABITAT: Introduced in North America; it is often planted along streets and in parks.

RANGE: A native of Southern Europe and Southwest Asia; it is widely cultivated.

USES: This species is a popular ornamental plant. Its wood is dense and useful for tools. The edible fruit is used for jams, distilling vodka, and other culinary purposes in Europe. It has also been used in traditional medicine.

SIMILAR SPECIES: The yellow flowers help to distinguish this species from other dogwoods in North America.

CONSERVATION STATUS:
Because of its wide distribution and high resistance to diseases and pests in the wild, Cornelian-cherry is believed to have a large and stable population, although it has not been formally assessed. Cultivated Cornelian-cherry trees have shown some vulnerability to fungal infections in orchards. The biggest potential threat to the species is possible genetic degradation resulting from deforestation and vegetative cultivation using tubers and cuttings.

Diameter: 6–12" (15–30 cm)
Height: 15–25' (4.6–7.6 m)

■ DENSE POPULATION
■ LIGHT POPULATION
■ NATURALIZED

Pacific Dogwood is one of the most handsome native ornamental trees on the Pacific Coast, with showy flowers and fruit. The apparently large, white-petaled flowers are somewhat misleading. The large white "petals" are actually long, broad bracts that surround the true small and inconspicuous flowers, produced in a dense, rounded, greenish-white flowerhead. American ornithologist and artist John James Audubon (1780–1851) painted this tree in his famous work *Birds of America* and named it for its collector, Thomas Nuttall (1786–1859), the British-American botanist and ornithologist.

DESCRIPTION: This tree has a dense, conical or rounded crown of often horizontal branches and beautiful white flower clusters. Its leaves are opposite, 2½–4½" (6–11 cm) long, 1¼–2¾" (3–7 cm) wide, and elliptical, with slightly wavy edges and five or six long, curved veins on each side of the midvein. They are shiny green and nearly hairless above and paler with woolly hairs beneath, turning orange and red in autumn. The bark is reddish-brown, thin, and smooth or scaly. The twigs are slender, and light green and hairy when young, becoming dark red or blackish.

FLOWERS: Spring and early summer, often again in late summer or fall. ¼" (6 mm) wide flowers with four greenish-yellow petals, with many crowded together in a head 1" (2.5 cm) wide, usually bordered by six (sometimes four to seven) large, elliptical, short-pointed, white (sometimes pinkish) petal-like bracts.

FRUIT: ½" (12 mm) long elliptical, shiny red or orange fruit with thin, mealy, bitter pulp and a stone containing one or two seeds, with many crowded together in a head 1½" (4 cm) across; it matures in fall.

HABITAT: Moist soils in mountains in understory of coniferous forests.

RANGE: Southwest British Columbia south to western Oregon and in the mountains to southern California; to 6000' (1829 m).

USES: This dogwood is a popular ornamental tree in western North America.

SIMILAR SPECIES: Flowering Dogwood.

CONSERVATION STATUS:

The Pacific Dogwood population is stable, and it is common throughout its discontinuous native range. Although no significant current or future threats to the species have been identified, a small, disjunct population in Idaho is listed as critically imperiled.

Diameter: 1' (0.3 m), rarely larger
Height: 50' (15 m)

GRAY DOGWOOD *Cornus racemosa*

- ■ DENSE POPULATION
- ■ LIGHT POPULATION
- ■ NATURALIZED

This small, shrubby dogwood is native to southeastern Canada and the eastern United States. These plants are natural early successional components of many woodland ecosystems in North America, and are adapted to take advantage of open areas with rapid growth and vigorous seeding. Birds and other wildlife commonly eat the fruit. It sometimes hybridizes with Silky Dogwood.

DESCRIPTION: Gray Dogwood is a thicket-forming, deciduous shrub, sometimes a small tree, with greenish-white blossoms in open, terminal clusters. Fall foliage can be reddish or purple. Young twigs are reddish and the fruit pedicels remain conspicuously red into late fall and early winter.

FLOWERS: Small, white flowers with four petals, occurring together in rounded clusters 1–2" (2.5–5 cm) wide.

FRUIT: A white ¼" (.64 cm) drupe that usually does not remain on the plant for long, attached by conspicuous red pedicels.

HABITAT: Thickets, riverbank woods, and low, open areas.

RANGE: Maine to Ontario and Manitoba, south to South Carolina and Arkansas.

USES: This dogwood is sometimes used in landscaping, particularly along ponds or other wet areas. It is an important food source for many birds and other wildlife.

SIMILAR SPECIES: Roughleaf Dogwood, Red Osier Dogwood.

CONSERVATION STATUS: **LC**
The Gray Dogwood has a stable population. There are no known threats to the species. A vigorous grower, its invasion into prairies and wetlands can be problematic for those sensitive ecosystems.

Height: 16' (4.9 m)

■ DENSE POPULATION
■ LIGHT POPULATION
■ NATURALIZED

This plant is usually a large shrub found growing in moist soils throughout much of Canada and across the northern United States. It can rarely take the form of a small tree. It is useful for erosion control on stream banks. The branch tips of this plant will root upon touching the ground and form new shoots. The common name recalls the resemblance of the reddish twigs to those of some willows (also called osiers).

DESCRIPTION: This is a large, spreading, thicket-forming shrub with several stems, clusters of small white flowers, and small whitish fruit; it is rarely a small tree. Its leaves are opposite, 1½–3½" (4–9 cm) long, and ⅝–2" (1.5–5 cm) wide; they are elliptical or ovate, short- or long-pointed, and without teeth, with five to seven long, curved, sunken veins on each side of the midvein. They are dull green above and whitish green and covered with fine hairs beneath, turning reddish in fall. The bark is gray or brown, and smooth or slightly furrowed into flat plates. The twigs are purplish-red, slender, and hairy when young, with rings at nodes.

FLOWERS: Late spring–early summer. ¼" (6 mm) wide with four spreading, white petals, with many crowded in upright, flattish clusters 1¼–2" (3–5 cm) wide.

FRUIT: Whitish, juicy berry ¼–⅜" (6–10 mm) in diameter, with a stone with two seeds, maturing in late summer.

HABITAT: Moist soils, especially along streams; it is found forming thickets and in understory of forests.

RANGE: Central Alaska east to Labrador and Newfoundland, south to northern Virginia, and west to California; it is also found in northern Mexico. It occurs to 5000' (1524 m), and to 9000' (2743 m) in the Southwest.

USES: Red Osier Dogwood is often used for erosion control on stream banks. Some Native Americans used the berries for medicinal purposes.

SIMILAR SPECIES: Roundleaf Dogwood.

CONSERVATION STATUS: LC
This species is widespread and globally secure. Its numbers are increasing globally but it may be declining in parts of its range in the eastern U.S.

Diameter: 3" (7.5 cm)
Height: commonly 3–10' (0.9–3 m),
rarely to 15' (4.6 m)

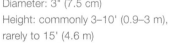

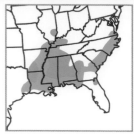

- ■ DENSE POPULATION
- ■ LIGHT POPULATION
- ■ NATURALIZED

Water Tupelo is a large tree that grows in the swamps and floodplains of the southeastern United States. The word Tupelo is from Muscogee (Creek) Native American words meaning "swamp tree." The tree grows in situations in which its roots are periodically under water. The spongy wood of the roots has served locally as a substitute for cork in floats of fish nets. This genus of aquatic trees was named *Nyssa* after one of the ancient Greek water nymphs, or spirits of lakes and rivers.

DESCRIPTION: This is a large aquatic tree with a swollen base; a long, straight trunk; a narrow, open crown of spreading branches; and large, shiny leaves. Those leaves are 5–8" (13–20 cm) long and 2–4" (5–10 cm) wide (sometimes larger). They are ovate, often with a few large teeth, and slightly thickened, with long hairy leafstalks. They are shiny dark green above and paler and hairy beneath. The bark is dark brown or gray and furrowed into scaly ridges. The twigs are reddish-brown, stout, and hairy when young.

FLOWERS: Early spring. Greenish, with many male flowers ¼" (6 mm) long in heads ⅝" (15 mm) wide; solitary female flowers are ⅜" (10 mm) long; male and female are usually on separate trees.

FRUIT: 1" (2.5 cm) long, oblong, berrylike, and dark purple, with thin sour pulp, and a stone with 10 winglike ridges; they mature in early fall.

HABITAT: Swamps and flood plains of streams, close to the water, where submerged a few months each winter and spring; often in pure stands.

RANGE: Southeastern Virginia south to northern Florida, west to southeastern Texas, and north to southern Illinois; to 500' (152 m).

USES: The wood from this tree is sometimes used for furniture and other purposes. The fruit attracts a variety of wildlife.

SIMILAR SPECIES: Ogeechee Tupelo.

CONSERVATION STATUS: LC

With a wide native range, the Water Tupelo is considered stable. The species has no known threats.

Diameter: 3' (0.9 m)
Height: 100' (30 m)

■ DENSE POPULATION
■ LIGHT POPULATION
■ NATURALIZED

This small tree has a limited range in the southeastern United States, found in southernmost South Carolina, southeastern Georgia, and northern Florida. The fruit is juicy and edible but sour, and is used to make Ogeechee lime preserves. Its natural propagation is by sprouts from roots and stumps as well as from seeds. It is known to attract various wildlife, notably pollinators, and is an excellent honey tree often planted in bee farms. The species is named for the Ogeechee River in southeastern Georgia, where the tree is found in abundance and was first described.

DESCRIPTION: This small tree has one or more crooked, leaning trunks and a narrow, rounded crown; it is sometimes a much-branched shrub. The leaves are 4–5½" (10–14 cm) long, 2–2½" (5–6 cm) wide, oblong to elliptical, blunt-pointed, not toothed (rarely with a few teeth), and slightly thickened. They are shiny dark green and slightly hairy above, and paler beneath, with velvety hairs on veins. The bark is dark brown, thin, and irregularly fissured into scaly plates. The twigs are light reddish-brown, covered with rust-colored hairs, and slender.

FLOWERS: Early spring. Greenish and hairy, occurring on short stalks in back of new leaves. Many tiny male flowers are clustered in heads ½" (12 mm) wide; solitary female flowers are ³⁄₁₆" (5 mm) long; male and female are usually on separate trees.

FRUIT: 1–1½" (2.5–4 cm) long, berrylike, oblong, and shiny or dull red, with thick juicy sour pulp, and a stone with 10–12 papery wings; it matures in late summer and remains attached in fall.

HABITAT: Permanently wet soils bordering streams, swamps, and lakes, and in swamp forests; it is usually found within 2' (0.6 m) of water level and in areas flooded for long periods.

RANGE: Extreme southern South Carolina, southern Georgia, and northern Florida; to 250' (76 m).

USES: The fruit, known as Ogeechee lime, is made into preserves and is used as a lime juice substitute.

SIMILAR SPECIES: Water Tupelo.

CONSERVATION STATUS: LC
The Ogeechee Tupelo population is stable and faces no significant threats.

Diameter: 1' (0.3 m)
Height: 30–40' (9–12 m)

■ DENSE POPULATION
■ LIGHT POPULATION
■ NATURALIZED

Black Tupelo occurs in a wide range in eastern North America, but it grows best in the Southeast. A handsome ornamental and shade tree, Black Tupelo is also a honey plant. Many birds and mammals consume the juicy fruit, and the flowers attract pollinators. Other common names for this tree include Sour Gum, Black Gum, Pepperidge, or simply Tupelo. A variety called Swamp Tupelo (var. *biflora*) with narrower, oblong leaves occurs in swamps in the Coastal Plain from Delaware to eastern Texas.

DESCRIPTION: This tree has a dense, conical or sometimes flat-topped crown, many slender, nearly horizontal branches, and glossy foliage turning scarlet in autumn. The leaves are 2–5" (5–13 cm) long and 1–3" (2.5–7.5 cm) wide. They are elliptical or oblong, not toothed (rarely with a few teeth), slightly thickened, and often crowded on short twigs. They are shiny green above and pale and often hairy beneath, turning bright red in early fall. The bark is gray or dark brown, thick, rough, and deeply furrowed into rectangular or irregular ridges. The twigs are light brown, slender, and often hairy, with some short spurs.

FLOWERS: Greenish and occurring at the end of long stalks at the base of new leaves in early spring, with many tiny male flowers in heads ½" (12 mm) wide, or two to six female flowers ³⁄₁₆" (5 mm) long. Male and female flowers are usually on separate trees.

FRUIT: ⅜–½" (10–12 mm) long, berrylike, elliptical, blue-black fruits with thin bitter or sour pulp, and a stone slightly 10- to 12–ridged; they mature in fall.

HABITAT: Moist soils of valleys and uplands in hardwood and pine forests.

RANGE: Extreme southern Ontario east to southwestern Maine, south to southern Florida, west to eastern Texas, and north to central Michigan; it is also local in Mexico. It occurs to 4000' (1219 m), sometimes higher in the southern Appalachians.

USES: Black Tupelo is commonly planted as an ornamental or shade tree. It is a major source of wild honey and the wood is sometimes used for pallets, pulpwood, firewood, and other purposes.

SIMILAR SPECIES: Water Tupelo, Ogeechee Tupelo, Common Persimmon, Bigleaf Snowbell.

CONSERVATION STATUS: Ⓛ🄲
Black Tupelo is common and widespread, and its population is presumed to be stable, with no significant threats identified.

Diameter: 2–3' (0.6–0.9 m)
Height: 50–100' (15–30 m)

■ DENSE POPULATION
■ LIGHT POPULATION
■ NATURALIZED

Ocotillo is an unusual cactus-like plant found in the Sonoran and Chihuahuan deserts in the southwestern United States. During most of the year the plant appears to consist of large, dead stems, but after rainfall the plant sprouts copious small green leaves that turn it a vibrant green. The leaf stalks harden into blunt spines, and new leaves sprout from the base of the spine. It is the only species in its family found north of Mexico.

DESCRIPTION: This is a large, erect, funnel-shaped shrub with several woody, almost unbranched, spiny, commonly straight stems; it is leafless most of the year, and has a tight cluster of red flowers at each branch tip. The leaves are up to 2" (5 cm) long, narrowly ovate, broader above the middle, and nearly stalkless, occurring in bunches above the spines shortly after rainfall.

FLOWERS: March–June, sometimes later. The corolla is ½–1" (1.5–2.5 cm) long and tubular, with five short, curled-back lobes; they occur in clusters to 10" (25 cm) long.

HABITAT: Open stony slopes in deserts.

RANGE: Southeastern California east to western Texas and south to northern Mexico.

USES: Ocotillo is sometimes used in native gardens and landscaping. The branches are sometimes used for canes or walking sticks. It has also been used for medicinal purposes, to treat a variety of ailments including urinary tract infections.

SIMILAR SPECIES: Compare Ocotillo to other species within the genus, such as Boojum Tree, all native to Mexico.

CONSERVATION STATUS: LC
The Ocotillo is a common species in its range. Although relatively robust, potential threats to these plants include habitat loss from urbanization, as well as severe wildfires that can damage the root crown and inhibit resprouting.

Height: to 30' (9 m)

■ DENSE POPULATION
■ LIGHT POPULATION
■ NATURALIZED

Satinleaf is native to Florida and elsewhere in the American tropics, and can grow in a variety of soil types but is sensitive to cold. The plant is named for the lower surfaces of its leaves, which have a smooth, satiny feel. Satinleaf is listed as an endangered species in Florida, but is widely cultivated and regularly used in landscaping for its attractive foliage. The fruit is edible but chewy.

DESCRIPTION: This shrub or small tree has evergreen, simple leaves. It has a straight trunk and an open, rounded crown. The leaves are 2–3" (5–7.6 cm) long and elliptic; they are shiny, dark green above and coppery brownish and densely hairy beneath. The bark is reddish brown.

FLOWERS: Year-round. ⅛" (3.2 mm) wide, white flowers with five spreading lobes and borne in leaf axils.

FRUIT: ½–¾" (1.3–1.9 cm) long, one-seeded berry; it is oblong to ovoid, fleshy, and purple when ripe; it is ripe year-round.

HABITAT: Tropical and coastal hammocks.

RANGE: It is found in central and southern Florida and the Florida Keys at sea level.

USES: This species is widely cultivated as an ornamental. The strong wood is sometimes used in construction and for fence posts and other purposes. The fruit is edible and usually eaten fresh.

SIMILAR SPECIES: In southern Florida, the tree is similar only to Star-apple (*C. cainito*), a strictly cultivated species that has much larger, 2" (5 cm) fruit.

CONSERVATION STATUS: Ⓥ

In Florida, this species is listed as endangered in the wild, although many cultivated specimens can still be found in south Florida and elsewhere. It is introduced and invasive in Hawaii.

Diameter: 3–6" (7.6–15 cm)
Height: 6–25' (1.8–7.6 m)

■ DENSE POPULATION
■ LIGHT POPULATION
■ NATURALIZED

This flowering plant, commonly known as Saffron-plum or Coma, is native to Texas and Florida in the United States. It is also found in parts of Mexico, Central America, and South America. Deer often browse the leaves of this plant, and other wildlife is attracted to the fruit.

DESCRIPTION: This species has the largest fruit of the tree bullies. The leaves are 1–1½" (2.5–3.8 cm) long and hairless beneath. The twigs are armed with ½" (1.3 cm) thorns; otherwise similar to Gum Bumelia.

FLOWERS: Creamy white flowers with five stamens.

FRUIT: ⅜–1" (1–2.5 cm) long, elliptical, one-seeded, blue-black berry. It is ripe year-round.

HABITAT: Coastal hammocks and ridges.

RANGE: It is found in southern Florida and southern Texas near sea level. It is also widely distributed in Mexico and Central America south to northern Colombia and Venezuela.

USES: This plant attracts a variety of wildlife, particularly White-tailed Deer (*Odocoileus virginianus*).

SIMILAR SPECIES: This species is related to Gum Bumelia. Both are similar to Tough Bully (*S. tenax*), a shrub or tree endemic to the outer Coastal Plain of the southeastern states.

CONSERVATION STATUS:

Although its numbers are limited in Texas and Florida, Saffron-plum is widely distributed in Mexico and Central America. No known threats to the species have been identified.

Diameter: 4–6" (10–15 cm)
Height: 10–20' (3–6 m)

■ DENSE POPULATION
■ LIGHT POPULATION
■ NATURALIZED

This species is native to the south-central and southeastern United States. Gum from cuts in the trunk is sometimes chewed. The fruit is showy and attracts birds, and is said to be sweet and edible but may be mildly toxic and cause stomach upset and dizziness if eaten in large quantities. The species name *lanuginosum*, means "woolly," in reference to the young leaves.

ALTERNATE NAME: *Bumelia lanuginosa*

DESCRIPTION: This is a tree or thicket-forming shrub with a straight trunk and a narrow crown of short, stiff branches, often with stout spines. The leaves are alternate or clustered on short side twigs; they are 1–3" (2.5–7.5 cm) long and ⅜–1" (1–2.5 cm) wide. They are elliptical or obovate, rounded at the tip, widest beyond the middle, and tapering toward a long-pointed base, without teeth and slightly thickened. They are shiny dark green above and densely covered with gray or rust-colored hairs beneath, falling irregularly in late autumn. The bark is dark gray and furrowed into narrow scaly ridges. The twigs are slender, often zigzag, and covered with gray or rust-colored hairs when young; they often end in straight spines, with single spines to ¾" (19 mm) long at the base of some leaves and with gummy, milky sap.

FLOWERS: Summer. ⅛" (3 mm) wide flowers with bell-shaped five-lobed white corolla, clustered on slender stalks at the leaf bases.

FRUIT: ⅜–½" (10–12 mm) long, elliptical, purplish-black berry with sweetish pulp and one seed, maturing in fall.

CAUTION: Poisonous parts: While some parts of this species are edible, other parts are toxic to humans. Eat only if you are absolutely sure which parts are edible.

HABITAT: Valleys and rocky slopes of uplands in hardwood forests; forming shrubby thickets in the Southwest.

RANGE: Eastern and southern Kansas and central Missouri, southeast to cylindrical Florida and west to southern and western Texas; local in southwestern New Mexico and southeastern Arizona; also northern Mexico; to 2500' (762 m).

USES: Gum from the trunk of this plant is sometimes chewed by children. The wood is sometimes used locally to make tool handles and cabinets.

SIMILAR SPECIES: This species is related to Saffron-plum. Both are similar to Tough Bully (*S. tenax*), a shrub or tree endemic to the outer Coastal Plain of the southeastern states.

CONSERVATION STATUS: NT
The Gum Bumelia is presumed to be stable but faces potential threats from loss of habitat.

Diameter: 1' (0.3 m)
Height: 50' (15 m)

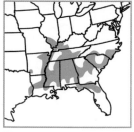

■ DENSE POPULATION
■ LIGHT POPULATION
■ NATURALIZED

This small tree (or shrub) is typically found along water in the southeastern United States. Unlike related species, Buckthorn Bumelia is usually thornless, and has narrow, pointed leaves instead of rounded ones. The thin fruit pulp is edible but is too bitter to be palatable. Birds, however, readily consume the berries, and livestock will browse the foliage.

ALTERNATE NAME: *Bumelia lycioides*

DESCRIPTION: This spiny shrub or small tree has a spreading, open crown; it is evergreen southward. The leaves are alternate or clustered on short side twigs, usually 2–4" (5–10 cm) long and ½–1¼" (1.2–3 cm) wide. They are elliptical or reverse lanceolate, widest beyond the middle, tapering to a long-pointed base, without teeth, thin, and with a prominent network of veins. They are shiny green above and paler and hairless or nearly so beneath. The bark is reddish-gray, thin, and smooth or scaly. The twigs are red-brown and slender, often with short straight or curved spines; the sap is slightly milky.

FLOWERS: Summer. ⅛" (3 mm) wide flowers with a calyx covered with rust-colored hairs, and a bell-shaped, five-lobed white corolla, clustered on short stalks at the leaf base.

FRUIT: ⅝" (15 mm) long, elliptical black berry with thin, bittersweet pulp and one large seed, maturing in fall.

HABITAT: Borders of streams, swamps, and lakes in forests; also rocky bluffs and dunes.

RANGE: Southeastern Virginia south to northern Florida, west to southeastern Texas, and north to southern Indiana; to 1000' (3–5 m).

USES: The fruit is edible but bitter.

SIMILAR SPECIES: Gum Bumelia.

CONSERVATION STATUS: LC
Buckthorn Bumelia faces no range-wide threats and is considered stable. It is classified as state-endangered at the margins of its range in Florida and Indiana.

Diameter: 6" (15 cm)
Height: 20' (6 m)

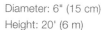

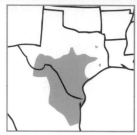

■ DENSE POPULATION
■ LIGHT POPULATION
■ NATURALIZED

This native persimmon can be found in riparian zones, prairie margins, and rocky slopes in central and southern Texas and northern Mexico. Many birds and mammals readily consume this tree's fruit. The edible, black persimmons from this tree stain teeth, lips, and hands, and have been used as a dye. They are sweet when ripe, but immature fruits are strongly astringent. The hard, dense heartwood from this species—found only in large trunks—is black and the sapwood is clear yellow.

DESCRIPTION: This is a much-branched shrub or tree with a short trunk and a narrow crown. Its leaves are ¾–1½" (2–4 cm) long, ⅜–¾" (10–19 mm) wide, obovate or oblong, rounded or slightly notched at the tip, widest above the middle, and tapering to a long-pointed base; they are thick, with the edges rolled under. They are shiny dark green and hairless (or nearly so) above and paler and hairy beneath, falling in winter. The bark is light reddish-gray, thin, smooth, and mottled and peeling, exposing gray inner layers. The twigs are gray, slender, slightly zigzag, and covered with fine hairs when young.

FLOWERS: Early–late spring. ¼" (6 mm) long and wide, bell-shaped flower with a five-lobed white, finely hairy corolla; it is short-stalked on the previous year's twigs back of new leaves. Male and female occur on separate trees, with one to three male flowers together; the female flowers are solitary.

FRUIT: A black berry ¾" (19 mm) in diameter, with black, juicy, sweet pulp and three to eight seeds, maturing in late summer.

HABITAT: Dry rocky uplands and in canyons, plains and open woodlands.

RANGE: Southeastern to central and Trans-Pecos Texas, south to northeastern Mexico; to 4000' (1219 m).

USES: The wood from Texas Persimmon is used for tools and other purposes, and the edible fruit is used in puddings and custards.

SIMILAR SPECIES: Ebony, Common Persimmon.

CONSERVATION STATUS: LC

Texas Persimmon has no known significant threats and the population is considered stable.

Diameter: 1' (0.3 m)
Height: 40' (12 m)

■ DENSE POPULATION
■ LIGHT POPULATION
■ NATURALIZED

This common tree of the eastern United States is important both in the wild and in cultivation. It is most abundant in the South Atlantic and Gulf states, and the largest trees occur in the Mississippi River basin, where individuals can reach more than 100' (30.5 m) in height. It is believed to have been cultivated by Native Americans for its fruit and wood. The fruit is sweet when ripe and recalls the flavor of dates; immature fruit contains tannin and is strongly astringent. Persimmons can be eaten fresh or used to make sweet desserts and beverages. Wildlife including opossums, raccoons, skunks, deer, and birds also eat the fruit. The word persimmon is of Algonquian origin, while the genus name *Diospyros* is from Greek and means "divine fruit."

DESCRIPTION: This is a tree with a dense cylindrical or rounded crown, or sometimes a shrub, best-known by its sweet, orange fruit in autumn. The leaves are 2½–6" (6–15 cm) long, 1½–3" (4–7.5 cm) wide, ovate to elliptical, long-pointed, without teeth, and slightly thickened. They are shiny dark green above and whitish-green and hairless to densely hairy beneath, turning yellow in autumn. The bark is brown or blackish, thick, and deeply furrowed into small square scaly plates. The twigs are brown to gray, slightly zigzag, and often hairy.

FLOWERS: The flower has a bell-shaped, four-lobed white corolla; it is fragrant, and scattered and almost stalkless at the leaf bases. Male and female flowers occur on separate trees in spring. Male flowers occur with two or three together, each ⅜" (10 mm) long. Female flowers are solitary and ⅝" (15 mm) long.

FRUIT: A rounded or slightly flat, orange to purplish-brown berry ¾–1½" (2–4 cm) in diameter, with four to eight large, flat seeds; maturing in fall before the frost and often remaining attached into winter. The orange pulp becomes soft and juicy at maturity.

HABITAT: Moist alluvial soils of valleys and in dry uplands; it is also found at roadsides and in old fields, clearings, and mixed forests.

RANGE: Connecticut south through Florida, west to central Texas, and north to extreme southeastern Iowa; to 3500' (1067 m).

USES: The edible fruits of this tree are eaten fresh or used to make puddings, cakes, and drinks. Native Americans made persimmon bread and stored the dried fruit like prunes. The wood is used for golf-club heads, shuttles for textile weaving, and furniture veneer.

SIMILAR SPECIES: Sweetleaf.

CONSERVATION STATUS: LC

The Common Persimmon population is considered stable. There are no known threats to the species.

Diameter: 1–2' (0.3–0.6 m)
Height: 20–70' (6–21 m)

■ DENSE POPULATION
■ LIGHT POPULATION
■ NATURALIZED

Commonly called Island Marlberry, this plant is native to the Caribbean, parts of Central America, and Florida. It is found in coastal strands and hammocks and in pine rocklands throughout central and southern Florida. It produces fragrant flowers and abundant fruit intermittently throughout the year. The berry-like fruit attracts a variety of birds and other wildlife.

DESCRIPTION: Marlberry is a large shrub to a small tree with a narrow, rounded crown and slender, upright branches. The leaves are 6" (15 cm) long, elliptical to obovate, wavy-edged, leathery, and evergreen. The bark is light gray, thin, and scaly.

FLOWERS: April–May. ⅓" (25 mm), five-petaled, white or lavender flowers in branched clusters.

FRUIT: ½" (1.3 cm), berrylike, and purple-black when mature, with one seed. Peak fruit production is in summer.

HABITAT: Hammocks, pinelands.

RANGE: Central Florida and southward, generally found on the coast.

USES: The fruit of this plant is edible, though some people consider the taste to be unpleasant. They are sometimes used locally as native landscape plants.

SIMILAR SPECIES: Myrsine.

CONSERVATION STATUS: (NT)

Marlberry has a relatively limited native range in Florida, and the loss of habitat due to coastal development has led to a population decline in the wild. This plant is available from local nurseries for use as a landscape plant, but care must be taken not to confuse it with two related, non-native and invasive *Ardisia* species, Coral Ardisia (*A. crentata*) and Shoebutton Ardisia (*A. elliptica*).

Height: 25' (7.6 m)

■ DENSE POPULATION
■ LIGHT POPULATION
■ NATURALIZED

This small evergreen plant is known by several names, including Myrsine, Rapanea, and Colicwood. It is native to coastal areas from Florida south through Central America.

ALTERNATE NAME: *Myrsine cubana, Rapanea punctata*

DESCRIPTION: This is a shrub to a small tree with an open, irregular crown and a few slender branches. The leaves are 4" (10 cm) long and oblanceolate to elliptical with rolled-under edges; they are leathery, often crowded at the branch ends, and evergreen. The bark is light gray, and smooth.

FLOWERS: October–January. Tiny, greenish white flowers with reddish streaks, arranged in dense clusters.

FRUIT: Tiny, berrylike, and blue-black.

HABITAT: Hammocks and pinelands.

RANGE: Central peninsular Florida and southward throughout the American tropics, mainly coastal.

USES: This species is cultivated as an ornamental plant.

SIMILAR SPECIES: Marlberry.

CONSERVATION STATUS:
The Myrsine has a wide native range and large population. It is considered stable with no threats identified.

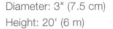
Diameter: 3" (7.5 cm)
Height: 20' (6 m)

■ DENSE POPULATION
■ LIGHT POPULATION
■ NATURALIZED

Typically a shrub, but sometimes a small tree, Joewood is native to the West Indies and found in limited areas of southern Florida. It is a slow-growing plant and somewhat rare. Attractive and fragrant, it is sometimes planted as an ornamental.

DESCRIPTION: This is a shrub to a small tree with a rounded crown. The leaves are 3" (7.6 cm) long, yellow-green, evergreen, and elliptical to obovate, with notched tips and rolled-under edges, often crowded at branch ends. The bark is smooth, thin, and blue-gray with pale patches.

FLOWERS: Year round. ½" (1.3 cm), white or yellowish, and fragrant, occurring in long clusters.

FRUIT: ½" (1.3 cm), green to orange-red berries.

HABITAT: Coastal hammocks and scrub; it is also found in pinelands.

RANGE: Found along the coast in limited areas of southern Florida and on many islands in the West Indies.

USES: This species is sometimes used as an ornamental.

SIMILAR SPECIES: Cudojewood.

CONSERVATION STATUS: NT

Loss of habitat has led to population decline in the state of Florida, where it is listed as threatened. Although still widespread in the West Indies, the species faces the potential for future habitat loss due to climate change and human activity, which could impact the global population.

Height: 20' (6 m)

■ DENSE POPULATION
■ LIGHT POPULATION
■ NATURALIZED

Loblolly Bay, also commonly known as Holly Bay or Gordonia, is native to the coastal southeastern United States. It is a showy, slender, flowering tree with evergreen leaves. Its size at maturity depends on soil moisture, but it can normally reach 30–60' (9–18 m). The Latin species name means "hairy-flowered," and the genus is named in honor of British nurseryman James Gordon (1728–1791).

DESCRIPTION: This is an evergreen tree or shrub with showy, large, white, fragrant flowers; a narrow, compact crown of upright branches; and shiny, leathery foliage. The leaves are evergreen, 4–6" (10–15 cm) long, and 1–2" (2.5–5 cm) wide; they are narrowly elliptical or lance-shaped, short-pointed, finely saw-toothed, thick and leathery, short-stalked, and shiny dark green, turning red before falling irregularly throughout the year. The bark is dark red-brown or gray, thick, and deeply furrowed into narrow flat ridges. The twigs are dark brown, stout, and hairless.

FLOWERS: Summer. 2½" (6 cm) wide and cup-shaped, with five large rounded white petals; they are waxy and silky on the outer surface, and have many yellow stamens; they are fragrant and borne singly on long reddish stalks at the leaf bases.

FRUIT: A pointed, gray, hairy, hard, egg-shaped capsule ½" (12 mm) in diameter; it is five-celled and splitting along five lines to below the middle, with several long-winged brown seeds, maturing in fall.

HABITAT: Wet soil of bays and edges of swamps; also in sandhills; with various hardwoods and conifers.

RANGE: Eastern North Carolina south to central Florida and west to southern Mississippi; to 500' (152 m).

USES: This plant is sometimes grown as an ornamental. The bark was once used locally for tanning leather.

SIMILAR SPECIES: Sweetbay, Swamp Bay.

CONSERVATION STATUS: LC
Loblolly Bay is common and widespread and its population is stable. The primary threats to the species are land conversion and fragmentation of habitat due to development.

Diameter: 1½' (0.5 m)
Height: 30–60' (9–18 m)

MOUNTAIN CAMELLIA *Stewartia ovata*

■ DENSE POPULATION
■ LIGHT POPULATION
■ NATURALIZED

This flowering shrub or small tree, also called Mountain Stewartia, is a member of the tea family. This uncommon species is mostly confined to the low to mid-elevations of the southern Appalachian Mountains, although small populations occur scattered throughout the southeastern United States. It is also cultivated as an ornamental for its showy flowers and red fall foliage. The scientific name refers to the egg-shaped leaves.

DESCRIPTION: This is a shrub or small tree with gray, hairless twigs, bright green foliage, and large white flowers. The leaves are 2½–4½" (6–11 cm) long, 1½–2¼" (4–6 cm) wide, ovate to elliptical, long-pointed; finely saw-toothed edges, and short-stalked. They are dark green and hairless above, and light gray-green and slightly hairy beneath, turning orange and red in fall. The bark is gray-brown and furrowed. The twigs are gray, slender, and hairless.

FLOWERS: Early summer. 4" (10 cm) wide and cup-shaped, with five unequal, rounded, waxy white petals with wavy edges, many whitish or yellowish stamens (rarely purple), and five styles; they are borne singly on short stalks at the leaf bases.

FRUIT: An egg-shaped pointed capsule ¾" (19 mm) in diameter; it is brown, hairy, hard, deeply five-angled, five-celled, and splitting into five parts, with several angled or winged seeds with dull surfaces; it matures in fall.

HABITAT: Moist soils of stream borders and bluffs; in understory of hardwood forests, chiefly in mountains.

RANGE: Virginia and Kentucky, south to extreme northern Georgia and northeastern Mississippi; usually at 1000–2500' (3–5–762 m); sometimes lower.

USES: This species is sometimes used as an ornamental or shrub border.

SIMILAR SPECIES: Silky Camellia.

CONSERVATION STATUS: LC
This species has a wide range but is not common and appears to struggle to become more widely established. It is threatened by land-use conversion, habitat fragmentation, and interspecific factors, but has not shown a population decline.

Diameter: 3" (7.5 cm)
Height: 20' (6 m)

■ DENSE POPULATION
■ LIGHT POPULATION
■ NATURALIZED

This small tree, or sometimes a shrub, is found primarily in the undergrowth of high forests of the southern United States. Also known as Horse-sugar, the sweet-tasting foliage of this plant is a favorite among browsing wildlife and livestock. Another common name for the plant, Yellowwood, is a reference to a yellow dye that was once obtained from the bark and leaves.

DESCRIPTION: This is a shrub or small tree with a short trunk, an open crown of spreading branches, and foliage with a sweetish taste. The leaves are deciduous, or evergreen southward; they are 3–5" (7.5–13 cm) long, 1–2" (2.5–5 cm) wide, elliptical or lance-shaped, slightly saw-toothed or without teeth, and slightly thickened. They are shiny dark green above, and paler and covered with fine hairs beneath. The bark is gray and smooth or slightly fissured. The twigs are light green, slender, and often hairy when young.

FLOWERS: Early spring. ⅜" (10 mm) wide and bell-shaped, with deeply five-lobed white or light yellow corolla; they are fragrant, and crowded and nearly stalkless in clusters on twigs behind new leaves.

FRUIT: ¼" (6 mm) long, elliptical, with five teeth at the tip; it is berrylike and green, turning brown, with thin pulp and a large one-seeded stone, with several arranged in stalkless clusters along the twig; it matures in summer and fall.

HABITAT: Moist valley soils in understory of hardwood forests.

RANGE: Southern Delaware to central Florida and west to eastern Texas. It generally occurs below 1000' (305 m), and to 3500' (1067 m) in western North Carolina.

USES: A yellow dye was once made from the bark and leaves of this plant. The bark was used as a tonic by early settlers.

SIMILAR SPECIES: Carolina Silverbell, Loblolly Bay, Common Persimmon.

CONSERVATION STATUS: ⓁⒸ

The Sweetleaf has a largely distributed range with a stable population. There are no known major threats to the species. Limited natural populations of this plant in Tennessee have earned a designation of Special Concern in the state.

Diameter: 4" (10 cm)
Height: 20' (6 m)

- ■ DENSE POPULATION
- ■ LIGHT POPULATION
- ■ NATURALIZED

This flowering plant is native to the southeastern United States, typically found growing in moist soils along riverbanks and other waterways. It is common in the southern Appalachians. The taxonomy surrounding this species is confusing, with the four-winged American silverbells having been assigned as many as four species names, with some disagreement remaining over their categorization. Little Silverbell (*H. parviflora*) has been incorporated into *H. carolina*. Mountain Silverbell (*H. monticola*) is sometimes considered a subspecies, but some authorities consider it a full species; it is much larger than the other silverbells and is common in the Great Smoky Mountains National Park.

ALTERNATE NAMES: *Halesia parviflora, Halesia tetraptera*

DESCRIPTION: This is a shrub or a small tree with an irregular, spreading, open crown and drooping, bell-shaped white flowers; in the southern Appalachians it becomes a large tree with a straight axis and a rounded crown. The leaves are 3–6" (7.5–15 cm) long, 1–2½" (2.5–6 cm) wide, elliptical, abruptly long-pointed, and finely saw-toothed or wavy. They are dull dark green and becoming hairless above, and covered with white hairs when young and often with tiny star-shaped hairs on the veins beneath, turning yellow in fall. The bark is reddish-brown to dark brown and furrowed into loose, scaly ridges. The twigs are brown and slender, with star-shaped hairs when young.

FLOWERS: Spring. ½–1" (1.2–2.5 cm) long with a bell-shaped, four-lobed white corolla (rarely pink) covered with fine hairs, opening before reaching full size, arranged in drooping clusters of two to five flowers on long stalks on the previous year's twig back of new leaves.

FRUIT: 1–2" (2.5–5 cm) long, oblong or club-shaped, and pod-like, with four long wings; it is pointed, dark brown, and dry, with a stone containing one to three seeds. It matures in late summer and fall and remains closed and attached into winter.

HABITAT: Moist soils along streams in the understory of hardwood forests; moist sandy valleys.

RANGE: Southern West Virginia south to northern Florida, northwest to southern Illinois; it is local in Arkansas and southeastern Oklahoma. It occurs to 5500' (1676 m) in the southern Appalachians.

USES: This plant has been cultivated as an ornamental, and the wood is sometimes used for lumber.

SIMILAR SPECIES: Sweetleaf.

CONSERVATION STATUS: LC
Carolina Silverbell has a wide distribution and large population, and its numbers appear to be stable. It does not currently face any known range-wide threats. Small natural populations in Illinois and Ohio are critically endangered and presumed extirpated, respectively.

Diameter: 6"–1' (15–30 cm); and 2' (0.6 m) in the south
Height: 25–30' (7.6–9 m); in the southern Appalachians 80' (24 m)

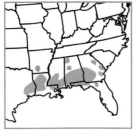

■ DENSE POPULATION
■ LIGHT POPULATION
■ NATURALIZED

Native to the southeastern United States, this plant prefers wet areas in ravines, swamps, and along streams. The common and scientific names both refer to the two-winged fruit. Wildlife—notably squirrels—readily consume the unripe, sour, green fruits. Another common name for this species is Two-winged Snowdrop Tree, a reference to the white flowers produced in clusters in the spring.

DESCRIPTION: This is a shrub or small tree with a spreading crown and white, bell-shaped, drooping flowers. The leaves are 2½–4½" (6–11 cm) long, 1½–2½" (4–6 cm) wide, broadly elliptical, abruptly long-pointed, finely saw-toothed, and short-stalked. They are light green and becoming nearly hairless above, and paler with tiny star-shaped hairs beneath. The bark is reddish-brown, and furrowed into irregular ridges, shedding in narrow plates. The twigs are light green, slender, and hairy when young.

FLOWERS: Early- to mid-spring. ⅝–1¼" (1.5–3.2 cm) long flowers with a bell-shaped, deeply four-lobed, white corolla; they are covered with fine hairs and occur in drooping clusters of three to six on long stalks on the previous year's twigs back of new leaves.

FRUIT: 1½–2" (4–5 cm) long, oblong, and podlike, with two long broad wings; it is dark brown and dry with a large stone, maturing in late summer and fall, and remaining closed.

HABITAT: Wet soils bordering streams and swamps in understory of hardwood and pine forests.

RANGE: Extreme southern South Carolina south to northwestern Florida, west to southeastern Texas; to 500' (152 m).

USES: Two-winged Silverbell is cultivated as an ornamental tree.

SIMILAR SPECIES: Carolina Silverbell.

CONSERVATION STATUS: (LC)
The Two-winged Silverbell population is stable with no known threats identified.

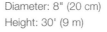

Diameter: 8" (20 cm)
Height: 30' (9 m)

■ DENSE POPULATION
■ LIGHT POPULATION
■ NATURALIZED

Native to the southeastern United States, this flowering shrub or small tree produces showy, white, bell-shaped, pleasantly fragrant flowers in the spring. These abundant flowers provide an important source of nectar for bees and butterflies. This plant prefers floodplains, swamps, and other low, wet areas. It has thin branches covered in bright green, glossy foliage.

DESCRIPTION: This is a shrub to a small, arching tree with an open crown. The leaves are 2½" (6.4 cm) long, elliptical or oval, with variable margins either toothed or untoothed. The bark is gray, thin, and smooth.

FLOWERS: April–June. ½", white flowers; the petals are strongly recurved and the anthers are bright yellow; they are borne in pairs from leaf axils.

FRUIT: ½" hairy capsules.

HABITAT: Bottomlands, moist woods, swamps.

RANGE: Southern Great Lakes and Virginia south to Florida and west to Texas and Missouri.

USES: American Snowbell is sometimes planted as an ornamental.

SIMILAR SPECIES: Bigleaf Snowbell, Japanese Snowbell.

CONSERVATION STATUS: LC

The American Snowbell population is believed to be stable, and no range-wide threats have been identified. It is classified as state-endangered in several Midwestern U.S. states where limited natural populations occur.

Height: 12' (4 m)

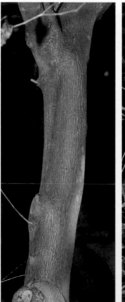

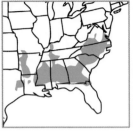

■ DENSE POPULATION
■ LIGHT POPULATION
■ NATURALIZED

Bigleaf Snowbell is native to the southeastern United States. It is most common in the Piedmont plateau region, where it is found in moist or wet soils of the valleys and uplands, often in the understory of hardwood forests. Many birds and other wildlife are attracted to the fruit and flowers as an important food source.

DESCRIPTION: Bigleaf Snowbell is a shrub, or sometimes a small tree, with a rounded crown; large, nearly round leaves; and showy, bell-shaped, white flowers. The leaves are 2½–5½" (6–14 cm) long, 1½–4" (4–10 cm) wide, broadly elliptical to nearly round, and abruptly short-pointed at the tip; they are not toothed or have only a few teeth. They are dark green and hairless or nearly so above, and paler and usually densely coated with gray, tiny, star-shaped hairs beneath. The bark is dark gray and smooth. The twigs are gray, slender, slightly zigzag, and scurfy with star-shaped hairs.

FLOWERS: Early spring. ¾–1" (2–2.5 cm) wide flowers with five long, narrow, white corolla lobes; they are short-stalked and fragrant, with many slightly drooping along an unbranched axis to 5" (13 cm) long, occurring at the end of short side twigs.

FRUIT: ¼" (6 mm) in diameter, rounded, hairy, brown, hard, and dry, with one large seed, maturing in late summer and fall and remaining closed.

HABITAT: Moist or wet soils of valleys and uplands; in understory of hardwood forests.

RANGE: Virginia to northern Florida, and west to eastern Texas; it can be found locally to Indiana and Illinois. It occurs to 1000' (305 m).

USES: This species is planted as an ornamental and to attract wildlife.

SIMILAR SPECIES: American Snowbell.

CONSERVATION STATUS: LC
The Bigleaf Snowbell is believed to be stable and faces no known range-wide threats. It is classified as state-endangered in several Midwestern U.S. states where limited natural populations occur.

Diameter: 4" (10 cm)
Height: 20' (6 m)

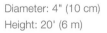

■ DENSE POPULATION
■ LIGHT POPULATION
■ NATURALIZED

This flowering shrub, or small tree, is native to the Appalachian Mountains, especially in the southeastern United States. It is also commonly called Cinnamon Clethra, named for the rich color of the inner bark, which is often visible as the outer layers naturally peel away. The name Pepperbush comes from the spicy taste of the seeds.

DESCRIPTION: This is usually a multi-stemmed shrub with exfoliating, red-brown bark. It can grow treelike to 20' (6 m). The dark-green foliage turns yellow-orange in fall.

FLOWERS: Dense, terminal, drooping spikes of white flowers are followed by brown capsules which persist through winter.

HABITAT: Wooded mountain bluffs and ravines.

RANGE: Found in the Appalachian Mountains from southern Pennsylvania and West Virginia to extreme northern Georgia.

USES: This attractive plant is sometimes used as a landscaping ornamental. Some early settlers used the spicy seeds as a substitute for black pepper.

SIMILAR SPECIES: Sweet Pepperbush.

CONSERVATION STATUS: NT
This plant has a somewhat limited range, and is somewhat threatened by land-use conversion, habitat fragmentation, and forest management practices.

Height: 12–20' (3.7–6 m)

■ DENSE POPULATION
■ LIGHT POPULATION
■ NATURALIZED

This tall shrub can form sizable patches in eastern and south-eastern North America. This plant is commonly used for natural gardens or erosion control. Sweet Pepperbush is sometimes used to halt the succession of trees along pathways such as roadcuts and forest clearings for utility lines, railroads, and pipe-lines. Its fragrant white flowers and nectar attract humming-birds and butterflies. Its dry fruiting capsules remain long after flowering and help to identify this plant in winter. It is named because the mature fruits vaguely resemble peppercorns, but they are not spicy.

DESCRIPTION: This is a tall, many-branched, leafy shrub with spike-like, upright clusters of fragrant white flowers. The leaves are up to 3" (7.5 cm) long, wedge-shaped, sharply toothed above the middle, untoothed at the base, and blunt or broadly pointed at the tip.

FLOWERS: July–September. Each flower is about ⅓" (8 mm) wide with 10 stamens and the style protruding.

FRUIT: Small, globular capsules with a persistent style.

HABITAT: Wetlands, especially swamps, and sandy woods.

RANGE: Coastal, from southern Maine south to Florida; west to eastern Texas.

USES: This shrub is cultivated as an ornamental and to stabilize streambanks.

SIMILAR SPECIES: Mountain Pepperbush has more pointed leaves and is found in south-ern mountains.

CONSERVATION STATUS: 🔘 LC
The Sweet Pepperbush is believed to have a stable population. There are no current or future threats identified for the species.

Height: 3–10' (90–300 cm)

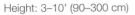

LEATHERWOOD   *Cyrilla racemiflora*

■ DENSE POPULATION
■ LIGHT POPULATION
■ NATURALIZED

This tree is native to the southeastern United States and throughout the American tropics. This tree is deciduous at the northern part of its range in the southeastern United States, and is nearly evergreen farther south into subtropical and tropical climates. It is known by several common names including Swamp Cyrilla, Titi, Swamp Titi, and several variations thereof. In the upper mountain forests of Puerto Rico, Leatherwood is a large, dominant tree known as *palo colorado* ("red tree") because of its reddish-brown bark and wood. Bees produce a dark honey from the flowers.

DESCRIPTION: Leatherwood is a small tree with a short, stout, crooked trunk and a spreading crown, or a much-branched shrub, with glossy foliage; profuse, tiny, whitish flowers; and clusters of tiny, brown or yellow fruit. The leaves are clustered near the end of the twigs; they are 1½–3" (4–7.5 cm) long, ⅜–1" (1–2.5 cm) wide, narrowly oblong, usually widest beyond the middle, blunt or slightly notched at the tip; they do not have teeth, and are slightly thickened and hairless. They are shiny green above and paler beneath, turning orange and red in fall. The bark is gray and smooth, becoming reddish-brown, thin, and scaly; it is whitish-pink and spongy at the base. The twigs are brown, slender, and hairless.

FLOWERS: Early summer. ⅛" (3 mm) wide with five pointed white petals, sometimes pinkish-tinged; they are fragrant and short-stalked, and crowded in upright narrow clusters 4–6" (10–15 cm) long.

FRUIT: ⅛" (3 mm) long, beadlike or egg-shaped, pointed or rounded, brown or yellow, and spongy, occurring in clusters; they are two-celled, with four or fewer tiny seeds, and mature in late summer, not splitting open.

HABITAT: Moist soils of river flood plains and riverbanks, flatwoods, and borders of sandy swamps and ponds.

RANGE: Southeastern Virginia south to central Florida and west to southeastern Texas; to 500' (152 m).

USES: This species is grown as an ornamental and to attract wildlife.

SIMILAR SPECIES: Sweet Pepperbush.

CONSERVATION STATUS: LC

Leatherwood has a large, widespread population and faces no major threats.

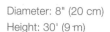
Diameter: 8" (20 cm)
Height: 30' (9 m)

■ DENSE POPULATION
■ LIGHT POPULATION
■ NATURALIZED

Florida Hobblebush, sometimes called Florida Leucothoe, is named for the impenetrably dense thickets it can form. The natural range of this species is concentrated in northern Florida, with a few small populations in wet hammocks and swampy woodlands elsewhere in the southeastern United States. The plant tends to sucker and can be used for stabilization along moist, shady streambanks. It is typically a large shrub in the wild. It is sometimes planted as an ornamental, and can be maintained at any height with proper pruning.

DESCRIPTION: Florida Hobblebush is a large, broadleaf evergreen shrub with a lax, arching, multi-stemmed habit. Its rich, glossy-green foliage, tinged with red on new growth, remains green through the winter.

FLOWERS: This plant's cream-colored, fragrant flowers are borne in profuse axillary clusters.

HABITAT: Wet hammocks and woodlands.

RANGE: South Carolina and Florida.

USES: This species is planted as an ornamental and to stabilize streambanks.

SIMILAR SPECIES: Coastal Doghobble, Highland Doghobble.

CONSERVATION STATUS: NT
The limited range and specific habitat needs of Florida Hobblebush are the main threats to this uncommon plant in the wild.

Height: 8–12' (2.4–3.7 m)

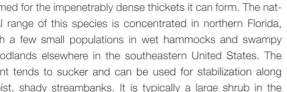

■ DENSE POPULATION
■ LIGHT POPULATION
■ NATURALIZED

This tree is native to the mountains of southeastern Arizona and southwestern New Mexico and farther south through the Sierra Madre Occidental mountain range of western Mexico. It is well represented in the moist montane forests at elevations between 4000–8000' (1219–2438 m) in canyon bottoms and hillsides in the oak-pine zone. Spanish settlers applied the term *madrono*—the Spanish name of Strawberry-tree (*Arbutus unedo*), a related species of southern Europe—to similar trees they encountered in Mexico and California.

DESCRIPTION: This is a medium-sized evergreen tree with a rounded crown of stout, crooked, smooth red branches; showy white flowers; and orange-red fruit. The leaves are evergreen, 1½–3" (4–7.5 cm) long, ½–1" (1.2–2.5 cm) wide, lance-shaped, short-pointed at both ends, without teeth or sometimes saw-toothed, and thick and stiff with short, slender stalks; they are shiny light green above and paler beneath. The bark is red-brown, smooth, and thin, and peeling off in thin, papery scales on the branches, and light gray or whitish and divided into square plates on the trunk. The twigs are whitish when young, turning red-brown, and finely hairy.

FLOWERS: Spring and summer. ¼" (6 mm) long with jug-shaped or urn-shaped white or pink corolla, short-stalked, with many arranged together in branched clusters about 2½" (6 cm) long and wide at twig ends.

FRUIT: ⅜" (10 mm) in diameter, berrylike, orange-red, finely warty, pulp mealy and sweetish, with a large stone and many flattish seeds; maturing in fall.

HABITAT: Oak woodland with other evergreen trees.

RANGE: Arizona, extreme southwestern New Mexico, and northwestern Mexico; at 4000–8000' (1219–2438 m).

USES: The edible fruit was eaten by some Native Americans and is still eaten locally.

SIMILAR SPECIES: Pacific Madrone, Texas Madrone.

CONSERVATION STATUS:

Although its range is limited in the southwestern U.S., Arizona Madrone has a large population and widespread distribution in Mexico. No significant threats have been identified for the species.

Diameter: 1½' (0.5 m)
Height: 40' (12 m)

■ DENSE POPULATION
■ LIGHT POPULATION
■ NATURALIZED

Pacific Madrone is native to the western coastal areas of North America, ranging as far north as British Columbia. This tree is one of the most beautiful broadleaf flowering evergreens, with its glossy foliage, large clusters of small white flowers, orange-red fruits, and showy, reddish, peeling bark. It is an important food plant for wildlife; many mammals and birds consume the fruit, and the flowers attract pollinators and are a source of honey. The scientific name honors Scottish physician and naturalist Archibald Menzies (1754–1842), who described the species during the Vancouver Expedition in the 1790s.

DESCRIPTION: Pacific Madrone is a handsome evergreen tree with a tall, reddish-brown trunk and an open, narrow, rounded or irregular crown of stout, smooth red branches. Its leaves are evergreen, 2–4½" (5–11 cm) long, 1–3" (2.5–7.5 cm) wide, elliptical, blunt at the tip, not toothed or sometimes saw-toothed, thick and leathery, and hairless except when young. They are shiny dark green above and paler or whitish beneath, turning red before falling. The bark is red, smooth, and thin, and peeling off in thin, papery scales on the branches, and dark reddish-brown and divided into square plates on the trunk. The twigs are light red or green, turning reddish-brown, and hairless.

FLOWERS: Early spring. ¼" (6 mm) long and jug-shaped or urn-shaped, with a white or pink-tinged corolla; they are short-stalked and arranged in branched clusters 2–6" (5–15 cm) long and wide at the twig ends.

FRUIT: ⅜–½" (10–12 mm) in diameter, berrylike, orange-red, and finely warty, with mealy pulp, a large stone, and many flattened seeds; it matures in fall. The fruit is edible but can cause stomach aches.

HABITAT: Upland slopes and canyons; in oak and coniferous forests, often in understory.

RANGE: The Pacific Coast from southwest British Columbia south to western Oregon and in the Coast Ranges to southern California; it is also found in the Sierra Nevada of central California and on Santa Cruz Island. It occurs to 5000' (1524 m), sometimes to 6000' (1829 m).

USES: This species is sometimes cultivated as an ornamental tree. Native Americans ate the fruits raw or cooked and made medicinal tea from the bark. The wood is durable and has a warm color after finishing, making it a popular choice for wood flooring and veneer, but it is difficult to work.

SIMILAR SPECIES: Arizona Madrone.

CONSERVATION STATUS: LC
The Pacific Madrone faces no major threats. The species is widespread and stable.

Diameter: 2' (0.6 m)
Height: 20–80' (6–24 m)

■ DENSE POPULATION
■ LIGHT POPULATION
■ NATURALIZED

This uncommon small tree is native to Central America, Mexico, and the southwestern United States. Some sources treat the variety that occurs in northernmost Mexico and the United States as a distinct species (*A. texana*). These northern representatives of the species tend to be arborescent shrubs in dry habitats, but the plant grows as a tree in the moist, tropical parts of Mexico and Central America. This plant is known by the colloquial name "Lady's Leg," referring to the smooth, pinkish to reddish-brown bark.

DESCRIPTION: This small evergreen tree has a short, often crooked trunk and a rounded crown of stout, crooked, spreading branches; it is sometimes a shrub. The leaves are evergreen, 1–3½" (2.5–9 cm) long, ⅝–1½" (1.5–4 cm) wide, and elliptical or ovate; they are usually without teeth, but sometimes wavy- or saw-toothed; they are thick and stiff with slender, hairy leafstalks. They are shiny green above, and paler and slightly hairy beneath. The bark on branches is pinkish to reddish-brown, smooth, thin, and peeling off in thin papery scales; on trunks, it is dark brown and divided into square plates. The twigs are red, and densely covered with hairs when young, becoming dark red-brown and scaly.

FLOWERS: Early spring. ¼" (6 mm) long and jug-shaped or urn-shaped, with a white or pink-tinged corolla; they are short-stalked and occur in upright branched clusters about 2½" (6 cm) long and wide.

FRUIT: ⅜" (10 mm) in diameter, berrylike, dark red to yellowish-red, and finely warty, with mealy sweetish pulp and a large stone containing many flat seeds, maturing in fall.

HABITAT: Canyons and rocky slopes of mountains, in oak woodlands, and on rocky plains.

RANGE: Central Texas (Edwards Plateau) to Trans-Pecos Texas, southeastern New Mexico (Guadalupe Mountains), and northeastern Mexico; at 2000–6000' (610–1829 m).

USES: The fruits of Texas Madrone may be edible, but those of some related European species are known to have narcotic properties. The wood has been used locally for tool handles.

SIMILAR SPECIES: Arizona Madrone.

CONSERVATION STATUS: LC

With a large population that is widely distributed across its native range, the Texas Madrone is a stable species with no identified threats.

Diameter: 8" (20 cm)
Height: 20' (6 m)

Arctostaphylos columbiana **HAIRY MANZANITA**

■ DENSE POPULATION
■ LIGHT POPULATION
■ NATURALIZED

Whether manzanitas should be considered trees is debatable; a few of the approximately 40 native species, including this one, reach tree size. However, they generally branch or fork near the ground, thus lacking the single trunk of a tree. Bears, deer, coyotes, and other mammals eat the red fruits of this species, which is most abundant in southern Oregon.

DESCRIPTION: Hairy Manzanita is usually a gnarled shrub of the coniferous forest understory. Its leaves are 2" (5 cm) long, elliptical, gray-green, hairy below, and evergreen. The bark is red and smooth, with peeling gray flakes.

FLOWERS: May–July. Tiny, pinkish to white, jug-shaped flowers occurring in 1" (2.5 cm) clusters.

FRUIT: Dry, red berries, ⅓" (1 cm) in diameter.

HABITAT: Hot rocky sites, areas of poor soil in wet forests, clearings.

RANGE: From Vancouver, British Columbia, south to California, mainly in the Coast Ranges.

USES: This species is sometimes grown as an ornamental. It is an important food source for a variety of wildlife.

SIMILAR SPECIES: Hoary Manzanita (*A. canescens*), a variable shrub that also occurs in the coastal mountain ranges of southwestern Oregon and northern California.

CONSERVATION STATUS: (LC)
Hairy Manzanita has a widespread, stable population. The species has no significant threats.

Height: 10' (3 m)

■ DENSE POPULATION
■ LIGHT POPULATION
■ NATURALIZED

Common Manzanita is endemic to California, where it is also commonly called Whiteleaf Manzanita. *Manzanita* is a Spanish word meaning "little apple." Wildlife of many kinds eat the mealy berries in great quantities. The dense evergreen foliage provides shelter for many birds and small mammals, and deer and livestock browse the leaves and twigs. These plants can be identified as manzanitas by the reddish bark on long, crooked branches.

DESCRIPTION: This is a large evergreen shrub, sometimes a small tree commonly branching near the base, with stout, crooked, twisted trunks and branches and a dense, rounded crown as broad as high. The leaves are evergreen, 1–1¾" (2.5–4.5 cm) long, ¾–1¼" (2–3 cm) wide, elliptical or nearly round; they are without teeth, thick, and short-stalked, and shiny or dull green on both surfaces, sometimes finely hairy. The bark is dark reddish-brown and smooth. The twigs are crooked, and densely covered with gray hairs.

FLOWERS: Late winter–early spring. ⁵⁄₁₆" (8 mm) long with a jug-shaped white or pale pink corolla ending in five tiny lobes; they are numerous and arranged in many-branched clusters, drooping at the end of the twig.

FRUIT: ⁵⁄₁₆–½" (8–12 mm) in diameter, berrylike, round or slightly flattish, and white, turning deep reddish-brown, with mealy pulp and several nutlets; it matures in late summer.

HABITAT: Dry slopes and in mountain canyons, chaparral, foothill and oak woodlands, and in Ponderosa Pine forest.

RANGE: Northern and central California in the northern Coast Ranges and the foothills of the Sierra Nevada; at 300–4000' (91–1219 m).

USES: The hard wood of this plant is used for making tool handles and as firewood. The tart berries are edible in small quantities—eating too many can cause stomach pain—and can be made into manzanita cider. The handsome, reddish-brown branches often become twisted into odd shapes, which are trimmed into collectors' items called "mountain driftwood."

SIMILAR SPECIES: This plant is not likely to be confused with other species in its native Californian range.

CONSERVATION STATUS: 🔘 LC
The Common Manzanita has a stable population. There are no known current or future threats to the species.

Diameter: 6" (15 cm)
Height: 20' (6 m) or more

■ DENSE POPULATION
■ LIGHT POPULATION
■ NATURALIZED

Elliottia, also commonly called Georgia Plume, is one of North America's rarest native trees. It is the only species in its distinct genus and is endemic to the U.S. state of Georgia. It is found at scattered locations in eastern and southern Georgia, and is now extinct at the single known site in South Carolina where it had last been collected in 1853. A management regime is now in place to preserve the remaining plants, with about 70 natural populations known. Elliottia is named for botanist Stephen Elliott (1771–1830).

DESCRIPTION: This is a shrub or small tree with a conical crown and showy, white clusters of flowers. The leaves are 2¾–4" (7–10 cm) long, 1–1½" (2.5–4 cm) wide, and oblong or elliptical; they are without teeth and short-stalked; they are dull green above and paler and slightly hairy beneath. The bark is light gray, smooth, and thin. The twigs are light brown, upright, slender, and hairy when young.

FLOWERS: Early summer. ¾" (19 mm) wide with four or five narrow, curved, white petals; they are fragrant and occur on long slender stalks, in upright narrow clusters 4–10" (10–25 cm) long.

FRUIT: A slightly flat, rounded capsule ⅜–½" (10–12 mm) in diameter, splitting open on four or five lines, with many winged seeds.

HABITAT: Moist, sandy, acid soils near streams and sometimes on dry ridges, in woods.

RANGE: Local in eight counties of eastern and southeastern Georgia, at 100–400' (30–122 m).

USES: This species is sometimes planted as an ornamental or specimen tree but is not widely used due to its rarity and limited range.

SIMILAR SPECIES: This plant is not likely to be confused with other species in its range.

CONSERVATION STATUS: EN
Many areas of suitable habitat have been lost due to conversion to agriculture and timber cultivation. The species requires periodic fire to reduce the density of the surrounding shrubs and prevent the plants from being shaded out by taller trees. Most of the known populations are sterile (they do not produce seeds), but the plants reproduce asexually by producing new stems from the existing underground roots; they produce new shoots especially vigorously after fire has killed the above-ground stems. It is unknown whether these remaining populations are stable in the absence of human intervention and management.

Diameter: 4" (10 cm)
Height: 20' (6 m)

■ DENSE POPULATION
■ LIGHT POPULATION
■ NATURALIZED

Mountain Laurel is native to the eastern United States and is common in the Appalachian Mountains, where it is most likely to attain tree size. It is found on rocky slopes and in montane forests. It is among the most beautiful native flowering plants and is widely planted as an ornamental in many parks. The stamens of the flowers have an unusual, springlike mechanism that spreads pollen when tripped by a bee or other pollinator.

DESCRIPTION: This is an evergreen, many-stemmed, thicket-forming shrub, or sometimes a small tree with a short, crooked trunk; stout, spreading branches; a compact, rounded crown; and beautiful, large, pink flower clusters. The leaves are evergreen, and alternate or sometimes opposite or in threes; they are 2½–4" (6–10 cm) long, 1–1½" (2.5–4 cm) wide, and narrowly elliptical or lance-shaped, with a hard whitish point at the tip; they are without teeth, and thick and stiff. They are dull dark green above and yellow-green beneath. The bark is dark reddish-brown and thin, fissured into long narrow ridges and shredding. The twigs are reddish-green with sticky hairs when young, later turning reddish-brown, peeling, and exposing a darker layer beneath.

FLOWERS: Spring. ¾–1" (2–2.5 cm) wide and saucer-shaped, with a five-lobed pink or white corolla with purple lines, from pointed deep pink buds, occurring on long stalks covered with sticky hairs in upright branched flat clusters 4–5" (10–13 cm) wide.

FRUIT: A rounded dark brown capsule ¼" (6 mm) wide, with long threadlike style at the tip, covered with sticky hairs; it is five-celled and splitting open along five lines, with many tiny seeds, maturing in fall and remaining attached.

CAUTION: Poisonous: Do not eat or ingest any part of this species. If accidentally ingested, contact your doctor or poison control center.

HABITAT: Dry or moist acid soils; it is found in the understory of mixed forests on upland mountain slopes and in valleys, and also in shrub thickets called "heath balds" or "laurel slicks."

RANGE: Southeastern Maine south to northern Florida, west to Louisiana, and north to Indiana; it occurs to 4000' (1219 m), and higher in the southern Appalachians.

USES: This species is cultivated as an ornamental, and has been introduced in Europe for this purpose. Although the plants do not grow large enough to be a commercially viable source of lumber, the strong wood is sometimes used for tool handles, spoons, bowls, and other small items.

SIMILAR SPECIES: Catawba Rosebay, Great Laurel.

CONSERVATION STATUS:

The Mountain Laurel has a large population that is considered stable, with no major threats identified. It is believed to have increased in numbers following the decline of the American Chestnut.

Diameter: 6" (15 cm)
Height: 20' (6 m)

■ DENSE POPULATION
■ LIGHT POPULATION
■ NATURALIZED

Dragon Tree is native to the southeastern United States, limited to southernmost South Carolina, southeastern Georgia, and the northern two-thirds of Florida. This tree is sometimes called Rusty Lyonia for the rust-colored lower surfaces of its leaves, and is called Rusty Staggerbush because of its crooked trunks. It is the only species in its genus that forms a tree.

DESCRIPTION: This evergreen shrub or small tree has a crooked trunk, an irregular crown, and tiny, nearly round flowers. The leaves are evergreen, 1–3½" (2.5–9 cm) long, ½–1" (1.2–2.5 cm) wide, thick and stiff, and oblong or narrowly elliptical, with a hard point at the tip and wavy edges turned under. They are shiny green above and gray or rusty scurfy with tiny scales beneath. The bark is reddish-brown and furrowed into long, narrow, scaly ridges. The twigs are slender, and gray or rusty scurfy, becoming dark red and smooth.

Diameter: 6" (15 cm)
Height: 20' (6 m)

FLOWERS: Early spring. ⅛" (3 mm) wide with a rounded, hairy, pinkish or white corolla with five teeth around a small opening, with several clustered on slender curved stalks at the base of the leaves.

FRUIT: A ¼" (6 mm) long, brown, egg-shaped capsule; it is long-stalked, splitting along lines into five parts, and many-seeded, maturing in early autumn and remaining attached.

HABITAT: Sandy soils of mixed woods and on dunes.

RANGE: Extreme southern South Carolina, southeastern Georgia, and Florida; to 300' (91 m).

USES: This shrub or tree is sometimes planted as an ornamental.

SIMILAR SPECIES: Malebarry and other laurels, which are mainly shrubs.

CONSERVATION STATUS: 🔵 **LC**
The Dragon Tree population appears to be secure, and does not face major threats from diseases or pests.

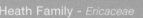

■ DENSE POPULATION
■ LIGHT POPULATION
■ NATURALIZED

These blueberry-like shrubs can grow quite tall and are frequently associated with shrubby or wooded swamps. It can thrive in a variety of habitats and conditions and is often found in transitional zones between biomes. Also sometimes called "He Huckleberry," the plant gets its name from the hard, dry, inedible capsules it produces (compared to the softer, fleshy fruits of related species such as blueberries, cranberries, and huckleberries). The genus name honors the early American botanist and explorer John Lyon (1765–1814).

DESCRIPTION: Maleberry is a much-branched, deciduous shrub with terminal clusters of globular white or pale rose flowers, constricted at the tip. The leaves are 1–3" (2.5–7.5 cm) long, alternate, oblong-oval, and finely toothed.

FLOWERS: May–July. About ⅛" (3 mm) wide.

FRUIT: A brown, rounded, five-angled capsule, persisting through winter.

HABITAT: Wet thickets, swamps.

RANGE: New England south to Florida; west to Texas; north to Kentucky and Oklahoma.

USES: This flowering plant is often cultivated as an ornamental shrub.

SIMILAR SPECIES: Other *Lyonia* species. At least three related species occur in eastern North America as shrubs: Staggerbush (*L. mariana*) has urn- or bell-shaped capsules. A southern evergreen species, Fetterbush (*L. lucida*), occurs from Virginia to Florida and Louisiana and has white to pale pink flowers, three-angled branches, and dark green lustrous leaves with a conspicuous vein next to the rolled margin. A third species (*L. ferruginea*) is evergreen, with white flowers and dull green leaves.

CONSERVATION STATUS: LC
The Maleberry has no known threats and is considered stable.

Height: 3–12' (1–3.6 m)

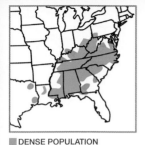

■ DENSE POPULATION
■ LIGHT POPULATION
■ NATURALIZED

Sourwood is native to a wide range in the eastern United States and is particularly abundant in Great Smoky Mountains National Park and elsewhere in the central Appalachian Mountains. The tree is hardy farther north of its natural range and makes an attractive ornamental throughout the year. Although it is named for the acidic taste of its foliage, the honey produced from its nectar is esteemed.

DESCRIPTION: This tree has a conical or rounded crown of spreading branches, clusters of flowers recalling Lily-of-the-valley, and glossy foliage that turns red in fall. The leaves are 4–7" (10–18 cm) long, 1½–2½" (4–6 cm) wide, elliptical or lance-shaped, and finely saw-toothed, with a sour taste. They are shiny yellow-green above and paler and slightly hairy on the veins beneath, turning red in fall. The bark is brown or gray, thick, and fissured into narrow, scaly ridges. The twigs are light yellow-green, slender, and hairless.

FLOWERS: Summer. The buds and young flowers hang down, short-stalked on one side of slender axes, with an urn-shaped white corolla, ¼" (6 mm) long, and slightly five-lobed, occurring in terminal drooping clusters 4–10" (10–25 cm) long.

FRUIT: A narrowly egg-shaped capsule ⅜" (10 mm) long, gray and covered with fine hairs, upright on curved stalks along drooping axes; it is five-celled, splitting along five lines, and many-seeded, maturing in fall and remaining attached into winter.

HABITAT: Moist soils in valleys and uplands with oaks and pines.

RANGE: Southwest Pennsylvania and southeastern Maryland, south to northwestern Florida, west to Louisiana, north to southern Indiana; it occurs to 5000' (1524 m) or slightly higher in the southern Appalachians.

USES: This species is grown as an ornamental tree. It is known for its nectar, which is used to make honey.

SIMILAR SPECIES: Sourwood is the sole species in its genus, unlikely to be confused with other trees within its range.

CONSERVATION STATUS: (LC)
The Sourwood has a large population with a wide range. There are no major threats known to the species.

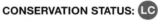

Diameter: 1' (0.3 m)
Height: 50' (15 m)

■ DENSE POPULATION
■ LIGHT POPULATION
■ NATURALIZED

Catawba Rhododendron is native to the eastern United States and is most plentiful in the southern Appalachian Mountains. On high mountain ridges, the plants usually flower in June; at lower altitudes, flowers appear earlier, although the times may vary slightly. Its dense thickets and purple flowers in May and June attract thousands of visitors each year to Great Smoky Mountains National Park. Frequently used as an ornamental, it has been crossed with some Himalayan species to produce some of our most spectacular cultivated rhododendrons. It is named for the Catawba River in North Carolina, which itself was named for the prominent Catawba tribe of Native Americans in the southeastern United States. This species goes by many common names including Purple Rhododendron, Purple Laurel, Mountain Rosebay, and several variations thereof.

DESCRIPTION: This is an evergreen, thicket-forming shrub or small tree with a broad, rounded crown and spectacular displays of large, pink to purplish blossoms. The leaves are evergreen, 3–6" (7.5–15 cm) long, 1½–2½" (4–6 cm) wide, elliptical, blunt at the tip, rounded at the base, thick and leathery with the edges often rolled under, and hairless, with long stout leafstalks. They are shiny dark green above and whitish beneath. The bark is brown or gray and fissured into narrow scaly ridges. The twigs are greenish and stout.

FLOWERS: Late spring–early summer. 2½" (6 cm) wide flowers with a bell-shaped corolla of five rounded lobes; they are waxy lilac-purple (sometimes pink), slender-stalked, and arranged in upright branched rounded clusters.

FRUIT: ½–⅝" (12–15 mm) long, a long-stalked narrowly egg-shaped capsule, densely covered with reddish-brown hairs; it is five-celled and splitting along five lines, and many-seeded; it matures in late summer and fall.

CAUTION: Poisonous: Do not eat or ingest any part of this species. If accidentally ingested, contact your doctor or poison control center.

HABITAT: Rocky slopes and ridges in understory of mountain forests, and in shrub thickets called "heath balds" or "laurel slicks."

RANGE: Western Virginia south to northern Georgia and northeastern Alabama, north to eastern Kentucky; it is usually found at 1500–6500' (457–1981 m), and locally down to 200' (61 m).

USES: This species is a popular ornamental plant, both in North America and in Europe.

SIMILAR SPECIES: Rosebay Rhododendron.

CONSERVATION STATUS:
The Catawba Rhododendron has a stable population with no known current or future threats.

Diameter: 4" (10 cm)
Height: 20' (6 m)

Pacific Rhododendron is a showy flowering shrub in the forests of western North America, with flower clusters that nearly rival in size and number those of its cultivated relatives. It is widely distributed in the Cascade Range and otherwise has a mainly coastal distribution. The plant is most abundant in Oregon, where mass displays provide a spectacular show in late spring and early summer.

DESCRIPTION: This evergreen shrub can grow tall and treelike under ideal conditions. It has large, deep green, leathery leaves and rounded clusters of large, pink, tubular flowers. The leaves are evergreen, alternate but often so closely spaced as to seem whorled, and 2½–8" (6–20 cm) long, with stout petioles. They are leathery, thick, not toothed, oblong to elliptical, and dark green and hairless above, and paler and sometimes rusty beneath.

FLOWERS: May–June. Pink to rose-purple, rarely white flowers about 1½" (3.8 cm) long, with five wavy-margined petals united to form a broadly bell-shaped corolla about 1½" (3.8 cm) long, arranged in rounded clusters up to 6" (15 cm) across.

FRUIT: Capsule about ¾" (2 cm) long.

CAUTION: Poisonous: Do not eat or ingest any part of this species. If accidentally ingested, contact your doctor or poison control center.

HABITAT: Acidic soils in open or shady forest.

RANGE: Southwest British Columbia south along the coast and in the Cascades to central California.

USES: This flowering shrub is sometimes used as an ornamental, though not as often as other *Rhododendron* species.

SIMILAR SPECIES: Western Azalea.

CONSERVATION STATUS: LC
The Pacific Rhododendron is considered stable and without major threats.

Height: 4–10' (1.2–3 m);
less common to 20' (6 m) nearly treelike

■ DENSE POPULATION
■ LIGHT POPULATION
■ NATURALIZED

Rosebay Rhododendron, also called Great Laurel or Great Rhododendron, is native to the Appalachian Mountains and north through New England to coastal Nova Scotia. It is the most common and dominant understory species of *Rhododendron* in the southern Appalachian region and is abundant in Great Smoky Mountains National Park. Often grown as an ornamental, it is one of the hardiest and largest evergreen rhododendrons. It is the state flower of West Virginia.

DESCRIPTION: This evergreen, thicket-forming shrub or tree has a short, crooked trunk, a broad, rounded crown of many stout, crooked branches, and large white blossoms. Its leaves are evergreen, 4–10" (10–25 cm) long, 1–3" (2.5–7.5 cm) wide, oblong or narrowly elliptical, and short-pointed at both ends; they are thick and leathery with the edges rolled under, and have short, stout leafstalks. They are shiny dark green above and whitish and covered with fine hairs beneath. The bark is red-brown, scaly, and thin. The twigs are green with reddish gland-hairs, becoming reddish-brown and scaly, and stout.

FLOWERS: Summer. 1½" (4 cm) wide with a bell-shaped corolla of five rounded lobes; they are waxy white or sometimes light pink (rarely reddish); the largest or upper lobe has many green spots; they are arranged in upright, branched, rounded clusters.

FRUIT: ½" (12 mm) long, long-stalked, narrowly egg-shaped capsule; it is dark reddish-brown, with gland-hairs; it is five-celled and splitting open along five lines, with many seeds, maturing in fall and remaining attached.

CAUTION: Poisonous: Do not eat or ingest any part of this species. If accidentally ingested, contact your doctor or poison control center.

HABITAT: Moist soils, especially along streams in understory of mountain forests, forming dense thickets.

RANGE: Maine southwest to western New York and south, mostly in mountains, to northern Georgia; to 6000' (1829 m) in southern Appalachians.

USES: This species is planted as an ornamental. The wood is occasionally used for tool handles and other small items, and a home remedy has been prepared from the leaves.

SIMILAR SPECIES: Catawba Rhododendron.

CONSERVATION STATUS: **LC**
The Rosebay Rhododendron has a widespread population that is stable. The species faces no current or future threats.

Diameter: 6" (15 cm)
Height: 20' (6 m)

■ DENSE POPULATION
■ LIGHT POPULATION
■ NATURALIZED

This flowering shrub is native to the northwestern United States, mainly in the coastal mountain ranges and east to the Sierra Nevada and Cascades. There is considerable diversity in the form and appearance of this species, with flower variations including mixtures of pale pink, deep pink, and yellow-orange. It is a popular ornamental and has been hybridized with other azaleas to produce numerous horticultural hybrids.

DESCRIPTION: This is a shrub and sometimes grows quite large with big, white to deep pink, fragrant flowers in large clusters at the stem ends. The leaves are 1¼–3½" (3–9 cm) long, thin, bright green, and elliptical.

FLOWERS: April–August. The corolla is 1½–2½" (4–6.5 cm) wide, with a narrow, tubular base and five wavy, pointed lobes, upper lobe with a yellow-orange patch; it has five stamens, with hairy filaments.

CAUTION: Poisonous: Do not eat or ingest any part of this species. If accidentally ingested, contact your doctor or poison control center.

HABITAT: Moist places; in open areas near the coast, otherwise where partly shaded.

RANGE: Southwestern Oregon south to southern California.

USES: Western Azalea is popularly planted as an ornamental.

SIMILAR SPECIES: California Rosebay, Pacific Rhododendron.

CONSERVATION STATUS: 🔵 LC
The Western Azalea population faces no known threats and is considered stable.

Height: 4–16½' (1.2–5.1 m)

■ DENSE POPULATION
■ LIGHT POPULATION
■ NATURALIZED

Sparkleberry is found throughout the southeastern and south-central United States. Its leaves are evergreen in the south of its range, but deciduous in colder climates farther north. The fruit has thin, slightly sweet pulp and large seeds. Although they are bitter, tough, and generally unpalatable to humans, many forms of wildlife readily consume the berries. This species, also commonly known as Farkleberry, is the tallest plant in this genus that includes blueberries, cranberries, and huckleberries.

DESCRIPTION: This is a shrub or tree with a short trunk; an irregular crown of crooked branches; small, glossy, elliptical leaves; and shiny black berries. The leaves are deciduous and, in the southern part of the range, evergreen or nearly so. They are ½–1¾" (1.2–4.5 cm) long, ¼–1" (0.6–2.5 cm) wide, elliptical or obovate, blunt or rounded at the tip, with tiny teeth or not toothed, a prominent network of fine veins, and slightly thickened. They are shiny dark green above and paler and sometimes slightly hairy on the veins beneath. The bark is light reddish-brown, thin, and covered with fine scales. The twigs are reddish, slender, stiff, and crooked, and hairy when young.

FLOWERS: Spring. ¼" (6 mm) long and wide with a bell-shaped, slightly five-lobed white corolla, occurring on drooping stalks in unbranched clusters 2–3" (5–7.5 cm) long.

FRUIT: ¼" (6 mm) in diameter, shiny black berries, with thin, dry, mealy pulp and eight to ten seeds, maturing in fall and remaining attached into winter.

HABITAT: Sandy and rocky dry uplands, in forest understory and clearings.

RANGE: Virginia south to central Florida, west to southeastern Texas, and north to southeastern Kansas; to 2500' (762 m).

USES: This species is often used as an ornamental and to attract wildlife.

SIMILAR SPECIES: Deerberry.

CONSERVATION STATUS: LC

The Sparkleberry population is considered stable and faces no major threats from pests or diseases.

Diameter: 6" (15 cm)
Height: 25' (7.6 m)

■ DENSE POPULATION
■ LIGHT POPULATION
■ NATURALIZED

Wavyleaf Silktassel is native to the coastal mountain ranges of California and Oregon. This is the only native species of the genus *Garrya* that reaches tree size. Although various parts of the plant have a bitter taste and are usually ignored by herbivores such as deer and rabbits, goats sometimes browse the foliage. As a landscaping ornamental, Wavyleaf Silktassel is a low-maintenance plant with a neat growing habit.

DESCRIPTION: This is an evergreen shrub or small tree with tassel-like clusters of flowers and fruit and paired, leathery, wavy-edged leaves. The foliage and other parts have a bitter taste. Its leaves are evergreen, opposite, 2–3¼" (5–8 cm) long, elliptical, and thick; they are shiny green and nearly hairless above and paler with a thick coat of woolly hairs beneath. The bark is gray and smooth, becoming finely fissured and slightly scaly. The twigs are four-angled and densely hairy when young, and greenish, becoming brown or blackish.

FLOWERS: Late winter–early spring. Tiny, scaly, greenish, and without petals, with many crowded together in drooping, narrow, catkin-like clusters 2–5" (5–13 cm) long; male and female flowers occur on separate plants.

FRUIT: ⅜" (10 mm) in diameter, rounded, berrylike, dark purple to black, and densely covered with white hairs, becoming dry; it has bitter pulp, is one-seeded, and matures in summer and splits open.

HABITAT: Dry slopes and ridges; often in thickets, chaparral, and mixed evergreen forests.

RANGE: Western Oregon south to southern California and Santa Cruz Island; to 2000' (610 m).

USES: This species is widely cultivated as a landscaping ornamental.

SIMILAR SPECIES: Ashy Silktassel, Fremont Silktassel.

CONSERVATION STATUS: LC
The Wavyleaf Silktassel is extremely resistant to cold temperatures and fairly resistant to drought. The species is considered stable and faces no known, significant threats.

Diameter: 4" (10 cm)
Height: 20' (6 m)

■ DENSE POPULATION
■ LIGHT POPULATION
■ NATURALIZED

Buttonbush is native to eastern North America and the southern United States. It is common as a shrub in many types of wetland habitats, but sometimes occurs as a small tree. This plant is a handsome ornamental suited to wet soils; it is also a honey plant. Waterfowl and other wildlife readily consume the seeds and hummingbirds and insects feed on the nectar. Some authorities classify the southwestern U.S. population of this plant as a distinct variety (var. *californicus*).

DESCRIPTION: Buttonbush is a spreading, much-branched shrub, or sometimes a small tree, with many branches (often crooked and leaning), an irregular crown, balls of white flowers resembling pincushions, and button-like balls of fruit. The leaves are opposite or whorled, 2½–6" (6–15 cm) long, 1–3" (2.5–7.5 cm) wide, ovate or elliptical, pointed at the tip, rounded at the base, and without teeth. They are shiny green above and paler and sometimes hairy beneath; at their southern limit they are nearly evergreen. The bark is gray or brown, becoming deeply furrowed into rough scaly ridges. The twigs occur mostly in threes; they are reddish-brown, stout, and sometimes hairy, with rings at the nodes.

FLOWERS: Late spring through summer. ⅝" (15 mm) long, with a narrow, tubular, white four-lobed corolla and long threadlike style; they are fragrant, stalkless, and crowded in upright long-stalked white balls of many flowers each, 1–1½" (2.5–4 cm) in diameter.

FRUIT: Compact rough balls ¾–1" (2–2.5 cm) in diameter, composed of many small, narrow, dry nutlets ¼" (6 mm) long, each two-seeded; it matures in autumn.

CAUTION: Poisonous: Do not eat or ingest any part of this species. If accidentally ingested, contact your doctor or poison control center.

HABITAT: Wet soils bordering streams and lakes.

RANGE: Southern Quebec and southwestern Nova Scotia, south to southern Florida, west to Texas, and north to southeastern Minnesota; to 3000' (914 m); in Arizona and California to 5000' (1524 m); also Mexico, Central America, and Cuba.

USES: Buttonbush is often planted as an ornamental and to help to control erosion. The bitter bark of Buttonbush has been used in traditional home remedies, but its medicinal value is uncertain.

SIMILAR SPECIES: Pinckneya (Georgia Fevertree).

CONSERVATION STATUS: 🅛🅒
Although listed as vulnerable or imperiled in some areas of the range due to limited numbers or local declines, Buttonbush is considered globally stable. The widespread species does not have any known current or future risks that pose a significant threat to the global population.

Diameter: 4" (10 cm)
Height: 20' (6 m)

■ DENSE POPULATION
■ LIGHT POPULATION
■ NATURALIZED

Also commonly known as Firebush or Hummingbird Bush, Scarletbush is often grown as an annual in areas north of its native Florida range; in northern areas it reaches only 2–3' (60–90 cm) in height. It is named for its leaves, which turn red in fall, especially farther north. In the subtropics it may remain evergreen.

DESCRIPTION: This is a shrub to a small tree with red stalks bearing clusters of bright, firecracker-shaped flowers. The leaves are up to 6" (15 cm) long, occurring in whorls of three to seven; they are elliptical to ovate, red-veined, red-stalked, and evergreen to the south. The bark is brownish or grayish, and smooth.

FLOWERS: Individual flowers are ¾" (2 cm) long, narrow tubes; they are red to orange, occurring in one-sided clusters on red stems; they bloom year-round in their native range.

FRUIT: Berries, ½" (12 mm) in diameter, ripening from green to yellow to red to purple-black, with a conspicuous ring on top; all colors may be present on the plant.

HABITAT: Hammocks, shell mounds, open disturbed areas, coastal dunes.

RANGE: Native to central Florida and southward throughout tropical America; it is an introduced garden plant elsewhere.

USES: This species is often planted to attract hummingbirds and other wildlife. It has also been used in traditional medicine to treat a variety of ailments.

SIMILAR SPECIES: Scarletbush is unlikely to be confused with other species within its native range.

CONSERVATION STATUS: LC
Scarletbush is widely distributed and has a large, stable population. The species has no significant current or future threats.

Height: to 12' (3.6 m), usually much shorter, especially farther north

■ DENSE POPULATION
■ LIGHT POPULATION
■ NATURALIZED

This small tree is native to the southeastern United States, where it is found in poorly drained, acidic soils such as those found along swamp margins. It is among the most beautiful native trees and is grown as an ornamental. The showy, pinkish, enlarged calyx lobes remain attached for several weeks. Also commonly known as Georgia Bark or Fevertree, the bitter inner bark of this species was once used in traditional medicine to reduce fever.

Pinckneya is the single, rare species of its genus.

ALTERNATE NAME: *Pinckneya pubens*

DESCRIPTION: This is a shrub or a small tree, usually with several trunks, and a narrow, rounded crown, and greenish-yellow, tubular flowers, some with a beautiful, pink, petal-like lobe. The leaves are opposite, 4–8" (10–20 cm) long, 2–4" (5–10 cm) wide, elliptical, short-pointed, and without teeth. They are dark green and covered with fine hairs above, and paler and with soft hairs beneath. The slender leafstalks are about 1" (2.5 cm) long. The bark is light brown, scaly, and bitter. The twigs are slender, covered with gray hairs, and turning light brown, with rings at the nodes.

FLOWERS: About 1½" (4 cm) long and narrow, with a tubular greenish-yellow corolla with red spots inside and five narrow lobes, arranged in open clusters of several flowers, each 6" (15 cm) wide, occurring at or near the ends of leafy twigs, blooming in late spring.

FRUIT: ¾" (19 mm) in diameter, slightly hairy rounded capsules; they are brown with whitish dots, thin-walled, two-celled, and splitting in two parts; they have many flat, light brown, broad-winged seeds, maturing in fall and remaining attached in winter.

HABITAT: Wet soils of swamps and stream borders; in understory of forests.

RANGE: Extreme southern South Carolina, Georgia, and northern Florida; to 300' (91 m).

USES: This species is cultivated as an ornamental. The bitter bark was once used as a home remedy to reduce fever.

SIMILAR SPECIES: Common Buttonbush.

CONSERVATION STATUS: NT

This species has a restricted range and is a regional endemic, but it is generally easy to find in small numbers in appropriate habitat. The species faces threats from habitat loss primarily due to conversion to agriculture and pine plantations, and few of its natural populations are protected.

Diameter: 6" (15 cm)
Height: 20' (6 m)

■ DENSE POPULATION
■ LIGHT POPULATION
■ NATURALIZED

The large leaves of this widespread tropical tree found in southern Florida are evergreen, but may be shed in the winter if the tree is in an unprotected site. Although slow-growing, this species is widely used as an ornamental plant throughout its native range. Geiger Tree is native throughout the American tropics. In Jamaica this plant is commonly known as Scarlet Cordia. Some sources list the tree as introduced in Florida, but others consider it native.

DESCRIPTION: This is a small tropical tree or slender shrub with a rounded crown and upright branches. The leaves are 6" (15 cm) long, heart-shaped, hairy, and evergreen. The bark is dark brown to blackish, furrowed, and scaly.

FLOWERS: March–November. 1½" (4 cm) long, trumpet-shaped, and orange-red, occurring in branched clusters 5–6" (12.5–15 cm) wide.

FRUIT: 1¼" (3 cm), berrylike, fleshy fruit that ripens to an ivory-white color, arranged in clusters.

HABITAT: Subtropical forests, hammocks.

RANGE: Through the American tropics; it is found in southern Florida.

USES: This species is widely planted as an ornamental, valued for its showy flowers.

SIMILAR SPECIES: Fragrant Manjack.

CONSERVATION STATUS: LC
Common throughout its native range, the Geiger Tree population is considered stable. Although the species faces no significant threats, urbanization has led to some habitat loss. The species faces a minor threat of defoliation from the Geiger Tortoise Beetle (*Eurypepla calochroma*), which is found in Florida as well as Central America and the Caribbean. As the beetle's name suggests, the Geiger Tree is its only food source.

Height: 20–30' (6–9 m)

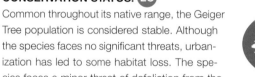

■ DENSE POPULATION
■ LIGHT POPULATION
■ NATURALIZED

A native of South America, Tree Tobacco has been introduced widely in the southern United States, and is a common and conspicuous weed along the roadsides of southern California. All species of *Nicotiana* contain the highly toxic alkaloid nicotine; this plant contains a more potent toxin called anabasine. An effective insecticide against aphids can be prepared by steeping tobacco in water and spraying the solution on affected parts of the plant. Consuming the leaves of this plant, however, can be fatal.

DESCRIPTION: This is an open shrub or small tree with few branches; the yellow, trumpet-shaped flowers tend to spread or hang on slender branches. The leaves are 2–7" (5–17.5 cm) long, smooth, ovate, and gray-green.

FLOWERS: April–November. The corolla is 1¼–2" (3.1–5 cm) long, with five stamens.

CAUTION: Poisonous: Do not eat or ingest any part of this species. If accidentally ingested, contact your doctor or poison control center.

HABITAT: Roadsides, slopes, and washes.

RANGE: Native to South America; it is introduced in central California, southern Arizona, and western Texas, south into Mexico.

USES: Tree Tobacco has been smoked and used for medicinal purposes by some Native American groups.

SIMILAR SPECIES: Green Cestrum.

CONSERVATION STATUS:
The conservation status of this species has not been formally evaluated in its native range.

Height: to 26' (8 m)

This tropical species is native to southern North America and northern South America. It has been introduced in many parts of the world; it is often cultivated in Florida gardens and landscapes as a buffer plant. It is quite fast-growing, and in some places has become a nuisance weed.

DESCRIPTION: This shrub or small tree has a low, flat crown and spreading branches. The leaves are 10" (25 cm) long, ovate to obovate, often wavy-edged, hairy, and evergreen. The bark is light green to grayish brown, and warty.

FLOWERS: Year-round. ½" (12 mm), star-shaped, and white, arranged in rounded clusters.

FRUIT: ¾" (2 cm) round, yellow berries.

HABITAT: Hammocks, thickets, disturbed areas.

RANGE: Florida and Texas and south into Mexico and the Caribbean.

USES: This species is grown as an ornamental plant, and has been used in traditional medicine in many countries. The leaves are sometimes used in the Philippines to clean dishes.

SIMILAR SPECIES: Turkey Berry, Earleaf Nightshade.

CONSERVATION STATUS:
The potato tree has not been formally assessed for conservation status.

Height: 10–16' (3–4.8 m)

■ DENSE POPULATION
■ LIGHT POPULATION
■ NATURALIZED

Fringe Tree is native to the United States where it is found primarily in the lowlands of the Southeast and less commonly north to New Jersey and west to Oklahoma and Texas. One of the last trees to bear new leaves in spring, Fringe Tree appears dead until the leaves and flowers appear. The genus name *Chionanthus* means "snow" and "flower," and describes the drooping clusters of whitish blossoms. This species is sometimes called White Fringetree and Old Man's Beard.

DESCRIPTION: This is a shrub or small tree with a short trunk; a narrow, oblong crown; and showy masses of fragrant, lacy, white flowers. The leaves are opposite, 4–8" (10–20 cm) long, 1–3" (2.5–7.5 cm) wide, narrowly elliptical, and without teeth; they are slightly thickened, with short, stout purplish leafstalks. They are shiny dark green above and paler and slightly hairy beneath, turning yellow in fall. The bark is brown with reddish-tinged scales. The twigs are light green, purplish at the nodes, stout, and often slightly angled; they are hairy when young, and slightly rough and warty.

FLOWERS: 1" (2.5 cm) long with a delicate corolla of four narrow, whitish lobes with purplish dots inside at the base; they are fragrant, with many hanging loosely on slender stalks in threes, arranged in branched drooping clusters 4–8" (10–20 cm) long.

FRUIT: ¾" (19 mm) long, elliptical, dark blue or blackish and often with whitish bloom; it has thin juicy pulp and a large stone, arranged in drooping grapelike clusters on long stalks, maturing in fall.

HABITAT: Moist soils of valley and bluffs; in understory of hardwood forests.

RANGE: Southern New Jersey south to central Florida, west to eastern Texas, and north to southern Missouri; to 4500' (1372 m) in the southern Appalachians.

USES: This species is often grown as an ornamental plant, valued for its white flowers and yellow fall foliage. Some Native Americans used the roots and bark for medicinal purposes.

SIMILAR SPECIES: Pygmy Fringetree, Wild Olive.

CONSERVATION STATUS:

The Fringe Tree population is considered stable. There are no known current threats to the species.

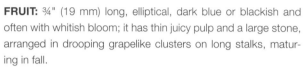
Diameter: 6" (15 cm)
Height: 30' (9 m)

■ DENSE POPULATION
■ LIGHT POPULATION
■ NATURALIZED

Sometimes called Eastern Swamp Privet, this flood-tolerant relative of true privets (genus *Ligustrum*) grows in swamps and other wetlands, and is most common in the Mississippi Valley. These plants are available commercially because they are useful for erosion control and attractive to wildlife. Bees and butterflies may visit the small, powder-puff flowers, and waterfowl sometimes consume the fruit.

DESCRIPTION: This is a shrub or small tree with slender, often leaning trunks, forming thickets at water edges, and with paired, diamond-shaped leaves. The leaves are opposite, 1½–4" (4–10 cm) long, ½–1" (1.2–2.5 cm) wide, diamond-shaped or ovate, and long-pointed at both ends, with small teeth beyond the middle; they are almost hairless and long-stalked. They are yellow-green above, and paler beneath. The bark is dark brown, thin, and smooth. The twigs are light brown and slender.

FLOWERS: Early spring, before leaves. Tiny, greenish-yellow flowers without petals in small clusters along the twig. Male and female are on separate plants.

FRUIT: ⅜–⅝" (10–15 mm) long, narrowly oblong or slightly curved, dark purple or blackish, with thin pulp and a large stone, arranged in short clusters of several flowers each, maturing in summer.

HABITAT: Wet soils bordering streams, swamps, and lakes; at edge of swamp forests.

RANGE: Eastern South Carolina to northern Florida, west to eastern Texas, and north to central Illinois and southwestern Indiana; to 500' (152 m).

USES: This species is sometimes planted to help with erosion control.

SIMILAR SPECIES: Florida Privet, Desert Olive.

CONSERVATION STATUS:
The Swamp-privet population appears secure and has no known threats.

Diameter: 4" (10 cm)
Height: 25' (7.6 m)

■ DENSE POPULATION
■ LIGHT POPULATION
■ NATURALIZED

Sometimes called Florida Swamp-privet or Southern Privet, the natural range of this species extends throughout the Caribbean to Puerto Rico and the Virgin Islands, where it is called "Ink-bush" because the berries are known to stain cloth and skin. In the United States, it is found in moist soils in coastal Georgia and Florida. The flowers attract bees, and birds often eat the fruit and use the plant as cover.

DESCRIPTION: This is an evergreen shrub or small tree with one or more short, much-branched trunks, forming dense, irregular crowns and thickets. The leaves are evergreen, opposite, ¾–2½" (2–6 cm) long, ½–¾" (12–19 mm) wide, narrowly elliptical or diamond-shaped, blunt-pointed, and without teeth, with few side veins and slightly curved up on the sides; they are some-what thickened and very short-stalked. They are shiny green or gray-green above, and dull light green with tiny dots beneath. The bark is whitish-gray and smooth. The twigs are greenish and covered with fine hair when young, becoming light gray, slender, and stiff.

FLOWERS: Spring. The flowers are ⅛" (3 mm) long and greenish-yellow, without a calyx or corolla, and short-stalked, arranged in small clusters at the leaf bases or on the previous year's twigs. Male and female flowers are generally on separate plants.

FRUIT: ¼–⅜" (6–10 mm) long, elliptical, and purplish or black-ish, with thin butter pulp and a large stone, maturing in summer.

HABITAT: Moist soil, including sand dunes, hammocks, and understory of pine forests.

RANGE: Along and near coasts of Georgia south through Florida Keys, also Bermuda and West Indies.

USES: Florida Privet is cultivated as a hedge plant.

SIMILAR SPECIES: Swamp-privet.

CONSERVATION STATUS: NT
The Florida Privet population is limited but apparently secure; however, more research is needed to assess possible threats.

Diameter: 4" (10 cm)
Height: 20' (6 m)

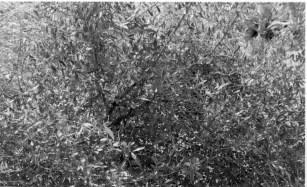

■ DENSE POPULATION
■ LIGHT POPULATION
■ NATURALIZED

Also known as American Ash, this tree is native to much of the eastern and central United States, where it is usually found in moist soils. Its strong, straight-grained wood is of high commercial value and the species is widely cultivated as a street or landscaping tree. Biltmore Ash is sometimes treated as a variation of White Ash and sometimes as a distinct species (*F. biltmoreana*); it has hairy twigs, leafstalks, and underleaf surfaces.

DESCRIPTION: This is a large tree with a straight trunk and a dense, conical or rounded crown of foliage with whitish lower surfaces. The leaves are opposite, pinnately compound, and 8–12" (20–30 cm) long, usually with seven (sometimes five to nine) leaflets 2½–5" (6–13 cm) long and 1¼–2½" (3–6 cm) wide; they are paired (except at the end), ovate or elliptical, and finely saw-toothed or almost without teeth. They are dark green above and whitish and sometimes hairy beneath, turning purple or yellow in fall. The bark is dark gray and thick, with deep diamond-shaped furrows and forking ridges. The twigs are gray or brown, stout, and mostly hairless.

FLOWERS: Early spring, before leaves. ¼" (6 mm) long and purplish, without a corolla, with many arranged in small clusters. Male and female are on separate trees.

FRUIT: 1–2" (2.5–5 cm) long brownish key with a narrow wing not extending down a cylindrical body; they hang in clusters and mature in late summer and autumn.

HABITAT: Moist soils of valleys and slopes, especially deep well-drained loams; in forests with many other hardwoods.

RANGE: Southern Ontario east to Cape Breton Island, south to northern Florida, west to eastern Texas, and north to eastern Minnesota; it occurs to 2000' (610 m) in the north and to 5000' (1524 m) in the south.

USES: White Ash is widely cultivated and used for a variety of purposes. The dense wood is used to make various sporting goods such as baseball bats, hockey sticks, oars, and tennis rackets, as well as tool handles.

SIMILAR SPECIES: Texas Ash, Biltmore Ash.

CONSERVATION STATUS: CR

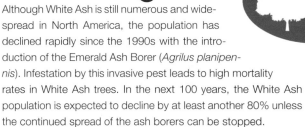

Although White Ash is still numerous and widespread in North America, the population has declined rapidly since the 1990s with the introduction of the Emerald Ash Borer (*Agrilus planipennis*). Infestation by this invasive pest leads to high mortality rates in White Ash trees. In the next 100 years, the White Ash population is expected to decline by at least another 80% unless the continued spread of the ash borers can be stopped.

Diameter: 2' (0.6 m)
Height: 80' (24 m)

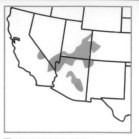

■ DENSE POPULATION
■ LIGHT POPULATION
■ NATURALIZED

Singleleaf Ash is native to the southwestern United States, and is particularly common in Grand Canyon National Park. Although most species in the genus *Fraxinus* are characterized by pinnately compound leaves, this species is distinct among ashes in that it usually has simple leaves.

DESCRIPTION: This is a shrub or small tree with a short trunk and a rounded crown of stout, curved branches. The leaves are opposite and simple or occasionally pinnately compound, with two or three leaflets; they are 1½–2" (4–5 cm) long, 1–2" (2.5–5 cm) wide (leaflets are slightly smaller), broadly ovate or nearly round, rounded or short-pointed at the tip, blunt or notched at the base, inconspicuously wavy-toothed or without teeth, slightly thick and leathery, and becoming hairless; long-stalked. Dark green above, paler beneath. The bark is dark brown and furrowed into narrow ridges. The twigs are brown, hairless, and four-angled or slightly winged.

FLOWERS: ⅛" (3 mm) long, greenish, and without petals, with many occurring together in small, hairy clusters on the previous year's twigs at the back of the leaves in spring. The flowers may be bisexual or female.

FRUIT: A flattened key ¾" (19 mm) long with an elliptical, rounded wing ⅜" (10 mm) wide, extending to the base, maturing in summer.

HABITAT: Dry canyons and hillsides including rocky slopes in upper desert; woodland and Ponderosa Pine forest.

RANGE: Western Colorado to eastern California, Arizona, and extreme northwestern New Mexico. It usually occurs to 2000–6500' (610–1981 m), and in California to 11,000' (3353 m).

USES: This species is sometimes planted as an ornamental but is not widely used.

SIMILAR SPECIES: This species is distinguished from other ashes by its (usually) simple leaves.

CONSERVATION STATUS: LC
Singleleaf Ash has a large, fairly widespread population that is considered stable. Fire poses a threat in its native range, but the species is believed to resprout quickly. Although the species has no significant current threats, the continued spread of the devastating Emerald Ash Borer (*Agrilus planipennis*) is a serious potential future threat, depending on how far west and south this invasive pest reaches.

Diameter: 6" (15 cm)
Height: 25' (7.6 m)

■ DENSE POPULATION
■ LIGHT POPULATION
■ NATURALIZED

Also commonly known as Mexican Ash, this species is native to the south-central United States and south into Mexico. This tree is a smaller southwestern relative of Green Ash, distinguished by fewer and smaller leaflets and smaller fruit. It is adapted to a warmer, less humid climate.

DESCRIPTION: This is a small tree with a short trunk and a rounded crown of spreading branches. Its leaves are opposite, pinnately compound, and 3–7" (7.5–18 cm) long, with three or five leaflets 1½–4" (4–10 cm) long and ⅝–1½" (1.5–4 cm) wide; they are paired (except at the end), lance-shaped, and sometimes elliptical; they have few teeth or are coarsely saw-toothed, and they are shiny green above, pale and sometimes hairy in vein angles beneath. The bark is gray, thick, and furrowed into narrow ridges. The twigs are light green, slender, and hairless.

FLOWERS: ⅛" (3 mm) long and without a corolla, occurring in small clusters of many flowers each, before the leaves in early spring. Male flowers are yellowish and female flowers are greenish, occurring on separate trees.

FRUIT: 1–1¼" (2.5–3 cm) long, light brown key with a narrow wing extending nearly to the base of the narrow body, sometimes three-winged, and hanging in clusters; it matures in late spring or summer.

HABITAT: Moist soils along streams and in canyons.

RANGE: Central and southern Texas and northeastern Mexico; to 1000' (3–5 m).

USES: This species is used as landscaping ornamental in southern Texas. The wood is sometimes used as firewood.

SIMILAR SPECIES: Green Ash, Velvet Ash, Fragrant Ash.

CONSERVATION STATUS: LC

The Berlandier is considered stable and widespread. It faces no current major threats. Like all North American ash trees, however, it faces a possible future threat from the Emerald Ash Borer (*Agrilus planipennis*) if the invasive pest's spread reaches its range.

Diameter: 1' (0.3 m)
Height: 30' (9 m).

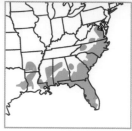

■ DENSE POPULATION
■ LIGHT POPULATION
■ NATURALIZED

Pop Ash is native to the southeastern United States, where it is typically found in coastal swamps and subtropical lowlands. This small tree's large, broadly winged, flat keys are distinctive. This tree is known by several common names, including Carolina Ash, Florida Ash, Swamp Ash, and Water Ash.

DESCRIPTION: Pop Ash is a tree with one (or sometimes more than one) trunk, often enlarged at the base and leaning, and a rounded or narrow crown. Its leaves are opposite, pinnately compound, and 6–12" (15–30 cm) long, with five or seven (sometimes nine) leaflets 2–4½" (5–11 cm) long and 1–2" (2.5–5 cm) wide; they are paired (except at the end), elliptical or ovate, long-pointed, coarsely saw-toothed, slightly thickened, and slender-stalked. They are green above and paler or whitish and often slightly hairy beneath. The bark is light gray, and thin and scaly, becoming rough and furrowed. The twigs are light green to brown, usually hairless, and slender.

FLOWERS: ⅛" (3 mm) long and without a corolla, occurring in small clusters of many flowers each, before the leaves in early spring. Male flowers are yellowish and female flowers are greenish, occurring on separate trees.

FRUIT: 1¼–2" (3–5 cm) long yellow-brown key with a broad elliptical wing extending to the base of the flat body; it is sometimes three-winged, hanging in clusters and maturing in summer and fall.

HABITAT: Wet soils of swamps and riverbanks that are flooded for part of the year; it is often found in swamp forests.

RANGE: Northeastern Virginia to southern Florida and west to southeastern Texas; to 500' (152 m).

USES: The soft, lightweight wood of the species is generally weak and not used commercially. This tree is sometimes planted as an ornamental, but it is not widely used.

SIMILAR SPECIES: Pumpkin Ash, White Ash.

CONSERVATION STATUS: EN

Some populations of Pop Ash have declined as a result of deforestation and habitat fragmentation. Like other trees in this genus, however, the most critical threat to this species is the invasive Emerald Ash Borer (*Agrilus planipennis*), which has spread rapidly throughout the Pop Ash's native range. The pest is expected to lead to the total mortality of virtually all Pop Ash trees in at least half of the species' range. With no known remediation for the invasion, the species is at risk of extinction.

Diameter: 1' (0.3 m)
Height: 30–50' (9–15 m)

■ DENSE POPULATION
■ LIGHT POPULATION
■ NATURALIZED

Fragrant Ash is native to the southwestern United States and northern Mexico. It grows on dry stream beds, rocky slopes, and ravines, and occurs in Grand Canyon National Park and Big Bend National Park. The scientific name refers to the sharp points or cusps of the leaflets. It is planted as an ornamental for the abundant showy flowers and attractive foliage.

DESCRIPTION: This is a many-branched shrub or small tree with fragrant, showy white flowers and long-pointed, sharply saw-toothed leaflets. The leaves are opposite, pinnately compound, and 3–7" (7.5–18 cm) long, usually with seven leaflets (sometimes three to nine), paired except at the end, and long-stalked; they are 1½–2½" (4–6 cm) long, ¼–¾" (6–19 mm) wide, lance-shaped or ovate, long-pointed, and sharply saw-toothed or not toothed. They are shiny dark green above and paler and slightly hairy when young beneath. The bark is gray and smooth, becoming fissured into scaly ridges. The twigs are gray, slender, and hairless.

FLOWERS: Spring. About ⅝" (15 mm) long with 4 very narrow, white corolla lobes; the flowers are fragrant, and droop on slender stalks, with many arranged in branched clusters 3–4" (7.5–10 cm) long at the end of side twigs.

FRUIT: An oblong key ¾–1" (2–2.5 cm) long with a rounded or slightly notched wing extending nearly to the base of the flattened body, maturing in summer.

HABITAT: Rocky slopes of canyons and mountains in oak woodlands.

RANGE: Southwest and Trans-Pecos Texas, New Mexico, Arizona, and northern Mexico; at 4500–7000' (1372–2134 m).

USES: This species is sometimes used as an ornamental plant.

SIMILAR SPECIES: Berlandier Ash, Velvet Ash.

CONSERVATION STATUS: (LC)
The Fragrant Ash has no known current threats and is considered widespread and stable. Like others in its genus, the invasive pest Emerald Ash Borer (*Agrilus planipennis*) is a possible future threat to the species.

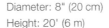

Diameter: 8" (20 cm)
Height: 20' (6 m)

TWO-PETAL ASH *Fraxinus dipetala*

■ DENSE POPULATION
■ LIGHT POPULATION
■ NATURALIZED

This species, also known as California Ash, is characterized by its clusters of sweet-smelling flowers with two long, narrow lobes divided almost into two separate petals. It is native to the southwestern United States and northern Baja California and typically grows in chaparral and foothill woodlands. It is often planted as an ornamental shrub.

DESCRIPTION: This is a shrub or small tree with showy white flowers composed of two petal-like lobes. The leaves are opposite, pinnately compound, and 1½–4½" (4–11 cm) long, with three to seven (sometimes nine) leaflets; they are paired except at the end, ¾–1½" (2–4 cm) long, ¼–⅝" (6–15 mm) wide, elliptical or obovate, blunt or short-pointed at the tip, short-pointed at the base, and coarsely saw-toothed, becoming hairless. They are dark green above and paler beneath. The bark is light gray, rough, and scaly. The twigs are green, slender, slightly four-angled when young, and usually hairless.

FLOWERS: Spring. ³⁄₁₆" (5 mm) long with two broad, white corolla lobes, drooping on slender stalks, with many in branched clusters to 4" (10 cm) long on twigs of the previous year.

FRUIT: An oblong key ¾–1" (2–2.5 cm) long with a broad wing often slightly notched and extending nearly to the base of the flattish body, maturing in early summer.

HABITAT: Dry slopes in foothills, chaparral, and woodland.

RANGE: California; also in northern Baja California; to 3500' (1067 m).

USES: This species is planted as an ornamental, valued for its showy white flowers.

SIMILAR SPECIES: Oregon Ash.

CONSERVATION STATUS: LC

The Two-petal Ash population is believed to be secure, but faces a potential future threat from the spread of the Emerald Ash Borer (*Agrilus planipennis*), which has significantly impacted other species in the *Fraxinus* genus.

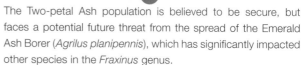

Diameter: 4" (10 cm)
Height: 20' (6 m)

■ DENSE POPULATION
■ LIGHT POPULATION
■ NATURALIZED

Oregon Ash is found west of the Cascade Range in the Pacific Northwest, and south through the mountains into central California. It is most abundant in southwestern Oregon and is planted as a shade tree along the Pacific Coast. This tree's hard wood resembles that of eastern ash species. An old superstition in the Northwest credits this tree with warding off poisonous snakes, suggesting that rattlesnakes will not crawl over a branch or stick from the tree.

DESCRIPTION: This tree has a long, straight trunk and usually a narrow, dense crown. The leaves are opposite, pinnately compound, and generally 5–12" (13–30 cm) long; the axis is usually hairy. It has five or seven (sometimes nine) leaflets that are paired except at the end; they are 2–5" (5–13 cm) long, 1–1½" (2.5–4 cm) wide, elliptical, short-pointed at the ends, and without teeth or slightly saw-toothed, with a prominent network of veins; they are stalkless or nearly so, and light green above, and paler and hairy beneath, turning yellow or brown in fall. The bark is dark gray or brown and thick, furrowed into forking, scaly ridges. The twigs are stout and covered with soft hairs.

FLOWERS: Early spring, before leaves. ⅛" (3 mm) long without a corolla; the male is yellowish and the female greenish, occurring on separate trees, with many clustered together on twigs.

FRUIT: A light brown key 1¼–2" (3–5 cm) long with a broad, rounded wing extending nearly to the base of the slightly flattened body, with many hanging in dense clusters; they mature in early fall.

HABITAT: Wet soils along streams and in canyons; it is often found with Red Alder, Black Cottonwood, willows, and Oregon White Oak.

RANGE: Western Washington, western Oregon, and south in Coast Ranges and Sierra Nevada to central California; to 5500' (1676 m).

USES: Oregon Ash wood is used for a variety of purposes, including furniture, flooring, and paneling. The species is also used as an ornamental tree.

SIMILAR SPECIES: Two-petal Ash, Velvet Ash.

CONSERVATION STATUS:
The Oregon Ash is a common species in its native range. Logging and habitat loss due to urbanization are minor threats impacting the species, but it is expected to face a more serious population decline of almost 30% due to the spread of the Emerald Ash Borer (*Agrilus planipennis*).

Diameter: 2' (0.6 m)
Height: 80' (24 m)

■ DENSE POPULATION
■ LIGHT POPULATION
■ NATURALIZED

Black Ash is widespread in eastern Canada and the northeastern United States. Our northernmost native ash, Black Ash is named for its dark brown heartwood. This tree is an important food source for a variety of wildlife, including frogs, which feed on fallen leaves as tadpoles. Various birds and small mammals eat the seeds and deer and moose browse the foliage. It was formerly abundant but has undergone a dramatic decline since the introduction of the Emerald Ash Borer (*Agrilus planipennis*), and now faces the possibility of extinction.

DESCRIPTION: This tree has a narrow, rounded crown of upright branches. The leaves are opposite, pinnately compound, and 12–16" (30–41 cm) long. It has seven to eleven leaflets 3–5" (7.5–13 cm) long and 1–1½" (2.5–4 cm) wide; they are paired (except at the end), broadly lance-shaped, finely saw-toothed, and stalkless. They are dark green above and paler beneath with tufts of rust-colored hairs along the midvein, turning brown in fall. The bark is gray, corky, and fissured into soft scaly plates that rub off easily. The twigs are gray and stout, becoming hairless.

FLOWERS: Before leaves in early spring. ⅛" (3 mm) long, purplish, and without a calyx or corolla, occurring in small clusters of many flowers each; male and female flowers occur on separate trees.

FRUIT: 1–1½" (2.5–4 cm) long key with a broad oblong wing extending to the base of the flat body, hanging in clusters and maturing in late summer.

HABITAT: Wet soils of swamps, peat bogs, and streams, especially cold swamps where drainage is poor; in coniferous and hardwood forests.

RANGE: Manitoba east to Newfoundland, south to West Virginia, and west to Iowa; it is local in northeastern North Dakota and northern Virginia. it occurs to 3500' (1067 m).

USES: Black Ash was once widely used for wooden items such as baskets, barrels and chairs. It is an important food source for native wildlife.

SIMILAR SPECIES: Green Ash, White Ash.

CONSERVATION STATUS: 🆅🆁
Black Ash is widespread but its populations are collapsing due to the infestation of the Emerald Ash Borer (*Agrilus planipennis*). Infected trees are essentially girdled by the insect's larval feeding habits, which leads to a nearly 100% mortality rate. No known preventative measures exist for controlling the spread of this invasive species, making the pest a critical and imminent threat to the survival of this and other ash species.

Diameter: 1' (0.3 m)
Height: 30–50' (9–15 m)

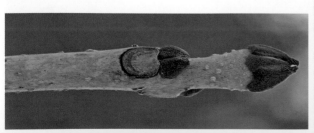

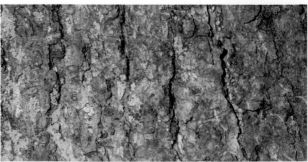

■ DENSE POPULATION
■ LIGHT POPULATION
■ NATURALIZED

Also commonly known as Red Ash, this is the most widespread native ash, extending westward into the Plains and nearly to the Rocky Mountains. It is one of the more adaptable ash species and thrives in a broad range of temperatures. It is an important food source for frogs and other native wildlife. Because it is usually found in moist soils along streams and other waterways, this species is sometimes called Swamp Ash or Water Ash.

DESCRIPTION: This tree has a dense, rounded or irregular crown of shiny green foliage. The leaves are opposite, pinnately compound, and 6–10" (15–25 cm) long, with five to nine (usually seven) leaflets 2–5" (5–13 cm) long and 1–1½" (2.5–4 cm) wide; they are paired (except at the end), lance-shaped or ovate, and coarsely saw-toothed or almost without teeth, and mostly hairless. They are shiny green above and green or paler and slightly hairy beneath, turning yellow in fall. The bark is gray and furrowed into scaly ridges, with a reddish inner layer. The twigs are green, becoming gray and hairless, and slender.

FLOWERS: ⅛" (3 mm) long and greenish, without a corolla, arranged in small clusters of many flowers each, appearing before the leaves in early spring. Male and female flowers occur on separate trees.

FRUIT: 1¼–2¼" (3–6 cm) long yellowish key with a narrow wing extending nearly to the base of the narrow body; they hang in clusters and mature in late summer and autumn.

HABITAT: Moist alluvial soils along streams in floodplain forests.

RANGE: Southeastern Alberta east to Cape Breton Island; south to northern Florida, west to Texas; to 3000' (914 m) in the southern Appalachians.

USES: This species is widely planted as an ornamental tree, particularly in urban areas. It is also used on spoil banks after strip mining, and the wood is used for a variety of purposes.

SIMILAR SPECIES: Black Ash, White Ash, Pumpkin Ash.

CONSERVATION STATUS: Ⓒᴿ
Although still common and widespread, Green Ash has been significantly impacted by the Emerald Ash Borer (*Agrilus planipennis*). All Ash trees are susceptible to this invasive species, but the pest strongly prefers the Green Ash, which suffers a mortality rate of nearly 100% after infestation. Although the southernmost population of Green Ash may be spared from the ash borers due to an unsuitable warm climate for the pest, this represents an extremely small portion of its current population.

Diameter: 1½' (0.5 m)
Height: 60' (18 m)

PUMPKIN ASH *Fraxinus profunda*

■ DENSE POPULATION
■ LIGHT POPULATION
■ NATURALIZED

Pumpkin Ash is found on wet, poorly drained sites at the margins of swamps or the flats of river floodplains where surface water tends to stand well into the growing season. Even before the invasion of the Emerald Ash Borer (*Agrilus planipennis*), this species was uncommon and occurred in small, scattered populations across the eastern United States and into southern Ontario. This species is similar to Green Ash but has large leaves and fruit. The scientific name, which means "deep," refers to the tree's swampy, often submerged habitat. It is an important food source for several moths and butterflies.

DESCRIPTION: This is a large tree with an enlarged and buttressed base and a narrow crown of spreading branches. The leaves are opposite, pinnately compound, and 8–16" (20–41 cm) long, with a hairy axis. There are usually seven or nine leaflets 3–7" (7.5–18 cm) long and 2–3" (5–7.5 cm) wide, sometimes larger; they are paired (except at the end), elliptical or narrowly ovate, without teeth or slightly saw-toothed, and slightly thickened. They are dark green and becoming nearly hairless above, and yellow-green and covered with soft hairs beneath. The bark is gray and furrowed into forking ridges forming a diamond pattern. The twigs are light gray and stout, and densely covered with hairs when young.

FLOWERS: ¼" (6 mm) long and without a corolla, arranged in small clusters of many flowers each, appearing before the leaves in early spring. Male flowers are yellowish and female flowers are greenish, occurring on separate trees.

FRUIT: 2–3" (5–7.5 cm) long yellowish key with a broad elliptical wing, extending nearly to the base of the thick body, hanging in clusters, and maturing in late summer and fall.

HABITAT: Wet soils of swamps and river valleys submerged part of the year; it is found in swamp forests.

RANGE: Southern Maryland south to northern Florida, west to Louisiana, and north to southern Illinois and southwest Ohio; to 500' (152 m).

USES: The heavy, strong wood of this species has been used for tool handles and other items.

SIMILAR SPECIES: Pop Ash, Green Ash.

CONCERVATION STATUS: CR

The population is declining, and the species is believed to have been extirpated from Florida and the Carolinas due to habitat destruction. Like other ashes, Pumpkin Ash's most significant threat is from the Emerald Ash Borer (*Agrilus planipennis*). The invasive species leads to almost 100% mortality of Pumpkin Ash trees within five years of infestation. Although the insect has not yet spread to all parts of this tree's range, the pest is a serious concern; Pumpkin Ash populations are projected to decline by at least 80% in the next century.

Diameter: 2' (0.6 m)
Height: 80' (24 m)

■ DENSE POPULATION
■ LIGHT POPULATION
■ NATURALIZED

Texas Ash is adapted to drier habitats than many of its relatives, often found growing on limestone canyon bluffs, rocky slopes in open woods, and along lakes. This tree's native range is confined to Texas except for a few sites in the Arbuckle Mountains in southern Oklahoma. Texas Ash is planted as a street tree in Texas and Oklahoma. This southwestern relative of White Ash has fewer and smaller leaflets and smaller fruit, and is adapted to a warmer, less humid climate. Some sources treat Texas Ash as a variety of White Ash.

DESCRIPTION: This tree has a short trunk and a spreading crown of stout, often crooked branches. Its leaves are opposite, pinnately compound, and 5–8" (13–20 cm) long, with three to seven leaflets 1¼–3" (3–7.5 cm) long and ¾–2" (2–5 cm) wide; they are paired (except at the end), elliptical, rounded or blunt

at the tip, coarsely wavy-toothed (except near the base), slightly thickened, and short-stalked. They are dark green above, and whitish beneath, with a network of veins; they are sometimes slightly hairy. The bark is gray or reddish-brown and deeply furrowed into broad scaly ridges. The twigs are greenish, slender, and hairless or nearly so.

FLOWERS: Early spring, before leaves. ⅛" (3 mm) long, purplish, and without a corolla, occurring in small clusters of many flowers each. Male and female flowers are on separate trees.

FRUIT: ⅝–1¼" (1.5–3 cm) long key with a narrow wing not extending down the cylindrical body; they hang in clusters and mature in late spring.

HABITAT: Dry rocky slopes and bluffs of canyons, especially limestone; in open forests.

RANGE: Southern Oklahoma and Texas; at 400–1600' (122–488 m).

USES: Texas Ash is planted as a shade, street, and landscaping ornamental tree.

SIMILAR SPECIES: White Ash, Biltmore Ash.

CONSERVATION STATUS: NT
The Texas Ash population is stable and it remains common where it is found. The future impact of the Emerald Ash Borer (*Agrilus planipennis*) on this species remains unknown; this tree's southern distribution may spare it from the brunt of the invading pests' destruction. It is likely that this tree, like other ashes, would experience an extremely high mortality rate if the insect were to spread to its range, however.

Diameter: 1½" (0.5 m)
Height: 40' (12 m)

VELVET ASH *Fraxinus velutina*

■ DENSE POPULATION
■ LIGHT POPULATION
■ NATURALIZED

Also commonly known as Arizona Ash, this variable species is the common ash in the southwestern United States and western Mexico. It is hardy in alkaline soils and fast-growing. In the desert, the presence of ash trees indicates a permanent underground water supply. It occurs in Grand Canyon National Park and Big Bend National Park. The leaflets of this small tree are of variable shapes and are often covered with velvety hairs beneath, as the species' name suggests, but they also may be hairless. A rapidly growing, cultivated variety called Modesto Ash is widely planted as a street tree in dry areas (including in alkaline soils) in the Southwest.

DESCRIPTION: Velvet Ash is a fairly small tree with an open, rounded crown of spreading branches and leaflets quite variable in shape and hairiness. Its leaves are opposite, pinnately compound, and 3–6" (7.5–15 cm) long, commonly with five leaflets (to 9) 1–3" (2.5–7.5 cm) long and ⅜–1¼" (1–3 cm) wide; they are paired except at the end, lance-shaped to elliptical, pointed at the ends, slightly wavy-toothed or sometimes without teeth, often slightly thickened and leathery, and short-stalked. They are shiny green above, and paler and densely covered with soft hairs or sometimes hairless beneath, turning yellow in fall. The bark is gray and deeply furrowed into broad, scaly ridges. The twigs are gray or brown, and often hairy when young.

FLOWERS: Early spring, before leaves. ⅛" (3 mm) long and without a corolla; the male is yellowish and the female is greenish, occurring on separate trees with many clustered together on twigs.

FRUIT: A light brown, narrow key ¾–1¼" (2–3 cm) long with a long wing not extending to the base, with many hanging in dense clusters and maturing in summer and early fall.

HABITAT: Moist soils of stream banks, washes, and canyons, mainly in mountains, desert, desert grassland; in oak woodland and Ponderosa Pine forest.

RANGE: Trans-Pecos Texas west to extreme southwestern Utah, southern Nevada, and southern California and south to northern Mexico; it occurs at 2500–7000' (762–2134 m).

USES: This tree is often planted as a shade or street tree in the Southwest.

SIMILAR SPECIES: Oregon Ash, Green Ash.

CONSERVATION STATUS:

Velvet Ash is currently widespread and faces no major current threats. Invasive *Tamarix* spp. in parts of the Rio Grande have outcompeted these trees and resulted in some habitat loss. As with other ashes, the largest possible future threat is from the invading Emerald Ash Borer (*Agrilus planipennis*), which has not yet reached this species' range.

Diameter: 1' (0.3 m)
Height: 40' (12 m)

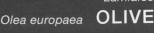

■ DENSE POPULATION
■ LIGHT POPULATION
■ NATURALIZED

Also called True Olive or European Olive, this species is the plant native to the Mediterranean Basin from which commercial olives and olive oil are derived. It is widely cultivated as an agricultural crop, including in the interior valleys of California and in several other warm-weather states. The bitter fruit becomes edible after curing or fermenting and thorough washing. Green and ripe fruit are eaten as pickles in brine; black olives are ripe fruits picked at full maturity. Olive oil, obtained by crushing and pressing the mature fruit, is used for salads, in cooking, and in soaps. Olive is a long-living and handsome ornamental and street tree with attractive gray-green foliage. The wild form has thorny twigs and small fruit with thin pulp.

DESCRIPTION: This is a small evergreen tree with a short, stout trunk, a rounded to irregular, dense crown, and the familiar olive as fruit. The leaves are evergreen, opposite, 1½–3" (4–7.5 cm) long, ⅜–¾" (10–19 mm) wide, lance-shaped or narrowly elliptical, and thick, with the edges slightly turned under. They are gray-green with veins not visible above, and silvery scales beneath. The bark is brownish-gray and furrowed. The twigs are light gray, slender, and covered with whitish, scaly hairs when young.

FLOWERS: Spring. ³⁄₁₆" (5 mm) long with a four-lobed, whitish corolla; they are fragrant and arranged in short, branched clusters at the leaf bases.

FRUIT: 1" (2.5 cm) long and elliptical, and olive green when immature, turning shiny black; it has bitter pulp and one seed, maturing in fall.

HABITAT: Subtropical or Mediterranean climates, especially hot, dry regions under irrigation.

RANGE: Native probably in the eastern Mediterranean region but cultivated from prehistoric times and widespread. Planted in California, Arizona, and Florida.

USES: This tree is the source for table olives and olive oil, commonly used in cooking. It is also planted as an ornamental tree. Olive wood is hard and durable, and useful for many small items, but is not often used because of the value of the live trees' fruits.

SIMILAR SPECIES: This tree is not likely to be confused with other species.

CONSERVATION STATUS:
The natural Olive population has not been fully assessed in the wild, but the population in Europe may be declining. Issues with genetic identity due to hybridization of the wild population with the cultivated olive poses a threat to the species, as does the risk of fires.

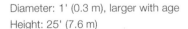

Diameter: 1' (0.3 m), larger with age
Height: 25' (7.6 m)

■ DENSE POPULATION
■ LIGHT POPULATION
■ NATURALIZED

Also known as American Olive, Wild Olive is native to the southeastern United States. It is also commonly called Devilwood, so named because the fine-textured wood is difficult to split and work. The fruit is a black drupe that resembles the cultivated Olive in the same family. The genus name *Osmanthus* is from the Greek words for "odor" and "flower," and refers to the strongly fragrant blossoms. This species is sometimes assigned to the genus *Cartrema*.

ALTERNATE NAME: *Cartrema americana*

DESCRIPTION: This is an evergreen shrub or small tree with a narrow, oblong crown of paired, glossy, leathery leaves, and with dark blue fruit like small olives. The leaves are opposite, 3½–5" (9–13 cm) long, ¾–1½" (2–4 cm) wide, lance-shaped to narrowly elliptical, and thick and leathery, with edges straight or turned under, and obscure side veins. They are shiny green above and dull and paler beneath. The stout leafstalks are about ⅝" (15 mm) long. The bark is gray, thin, and scaly, exposing dark red inner bark. The twigs are red-brown, slender, and slightly angled, becoming hairless.

FLOWERS: Early spring. ³⁄₁₆" (5 mm) wide with a bell-shaped, four-lobed, whitish or yellowish corolla; they are strongly fragrant and almost stalkless, arranged in branching clusters of many flowers each, at the leaf bases, usually with male and female flowers (sometimes bisexual) on separate plants.

FRUIT: ⅜–¾" (10–19 mm) long, elliptical, and dark blue, with thin pulp and a large stone, maturing in fall and remaining attached in winter.

HABITAT: Moist soils from river valleys to sandy uplands and dunes; in understory of forests.

RANGE: Southeastern Virginia to central Florida and west to southeastern Louisiana; it is also found in Mexico. It occurs to 500' (152 m).

USES: Wild Olive is often planted as an ornamental, valued for its fragrant flowers.

SIMILAR SPECIES: Fringe Tree, Common Sweetleaf.

CONSERVATION STATUS: 🅛🅒
The natural Wild Olive population is believed to be stable. It has no known significant threats.

Diameter: 1' (0.3 m)
Height: 30' (9 m), sometimes larger

Japanese Tree Lilac is one of the hardiest and most versatile of the lilacs. It is native to far-eastern Asia, especially northern China, Japan, and Korea. It is planted as an ornamental in North America, where it does best in climates with cooler summers. The widely planted Common Lilac (*S. vulgaris*) is a shrub known for its purple flowers.

DESCRIPTION: This is a small tree or large shrub with deciduous, simple leaves. It may have a single trunk, or be multi-stemmed, and it has a pyramidal crown, becoming open and broadly round-topped, with erect branches. The bark is red-brown and cherry-like; the showy white flowers and small fruits occur in large, loose clusters. The leaves are 2–5" (5–13 cm) long and ovate, with a pointed tip and untoothed margins. They are dark green.

FLOWERS: Summer. White flowers are arranged in 6–12" (15–30 cm) long, loose, branching clusters. They are slightly fragrant.

FRUIT: ¾" (2 cm) long, warty capsule maturing in late summer and persisting through winter.

HABITAT: Cultivated in North America.

RANGE: Native to Asia, it is introduced and naturalized in Ontario, Wyoming, New York, New Hampshire, and Massachusetts.

USES: This species is commonly planted as an ornamental.

SIMILAR SPECIES: Common Lilac (*S. vulgaris*) is a commonly planted shrub.

CONSERVATION STATUS: LC
The Japanese Tree Lilac is common and thought to have a large, stable population in its native range, with no known current or future threats.

Height: 20–30' (6–9 m)

■ DENSE POPULATION
■ LIGHT POPULATION
■ NATURALIZED

Black Mangrove is native to coastal Florida and the Gulf Coast, in addition to being widely distributed in the tropics around the world. It typically grows in the intertidal regions of sheltered tropical and subtropical coasts and can be dominant in the zones near open water. Black Mangrove is the hardiest of the four species forming the mangrove swamp forests of southern Florida, and it ranges the farthest inland and north along the Gulf Coast, where it becomes smaller and shrubby. At the limits of its range it is killed back in cold winters, but new seedlings persist for a few years, sprouting from seeds transported by currents. This species is an important honey plant that yields clear, whitish honey of high quality.

DESCRIPTION: Black Mangrove is an evergreen shrub or, in tropical regions, a tree with a rounded crown of spreading branches. The leaves are evergreen, opposite, 2–4" (5–10 cm) long, ¾–1½" (2–4 cm) wide, narrowly elliptical or lance-shaped, without teeth, and slightly thickened. They are yellow-green and often shiny above, and gray-green and covered with fine hairs beneath; both surfaces have scattered salt crystals. The bark is dark gray or brown and smooth, becoming thick, fissured, and scaly. The twigs are gray or brown, and covered with fine hairs when young, with rings at the enlarged nodes.

FLOWERS: Summer, or nearly year-round in the tropics. The flowers are ⅜" (10 mm) wide with four white corolla lobes, rounded or notched, stalkless, and crowded in upright clusters branched on four-angled stalks at the end of twigs.

FRUIT: 1–1¼" (2.5–3 cm) long, flat elliptical capsules with tiny points; they are yellow-green and covered with fine hairs, maturing throughout the year. They have one large, dark green seed often germinating on the tree, and splitting into two parts.

HABITAT: Silt seashores, in salt and brackish water; it is found in mangrove swamp forests.

RANGE: Coasts and islands from northern to southern Florida, west to southern Louisiana and south Texas; it is also found in Bermuda, the West Indies, and from Mexico to Brazil and Peru. It occurs at sea level.

USES: The wood of Black Mangrove is used for posts, fuel, and other purposes. Tannin from the bark has been used to tan leather.

SIMILAR SPECIES: The flat, pod-like fruit distinguishes this plant from other mangrove species. The other native mangrove species are tropical trees confined mostly to southern and coastal Florida.

CONSERVATION STATUS: LC
The Black Mangrove population is decreasing. The species is threatened throughout its coastal range by habitat conversion, pollution, hurricanes, and sea level rise associated with climate change.

Diameter: 1' (0.3 m)
Height: 40' (12 m)

■ DENSE POPULATION
■ LIGHT POPULATION
■ NATURALIZED

Southern Catalpa is believed to be native to the southeastern United States but has been widely cultivated since long before European settlers arrived in North America. The species is able to grow in many parts of North America, and has become widely naturalized outside of its natural range. It is planted as a shade tree and an ornamental for the abundant showy flowers, cigar-like pods, and coarse foliage. Catalpa is the name the Cherokee Native Americans gave to this valued plant.

DESCRIPTION: This short-trunked tree sports a broad, rounded crown of spreading branches; large, heart-shaped leaves; large clusters of showy white flowers; and long, beanlike fruit. The leaves are opposite, 5–10" (13–25 cm) long, 4–7" (10–18 cm) wide, ovate, abruptly long-pointed at the tip, notched at the base, and without teeth. They are dull green above, and paler and covered with soft hairs beneath, turning blackish in fall. They have an unpleasant odor when crushed. The slender leafstalk is 3½–6" (9–15 cm) long. The bark is brownish-gray and scaly, and the twigs are green, turning brown, stout, and hairless or nearly so.

FLOWERS: Late spring. 1½" (4 cm) long and wide with bell-shaped corolla of five unequal rounded fringed lobes; they are white with two orange stripes and many purple spots and stripes inside, slightly fragrant, and arranged in upright branched clusters to 10" (25 cm) long and wide.

FRUIT: 6–12" (15–30 cm) long, 5/16–⅜" (8–10 mm) in diameter, narrow, cylindrical, dark brown capsule; it is cigarlike, thin-walled, and splitting into two parts, with many flat light brown seeds with two papery wings; it matures in fall and remains attached in winter.

HABITAT: Moist soils in open areas such as roadsides and clearings.

RANGE: The original range is uncertain; it is probably native in southwestern Georgia, northwestern Florida, Alabama, and Mississippi. It is widely naturalized from southern New England south to Florida, west to Texas, and north to Michigan; it occurs at 100–500' (30–152 m).

USES: This species is often planted as an ornamental tree.

SIMILAR SPECIES: Northern Catalpa, Princesstree.

CONSERVATION STATUS:

Southern Catalpa is so widespread in cultivation across North America that its wild range is difficult to identify. It faces no significant threats but is susceptible to a variety of pests and fungal diseases. Catalpa Sphinx Moth (*Ceratomia catalpae*) larvae are a common minor pest to Southern Catalpa and can lead to defoliation, but generally do not threaten the plant.

Diameter: 2' (0.6 m)
Height: 50' (15 m)

■ DENSE POPULATION
■ LIGHT POPULATION
■ NATURALIZED

Northern Catalpa is native to the Midwest—probably only to a small zone along the Ohio and Mississippi Rivers—and is widely naturalized across the eastern two-thirds of the United States and in southern Ontario. This fast-growing tree has long been cultivated for its wood and has become a popular ornamental. It is hardier than Southern Catalpa, which blooms later and has slightly smaller flowers and narrower, thinner-walled capsules. Both are commonly called Cigar Tree because of the distinctive fruits. Northern Catalpa is also sometimes called Hardy Catalpa or Western Catalpa.

DESCRIPTION: This tree has a rounded crown of spreading branches; large, heart-shaped leaves; large, showy flowers; and long, beanlike fruit. The leaves are opposite, 6–12" (15–30 cm) long, 4–8" (10–20 cm) wide, ovate, long-pointed, and without teeth. They are dull green above, and paler and covered with soft hairs beneath, turning blackish in autumn. The slender leafstalk is 4–6" (10–15 cm) long. The bark is brownish-gray and smooth, becoming furrowed into scaly plates or ridges. The twigs are green, turning brown, and stout, and becoming hairless.

FLOWERS: Late spring. The flowers are 2–2¼" (5–6 cm) long and wide with a bell-shaped corolla of five unequal, rounded, fringed lobes; they are white with two orange stripes and purple spots and lines inside, arranged in branched upright clusters, 5–8" (13–20 cm) long and wide.

FRUIT: A narrow, cylindrical, dark brown capsule 8–18" (20–46 cm) long, ½–⅝" (12–15 mm) in diameter, cigarlike, thick-walled, and splitting into two parts, with many flat, light brown seeds with two papery wings; it matures in fall and remains attached in winter.

HABITAT: Moist valley soils by streams; it is naturalized in open areas such as roadsides and clearings.

RANGE: The original range is uncertain; it appears to be native from southwestern Indiana to northeastern Arkansas. It is widely naturalized throughout eastern North America; it occurs at 200–500' (61–152 m).

USES: Northern Catalpa is commonly planted as an ornamental tree. It was widely cultivated for its soft, light wood, which is used for furniture, cabinets, and other purposes.

SIMILAR SPECIES: Southern Catalpa.

CONSERVATION STATUS: LC

The Northern Catalpa has a widespread population and is considered stable. There are no known current or potential threats that pose a significant risk to the species.

Diameter: 2½' (0.8 m)
Height: 50–80' (15–24 m)

■ DENSE POPULATION
■ LIGHT POPULATION
■ NATURALIZED

Desert Willow is a native of the southwestern United States and is adapted to arid climates. It is highly tolerant of drought and heat, prefers full sun, and grows best on well-drained soil in areas with less than 30" of annual precipitation. It is most common in dry washes and along riverbanks, and is a popular ornamental thanks to its rapid growth and low maintenance requirements. Hummingbirds and native bees visit the flowers. It is so named because its leaves resemble those of willows, but it is not related to true willows in the genus *Salix*.

DESCRIPTION: This is a large shrub or small tree, often with a leaning trunk; an open, spreading crown; narrow, willow-like leaves; large, showy flowers; and very long, narrow, beanlike fruit. Its leaves are opposite and alternate, 3–6" (7.5–15 cm) long, ¼–⅜" (6–10 mm) wide, linear, straight or slightly curved, long-pointed at the ends, not toothed, drooping, and short-stalked; they are light green, and sometimes hairy or sticky. The bark is dark brown and furrowed into scaly ridges. The twigs are brown, slender, and sometimes hairy or sticky.

FLOWERS: Late spring–early summer. 1¼" (3 cm) long and wide bell-shaped corolla with five unequal lobes, whitish tinged with pale purple or pink and with yellow in the throat; they are fragrant and arranged in usually unbranched clusters to 4" (10 cm) long at the ends of twigs.

FRUIT: A dark brown, cigarlike capsule 4–8" (10–20 cm) long and ¼" (6 mm) in diameter, maturing in fall, splitting into two parts, and remaining attached in winter, with many flat, light brown seeds with two papery, hairy wings.

HABITAT: Moist soils of stream banks and drainages in plains and foothills, desert and desert grassland zones, often forming thickets.

RANGE: Texas and New Mexico west to extreme southwestern Utah and southern California; it is also found in northern Mexico. It occurs at 1000–5000' (305–1524 m).

USES: This species is planted as an ornamental and is also used in erosion control. Native Americans made bows from the stiff, durable wood, which is also suitable for fence posts.

SIMILAR SPECIES: Northern Catalpa, Southern Catalpa.

CONSERVATION STATUS: **LC**
Desert Willow is considered stable with a large, widespread population. The species faces no known threats. It can become weedy or invasive when planted outside of its natural range.

Diameter: 6" (15 cm)
Height: 25' (7.6 m)

■ DENSE POPULATION
■ LIGHT POPULATION
■ NATURALIZED

This drought-tolerant flowering shrub or small tree is native to the subtropical and tropical Americas, where it prefers dry, rocky areas. It is also called Yellow Bells because of its large, showy yellow flowers when in full bloom, and is a popular ornamental planting. The flowers attract many pollinators such as bees, butterflies, and hummingbirds.

DESCRIPTION: This is an irregularly branched shrub or small tree. Its leaves are 5" (13 cm) long and pinnately compound, usually with seven to nine leaflets, lanceolate to linear, and serrate margins.

FLOWERS: May–October. 1½" (3.9 cm) long, yellow, and tubular with irregular symmetry.

FRUIT: 6" (15 cm) long, narrow capsules with flat, winged seeds.

HABITAT: Arid, exposed, well-drained rocky or gravelly slopes.

RANGE: Arizona, New Mexico, Texas, and Florida, and throughout the American tropics.

USES: This species is a popular ornamental plant. It is also planted to attract butterflies and hummingbirds.

SIMILAR SPECIES: Golden Trumpet.

CONSERVATION STATUS:
Yellow Trumpetbush faces no known threats that pose a major concern. The species is considered stable due to its large and widespread global population. It is an aggressive plant that can invade rocky, sandy, or disturbed sites.

Height: 24' (7 m)

■ DENSE POPULATION
■ LIGHT POPULATION
■ NATURALIZED

This species is native to the Mediterranean region and is widely cultivated for its pleasant-smelling foliage and attractive, fragrant flowers. Lilac Chastetree is hardy as far north as Washington and New York, but it is more commonly grown in the southern United States, where it can top 20' (6 m) and is spectacular in full bloom. Sometimes simply called Chaste Tree, other common names for this plant are Vitex and Monk's Pepper.

DESCRIPTION: This is a small tree or shrub with deciduous, palmately compound leaves. It is multi-stemmed and dense, with a spreading crown as wide as it is high. The leaflets are lance-shaped and untoothed. The leaves and twigs have a spicy scent when crushed; the twigs are hairy.

FLOWERS: The fragrant, blue-lilac flowers are arranged in 5–7" (13–18 cm) spikes at the tips of the twigs.

FRUIT: ⅛" (3 mm) wide, round, green drupe, occurring in clusters; it matures in late summer and persists through winter.

HABITAT: Introduced in North America; tolerant of diverse conditions and widely naturalized. Native to Europe, Asia, and Africa.

RANGE: Native to the Mediterranean region; it is introduced in North America and grown mainly in the southern United States, but is also widely naturalized in the southern tier of the U.S. and up both coasts to Pennsylvania and Oregon.

USES: This species is a popular ornamental plant, and is often grown to attract butterflies.

SIMILAR SPECIES: Chinese Chastetree, Simpleleaf Chastetree.

CONSERVATION STATUS:
Collection in the wild is a suspected threat; however, the Lilac Chastetree has not been fully assessed. This non-native species can be weedy in North America but is not considered invasive.

Height: 10–20' (3–6 m)

■ DENSE POPULATION
■ LIGHT POPULATION
■ NATURALIZED

This flowering shrub is native to the southwestern deserts of the United States and northwestern Mexico. It is found in areas associated with water, commonly in dry washes. The flowers are fragrant and readily attract pollinators including bees, butterflies, and hummingbirds.

ALTERNATE NAME: *Condea emoryi*

DESCRIPTION: This species may be an erect or spreading shrub with an overall pale gray appearance and lavender-scented leaves. It can grow quite tall and tree-like in optimal conditions. The leaves are 2½" (6.5 cm) long, ovate, toothed, and whitish, with woolly hairs; they are semi-evergreen.

FLOWERS: January–May. Tiny, blue-violet, and two-lipped flowers are arranged in clusters at the leaf axils.

HABITAT: Arid, low-elevation rocky slopes, washes, canyons; it is often intermixed with other species.

RANGE: California deserts and eastward into Arizona and southernmost Nevada.

USES: This shrub is planted as an ornamental and to attract butterflies and other wildlife.

SIMILAR SPECIES: Brittlebush.

CONSERVATION STATUS: LC

The Desert Lavender population is apparently secure. There are no known threats to the species.

Height: 10' (3 m)

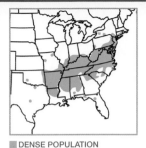

■ DENSE POPULATION
■ LIGHT POPULATION
■ NATURALIZED

This rapid-growing tree native to China resembles a catalpa, with vigorous shoots and enormous leaves 2' (0.6 m) or more in length and width. Introduced to North America as a garden plant, it has naturalized in the eastern and southern United States and is considered an exotic pest plant. It grows quickly in most soil types, aggressively colonizes disturbed sites, and is difficult to contain. The species is named for Queen Anna Paulowna of Russia (1795–1865), a queen consort of the Netherlands.

ALTERNATE NAMES: Empress Tree, Foxglove-Tree, Royal Paulownia

DESCRIPTION: This is a naturalized, invasive tree with a short trunk; a broad, open crown of stout, spreading branches; large leaves; and showy purple flowers. Its leaves are opposite, 6–16" (15–41 cm) long, 4–8" (10–20 cm) wide, broadly ovate, and long-pointed at the tip, with several veins from the notched base; they are sometimes slightly three-toothed or three-lobed. They are dull light green and slightly hairy above, and paler and densely covered with hairs beneath. The leafstalks are 4–8" (10–20 cm) long. The bark is gray-brown, with a network of irregular shallow fissures. The twigs are light brown and stout, and densely covered with soft hairs when young.

FLOWERS: Early spring. 2" (5 cm) long with a bell-shaped pale violet corolla ending in five rounded unequal lobes; they are fragrant, and arranged in upright clusters, 6–12" (15–30 cm) long on stout, hairy branches, from rounded brown hairy buds formed the previous summer.

FRUIT: 1–1½" (2.5–4 cm) long egg-shaped capsule; it is pointed, brown, and thick-walled, splitting into two parts, with many tiny winged seeds; it matures in fall and remains attached.

HABITAT: Wasteplaces, roadsides, streamsides, slopes, valleys, woodland edges, and open areas.

RANGE: Native of China. Cultivated and naturalized from southern New England, Montreal, and New York south to northern Florida, west to southern Texas, and north to Missouri.

USES: This tree has been widely planted as an ornamental, but is now considered an invasive pest and should be avoided in landscaping.

SIMILAR SPECIES: Southern Catalpa.

CONSERVATION STATUS:
The conservation status of this species in the wild has not been fully assessed. It is a persistent exotic invasive plant in North America.

Diameter: 2' (0.6 m)
Height: 50' (15 m)

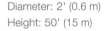

■ DENSE POPULATION
■ LIGHT POPULATION
■ NATURALIZED

Carolina Holly, or Sand Holly, is native to the southeastern and south-central United States, where it grows in many sandy habitats and woodlands. This holly is recognized by the usually elliptical, translucent fruit and by the petals, which sometimes number five instead of four. The Latin species name, meaning "ambiguous" or "doubtful," suggests uncertainty in the classification when the species was first described and named.

DESCRIPTION: This is a deciduous shrub, or sometimes a small tree, with an irregular or rounded crown. The leaves are deciduous, 1–2" (2.5–5 cm) long, ⅜–1" (1–2.3 cm) wide, elliptical, finely wavy-toothed, and hairless or slightly hairy. They are dull green above and paler beneath. The bark is dark brown or blackish, and smooth or scaly. The twigs are dark reddish-brown, slender, and hairless to densely hairy.

FLOWERS: Spring. ¼" (6 mm) wide with four (sometimes five) rounded white petals, occurring on short twigs. Male and female are on separate plants.

FRUIT: ¼" (6 mm) in diameter, berrylike, elliptical or sometimes round, red, translucent, and short-stalked, with four brown narrow grooved nutlets, maturing in late summer and soon shedding.

CAUTION: Poisonous: Do not eat or ingest any part of this species. If accidentally ingested, contact your doctor or poison control center.

HABITAT: Moist well-drained soils of upland forests.

RANGE: North Carolina to central Florida, west to eastern Texas and southeastern Oklahoma; to 1000' (3–5 m).

USES: Carolina Holly is sometimes planted as an ornamental.

SIMILAR SPECIES: Possumhaw Holly.

CONSERVATION STATUS: LC
Carolina Holly has a stable, widely distributed population. The species faces no known significant threats.

Diameter: 4" (10 cm)
Height: 18' (5.5 m)

■ DENSE POPULATION
■ LIGHT POPULATION
■ NATURALIZED

English Holly is an evergreen tree or shrub native to Europe, northwest Africa and southwest Asia. It can form dense thickets and dominate moist, shady environments such as those found in forests or gorges. Numerous horticultural varieties differ in leaf size, shape, spines, color, and in tree habit. Several varieties grown in orchards for Christmas decorations have larger berries and larger leaves than native hollies. To ensure fruit, a male plant is needed to pollinate the female. Cultivated since ancient times, it is propagated by cuttings and seeds.

DESCRIPTION: This cultivated evergreen tree has a dense, conical crown of short, spreading branches and shiny red berries. Its leaves are evergreen, 1¼–2¾" (3–6 cm) long, ¾–1½" (2–4 cm) wide, elliptical, spiny-pointed, and wavy-edged with large spiny teeth; they are blunt at the base, and stiff and leathery; they are shiny dark green above and paler beneath. The bark is gray and smooth or nearly so. The twigs are greenish or purplish, angled, and hairless or with short hairs.

FLOWERS: Late spring. ¼" (6 mm) wide with four rounded white petals, with several on short stalks at the base of the previous year's leaves; male and female occur on separate plants.

FRUIT: ¼–⅜" (6–10 mm) in diameter, berrylike, shiny red, clustered at leaf bases, and short-stalked, with four nutlets; it matures in fall and remains attached in winter.

CAUTION: Poisonous—do not eat or ingest any part of this species. If accidentally ingested, contact your doctor or poison control center.

HABITAT: Moist soils in humid temperate regions.

RANGE: A native of Southern Europe, Northern Africa, and Western Asia. It is widely planted across the United States, mainly in Atlantic, southeastern, and Pacific states. It is naturalized from British Columbia to California.

USES: Decorative varieties are grown for ornament and are widely included in parks and gardens. The wood is used for veneers and inlays. Holly berries contain alkaloids, caffeine, and a yellow pigment called ilexanthin. The berries, now thought to be toxic, were used in traditional medicine as a laxative and diuretic.

SIMILAR SPECIES: American Holly, Carolina Holly, Dahoon Holly.

CONSERVATION STATUS: LC
Common and widespread in its native range. English Holly is considered an invasive species in western North America and on the Hawaiian Islands.

Diameter: 1½' (0.5 m)
Height: 50' (15 m)

■ DENSE POPULATION
■ LIGHT POPULATION
■ NATURALIZED

Native to the far southeastern United States, this species is often planted as an ornamental for the glossy evergreen foliage and attractive, bright red fruit. Its natural range is mainly along the coast but it has been spread farther inland by planting. The name Dahoon comes from the Timucua language spoken by the region's Native Americans.

DESCRIPTION: This is an evergreen shrub or small tree with a rounded, dense crown and abundant, bright red berries. The leaves are evergreen, 1½–3½" (4–9 cm) long, ¼–1¼" (0.6–3.2 cm) wide, oblong or obovate, slightly thick and leathery, and usually without teeth or spines, with the edges often turned under. They are shiny dark green and becoming hairless above, and light green (and densely hairy when young) beneath. The bark is dark gray, thin, and smooth to rough and warty. The twigs are slender and densely covered with silky hairs, becoming brown.

FLOWERS: ³⁄₁₆" (5 mm) wide with four rounded white petals, occurring on short stalks mostly at the base of new leaves in spring, with male and female on separate plants.

FRUIT: ¼" (6 mm) in diameter, berrylike, round, shiny red (sometimes yellow or orange), and short-stalked, with mealy bitter pulp and four narrow grooved brown nutlets, maturing in fall, and remaining attached in winter.

CAUTION: Poisonous: Do not eat or ingest any part of this species. If accidentally ingested, contact your doctor or poison control center.

HABITAT: Wet soils along streams and swamps, sometimes sandy banks or brackish soils.

RANGE: North Carolina south through Florida, and west to southern Louisiana; to 200' (61 m); also Bahamas, Cuba, Puerto Rico and one variety in Mexico.

USES: Dahoon Holly is a popular ornamental plant, valued for its bright red berries and glossy green foliage.

SIMILAR SPECIES: Myrtle Dahoon, American Holly.

CONSERVATION STATUS: (LC)
Dahoon Holly is common and widespread. The species faces no current or anticipated threats.

Diameter: 1' (0.3 m)
Height: 30' (9 m)

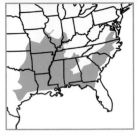

- DENSE POPULATION
- LIGHT POPULATION
- NATURALIZED

Also called Deciduous Holly or Swamp Holly, Possum Haw is a common plant native to the southern and eastern United States. It is conspicuous in winter, with its many, small, red berries along leafless, slender, gray twigs. Its leaf shape is distinctive among hollies—spoon-shaped or narrowly obovate and simple. Opossums, raccoons, other mammals, and a wide variety of birds readily eat the fruit of this and other hollies.

DESCRIPTION: This is a deciduous shrub or small tree with a spreading crown and bright red berries. The leaves are deciduous, mostly clustered on short spur twigs and alternate on vigorous twigs; they are 1–3" (2.5–7.5 cm) long, ⅜–1¼" (1–3 cm) wide, spoon-shaped or narrowly obovate, and finely wavy-toothed. They are dull green above, and paler and hairy on the veins beneath. The bark is light brown to gray, thin, and smooth or warty. The twigs are light gray, slender, and hairless.

FLOWERS: Spring. ¼" (6 mm) wide with four rounded white petals, occurring on slender stalks at the end of spur twigs, with male and female on separate plants.

FRUIT: ¼" (6 mm) in diameter, berrylike, and red, with bitter pulp and four narrow grooved nutlets; it is short-stalked and arranged in clusters, maturing in fall, and remaining attached in winter.

CAUTION: Poisonous: Do not eat or ingest any part of this species. If accidentally ingested, contact your doctor or poison control center.

HABITAT: Moist soils along streams and in swamps.

RANGE: Maryland south to central Florida, west to Texas, and north to southeastern Kansas; to about 1200' (366 m).

USES: Possum Haw is often planted as an ornamental and to attract birds and other wildlife.

SIMILAR SPECIES: Carolina Holly, Winterberry, Walter Viburnum.

CONSERVATION STATUS: LC
Possum Haw is considered stable because of its large population and wide range. There are no known current or future threats to the species.

Diameter: 6" (15 cm)
Height: 20' (6 m)

■ DENSE POPULATION
■ LIGHT POPULATION
■ NATURALIZED

Although a shrub rather than a tree, Gallberry is an abundant, large plant in its wide range, and is both ecologically and economically important. It is found on the coastal plain of the United States along both the Atlantic and Gulf coasts, becoming sporadic and less common into New England and to Nova Scotia. It is commonly called Inkberry, and the plant's black berries persist well into winter, making this a valuable food source for a variety of wildlife. The evergreen foliage varies in color from dark to light green both in summer and fall.

DESCRIPTION: This is a mound-shaped, colony-forming shrub, becoming somewhat open with age. The leaves are 2" (5 cm) long, lance-shaped, sparingly toothed, glossy, and leathery.

FLOWERS: February–July. Tiny, white, and inconspicuous.

FRUIT: Tiny, berrylike, and purple-black.

CAUTION: Poisonous: Do not eat or ingest any part of this species. If accidentally ingested, contact your doctor or poison control center.

HABITAT: Pinelands, thickets, bogs, wet woods of coastal plains.

RANGE: Coastal Plain from Nova Scotia to Florida, west to Louisiana and eastern Texas.

USES: Nectar from this plant is used to produce honey in parts of the southeastern United States. Some Native Americans used the leaves to make a tea-like drink.

SIMILAR SPECIES: Large Gallberry (*I. coriacea*), usually an understory shrub that shares this plant's range on the coastal plain in the southeastern United States.

CONSERVATION STATUS: LC

The overall Gallberry population is widespread and stable, but native subpopulations at the limits of its natural range in Texas and New England are seriously threatened. No significant range-wide threats are currently known.

Height: 6–12' (1.8–3.6 m) tall and wide

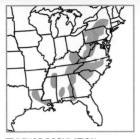

- DENSE POPULATION
- LIGHT POPULATION
- NATURALIZED

Also called Mountain Holly, Mountain Winterberry is native to the eastern United States, particularly in the Appalachian Mountain region. The showy, brightly colored berries persist into winter. It is closely related to Carolina Holly, and is sometimes classified as a variety of that species; Mountain Winterberry has larger leaves and fruit.

DESCRIPTION: This deciduous, spreading shrub or small tree has a narrow crown and relatively large orange-red berries. The leaves are 2½–5" (6–13 cm) long, ¾–2¼" (2–6 cm) wide, elliptical or ovate, abruptly pointed at the tip, and finely saw-toothed. They are dull green above, and paler and hairy on the veins beneath, turning yellow in fall. The bark is light brown to gray, thin, and smooth or warty. The twigs are brown to gray and slightly zigzag.

FLOWERS: Late spring. Nearly ¼" (6 mm) wide with four (sometimes five) rounded white petals, occurring on slender stalks; males and females are on separate plants.

Diameter: 8" (20 cm)
Height: 30' (9 m)

FRUIT: ⅜–½" (10–12 mm) in diameter, berrylike, and bright orange-red (sometimes red or yellow), with four narrow grooved nutlets; it is short-stalked and matures in fall, often remaining attached into winter.

CAUTION: Poisonous: Do not eat or ingest any part of this species. If accidentally ingested, contact your doctor or poison control center.

HABITAT: Moist soils in mixed hardwood forests.

RANGE: Western Massachusetts and New York south to northern Georgia; it is local to northwestern Florida and Louisiana. It occurs from 200' (61 m) to 6000' (1829 m) in the southern Appalachians.

USES: Mountain Winterberry is sometimes planted as an ornamental.

SIMILAR SPECIES: Carolina Holly, Winterberry, Smooth Winterberry.

CONSERVATION STATUS:
Mountain Winterberry has a broad range and is believed to be stable. The species faces no known current or future threats. It is listed as state-endangered in a few states at the northern limit of its natural range.

■ DENSE POPULATION
■ LIGHT POPULATION
■ NATURALIZED

This smallish plant is native to the southeastern United States, especially near the southern Atlantic and Gulf coasts, where it prefers moist, sandy soils. Also called Myrtle-leaved Holly, Myrtle Dahoon is named for the resemblance of its leaves to those of the unrelated true Myrtle (*Myrtus communis*), native to the Mediterranean region. Myrtle Dahoon is closely related to Dahoon Holly and has been considered a variety of that species; the latter has larger, broader leaves and grows in richer, wet soil.

DESCRIPTION: This is an evergreen shrub or small tree with a broad, dense crown of many crooked branches and small, narrow leaves. Those leaves are evergreen, ½–1¼" (1.2–3.2 cm) long, ⅛–⅜" (3–10 mm) wide, linear, bristle-tipped, short-pointed at the base, and thick and stiff, with the edges turned under and usually without teeth; they are crowded and short-stalked, becoming nearly hairless. They are dark green above and paler beneath. The bark is whitish-gray, and rough and warty. The twigs are brown, slender, and stiff, and hairy when young.

FLOWERS: Spring. ³⁄₁₆" (5 mm) wide with four rounded white petals, occurring on short stalks at the leaf bases, with males and females on separate plants.

FRUIT: ¼" (6 mm) in diameter, berrylike, and red (rarely orange or yellow), with thin bitter pulp and four narrow grooved nutlets; it is short-stalked and matures in fall, remaining attached in winter.

CAUTION: Poisonous: Do not eat or ingest any part of this species. If accidentally ingested, contact your doctor or poison control center.

HABITAT: Wet, mostly poor or acid sandy soils, bordering ponds and swamps; it is often found in pine or Bald Cypress forests.

RANGE: North Carolina to central Florida, west to southeastern Louisiana; to 200' (61 m).

USES: This species is grown as an ornamental plant.

SIMILAR SPECIES: Yaupon Holly, Dahoon Holly.

CONSERVATION STATUS: LC

Common and widespread, Myrtle Dahoon has a stable population. The species faces no significant risks.

Diameter: 6" (15 cm)
Height: 18' (5.5 m)

■ DENSE POPULATION
■ LIGHT POPULATION
■ NATURALIZED

American Holly is common in the eastern and south-central United States, where it typically grows as an understory tree in moist forest. Many improved varieties are grown for ornament, shade, and hedges. The evergreen fruiting branches from wild and planted American Holly trees are popular Christmas decorations. Many kinds of birds and mammals eat the bitter berries of this and other hollies, and the flowers attract a multitude of insect pollinators including bees, ants, and moths.

DESCRIPTION: This is an evergreen tree with a narrow, rounded, dense crown of spiny leaves; small white flowers; and bright red berries. The leaves are evergreen, spreading in two rows, 2–4" (5–10 cm) long, ¾–1½" (2–4 cm) wide, elliptical, spiny-pointed and coarsely spiny-toothed, thick, stiff and leathery, and dull green above and yellow-green beneath. The bark is light gray, thin, and smooth or rough and warty. The twigs are brown or gray and stout, and covered with fine hairs when young.

FLOWERS: Spring. ¼" (6 mm) wide with four rounded white petals, arranged in short clusters at the base of new leaves and along twigs. Male and female occur on separate trees.

FRUIT: ¼–⅜" (6–10 mm) in diameter, berrylike, and bright red (rarely orange or yellow), with bitter pulp and four brown nutlets; they are scattered and short-stalked, maturing in fall and remaining attached in winter.

CAUTION: Poisonous: Do not eat or ingest any part of this species. If accidentally ingested, contact your doctor or poison control center.

HABITAT: Moist or wet well-drained soils, especially flood plains; it is typically found in mixed hardwood forests.

RANGE: Eastern Massachusetts south to central Florida, west to south central Texas, and north to southeastern Missouri; it occurs to 4000' (1219 m), higher in the southern Appalachians.

USES: American Holly is a popular ornamental and also attracts a variety of birds and other wildlife. The whitish, fine-textured wood is especially suited for inlays in cabinetwork, handles, carvings, and rulers, and can be dyed various shades, even black.

SIMILAR SPECIES: English Holly.

CONSERVATION STATUS: 🔵 LC
American Holly has an extensive range and large population. The species has no major threats.

Diameter: 1–2' (0.3–0.6 m)
Height: 40–70' (12–21 m)

■ DENSE POPULATION
■ LIGHT POPULATION
■ NATURALIZED

Also known as Common Winterberry, Canada Holly, and Michigan Holly, this species is hardy farther north than most of its relatives. It is native to most of eastern North America, found as far north as Minnesota and Newfoundland and south to the Gulf Coast. This shrub is extremely showy in late fall and early winter when it is covered with bright fruits. Birds, particularly American Robins (*Turdus migratorius*), are readily attracted to them. A number of ornamental varieties have been selected for berry and leaf color.

DESCRIPTION: Winterberry can be a deciduous shrub or a small tree with clusters of small, white flowers in the leaf axils. The leaves are 2" (5 cm) long, elliptical, and toothed but not spiny.

FLOWERS: Clusters ¼–½" (6–13 mm) wide; the sepals, petals, and stamens number four to eight each.

FRUIT: Berrylike, showy red (rarely yellow), and less than ¼" (6 mm) wide, occurring on short stalks, either singly or in small clusters along branches.

CAUTION: Poisonous: Do not eat or ingest any part of this species. If accidentally ingested, contact your doctor or poison control center.

HABITAT: Swamps, damp to dry thickets, and pondsides.

RANGE: Ontario east to Newfoundland, south to Florida, west to Texas, and north through Arkansas to Minnesota.

USES: Native Americans used the berries and other parts of this species for medicinal purposes. Winterberry is commonly used as an ornamental plant and to attract birds and other wildlife.

SIMILAR SPECIES: Possum Haw, Mountain Winterberry. Smooth Winterberry (*I. laevigata*) is similar but is usually a shrub restricted to the easternmost United States and is less common in cultivation.

CONSERVATION STATUS: Winterberry is considered highly stable with the widest distribution of any holly native to North America. There are no known threats to the species.

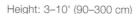

Height: 3–10' (90–300 cm)

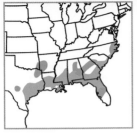

- ■ DENSE POPULATION
- ■ LIGHT POPULATION
- ■ NATURALIZED

Yaupon Holly is native to the southeastern United States. It is sometimes grown for ornament and trimmed into hedges; the ornamental twigs with shiny evergreen leaves and numerous red berries are a favorite Christmas decoration. The plant's shiny red berries are an important food source for numerous native birds and mammals.

DESCRIPTION: This evergreen, much-branched, thicket-forming shrub or small tree has a rounded, open crown; small shiny leaves; and abundant, round, shiny red berries. The evergreen leaves are usually ¾–1¼" (2–3 cm) long and ¼–½" (6–12 mm) wide; they are elliptical, blunt at the tip, rounded at the base, finely wavy-toothed, thick and stiff, and short-stalked. They are shiny green above and paler beneath. The bark is red-brown, thin, and finely scaly. The twigs are gray, branching at right angles, and slightly angled, and hairy when young, becoming rough.

FLOWERS: ³⁄₁₆" (5 mm) wide with 4 spreading rounded white petals, occurring on short stalks at the base of old leaves, with male and female on separate plants.

FRUIT: ¼" (6 mm) in diameter, berrylike, shiny red, clustered along twigs, and short-stalked, with bitter pulp and four narrow grooved nutlets; they mature in fall, often remaining attached in winter.

CAUTION: Poisonous: Do not eat or ingest any part of this species. If accidentally ingested, contact your doctor or poison control center.

HABITAT: Moist soils, especially along coasts and in valleys, sometimes in sandhills.

RANGE: Southeastern Virginia south to central Florida, west to Texas, and north to southeastern Oklahoma; to 500' (152 m).

USES: The caffeine-containing leaves of this species have been widely used by several Native American groups to brew a tea for medicinal and ceremonial purposes. Yaupon Holly is a popular ornamental plant.

SIMILAR SPECIES: Myrtle Dahoon.

CONSERVATION STATUS: LC
Yaupon Holly is widespread with a large population, and there are no known threats to the species. This plant can become weedy or invasive in some regions and habitats, and may displace other desirable vegetation if not managed.

Diameter: 6" (15 cm)
Height: 20' (6 m)

■ DENSE POPULATION
■ LIGHT POPULATION
■ NATURALIZED

Big Sagebrush is the dominant shrub over vast areas of the Great Basin region. Sagebrush is a valuable and highly nutritious forage plant for wildlife, particularly during the winter. Deer, moose, elk, antelope, and bighorn sheep browse this plant, especially in late winter and spring. Sage-grouse (*Centrocercus* spp.) are almost entirely reliant on sagebrush communities for their food and cover; sagebrush also provides nesting sites for a variety of native songbirds of the Intermountain West. Several subspecies have been identified, all more or less similar to the typical form; how to divide the species taxonomically remains the subject of considerable disagreement.

DESCRIPTION: This is a much-branched gray-green shrub with a pungent sagelike aroma; it has a short, thick trunk or a few stems rising from the base. The leaves are evergreen and alternate or in bundles, usually to ¾" (2 cm) long but range from ⅜" to 1½" (1–4 cm) long. The upper leaves may be linear or oblanceolate and untoothed; they are narrowly wedge-shaped, usually with three terminal lobes, and covered with fine gray hairs. The bark is gray and shreddy.

FLOWERS: July–November. The heads are ovoid, silver-green, and numerous, usually with three to six minute disk flowers (ray flowers absent) in erect, leafy panicles from the ends of branchlets.

FRUIT: Resinous achenes.

HABITAT: Deep, fine soils in arid basins and on mountain slopes to near the timberline; it is often found with pinyons and junipers.

RANGE: British Columbia east to North Dakota, south southern California, Arizona, and New Mexico.

USES: The primary use of this plant is firewood—the volatile oils responsible for its pungent aroma are so flammable that they can cause even green plants to burn. It has also been used for medicinal purposes by Native Americans.

SIMILAR SPECIES: California Sagebrush.

CONSERVATION STATUS:
Big Sagebrush is widespread with a large population that is stable or possibly increasing. The species is highly susceptible to fire, relying on seeds carried on the wind from unaffected areas to reseed after a burn. The species faces the most significant potential threat from future population fragmentation and habitat loss.

Height: 1½–15' (0.5–4.5 m)

■ DENSE POPULATION
■ LIGHT POPULATION
■ NATURALIZED

Florida Groundsel Bush is the only native eastern species of the aster family that reaches tree size. This species is native to the Atlantic Coastal Plain and has extended its range inland somewhat. Also called Eastern Baccharis, it is usually found in wetlands and is tolerant of saltwater spray, making it one of the few eastern shrubs suitable for planting as an ornamental near the ocean. Small songbirds often use the dense branches for cover and nest sites.

DESCRIPTION: Florida Groundsel Bush can be a shrub or a small, much-branched, spreading and rounded tree with fruit resembling silvery paint brushes. Its leaves are deciduous (evergreen in the far Southeast), 1–2½" (2.5–6 cm) long, ⅜–1¼" (1–3 cm) wide, and elliptical to obovate; usually with a few coarse teeth toward the short-pointed tip, tapering to the short-stalked base; they are slightly thick, and hairless or resinous. They are dull gray-green above and beneath, falling late in autumn. The bark is dark brown and thin, fissured into forking ridges. The twigs are green, gray, or brown, and slender, angled, and hairless or nearly so.

FLOWERS: Late summer–fall. Tiny and crowded in greenish bell-shaped heads less than ¼" (6 mm) long; the heads occur in large upright branching clusters, with the male and female flowers on separate plants.

FRUIT: ½" (12 mm) long and wide brush-like whitish head composed of many narrow seeds, each with tufts of whitish hairs or bristles to ⅜" (10 mm) long, maturing in late fall.

CAUTION: The leaves of this plant are toxic to livestock.

HABITAT: Moist soils, including salt marshes, borders of streams, roadside ditches, open woods, and waste places.

RANGE: Massachusetts south through Florida west through Texas, and north to Oklahoma and Arkansas; it is also found in the Bahamas and Cuba. It occurs to 800' (244 m).

USES: Florida Groundsel Bush is used as an ornamental plant, suited for planting near the ocean. It provides important habitat for coastal wildlife.

SIMILAR SPECIES: Bigleaf Marshelder.

CONSERVATION STATUS: LC

Florida Groundsel Bush has an extensive range with a large population. It is considered stable, with no current or future threats identified.

Diameter: 4" (10 cm)
Height: 16' (5 m)

■ DENSE POPULATION
■ LIGHT POPULATION
■ NATURALIZED

This is a widespread western plant, common in moist, open places and wetlands at elevations up to 10,000' (3,048 m). Blue Elderberry's sweetish fruits are used in preserves and pies but should never be eaten fresh or raw. Cooking degrades the alkaloid compounds found in elderberries that can cause nausea. Blue Elderberry is planted as an ornamental for the numerous whitish flowers and bluish fruits.

ALTERNATE NAME: *Sambucus cerulea, Sambucus mexicana*

DESCRIPTION: This large, many-branched, thicket-forming shrub or small tree often has several trunks; a compact, rounded crown; numerous small, whitish flowers in large clusters; and bluish fruit. The leaves are opposite, pinnately compound, and 5–7" (13–18 cm) long; they are sometimes nearly evergreen southward; there are five to nine leaflets 1–5" (2.5–13 cm) long, ⅜–1½" (1–4 cm) wide, paired except at the end, narrowly ovate or lance-shaped, long-pointed at the tip, short-pointed and unequal at the base, sharply saw-toothed, and short-stalked. They are yellow-green above and paler and often hairy beneath. The bark is gray or brown and furrowed. The twigs are green, stout, angled, and often hairy, with ringed nodes and thick, white pith.

FLOWERS: Summer. Nearly ¼" (6 mm) wide with a yellowish-white, five-lobed corolla; they are fragrant and arranged in upright, flat-topped, many-branched clusters 4–8" (10–20 cm) wide.

FRUIT: A dark blue berry nearly ¼" (6 mm) in diameter, with a whitish bloom and three one-seeded nutlets, arranged in many clusters and maturing in summer and fall.

CAUTION: Poisonous: Do not eat or ingest any part of this species. If accidentally ingested, contact your doctor or poison control center.

HABITAT: Moist soils along streams and canyons of mountains, in open areas in coniferous forests; also along roadsides, fence-rows, and clearings.

RANGE: Southern British Columbia south along coast to southern California, east in mountains to Trans-Pecos Texas, and north to western Montana; also in northwest Mexico; to 10,000' (3,048 m).

USES: This species is often planted as an ornamental. The berries are made into jelly, wine, and other food items, but should never be eaten when fresh and raw. A traditional remedy for fever has been made from the bark.

SIMILAR SPECIES: American Elderberry.

CONSERVATION STATUS:
Blue Elderberry is fairly common with no known threats to the species.

Diameter: 1' (0.3 m)
Height: 25' (7.6 m)

■ DENSE POPULATION
■ LIGHT POPULATION
■ NATURALIZED

This common, widespread plant sprouts from roots and grows in a variety of conditions. It is found in eastern Canada and throughout the United States except for the Great Basin and Pacific Northwest. Elderberries are inedible when fresh and raw, but can be used for making jelly, preserves, pies, and wine. Many species of native birds and mammals eat the berries. This plant is known by many common names including American Black Elderberry, Canada Elderberry, and Common Elderberry.

ALTERNATE NAMES: *Sambucus canadensis, Sambucus mexicana*

DESCRIPTION: This is a large shrub or small tree with an irregular crown of few, stout, spreading branches, clusters of white flowers, and many small black or purple berries. The leaves are opposite, pinnately compound, and 5–9" (13–23 cm) long, with a yellow-green axis and three to seven leaflets 1½–4" (4–10 cm) long and ¾–2" (2–5 cm) wide; they are paired (except at the end), elliptical, sharply saw-toothed, stalkless or nearly so, and shiny green above and dull light green and hairy along the midvein beneath. They are often evergreen and leathery in the South and Southwest. The bark is light gray or brown with raised dots, and smooth or becoming fissured and rough. The twigs are light green, stout, and angled, with ringed nodes and thick white pith.

FLOWERS: Late spring and early summer (year-round in the South). ¼" (6 mm) wide with a white corolla of four or five lobes; they are fragrant, with many arranged in upright flat-topped, much-branched clusters, 2–8" (5–20 cm) wide.

FRUIT: ¼" (6 mm) in diameter, black, purplish-black, or dark blue berry; it is juicy, with three to five one-seeded nutlets, maturing in late summer and fall (year-round in the South).

CAUTION: Poisonous: Do not eat or ingest any part of this species. If accidentally ingested, contact your doctor or poison control center. New growth of this plant can be fatal to livestock.

HABITAT: Wet soils, especially in open areas near water at forest edges; along streams and drainages.

RANGE: Manitoba east to Nova Scotia, south to Florida, and westward across Texas and the plains states to California and south into Mexico; to 5000' (1524 m).

USES: This species is often planted as an ornamental and to attract birds and other wildlife. The berries have been used to make jelly, wine, and other food items, but should never be eaten when fresh and raw. The bark, leaves, and flowers have served as traditional medicines but can be toxic.

SIMILAR SPECIES: Red Elderberry.

CONSERVATION STATUS: LC
American Elderberry is common and considered stable. The species has no known threats.

Diameter: 6"–1' (15–30 cm)
Height: 16–25' (5–7.6 m)

■ DENSE POPULATION
■ LIGHT POPULATION
■ NATURALIZED

Red Elderberry is native to to Europe, northern Asia, and throughout North America. It is widespread in North America, typically found along stream banks and in moist habitats from foothills to 3200' (975 m). The red fruit has a disagreeably bitter taste and is inedible when raw, but can be made into preserves or wine. Birds and mammals eat the fruit, and the plant is sometimes used as browse for livestock. There is considerable disagreement among experts over the classification of *Sambucus* species; several former species are now considered part of *Sambucus racemosa* as varieties or subspecies.

ALTERNATE NAMES: *Sambucus racemosa* var. *racemosa, Sambucus pubens, Sambucus callicarpa, Sambucus microbotrys*

DESCRIPTION: This is a clump-forming shrub or sometimes a small tree with many small, white flowers in concave or pyramidal clusters and bright red berries; the flowers and crushed foliage have an unpleasant odor. The leaves are opposite, pinnately compound, and 5–10" (13–25 cm) long, with five or seven leaflets 2–5" (5–13 cm) long and 1–2" (2.5–5 cm) wide; they are paired except at the end, lance-shaped or elliptical, and finely and sharply saw-toothed. They are green and nearly hairless above and paler and hairy beneath. The bark is light to dark gray or brown and smooth, becoming fissured into small, scaly or shaggy plates. The twigs are gray and stout, and hairy when young, with ringed nodes and thick, whitish pith becoming yellow-orange or brown.

FLOWERS: Spring–early summer. ¼" (6 mm) wide with a white, five-lobed corolla, arranged in upright, much-branched clusters to 4" (10 cm) long.

FRUIT: A round berry ⁵⁄₁₆" (8 mm) in diameter; it is bright red or sometimes orange, and juicy, with one-seeded, poisonous nutlets; it matures in summer.

CAUTION: Poisonous: Do not eat or ingest any part of this species. If accidentally ingested, contact your doctor or poison control center.

HABITAT: Moist soils in rich woods and clearings, such as cutover coniferous forests.

RANGE: Alaska east to Newfoundland, south to Georgia, and northwest to Tennessee, Missouri, and South Dakota; also throughout West south to California, Colorado, and New Mexico.

USES: This plant is cultivated as an ornamental plant. The fruits are an important food source for many bird species, and the flowers attract butterflies and hummingbirds. This species has also been used for medicinal purposes.

SIMILAR SPECIES: American Elderberry.

CONSERVATION STATUS:
Red Elderberry has a large population that is widely distributed. The species is not known to face any significant risks.

Diameter: 6" (15 cm)
Height: 20' (6 m)

■ DENSE POPULATION
■ LIGHT POPULATION
■ NATURALIZED

Southern Arrowwood is widespread in the easern United States, where it usually grows as a shrub. All arrowwoods are good plants for a backyard wildlife habitat: They have attractive flowers and fruit that attract birds and other wildlife, and the branching habit forms protective cover for birds and small mammals. Some botanists recognize two separate species for this highly variable plant, the other being Northern Arrowwood (*V. recognitum*), which differs from other arrowwoods in having smooth twigs, while Southern Arrowwood has downy twigs.

DESCRIPTION: This is usually a shrub with downy twigs, coarsely toothed leaves, and flat-topped clusters of small, white flowers. It sometimes grows rather tall and tree-like. The leaves are 1½–3" (3.8–7.5 cm) long and rounded or heart-shaped at the base; they are opposite, ovate, or egg-shaped, with saw-like teeth, and large-toothed, turning shiny red. The bark is smooth and gray, and the twigs are hairless.

FLOWERS: May–June. Small, white flowers are borne in 2–4" (5–10 cm), flat-topped clusters.

FRUIT: Tiny, purplish-black or blue-gray berrylike drupes, occurring in 3½" (9 cm) clusters of 30 or more.

HABITAT: Wet or dry thickets and borders of woods. Moist low areas, wet or dry thickets, woodland borders.

RANGE: Ontario to New Brunswick; south to Florida; west to Texas. Maritime Canada and Maine to Wisconsin and south to Georgia and Alabama.

USES: This shrub is planted for ornament and to attract birds and other wildlife. Arrow shafts were made from the straight young stems of this plant.

SIMILAR SPECIES: Nannyberry.

CONSERVATION STATUS:

Although it is susceptible to mild pests such as various moth larvae and the introduced Viburnum Leaf Beetle (*Pyrrhalta viburni*), Southern Arrowwood does not face any known significant threats. The population is thought to be large and secure.

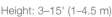

Height: 3–15' (1–4.5 m)

■ DENSE POPULATION
■ LIGHT POPULATION
■ NATURALIZED

Nannyberry, sometimes called Sweet Viburnum, is native to southern Canada from Saskatchewan east, and in the U.S. in the Midwest and Northeast. It is adaptable and can flourish in a variety of sites, but prefers moist soil. A popular ornamental grown for its attractive fruit and foliage, the plant can be readily propagated by softwood cuttings; when cut, the plants sprout from roots, and old branches will often arch down and take root. Many native birds and mammals eat the sweet fruit in winter.

DESCRIPTION: This is a large shrub or small tree with a short trunk; a compact, rounded crown of drooping branches; small white flowers in clusters; and small bluish-black fruit. The leaves are opposite, 2½–4" (6–10 cm) long, 1½–2½" (4–6 cm) wide, elliptical, long-pointed, and finely saw-toothed; with a prominent network of veins, and a broad, often hairy leafstalk. They are shiny green above, and yellow-green with tiny black dots beneath, turning purplish-red and orange in autumn. The bark is reddish-brown or gray, irregularly furrowed into scaly plates, and has an unpleasant skunk-like odor. The twigs are light green, slender, and slightly hairy when young, ending in a long-pointed hairy reddish bud.

FLOWERS: Late spring. ¼" (6 mm) wide with 5 rounded white corolla lobes; they are slightly fragrant and are arranged in branched upright, stalkless clusters of many flowers each, 3–5" (7.5–13 cm) wide.

FRUIT: ½" (12 mm) long, elliptical or sometimes nearly round, slightly flat, and blue-black with whitish bloom; it has sweet juicy pulp and a somewhat flat stone, drooping on slender reddish stalks, and maturing in fall and remaining attached in winter.

HABITAT: Moist soils of valleys and rocky uplands; at forest edges.

RANGE: Southeastern Saskatchewan east to New Brunswick and Maine, south to West Virginia, west to Nebraska and northeastern Wyoming; it is local in southwest Virginia. It occurs to 2500' (762 m), and to 5000' (1524 m) in the Black Hills.

USES: Nannyberry is often cultivated as an ornamental plant. The sweet fruit is edible, unlike most other *Viburnum* species. The bark and leaves have been used for medicinal purposes.

SIMILAR SPECIES: Southern Arrowwood.

CONSERVATION STATUS: LC

Nannyberry is considered stable due to its extensive range and large population. There are no current threats known to affect the species.

Diameter: 6" (15 cm)
Height: 20' (6 m)

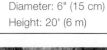

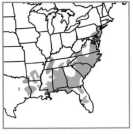

■ DENSE POPULATION
■ LIGHT POPULATION
■ NATURALIZED

This species is native to eastern North America, usually preferring swamps and other wet areas. As the common name suggests, opossums and other wildlife readily consume the fruit; deer also browse the foliage. It is also called Withe-rod or Withe-rod Viburnum. The scientific name of the species, from the Latin for "naked," refers to the plant's stalked, leafless flower clusters.

DESCRIPTION: This shrub or small tree has a spreading, open crown of irregular branches; many small, white or yellowish flowers; and small blue fruit. The leaves are opposite, 2–5" (5–13 cm) long, 1–2½ (2.5–6 cm) wide, and elliptical to narrowly elliptical, with the edges turned under (without teeth or slightly wavy); they are slightly thick with prominently curved side veins raised beneath. They are shiny green and becoming hairless above; and paler and rusty with tiny dots beneath, turning red in fall. The short leafstalk is covered with rust-colored hairs. The bark is gray to brown and smooth. The twigs are brown and slender, ending in long-pointed buds covered with rust-colored hairs.

FLOWERS: Spring. ½" (12 mm) wide with 5 rounded creamy white or yellowish corolla lobes, arranged in upright short-stalked flat clusters, 2½–5" (6–13 cm) wide.

FRUIT: ⁵⁄₁₆" (8 mm) in diameter and nearly round, turning from pink to deep blue or blue-black with a bloom; it has bitter pulp and a slightly flat stone, occurring on slender stalks and maturing in fall.

HABITAT: Moist soil near streams and swamps and less frequently in upland slopes; in open forests, pinelands, and thickets.

RANGE: Southwestern Connecticut to central Florida, west to eastern Texas, and north to central Arkansas and western Kentucky; to 3000' (914 m).

USES: This species has been used for medicinal purposes, and some people have used the fruit as a food source.

SIMILAR SPECIES: This plant is not to be confused with the species of holly commonly called Possum Haw (genus *Ilex*).

CONSERVATION STATUS: (LC)
Possumhaw Viburnum has a wide distribution and large population. It faces no known current or anticipated threats.

Diameter: 4" (10 cm)
Height: 16' (5 m)

■ DENSE POPULATION
■ LIGHT POPULATION
■ NATURALIZED

Walter Viburnum occurs in a limited area in the southeastern United States, where it is found in the coastal plains in hammocks, thickets, and swamp margins. This species is named for botanist Thomas Walter (1740–1789), who described this species in his 1788 book *Flora Caroliniana* detailing the plant life of South Carolina. Birds readily consume the shiny black fruit. Other common names for this plant include Small-leaf Viburnum and Small-leaf Arrowwood.

DESCRIPTION: This evergreen shrub or small tree has a few trunks; a broad, spreading crown; small, white flowers; and small, shiny black fruit. The leaves are evergreen, opposite, 1–2½" (2.5–6 cm) long, ⅜–1¼" (1–3 cm) wide, and spoon-shaped, with a few teeth toward the blunt tip (or not toothed); they are thick and leathery, with a short-stalked base. They are shiny dark green above, and light green with reddish-brown gland-dots beneath. The bark is blackish, thick, and furrowed into many angular blocks. The twigs are red-brown to gray, partly short, stiff side twigs or spurs, ending in a narrow reddish-brown hairy bud.

FLOWERS: Early spring. ¼" (6 mm) wide with five rounded white corolla lobes, arranged in upright flat, almost stalkless clusters 1½–2½" (4–6 cm) wide.

FRUIT: ⁵⁄₁₆" (8 mm) long, elliptical, and red turning shiny black, with thin, almost tasteless pulp, and maturing in summer or fall and remaining attached.

HABITAT: Moist valley soils, especially stream banks, flood plains, and swamps.

RANGE: Eastern South Carolina to central Florida; it occurs from sea level to 200' (61 m).

USES: This species is grown as an ornamental plant or to attract birds and butterflies.

SIMILAR SPECIES: Other *Viburnum* spp.

CONSERVATION STATUS: LC

The Walter Viburnum population has a restricted range but is believed to be secure. No threats to the species have been identified.

Diameter: 4" (10 cm)
Height: 20' (6 m)

 DENSE POPULATION
 LIGHT POPULATION
■ NATURALIZED

Also known as Smooth Blackhaw, Sweet Haw, or Stag Bush, this plant is native to the eastern and central United States, where it prefers moist valleys. A wide variety of native birds and mammals readily consume the fruit, which can be eaten raw or made into preserves. As an ornamental plant, it provides attractive fall color and provides food and cover for birds. The Latin species name refers to the leaves' resemblance to plum leaves.

DESCRIPTION: This is a shrub or small tree with a short trunk; a spreading, rounded or irregular crown; many showy, small, white flowers; and small, blue-black fruit. The leaves are opposite, 1½–3" (4–7.5 cm) long, ¾–2" (2–5 cm) wide, elliptical, finely saw-toothed, slightly thick, and hairless or nearly so. They are shiny green with a network of sunken veins above, and dull light green beneath, turning shiny red in fall. The bark is gray, rough, and furrowed into rectangular plates. The twigs are gray, slender, and stiff, ending in a flat, oblong, hairy brown bud.

FLOWERS: Spring. ¼" (6 mm) wide with five rounded white corolla lobes, arranged in upright flat, stalkless clusters, 2–4" (5–10 cm) wide.

FRUIT: ½" (12 mm) long, elliptical, and slightly flat; it is dark blue-black with whitish bloom, thin and slightly sweetish edible pulp, and a somewhat flat stone; it droops on long slender reddish stalks and matures in fall, remaining attached into early winter.

HABITAT: Moist soils, especially in valleys, and on slopes; in thickets and at borders of forests.

RANGE: Southwestern Connecticut south to Alabama, west to eastern Kansas, and north to southeastern Wisconsin and southwestern Iowa; it occurs to 3000' (914 m).

USES: Blackhaw is planted as an ornamental and to attract birds and other wildlife. The astringent bark was once used medicinally. The edible fruit is often made into preserves.

SIMILAR SPECIES: Rusty Blackhaw.

CONSERVATION STATUS: 🔵 LC
Blackhaw is believed to have a stable population with no known significant threats.

Diameter: 4" (10 cm)
Height: 20' (6 m)

■ DENSE POPULATION
■ LIGHT POPULATION
■ NATURALIZED

This plant is native to the southeastern and south-central United States, commonly found in upland forests. It is named for the rusty reddish-brown hairs on its foliage and other parts of the plant. These reddish-brown hairs help to distinguish Rusty Blackhaw from the more northerly Blackhaw, as do its slightly larger leaves and paler blue fruit. These two related, widespread species often intergrade in areas where both occur.

DESCRIPTION: This is a large shrub or small tree with a short trunk; a spreading, irregular crown; many small white flowers; leaves covered with rust-colored hairs beneath; and small blue fruit. The leaves are opposite, 2–4" (5–10 cm) long, 1–2½" (2.5–6 cm) wide, elliptical, finely saw-toothed, and slightly thick. They are shiny green with a network of sunken veins above, and covered with rust-colored hairs beneath, especially along the veins, and turning shiny red in fall. The short leafstalks are covered with rust-colored hairs. The bark is gray, rough, and furrowed into rectangular plates. The twigs are slender, stiff, and covered with rust-colored hairs when young, turning gray; they end in oblong, flat, reddish-brown hairy buds.

FLOWERS: ¼" (6 mm) wide with five rounded white corolla lobes, occurring in upright flat, stalkless clusters, 3–6" (7.5–15 cm) wide, and covered with rust-colored hairs.

FRUIT: ½" (12 mm) long, elliptical, and slightly flat; it is blue with whitish bloom and has slightly sweetish edible pulp and a somewhat flat stone; it droops on slender reddish stalks and matures in late summer and fall, remaining attached into early winter.

HABITAT: Uplands and less often in valleys; in forests and at edges of woods.

RANGE: Southeast Virginia, south to northern Florida, west to central Texas and north to central Missouri; it occurs to 2500' (762 m).

USES: This species is sometimes used as an ornamental plant.

SIMILAR SPECIES: Flowering Dogwood, Blackhaw.

CONSERVATION STATUS:
The Rusty Blackhaw has a large population that is widely distributed. The species does not currently face any current or anticipated threats.

Diameter: 6" (15 cm)
Height: 20' (6 m)

- ■ DENSE POPULATION
- ■ LIGHT POPULATION
- ■ NATURALIZED

This native of eastern North America is sometimes planted for its exotic appearance and lacy compound leaves. Devil's Walking-stick is widespread and grows at forest edges or in the understory, typically in deep, moist soil. It sometimes forms clonal thickets by sprouting from the roots. Early settlers used the aromatic, spicy roots and fruit in home remedies, including as a cure for toothaches. Devil's Walking-stick is sometimes called Hercules' Club, leading to confusion between this species and the unrelated *Zanthoxylum clava-herculis.*

DESCRIPTION: This is a spiny, aromatic, thicket-forming shrub or small tree with one (sometimes several) stout and usually unbranched trunks, very large compound leaves, and big clusters of tiny flowers; it sometimes has a few spreading branches and a thin crown. The leaves are clustered at the ends of twigs, bipinnately compound, and 15–30" (38–76 cm) long and nearly as wide, with a prickly branched axis. They have numerous leaflets 2–3½" (5–9 cm) long; they are ovate or broadly elliptical, finely saw-toothed, and nearly hairless. They are dark green above, and paler and often with prickles on the midvein beneath, turning light yellow in autumn. The bark is dark brown, thin, and fissured, often with scattered stout spines. The twigs are light brown, green inside, and very stout, with many straight slender sharp prickles and large pith.

FLOWERS: Late summer. Less than ⅛" (3 mm) long and wide flower with five tiny white petals, occurring in upright clusters 8–16" (20–41 cm) long, with many hairy branches.

FRUIT: ¼" (6 mm) in diameter and berrylike, with black skin, thin purplish juicy pulp, and three to five seeds, maturing in fall.

HABITAT: Moist soils mostly near streams in understory of hardwood forests; often forms dense groves from root sprouts.

RANGE: New Jersey and New York south to central Florida, west to eastern Texas, and north to Missouri; it is naturalized north to New England, southern Ontario, and Wisconsin. It occurs to 3500' (1067 m), sometimes to 5000' (1524 m) in the southern Appalachians.

USES: This species is sometimes cultivated as an ornamental, although not as often as its Asian relatives.

SIMILAR SPECIES: This species is related to the introduced species Japanese Angelica Tree or Chinese Angelica Tree (*A. elata*), which has become an invasive plant in northeastern North America.

CONSERVATION STATUS: LC

Devil's Walking-stick is common in much of its native range. The species faces no known threats.

Diameter: 8" (20 cm)
Height: 30' (9 m)

SITKA SPRUCE (*PICEA SITCHENSIS*)

Pinaceae — Pine

This family includes many of the familiar conifers, including cedars, firs, hemlocks, larches, pines, and spruces. The family consists of aromatic, resinous, coniferous trees and shrubs, most of the Northern Hemisphere, many of which are important for lumber, pitch, turpentine, oils, seeds, and ornamental plantings.

Araucariaceae — Monkey-puzzle

This ancient family of conifers once occurred worldwide during the Jurassic and Cretaceous periods, but largely disappeared from the Northern Hemisphere around the same time as the extinction of the dinosaurs. Today's members of this family are mainly native to the Southern Hemisphere, but some species are planted as ornamentals worldwide. They are typically extremely tall evergreen trees.

Cupressaceae — Cypress

This family includes many aromatic, resinous, coniferous trees and shrubs from temperate regions throughout the world, valued for ornamental beauty, good lumber, resins, and oils. There are about 140 species worldwide. Notable groups within the Cypress family in North America include junipers and redwoods.

Taxaceae — Yew

This family includes about 30 species of many-branched coniferous trees and shrubs, predominantly of the Northern Hemisphere. Many species in this family are important as ornamentals, but the seeds of these plants are generally toxic to humans.

MONKEY-PUZZLE TREE (*ARAUCARIA ARAUCANA*)

PACIFIC YEW (*TAXUS BREVIFOLIA*)

Ephedraceae — Ephedra

This family is composed of only a single genus, which mainly consists of gymnospermic shrubs but is sometimes climbing vines or small trees. These plants are adapted to extremely high-altitude, arid regions, including high in the Andes and Himalayas.

Zamiaceae — Zamiad

This family includes about 150 species found in the tropical and subtropical zones around the world. They are evergreen cycads—an ancient group of slow-growing, long-lived gymnosperms with cylindrical trunks—that superficially resemble palms or ferns, but are not closely related to either.

Ginkgoaceae — Ginkgo

This ancient family of gymnosperms evolved during the Mesozoic Era when diverse forests of these and related plants thrived for hundreds of millions of years. Today only a single species, *Ginkgo biloba*, remains.

Schisandraceae — Star-anise

This is a family of about 90 species of subtropical and tropical plants found in North America and East Asia that have symmetrical flowers. These are flowering, woody plants that produce essential oils, found from tropical to temperate latitudes. Star-anise (*Illicium verum*), native to Vietnam and China, is a well-known and commercially important member of this family.

Magnoliaceae — Magnolia

This is a family of more than 200 species of flowering shrubs and trees found in subtropical climates around the world. The flowers of these plants tend to be bisexual, fragrant, and attractive. Petals occur without distinct sepals, and stamens grow in a spiral around a central cone. These structures appear in fossilized plants believed to be an early form of angiosperm. Some species produce a fine-grained wood that is excellent for woodworking, while bark and buds are used in some traditional medicines.

Annonaceae — Custard-apple

Roughly 2400 species make up this family of flowering plants, occurring mainly as trees and shrubs in tropical regions around

CHINESE MAGNOLIA (*MAGNOLIA X SOULANGEANA*)

the world. Trees in this family tend to be aromatic, and many produce edible fruits including the custard apple, soursop, and pawpaw.

Lauraceae — Laurel

This family includes more than 2800 species of aromatic evergreen and deciduous trees and shrubs. Laurels are found in tropical and warm temperate regions. Several species are prized for their fragrant wood, while others produce common cooking ingredients, including bay leaves, cinnamon, and avocados.

Arecaceae — Palm

This family includes 2600 species of trees, shrubs, climbers, and stemless plants occurring in tropical to subtropical regions. Palms are well-known for their straight, branchless trunks and long, wide leaves called fronds. Palms are often associated with tropical paradises, and this has become one of the most cultivated plant families. Products from this commercially important family include dates, coconuts, palm oil, palm syrup, heart of palm, acai berries, and carnauba wax.

Papaveraceae — Poppy

The poppy family includes 775 species. The majority of species are herbaceous plants inhabiting temperate to subtropical climates in the Northern Hemisphere, but some species are shrubs and small trees. Poppies are widely cultivated as ornamental plants due to their colorful blossoms. The Opium

AMERICAN SYCAMORE (*PLATANUS OCCIDENTALIS*)

Poppy (*Papaver somniferum*) is used in the production of medical opiates.

Platanaceae — Sycamore

This is a small family made up of only eight species of large, flowering trees that occur primarily in temperate to subtropical climates in the Northern Hemisphere. Commonly called plane trees or sycamores, these trees are widely planted in urban areas and parks to provide shade.

Altingiaceae — Sweetgum

This family includes 15 species of deciduous trees occurring in Central and North America, Southeast Asia, and the Mediterranean. Commonly called gum trees, some species produce resin that can be used as chewing gum. Many species are cultivated as ornamentals or for their fine wood. They produce round, spiky fruit, sometimes called gumballs.

Hamamelidaceae — Witch-hazel

This family of flowering, deciduous shrubs and small trees consists of 140 species found mainly in eastern North America, Central America, Southeast Africa, and Southeast Asia. This family is distinguished by its unique floral characteristics. The taxonomic placement of this family and some of its subfamilies have been revisited several times and remain the subject of ongoing studies. An astringent extract commonly known as witch hazel is produced from a few species and used to soothe skin irritation.

Cercidiphyllaceae — Katsura-tree

This family contains a single genus of only two species of deciduous trees endemic to China and Japan. The flowers of these trees are unisex and appear on distinct male and female individuals. Katsura-tree (*Cercidiphyllum japonicum*) is the better known of these species; it is an extremely large hardwood often grown for ornamental purposes, prized for its heart-shaped leaves and strong wood.

Zygophyllaceae — Caltrop

This is a family of 285 species of trees, shrubs, and herbaceous plants found in warm temperate to tropical regions around the world. These plants are especially suited for dry environments, and many of its species have been cultivated for ornamental purposes. Many representatives of this family produce small seeds covered in sharp spines that easily pierce the skin or become entangled in animal hair. One genus produces fine hardwood lumber as well as chemical compounds with various medical applications.

Fabaceae — Pea

At 19,000 species, this is one of the largest families of terrestrial plants, and consists of trees, shrubs, woody vines, and herbaceous plants. Members of this family occur around the world, generally in tropical environments and dry forests. Cultivated globally, these plants are among the most important sources of food for humans and animals. Commonly called legumes,

SWEET ACACIA (*ACACIA FARNESIANA*)

the family includes beans, peas, lentils, alfalfa, clover, peanuts, and tamarind. Various species are used for timber, dyes, and natural gums.

Surianaceae — Bay-cedar

This is a small family containing eight species of flowering shrubs and trees occurring in tropical regions, generally in coastal areas. Most species are relegated to small distributions; however, one species—Bay-cedar (*Suriana maritima*)—inhabits coastal regions around the world. The tree's hard, buoyant fruit allows the ocean to distribute its seeds while protecting them from the salt water.

BRADFORD PEAR (*PYRUS CALLERYANA*)

Rosaceae — Rose

This family includes more than 4800 species of trees, shrubs, and herbaceous plants occurring nearly everywhere other than Antarctica. Many species have striking blossoms and are cultivated as ornamentals. A number of economically important food plants belong to this family, including apples, peaches, cherries, blackberries, almonds, and strawberries.

Elaeagnaceae — Oleaster

This family contains about 60 species of trees and shrubs occurring mainly in temperate regions of North America, Europe, and Asia. Many species thrive in dry areas and can tolerate soil with high salinity. Some are nitrogen-fixing plants that are useful in soil reclamation. Several species in this family produce delicious fruits that are eaten or used to make beverages and preserves.

Rhamnaceae — Buckthorn

This large family consists of 950 species, mainly flowering trees, shrubs, and woody vines. They are found in tropical to subtropical regions around the world. Many species in this family are popular as ornamentals and some produce compounds used to make green and yellow dyes.

Ulmaceae — Elm

This is a family of about 45 species of evergreen or deciduous trees and shrubs occurring in temperate zones throughout the Northern Hemisphere. Typical members of this family contain mucilaginous substances in the leaf and bark tissue. Elm trees have been prized for their excellent lumber for centuries, and many are grown as attractive ornamentals. In addition, the inner bark of some species has medicinal applications.

WINGED ELM (*ULMUS ALATA*)

Moraceae — Mulberry

This family includes about 1100 species of flowering plants ranging from tall trees to shrubs, vines, and herbaceous plants. They are mainly from tropical to subtropical regions worldwide. Many species are prized for their fruits and include jackfruit, mulberries, and figs. Some are cultivated as ornamentals and a few species were historical sources of latex for rubber.

Fagaceae — Beech

This is a widespread family of 927 trees and shrubs that occur throughout the temperate zones of North America, Europe, and Asia. This family includes oaks, beeches, and chestnuts. Fruit develops as one or more seeds partially or completely enveloped by a woody covering. Several species are economically important sources of hardwood timber. Seeds and nuts from this family are a major food source for humans and wildlife alike.

Myricaceae — Wax-myrtle

This family consists of about 55 species of trees and shrubs found naturally on every continent other than Australia and Antarctica. Several species commonly known as bayberries have a wide variety of uses. The waxy fruit is used to produce candles and as a food crop. The leaves contain compounds that are useful for repelling insects.

PIGNUT HICKORY (*CARYA GLABRA*)

Juglandaceae — Walnut

This is a family of about 50 aromatic trees and shrubs best known for producing valuable nuts and timber. These plants primarily range from temperate to tropical areas of the Americas and Asia. The drupe-like nuts of several species, including walnuts, pecans, and hickory nuts, are used extensively for food and commercially important. Many species are also prized for their excellent hardwood timber.

Casuarinaceae — Australian-pine

This family includes about 90 species of evergreen trees and shrubs occurring primarily in Madagascar, Australia, Southeast Asia, and Polynesia. They are typified by clusters of woody, ovoid, cone-like fruiting bodies. Many species are commonly known as she-oak due to their hardwood timber being similar to that of oak.

Betulaceae — Birch

This family consists of 167 species of deciduous trees and shrubs found mainly in temperate regions of the Northern Hemisphere with some species in the mountains of South America. Two species in this family produce edible nuts of great economic importance. Trees in this family provide extremely hard wood that makes valuable timber, and several species are grown as ornamentals.

SOUTHERN BAYBERRY (*MORELLA CERIFERA*)

Celastraceae — Staff Tree

This is a large and diverse family of 1350 species of trees, shrubs, vines, and herbaceous plants. Most of the plants in this family are tropical, but others are widespread around the world, ranging to temperate, arctic, and alpine environments. Many are used as ornamentals. The leaves of most of these plants are leathery, and their flowers are usually small. Several members of this family produce colorful fruit.

Rhizophoraceae — Red Mangrove

This family includes 147 species of flowering plants that grow throughout the tropics worldwide. Mangroves thrive in saltwater marshes due to an adaptation that allows them to remove excess salt from their systems, and they constitute an important part of the coastal ecosystem by protecting the land from erosion and damage from tropical storms. Mangrove bark produces tannins used for making leather, and its dense wood is used as pulp or timber.

Malpighiaceae — Malpighia

This is a family of more than 1300 species of trees, shrubs, and vines native to tropical and subtropical environments in the Americas, Africa, and Southeast Asia. A few species produce edible fruit that is high in vitamin C. Several species are cultivated ornamentally for their attractive, fragrant flowers.

Chrysobalanaceae — Coco-plum

This family consists of 533 species of trees and shrubs occurring in tropical regions around the world. Several species produce a small red or black fruit similar to a plum. The fruits contain a variety of flavonoids and are widely used for food.

Salicaceae — Willow

This is a family made up of more than 1000 species of flowering trees and shrubs ranging from temperate to tropical regions. The family is most common in the tropics but several commonly known species of poplars and willows are native to temperate zones. Several species in this family are cultivated for fruit, medicine, and timber, and many are grown as ornamentals. Willow and poplar wood have a number of commercial uses.

Euphorbiaceae — Spurge

This family includes about 7000 species, many of which are herbs with milky sap. Some species in this family, especially those in the tropics, are shrubs or trees. Among the valuable products made from members of this economically important family are rubber, castor and tung oils, and tapioca. Most members of the family are poisonous, and their milky sap will irritate the membranes of the eyes and mouth.

WHITE WILLOW (*SALIX ALBA*)

Combretaceae — White Mangrove

This family includes 530 species of trees, shrubs, and vines occurring in tropical and subtropical climates. Several species are economically important as sources of good timber, resins, gums, and edible fruits, as well as decorative ornamentals. A few members of this family provide yellow dyes and medicinal compounds used to treat a wide range of illnesses.

Lythraceae — Loosestrife

This is a family of about 620 species of trees, shrubs, and herbaceous plants found mostly in the tropics and temperate regions worldwide. Members of this family include several plants that are valued as sources of dyes and food crops, and some that are cultivated as ornamentals for their showy flowers. Among its members are Pomegranate (*Punica granatum*), which has been cultivated as food for thousands of years, and water caltrops (*Trapa* spp.), which are extensively harvested in China and India.

Myrtaceae — Myrtle

This family consists of at least 3000 species of woody trees and shrubs found in tropical to warm temperate zones. The plant family is valued as a source of lumber, gums, essential oils, spices, edible fruits, and ornamentals. Well-known members of the Myrtaceae family include clove, allspice, eucalyptus, and guava. Several species produce a crisp fruit with a sweet, floral flavor known as a rose-apple.

Staphyleaceae — Bladdernut

This family consists of about 45 species of trees and shrubs occurring in temperate to tropical regions in the Northern Hemisphere and South America. Several species are cultivated as ornamentals for their attractive flowers and interesting, papery fruits.

AMERICAN BLADDERNUT (*STAPHYLEA TRIFOLIA*)

Burseraceae — Torchwood

This is a family of 540 species of trees and shrubs with smooth bark and aromatic resin. All but a few representatives of the family occur in the tropical zones of the Americas, Africa, and Asia. Several species provide wood used in construction and manufacturing, some produce food crops of fruits and nuts, and others are well-known for their fragrant resins that are used to make incense and perfumes such as frankincense and myrrh.

Anacardiaceae — Cashew

This family consists of 860 species of trees and shrubs. Habitats range from tropical to temperate regions. The family includes several food-producing species that yield our commercial supply of cashew nuts, pistachios, and mangos. Several infamous plants in the appropriately named genus *Toxicodendron* produce an irritant called urushiol that can cause a severe skin rash upon contact—these include plants such as Poison Ivy (*T. radicans*) and Poison Sumac (*T. vernix*).

Sapindaceae — Soapberry

This widespread family includes more than 1800 species of trees, shrubs, vines, and herbaceous plants occurring in temperate to tropical zones worldwide. One genus includes maple trees, which are prized both for their fine hardwood and sap used to produce maple syrup. Several species produce well-known fruit such as lychee, while others contain compounds used in cosmetics.

CALIFORNIA BUCKEYE (*AESCULUS CALIFORNICA*)

Rutaceae — Citrus

This family includes about 1600 species of trees, shrubs, and herbaceous plants that are found in temperate to tropical regions around the world. The fruits of many species are used for culinary, medicinal, and ornamental purposes, as well as for their fragrant oils. Oranges, lemons, and grapefruits are among the most economically significant members of the family. Several species in the genus *Zanthoxylum* produce the popular spice commonly known as Sichuan pepper, which creates a numbing sensation in the mouth when eaten.

Simaroubaceae — Quassia

This family includes 170 species of trees and shrubs occurring in tropical and subtropical regions globally. Several species contain compounds that are valued for ornament and for medicinal extracts.

Meliaceae — Mahogany

This family consists of about 600 species of trees, shrubs, and herbaceous plants. Found mostly in tropical areas globally with a few species occurring in temperate zones. Several species of mahogany are prized for their rich, dark brown hardwood. Other species produce oils, soaps, and insecticides.

MAHOGANY (*SWIETENIA MAHAGONI*)

LITTLELEAF LINDEN (*TILIA CORDATA*)

Malvaceae — Mallow

This family includes more than 4000 species of mainly shrubs and herbaceous plants with some trees and vines. The leaves and the fruits of many of its species are edible. Several members of this family are economically important, producing major crops including cotton, cacao, okra, and hibiscus.

Caricaceae — Papaya

This is a family of about 35 species, mostly evergreen trees and shrubs found in the tropics in the Americas and Africa. This family gives us the widely cultivated Papaya (*Carica papaya*), whose large fruits are eaten worldwide. Several species in this family have culinary and medicinal uses.

Capparaceae — Caper

This family consists of about 700 species of trees, shrubs, and vines occurring in tropical to subtropical regions. Several species in this family produce edible fruit, while others are used in a variety of traditional medicines. The family is considered closely related to the mustard family (Brassicaceae) because both can produce pungent mustard oils.

TREE FAMILIES

Tamaricaceae — Tamarisk

This is a family of 78 species of trees and shrubs native mainly to drier areas and desert regions of Africa, Asia, and Europe. Members of the family thrive in soils with high levels of salinity. Some species are known to draw up salts dissolved in deep water sources before excreting the salts through their leaves, creating an inhospitable environment for any competing plants.

Polygonaceae — Buckwheat

This family includes about 1200 species of trees, shrubs, vines, and herbaceous plants. The highest concentrations of species are found in northern temperate zones. Many members of this family are cultivated for a variety of uses including for lumber, food, and ornament. Some bear edible fruit or leaves, and several species are economically important food crops.

Cornaceae — Dogwood

This family consists of two genera and about 85 species, most of which are trees and shrubs. They can be found from tropical to cool temperate zones worldwide. Several species are cultivated for ornament or for their wood and fruit. A few species are used extensively in traditional medicine.

FLOWERING DOGWOOD (*CORNUS FLORIDA*)

Fouquieriaceae — Ocotillo

The common name of this family (pronounced *o-ko-TEE-yo*) refers to a pitchy Mexican pine and calls attention to the resinous nature of the stems. There is only one genus with about 11 species, all found in North America and all but one restricted to Mexico. Ocotillo (*Fouquieria splendens*) is the northernmost species; the most unusual is Boojum Tree (*F. columnaris*) of Baja California, with its conical trunk and radiating spiky branches. Leaves appear quickly after rain and wither when the soil dries; a cycle commonly repeated several times during the warm season.

Sapotaceae — Sapodilla

This is a tropical family of about 800 species of evergreen trees and shrubs. Several species produce edible fruits, oil-rich nuts, or commercially valuable sap. Oils and sap extracted from members of this family are used for food, cleaning products, cosmetics, and skin treatments.

Ebenaceae — Ebony

This family includes 768 species of mostly tropical trees and shrubs. The family includes the genus *Diospyros*, commonly called ebony trees, which are prized for their exceptionally heavy, dark black hardwood. Other members of the family include persimmons, which are an economically important food crop.

Primulaceae — Primrose

The primrose family consists of 2790 species of trees, shrubs, and herbaceous plants found worldwide in tropical to temperate regions. A large number of species are cultivated ornamentally as garden flowers.

Theaceae — Tea

This family includes about 40 genera and perhaps 200–300 species of trees and shrubs occurring in tropical to temperate zones worldwide. The familiar cured leaves of one species in this family, *Camellia sinensis*, are the source of tea; the species is widely cultivated for this purpose. Other species in the genus *Camellia* are sometimes used to brew similar beverages and to make cooking oil, especially in Asia. Many other species in this family are valued as flowering ornamentals.

Symplocaceae — Sweetleaf

This family includes about 300 species of trees and shrubs primarily distributed in Asia and the Americas in humid, tropical, or montane forests. These plants are generally used for ornamental purposes and to make tonics. The bark of some species is used to produce a yellow dye, while the roots of others are used in traditional medicines.

Styracaceae — Snowbell

This is a family containing about 160 species of mostly large shrubs to small trees that are native to subtropical and warm temperate regions in the Northern Hemisphere. Several species are cultivated for their fragrant bark extracts and as ornamental trees with attractive white flowers. Benzoin resin is extracted from several species of the genus Styrax; this resin is commonly used in medicines, perfumes, and incense.

BIGLEAF SNOWBELL (*STYRAX GRANDIFOLIUS*)

Clethraceae — Clethra

This family consists of two genera and about 75 species of trees and shrubs that typically occur in tropical and warm temperate regions of Asia and the Americas. Some species are cultivated ornamentally or for erosion control. Many of these species feature clusters of small, bell-shaped flowers; others have large, showy flowers.

Cyrillaceae — Cyrilla

This is a small family of two genera of trees and shrubs occurring in the warm temperate to tropical zones of the Americas, typically in wetlands and rainforests. The number of species in one genus, *Cyrilla*, is the subject of debate among botanists; some sources recognize up to a dozen species while others contend that all of these represent a single widespread and variable species, *C. racemiflora*, commonly called Leatherwood or Swamp Titi. Although usually evergreen, some specimens become deciduous near the northern limits of their range.

Ericaceae — Heath

This is a large, diverse, and globally distributed family containing about 4250 species of trees, shrubs, and herbaceous plants. Many species thrive in acidic and infertile soils by forming a symbiosis with fungi to draw more nutrients from the soil. Several economically important members of this family of plants are often cultivated for their ornamental merit (azaleas, rhododendrons) and edible fruits (blueberries, cranberries).

TEXAS MADRONE (*ARBUTUS XALAPENSIS*)

WAVYLEAF SILKTASSEL (*GARRYA ELLIPTICA*)

Garryaceae — Silktassel

This is a small family with 28 species of trees and shrubs found predominantly in tropical and temperate regions in East Asia and the Americas. Several species are cultivated as ornamentals for their long, hanging clusters of flowers.

Rubiaceae — Madder

This is an extremely large and economically important family consisting of about 13,500 species of trees, shrubs, vines, and herbaceous plants that are distributed globally. The largest diversity occurs in the tropics but members of this family are found in all but the most extreme environments around the world. This family includes plants in the genus *Coffea*, the source of coffee beans, which are widely cultivated and among the most important trade commodities in the world. Important medicinal compounds are derived from the bark and leaves of some plants in this family, and a few species are used to produce red and yellow dyes.

Boraginaceae — Borage

This family consists of about 2000 species of trees, shrubs, and herbaceous plants distributed globally. Some members of the family are cultivated for a variety of applications including cosmetic dyes, wood stains, food seasonings, and traditional medicines. Several species are popular garden ornamentals. Most members of this family have hairy leaves, which can sometimes induce a skin reaction such as an itchy rash when the plants are handled.

Solanaceae — Nightshade

This is a widely distributed and economically important family that includes about 2700 species of trees, shrubs, vines, and herbaceous plants that can be found on every continent except Antarctica. This family produces a number of familiar and widely cultivated crops, including tomatoes, potatoes, eggplants, peppers, and tobacco. Many species are cultivated ornamentally, including petunias (genus *Petunia*).

Oleaceae — Olive

This family includes about 700 species of trees, shrubs, and woody vines that occur around the globe from subpolar to tropical zones. Olives, jasmines, and lilacs are well-known representatives of this family, many of which are prized for their oils, wood, or flowers. Many species are fragrant and several are grown as ornamentals. The family also contains the genus *Fraxinus*, which includes the ash trees.

SINGLELEAF ASH (*FRAXINUS ANOMALA*)

Acanthaceae — Acanthus

This family includes 2500 species of shrubs, vines, and herbaceous plants, mainly found across the tropical regions of Africa, Southeast Asia, and the Americas. Several species are widely used for the medicinal value of their leaves. At least one species is used in herbal teas due to its abundant antioxidants.

Bignoniaceae — Trumpet Creeper

This family consists of about 860 species of trees, shrubs, vines, and herbaceous plants. The majority of species are found in tropical regions of South America. Several species of trees from this family are important sources of timber. Many species are cultivated as ornamental. A few members of the Bignoniaceae family produce red, blue, or brown dyes.

Lamiaceae — Mint

This family contains about 7200 species of aromatic trees, shrubs, and herbaceous plants that occur globally. Many plants in this family are grown for their edible leaves, fragrance, or essential oils. Common seasonings and fragrances originating from plants in this family include oregano, basil, thyme, sage, rosemary, and lavender. It is closely related to the vervain family (Verbenaceae), and many species in this family were formerly classified there.

Paulowniaceae — Figwort

This is a small family of deciduous trees native to East Asia and cultivated worldwide as ornamentals. Its taxonomic status remains in flux, with various authorities recognizing between 7 and 17 species. The most widely distributed species is *Paulownia tomentosa*, commonly called Princesstree, which is considered invasive in North America. Paulownia trees are extremely fast-growing and have been known to grow up to 20 feet in a single year, and often regrow quickly after being harvested for wood.

Aquifoliaceae — Holly

This family consists of about 480 species of trees, shrubs, and vines that occur primarily in tropical to temperate climates worldwide. Many species in this family are poisonous, with every part of the plant being toxic. Some species are cultivated ornamentally or grown as garden hedges, and a few are grown for culinary purposes.

YAUPON HOLLY (*ILEX VOMITORIA*)

Asteraceae — Aster

This is probably the largest plant family with its 32,000 species of trees, shrubs, vines, and herbaceous plants. These plants occur in a wide variety of habitats from the subpolar to the tropical zones worldwide. The family includes many important food crops such as sunflowers, chicory, lettuce, and artichokes. Many of its species are cultivated ornamentally and can be used in herbal medicines and teas.

Adoxaceae — Moschatel

This family includes 200 species of mostly shrubs and herbaceous plants occurring globally in temperate to subtropical climates, mainly in the Northern Hemisphere. Members of the genus *Sambucus*, commonly known as elderberries, produce economically important fruits. Several species are cultivated as ornamentals.

Araliaceae — Ginseng

This family includes 1500 species of trees, shrubs, vines, and herbaceous plants that are widely distributed mainly in temperate and tropical regions of Asia and the Americas. Ivies belonging to this family are often cultivated as ornamental wall or ground covers. The genus *Panax* includes several species of ginseng, a root long used in traditional medicine for its stimulant qualities.

GLOSSARY

A

Achene – A small, dry, seed-like fruit with a thin wall that does not open.

Acorn – The hard-shelled, one-seeded nut of an oak, with a pointed tip and a scaly cup at the base.

Aggregate fruit – A fused cluster of several fruits, each one formed from an individual ovary, but all derived from a single flower, as in a magnolia.

Alternate – Leaves arranged singly along a twig or shoot, and not in whorls or opposite pairs.

Anther – The terminal part of a flower's stamen, containing pollen in one or more pollen sacs.

Axis – The central stalk of a compound leaf or flower cluster.

B

Bark – The outer covering of the trunk and branches of a tree, usually corky, papery, or leathery.

Berry – A fleshy fruit produced from the ovary of a single flower, typically with more than one seed.

Bipinnate – With leaflets arranged on side branches off a main axis. Syn. twice-pinnate; bipinnately compound.

Bisexual – Having both male and female organs in the same flower.

Blade – The broad, flat part of a leaf.

Bract – A modified and often scale-like leaf, usually located at the base of a flower, a fruit, or a cluster of flowers or fruits.

Bud – A young and undeveloped leaf, flower, or shoot, usually covered tightly with scales.

C

Calyx – Collective term for the sepals of a flower.

Capsule – A dry, thin-walled fruit containing two or more seeds and splitting along natural grooved lines at maturity.

Catkin – A compact and often drooping cluster of reduced, stalkless, and usually unisexual flowers.

Cell – A cavity in an ovary or fruit, containing ovules or seeds.

Compound leaf – A leaf whose blade is divided into three or more smaller leaflets.

Cone – A conical fruit consisting of seed-bearing, overlapping scales around a central axis.

Cone-scale – One of the scales of a cone.

Conifer – A cone-bearing tree of the pine family, usually evergreen.

Corolla – Collective term for the petals of a flower.

Critically Endangered – A Critically Endangered (CR) species is one that has been categorized by the International Union for Conservation of Nature (IUCN) as facing an extremely high risk of extinction in the wild. It is the most severe conservation status; species that are possibly extinct in the wild are listed as critically endangered until conclusively determined to be extinct.

Crown – The mass of branches, twigs, and leaves at the top of a tree, with particular reference to its shape.

Cultivated – Planted and maintained by humans.

D

Deciduous – Shedding leaves seasonally, and leafless for part of the year.

Drupe – A fleshy fruit with a central stone-like core containing one or more seeds.

E

Elliptical – Elongately oval, about twice as long as wide, and broadest at the middle; shaped like an ellipse.

Endangered – An Endangered (EN) species is a species which has been categorized by the International Union for Conservation of Nature (IUCN) as likely to become extinct in the near future.

Entire – Smooth-edged, not lobed or toothed.

Escaped – Spread from cultivation and now growing and reproducing without aid from humans.

Exotic – See Introduced.

Extant – Refers to a population with members that remain alive. Cf. Extinct.

Extinct – Refers to a species that is no longer alive, generally assigned after the last individual of a species dies. Species with only captive populations remaining may be classified as Extinct in the Wild. Cf. Extant.

F

Family – A group of related genera.

Filament – The threadlike stalk of a stamen.

Fleshy fruit – A fruit with juicy or mealy pulp.

Flower – The reproductive structure of a tree or other plant, consisting of at least one pistil or stamen, and often including petals and sepals.

Follicle – A dry, one-celled fruit, splitting at maturity along a single grooved line.

Fruit – The mature, fully developed ovary of a flower, containing one or more seeds.

G

Genus – A group of closely related species. Plural: genera.

Gland-dot – A tiny, dot-like gland or pore, usually secreting a fluid.

H

Habit – The characteristic growth form or general shape of a plant.

Herb – A plant with soft, not woody, stems, that dies to the ground in winter.

Hybrid – A plant or animal of mixed parentage, resulting from the interbreeding of two different species.

I

Intergradation – The gradual merging of two or more distinct forms or kinds, through a series of intermediate forms.

Introduced – Intentionally or accidentally established in an area by humans, and not native; Syn. exotic, foreign.

Irregular flower – A flower with petals of unequal size.

IUCN Red List of Threatened Species – The International Union for Conservation of Nature's Red List of Threatened Species. Version 2020–1. www.iucnredlist.org. Downloaded on 20 March 2020.

K

Keel – A sharp ridge or rib, resembling the keel of a boat, found on some fruits and seeds.

Key – A dry, one-seeded fruit with a wing; a samara.

L

Lanceolate – Shaped like a lance, several times longer than wide, pointed at the tip and broadest near the base.

Leader – The highest, terminal shoot of a plant.

Leaflet – One of the leaf-like subdivisions of a compound leaf.

Least Concern - A Least Concern (LC) species has been categorized by the International Union for Conservation of Nature (IUCN) as not being a focus of species conservation. These species do not qualify as threatened or conservation dependent.

Linear – Long, narrow, and parallel-sided.

Lobed – With the edge of the leaf deeply but not completely divided.

M

Male cone – The conical, pollen-bearing male element of a conifer.

Midvein – The prominent, central vein or rib in the blade of a leaf; midrib.

Multiple fruit – A fused cluster of several fruits, each one derived from a separate flower, as in a mulberry.

N

Native – Occurring naturally in an area and not introduced by humans; indigenous.

Naturalized – Successfully established and reproducing naturally in an area where not native.

Near Threatened – Near Threatened (NT) is a conservation status given by the International Union for Conservation of Nature (IUCN) to species or subspecies that may be threatened with extinction in the near future, but which is not yet elevated to threatened.

Needle – The long and narrow leaf of pines and related trees.

Node – The point on a shoot where a leaf, flower, or bud is attached.

Nut – A dry, one-seeded fruit with a thick, hard shell that does not split along a natural grooved line.

Nutlet – One of several small, nutlike parts of a compound fruit, as in a sycamore; the hard inner core of some fruits, containing a seed and surrounded by softer flesh, as in a hackberry.

O

Oblanceolate – Reverse lanceolate. Shaped like a lance, several times longer than wide, broadest near the tip and pointed at the base.

Oblong – With nearly parallel edges.

Obovate – Reverse ovate; oval, with the broader end at the tip.

Opposite – Leaves arranged along a twig or shoot in pairs, with one on each side, and not alternate or in whorls.

Ovary – The enlarged base of a pistil, containing one or more ovules.

Ovate – Oval with the broader end at the base.

Ovule – A small structure in the cell of an ovary, containing the egg and ripening into a seed.

P

Palmate – With leaflets attached directly to the end of the leafstalk and not arranged in rows along an axis; Syn. palmately compound, digitate.

Palmate-veined – With the principal veins arising from the end of the leafstalk and radiating toward the edge of the leaf, and not branching from a single midvein.

Persistent – Remaining attached, and not falling off.

Petal – One of the flower parts lying within the sepals and next to the stamens and pistil, often large and brightly colored.

Pinnate – With leaflets arranged in two rows along an axis; pinnately compound.

Pinnate-veined – With the principal veins branching from a single midvein, and not arising from the end of the leafstalk and radiating toward the edge of the leaf.

Pistil – The female structure of a flower, consisting of the stigma, style, and ovary.

Pith – The soft, spongy, innermost tissue in a stem.

Pod – A dry, one-celled fruit, splitting along natural grooved lines, with thicker walls than a capsule.

Pollen – Tiny grains containing the male germ cells and released by the stamens.

Pollen-sac – The part of an anther containing the pollen.

Pome – A fruit with fleshy outer tissue and a papery-walled, inner chamber containing the seeds.

R

Regular flower – A flower with petals all of equal size.

Resin – A plant secretion, often aromatic, that is insoluble in water but soluble in ether or alcohol.

Ring-scar – A ring-like scar left on a twig after a leaf falls away.

S

Samara – A dry, one-seeded fruit with a wing; a key.

Scale – One of the short, pointed, and overlapping leaves of some conifers; the leaflike covering of a bud or of the cup of an acorn.

Scurfy – Covered with tiny, broad scales.

Seed – A fertilized and mature ovule, containing an embryonic plant.

Sepal – One of the outermost flower parts, arranged in a ring outside the petals, and usually green and leaflike.

Sheath – A tubular, surrounding structure; in some conifers, the papery tube enclosing the base of a bundle of needles.

Shoot – A young, actively growing twig or stem.

Shrub – A woody plant, smaller than a tree, with several stems or trunks arising from a single base; a bush.

Simple fruit – A fruit developed from a single ovary.

Simple leaf – A leaf with a single blade, nor compound or composed of leatlets.

Sinus – The space between two lobes of a leaf.

Solitary – Borne singly, not in pairs or clusters.

Species – A kind or group of plants or animals, composed of populations of individuals that interbreed and produce similar offspring.

Spur – A short side twig, often bearing a cluster of leaves.

Stamen – One of the male structures of a flower, consisting of a threadlike filament and a pollen-bearing anther.

Stigma – The tip of a pistil, usually enlarged, that receives the pollen.

Stipule – A leaflike scale, often paired, at the base of a leaf stalk in some trees.

Style – The stalk-like column of a pistil, arising from the ovary and ending in a stigma.

T

Threatened – A term generally used to refer to the three International Union for Conservation of Nature (IUCN) categories that indicate the greatest conservation concern (Vulnerable, Endangered, and Critically Endangered).

Toothed – With an edge finely divided into short, tooth-like projections.

Tree line – The upper limit of tree growth at high latitudes or on mountains; also called the timberline.

Trunk – The major woody stem of a tree.

Tubular – With the petals partly united to form a tube.

U

Unisexual – With male or female organs, but not both, in a single flower.

V

Vein – One of the rib-like vessels in the blade of a leaf.

Vulnerable - Vulnerable (VU) is a conservation status designated by the International Union for Conservation of Nature (IUCN), referring to a species that is likely to become endangered unless the circumstances that threaten its survival improve.

W

Whorled – Arranged along a twig or shoot in groups of three or more at each node.

Wing – A thin, flat, dry, shelf-like projection on a fruit or seed, or along the side of a twig.

Wood – The hard, fibrous inner tissue of the trunk and branches of a tree or shrub.

INDEX

Please note that we have provided full names of the photographers where possible; some contributors are known by usernames, which we also wish to credit.

4028mdk09
A. Stewart
A. Barra
A. Abrahami
abdallahh
Adam Baker
Adam Jones, Ph.D.
Adbar
AfroBrazilian
Agaath
Agnieszka Kwiecień
Agrosylva
ahaislip
Al Fischer
Alan Cressler, Lady Bird Johnson Wildflower Center
Alan Kneidal
Alan Schmierer
Alec Cook via Calflora
Alfried H.
Alison Northup
Alpsdake
Alvesgaspar
Amada44
Andrew Butko
Andrew Smith
Andrew Stewart
Andrew Petro
Andrey Zharkikh
AnemoneProjectors
Anna reg
Annie Spratt
Anthony J. Valois
Аимаина хикари
Arches National Park
Arlington National Cemetery
Arnold Arboretum
Art Poskanzer
Arthur Chapman
Arthur Haines
Atamari
Atirador
Auburn University College of Science and Mathematics
Austin Pursley
Averater
Axel Kristinsson
B. Navez
Bananenfalter
Basotxerri
Beentree
Benjamin Smith
Benny White
Bernard Gagnon
Beyond My Ken
Bgautrea
Bill Cook
Bj.schoenmakers
Bjocar
Björn S

blahedo
bloodredrapture
BlueCanoe
Bob Peterson
Böhringer Friedrich
Bostonian13
BotBln
brdavids
Brewbooks
Bri Weldon
Brian Stansberry
Bruce Kirchoff
Bruce Marlin
Burr
C T Johansson
Celerylady
Charles Kane
Charles T. Bryson
Chhe
Chiselwit
Choess
Chris Light
Chris M.
Christian Ferrer
Christian Grenier
Chrumps
Chuck B.
cibomahto
Claire Secrist
Claudio Gioseffi
Coconuttree
Colsu
Congaree National Park
Consultaplantas
Costoa
Crosby
Crusier
Cwarneke
Daderot
Dalgial
Dan Keck
Daniel Case
Darrekk2
Darren Swim
Dave Creech
Dave Powell USDA Forest Service
David J. Stang
David Perez
Dcrjsr
D.E. Herman, USDA-NRCS PLANTS Database
Derek Ramsey
Dietmar Rabich
Dinesh Valke
Dinkum
Dmitriy Konstantinov
Dominic Sherony
Dominicus Johannes Bergsma

Donald Lee Pardue
Dr. George Rogers
Dr. Otto Wilhelm Thomé
Dustin Minialoff
dvs
Ebyabe
Ed Ogle
Eddie Welker
Emily Webb
Emőke Dénes
Eric Guinther
Eric Hunt
Eric Jones
ernie
Eugene Zelenko
EvaK
evan barker
Everglades NPS
Fagus_sylvatica_020
Famartin
Fepup
Filo gèn'
Forest & Kim Starr
Franco Folini
Franklin Bonner
Fritzflohrreynolds
Fungus Guy
gailhampshire
GatesofHell
Gene Sturla
Genet
Geoff Gallice
Georg Slickers
George Chernilevsky
Georges Jansoone
Gojaz Alkimson
Guettarda
Gustav Svensson
Gzzz
H. Zell
Hans Hillewaert
Harlan B. Herbert
harum.koh
Helen Lowe Metzman
hohey22
Homer Edward Price
International Oak Society
Ivar Leidus
J. Grimshaw via treesandshrubsonline.org
jacilluch
Jakec
James H. Miller & Ted Bodner
James H. Miller, USDA Forest Service
James Henderson
James Holland
James Steakley
Jason Hollinger
Jason Sturner

Jason Zhang
Jay Sturner
Jcomeau ictx
Jean Claude
Jean-Pol Grandmont
Jebulon
Jeff McMillian
Jeff Turner
Jerome, D., Wenzell, K. & Kenny, ucnredlist.org
JerryFriedman
Jerzystrzelecki
Jesse Rorabaugh
Jesse Taylor
jilllybean
JJ Harrison
jkirkhart35
JLPC
JMK
Joan Simon
João Medeiros
Joe Blowe
Joe Decruyenaere
joedecruyenaere
John Verive
John Bradford
John Comeau
John D. Byrd
John Game
John Gwaltney
John Morgan
John Rusk
John Verive
John Wagman
JoJan
Jon Peli Oleaga
Jonathon Jongsma
Joseph O'Brien, USDA Forest Service
Josh Jackson
Joshua Mayer
Jpaganpr
Jsayre64
Judy Gallagher
kallerna
Karelj
Katja Schulz
Katy Chayka
Keith Kanoti
Konpei
Kenraiz
Kent
Kerry Dressler
Kevin Keegan
Kim Scarborough
Kretyen
Kristine Paulus
Kritzolina
Krzysztof Golik
Krzysztof Ziarnek

Kymi
Ladislav Luppa
Larry Allain, U.S. Geological
    Survey
Lazaregagnidze
Lazare Gagnidze
Lee Adlaf
Lemur12
Léna
limkin
Liné1
Lokal_Profil
Louise Wolff
LRBurdak
Luis Fernandez Garcia
Luis Miguel Bugallo Sánchez
Lynden Gerdes
Maas Morton
Mac H. Alford
Magnus Manske
Manhattan Beach Botanical
    Garden
Marco Schmidt
MarcusObal
Marie Selby Botanical
    Gardens
Marija Gajić
MartinThoma
Masebrock
Mason Brock
Matrixupgrade
Matt Lavin
Matthieu Sontag
Max Licher
MBisanz
mefisher
Melissa McMasters
Meneerke bloem
Michael DePue
Michael Rivera
Michael Wunderli
Miguel Vieira
miheco
Mike Mulqueen
Mitch
Miwasatoshi
Mmcknight4
Mongo
MPF
MrPanyGoff
MSha
Mwanner
Mykola Swarnyk
Nadiatalent
Naples Botanical Garden
National Park Service
Neepster
New York Botanical Garden
Nicholas_T
Ninjatacoshell
NTNU Vitenskapsmuseet
Orjen
Paliienko Konstantin
Patrice78500
Paul Asman and Jill Lenoble
pellaea

Peter Halasz
Peter M. Dzuik
Peter O'Connor
Petr Filippov
pfly
Piotrus
Plant Image Library
Prairie Voyageur
Prenn
Przykuta
Ptelea trifoliata
PumpkinSky
Quartl
R. A. Nonenmacher
Radomil
Ragesoss
Randy A. Nonenmacher
Ranjith-chemmad
Rasbak
Rebekah D. Wallace
Renee
Reuven Martin
Richard Avery
Richard Sniezko, U.S. Forest
    Service
Richard Webb
Rison Thumboor
rmceoin
Rob Routledge
Rob Young
Robb Hannawacker
Robert Flogaus-Faust
Robert H. Mohlenbrock
Robert Vid
Robin Stott
Rogów
Rohit Naniwadekar
Ron Wiliams
Rosser1954
Rowan Adams
Roy Winkelman
Ryan Hodnett
S. Rae
Salicyna
Salvor
Sandy__R
SB_Johnny
Schzmo
Scott Loarie
Scott Zona
Sean Kelleher
Sgerbic
Sheri Hagwood
Sherief Saleh
Shih-Shiuan Kao
Si Griffiths
Siddharth Patil
Silversyrpher
SK Siddhartthan
Skuzbucket
Slarson789
snowpeak
SonoranDesertNPS
Southcoastwholesale
SriMesh
Stan Shebs

Sten Porse
Steve Hillebrand, U.S. Fish
    and Wildlife Service
Steve Law
Steve Ryan
Steven Katovich
Steven A. Wright, LSU
    School of Renewable
    Natural Resources
Stickpen
Superior National Forest
Susan McDougall
Syrio
TeunSpaans
Thayne Tuason
theforestprimeval
Themodoccypress
Thesupermat
Tom Nagy
Tomás Esparza
Tortie Tude
Trails LA County
Treegrow
Tree World Wholesale
Tyler Karaszewski
U.S. Fish and Wildlife Service
U.S. Forest Service
Ulf Mehlig
United States Botanic Garden
University of California
    Botanical Garden
University of Texas School of
    Biological Sciences
Uoaei1
USDA-NRCS PLANTS
    Database
U.S. Environmental Protection
    Agency
USFWS - Pacific Region
USGS Bee Inventory and
    Monitoring Lab
Valérie75
Van Den Berk, UK Ltd
Velo Steve
Vinayaraj
Vlad Butsky
Vojtěch Zavadil
Walter Siegmund
Whitney Cranshaw
Wikimedia Images
WildBoar
Willow
Wing-Chi Poon
Wlodzimierz
Wolfman SF
Wouter Hagens
Wynn Anderson
yamatsu
Yath
Yuriy Kvach
Zhangzhugang
Zolty Straczyn
Zooclub
Zoya Adulova

ACKNOWLEDGMENTS

Any work of this scope takes a dedicated team of talented experts, advisors, and editors to bring it to life. It also takes an inspirational spark, and for that, we want to thank the National Audubon Society, which since 1905 has championed the conservation of bird species and habitat throughout North America. **www.audubon.org**

The images used throughout this book reflect countless, patient hours in the field on the part of scores of photographers as well as the painstaking curation of those images. We want to particularly thank Susan McDougall and Gene Sturla, along with all of the photographers noted on pages 589–590, whose images bring these tree species to the printed page with grace, dynamism, and immediacy.

We are grateful to the following collaborators whose expertise informed the tree/dendrology discussions in the beginning of this book. We appreciate their work to further the cause of tree research, conservation, and education:

Marlyse Duguid, Ph. D., Thomas G. Siccama Lecturer in Environmental Field Studies and Associate Research Scientist at the Yale School of Forestry and Environmental Studies

Kyle Carlsen, writer, editor, and naturalist

Patrick Sweeney, Ph.D., Collections Manager at the Peabody Museum of Natural History at Yale University

Tyler Deal, writer and editor

Lawrence M. Kelly, Ph. D., Associate Vice President of Science Administration at the New York Botanical Garden (taxonomic consulting)

Warm thanks to Jane Cirigliano and Alan Smith for masterfully pulling it all together into the pages you see here, Heather Rowland of Add+Water for the book design, and to Kirill Litvinov and Alexandra Kovernzieva for rendering the range maps. We are also grateful to Eve Kalondu and Larry Farr for their meticulous work in indexing and managing the vast files of species data which are the heart of this book. Very special thanks to our publishing partners at Alfred A. Knopf for their ongoing encouragement, guidance, and support; most notably Andy Hughes and his editorial staff.

At the core of this effort is the team at Fieldstone Publishing, whose devotion to this project imbues every page of this book:

Shyla Stewart, President & CEO
Andrew Stewart, Publisher Emeritus
Jim Cirigliano, Editor-In-Chief
Katy Savage, Production Coordinator
Heather Coon, Finance
Shavar Dawkins, Photo Editor / Research
Sophia Foster, Social Media Coordinator

We thank all of the Audubon experts who contributed their knowledge to ensure this book represents the organization's mission: Audubon's Marketing Division, including Jose Carbonell, Chief Marketing Officer; Julisa Colón, Senior Manager of Brand Marketing; Holly Mascaro, Manager of Brand Marketing; Audubon's Science Division, including Geoff LeBaron, Director of the Christmas Bird Count; Audubon's Network Division, including John Rowden, Senior Director of Bird-Friendly Communities; Marlene Pantin, Partnerships Manager of Plants for Birds; and Roslyn Rivas, Program Coordinator for Plants for Birds; Audubon's Content Division, including Jennifer Bogo, Vice President of Content; Kristina Deckert, Art Director; Sabine Meyer, Photography Director; Melanie Ryan, Assistant Art Director; Hannah Waters, Senior Editor, Climate; and Camilla Cerea, Photo Editor and Photographer.

THIS IS A BORZOI BOOK PUBLISHED BY ALFRED A. KNOPF

www.aaknopf.com

Knopf, Borzoi Books, and the colophon are registered trademarks of Penguin Random House LLC.

Audubon™ is a licensed, registered trademark of the National Audubon Society. All rights reserved.

Based on *National Audubon Society Field Guide to North American Trees: Eastern Region* (Alfred A. Knopf, 1980) and *National Audubon Society Field Guide to North American Trees: Western Region* (Alfred A. Knopf, 1980).

For more information about Audubon, including how to become a member, visit www.audubon.org or call 1-844-428-3826.

Library of Congress Cataloging-in-Publication Data
Names: National Audubon Society, editor. | Cirigliano, Jim, 1981- editor.
Title: National Audubon Society trees of North America / edited by Jim Cirigliano, editor in chief, Fieldstone Publishing, Inc.
Description: First edition. | New York: Alfred A. Knopf, 2021. | Includes index.
Identifiers: LCCN 2020026120 (print) | LCCN 2020026121 (ebook) | ISBN 9780525655718 (hardcover) | ISBN 9780525655725 (ebook)
Subjects: LCSH: Trees—North America—Identification. | Forests and forestry—North America—Identification.
Classification: LCC SD140 .N38 2021 (print) | LCC SD140 (ebook) | DDC 634.9097—dc23
LC record available at https://lccn.loc.gov/2020026120
LC ebook record available at https://lccn.loc.gov/2020026121

Cover photograph of the Angel Oak, Johns Island near Charleston, South Carolina. Daniela Duncan / Moment / Getty Images
Cover design by Linda Huang
Frontispiece by Nazar Abbas Photography / Moment / Getty Images

Manufactured in China

First Edition